Symbol	*Meaning*
$\mathbf{A}^T$	the transpose of matrix **A**
$\mathbf{A}^{-1}$	the inverse of matrix **A**
p_{ij}	transition probability
$\mathbf{P}^n$	nth step transition matrix
$\mathbf{P}_0$	vector of initial conditions
$\{(x,y) \mid y = x + 2\}$	the set of all ordered pairs where the y-value is two more than the x-value
$a < b$	a is less than b
$a \leq b$	a is less than or equal to b
$a > b$	a is greater than b
$a \geq b$	a is greater than or equal to b

$$
\begin{array}{cccc|cccc|c}
x_1 & x_2 & \cdots & x^n & s_1 & s_2 & \cdots & s_m & \mathbf{B} \\
\hline
a_{11} & a_{12} & \cdots & a_{1n} & 1 & 0 & \cdots & 0 & b_1 \\
a_{21} & a_{22} & \cdots & a_{2n} & 0 & 1 & \cdots & 0 & b_2 \\
\cdot & & & & \cdot & \cdot & & & \\
\cdot & & & & \cdot & \cdot & & & \\
\cdot & & & & \cdot & \cdot & & & \\
a_{m1} & a_{m2} & \cdots & a_{mn} & 0 & 0 & \cdots & 1 & b_m \\
\hline
-c_1 & -c_2 & \cdots & -c_n & 0 & 0 & \cdots & 0 & 0
\end{array}
$$

the initial simplex tableau

Symbol	*Meaning*
$\mathbf{P} = (p_1 \quad p_2)$	row player's strategy
$\mathbf{Q} = \begin{pmatrix} q_1 \\ q_2 \end{pmatrix}$	column player's strategy
$E = \mathbf{PAQ}$	the value of a mixed-strategies game
r'	a recessive row
c'	a recessive column
$\mathbf{A}_r$	a reduced game matrix **A**
$\mathbf{P}_r$	row player's strategy to be applied to $\mathbf{A}_r$
$\mathbf{Q}_r$	column player's strategy to be applied to $\mathbf{A}_r$
$\bar{x}$	the mean of the x's
$\sum_{i=1}^{n} x_i$	$x_1 + x_2 + \cdots + x_n$
$\tilde{x}$	the median of the x's
v	variance
s	standard deviation
$\sqrt{6}$	the square root of six
z	a z-value
P_{32}	the thirty-second percentile
Q_3	the third quartile
D_9	the ninth decile
μ	the mean of the population
σ	the standard deviation of the population

Finite Mathematics
an elementary approach

Burton Rodin, Series Editor

Goodyear Series in Mathematics

Finite Mathematics: an elementary approach
Lawrence G. Gilligan and Robert B. Nenno

Introduction to Mathematics
Ramakant Khazanie and Daniel Saltz

Finite Mathematics
an elementary approach

Lawrence G. Gilligan
Robert B. Nenno

GOODYEAR PUBLISHING COMPANY, INC.,
Pacific Palisades, California

Library of Congress Cataloging in Publication Data
Gilligan, Lawrence G.
Finite mathematics.

(Goodyear series in mathematics)
Includes index.
1. Mathematics—1961- I. Nenno, Robert B., joint author. II. Title.
QA39.2.G533 510 74-11826
ISBN 0-87620-327-6

Current printing (last digit):
10 9 8 7 6 5 4 3 2 1

ISBN: 0-87620-327-6
Y- 3276-6

Designer: Don Fujimoto
Project Supervisor: Jeri Walsh
Library of Congress Catalog Card Number: 74-11826
Printed in the United States of America

This book is dedicated with love to
Corine and Sue

Contents

chapter 4 Probability

chapter 5 Vectors and Matrices

chapter 6 Linear Programming

chapter 7 The Theory of Games

chapter 8 Statistics

chapter 9 Models

Preface

To the Instructor

This project began in 1971 with one major goal in mind: *To create a Finite Mathematics text that is truly readable.* We think we have achieved that goal; we are proud of the text.

The intended audience for this text is the student with little or no background in mathematics, who is majoring in the social sciences, business, economics, or the liberal arts. With that in mind, the instructor should realize that rigor is minimal, and our approach is to present complete solutions to numerous examples.

Since "Finite Mathematics" is not a well-defined course and its content varies greatly from one school to the next, we have included most of the typical topics. Various selections of the topics can be made as suggested below to compromise a one-semester course, or all the topics can be included to make a two-semester course.

Chapters	Possible omissions	Time allotment
1–9	None	Two semesters
1–8	Sections 1.6–1.10, 3.5, 3.6, 4.7, 5.5, 5.6, 6.4, 6.5, 7.5, 7.6	One semester
1–4	None	One semester
1–5	Sections 1.8–1.10, 4.7	One semester
1–6	Sections 1.6–1.10, 3.5, 3.6, 4.7	One semester
1–5, 7	Sections 1.6–1.10, 3.5, 3.6, 4.7	One semester
1–4, 8	Sections 1.8–1.10, 4.7	One semester
3–7	*	One semester
4–8	**	One semester

(Models in Mathematics (Chapter 9) may be able to "work its way" into any of the above suggestions as you may see fit.)

We welcome criticism and communication from our colleagues; please feel free to contact us.

To the Student

This book was written for you; USE IT! There is more involved than just reading a text; you have to dissect it. We have included, to assist your understanding of the textual matter, more worked-out examples than any other text of this type. Read them carefully—they are the heart of the text.

* In order to omit chapters 1 and 2, some notation, tree diagrams, and the $r \cdot s$ rule from chapter 2 should first be introduced to assure a "smoothness" in chapter 3.

** To begin in chapter 4, some notation from chapters 1 through 3 should first be introduced. Also, some knowledge of counting is required.

When studying for exams or when notation becomes a problem, the chapter summaries and the glossary of symbols may be useful.

We have noticed in the past that the best critic is the student. So, at any point in your experience with this text, should you care to offer any criticism, or should you have any questions, or if you just want to make a comment, feel free to contact us.

Lawrence G. Gilligan
Robert B. Nenno
c/o Goodyear Publishing Co., Inc.
15115 Sunset Blvd.
Pacific Palisades, Calif. 90272

Acknowledgments

We wish to thank the Department of Mathematics and the students of Monroe Community College for helping us improve earlier editions of this text in manuscript form. Professors Bonnie Glickman and Nelson Rich deserve a special note of thanks.

For expressing the earliest interest and constant motivation, our thanks go to friend and colleague Calvin Lathan. And to William Setek, who we missed at lunches, poker, and paddleball because we were writing, we say "thanks" for understanding.

For his review of the final manuscript we want to thank Professor Gerald L. Bradley, Claremont Men's College, who did a brilliant job. We are also grateful to Professor Burton Rodin, University of California, San Diego, for his help. We thank the following people who reviewed the manuscript at various stages of its development: Burton Rodin, Gerald L. Bradley, Joan H. McCarter, and A. E. Halteman. Thanks to Kenneth C. Millett, Ken Goldstein, Carole Bauer, Linsley Wyant, William J. Claffey, Arnold L. Schroeder, and Harry D. Eylar, who responded to questionnaires sent to them.

We would like to thank those at Goodyear Publishing Company for having so much faith in us—especially Tom Paul, whose enthusiasm and interest in this project were unmatched, and Jack Pritchard, the editor who endured so many of our idiosyncracies.

For his concern about our desires regarding artwork, we sincerely thank Curt Cowan, who did a superb job. Our project supervisor, Jeri Walsh, diligently, unselfishly, and pleasantly devoted her time to the culmination of this book. For that we are indebted.

To the world's greatest mathematics secretary, Pam Dretto: thank you for such perfect service.

We probably should also thank each other and point out that this project was a joint effort with Nenno's name appearing after Gilligan's only because one had to appear first.

Finally and foremost, to our friends who tolerated our almost nonexistence at times, especially to Corine (Bob's daughter) and Sue (Larry's wife), we want you all to know we are alive and well.

Lawrence G. Gilligan
Robert B. Nenno

chapter 1
Logic

1.1 INTRODUCTION

Logic is the systematic study of the basic formal principles of reasoning. More meaningfully, logic deals with statements, with ways of combining statements to form "compound" statements, and with "proving" things.

First, let us examine exactly what is meant by a statement. Recall from elementary grammar that all sentences in the English language can be classified as one of the following types: declarative, exclamatory, imperative, or interrogative. In logic, a **statement** is a declarative sentence that is either true or false (but not both simultaneously). Exclamatory, imperative, and interrogative sentences will be of no use to us.

Today is Thursday.
Your nose is running.
Bob and Larry wrote this book.
The Dow Jones Industrial Average dropped three points today.

are all examples of statements, while

Are you for real?
Look out!
Make love, not war.

are not.

As in grammar, we have **simple statements** and **compound statements.** A *simple statement* is a statement that conveys exactly one thought, while a *compound statement* is a connection of simple statements. The word or words that perform this "connection" are called **connectives.** Consider the following statements:

1. Jesus wept.
2. Harold is an A student.
3. Ernie is an attorney.
4. Gloria and Vinnie are both attending school.
5. The Soytex Corporation will relocate or I will lose $10,000.
6. If I plan to love Edna, then Edna will have a headache.

We see that statements 1, 2, and 3 are simple statements; 4, 5, and 6 are compound. In 4 the connective is "and." In 5 the connective is "or." In 6 it is "if . . . , then." Let us examine further some of the connectives.

1.2 BASIC CONNECTIVES

When we wish to change a statement to its **negation,** one with the opposite meaning, we introduce the word "not." So,

Today is Thursday

can be negated to

Today is not Thursday

The word "not" is commonly called a connective even though it isn't really connecting simple statements here.

It is important to determine the **truth value** (the "trueness" or "falseness") of a statement. We can see that the negation of a statement has the opposite truth value of the given statement. To simplify matters we will use the lower case letter p to represent "Today is Thursday." The connective "not" will be symbolized by "$\sim$." So "$\sim p$" (not p) represents "Today is not Thursday."

In order to display truth values, it is common to list all possibilities in table form. Such a configuration is called a **truth table.** The figure below is the truth table for $\sim p$, where T stands for "true" and F for "false."

p	$\sim p$
T	F
F	T

A common connective in the English language is "and." Consider the following statements:

p: The card I am holding is a queen
q: The card I am holding is a heart

The **conjunction** of the two statements p and q, denoted $p \wedge q$, is read, "The card I am holding is a queen *and* the card I am holding is a heart." This can be shortened to, "The card I am holding is a queen and a heart.

It is important to realize just when $p \wedge q$ is true and when it is false. Common sense prevails. In order for $p \wedge q$ to be true, the card I am holding must actually be the queen of hearts. That is, statement p must be true *and* statement q must be true in order for $p \wedge q$ to be true. Otherwise, $p \wedge q$ will be false. The figure below is the truth table for $p \wedge q$.

p	q	$p \wedge q$
T	T	T
T	F	F
F	T	F
F	F	F

Note that there are now *four* rows in the truth table (as opposed to two previously). This is because all possibilities of truth values for p and q must be listed. Notice that in the column headed $p \wedge q$ there is just one "T" and that occurs in the first row of the table, where both p and q are true.

Another common connective in our language is "or." For example, consider these statements:

p: I will date Ann
q: I will date Marcia

The **disjunction** of these two statements is read, "I will date Ann or I will date Marcia." This can be shortened to "I will date Ann or Marcia." We will now examine the truth value of this compound statement.

When a person makes the above statement, he might mean that he will date just one of the two girls Ann, Marcia. However, it is also possible that he has in mind dating at least one of the two girls, possibly both. So we see that the word "or" can be used in two different senses. To be concise, we will distinguish between two disjunction processes, namely,

exclusive disjunction: the disjunction process that excludes the possibility of dating both Ann and Marcia. In symbols we write $p \veebar q$ (p or q, but not both).

inclusive disjunction: the disjunction process that includes the possibility of dating both Ann and Marcia. In symbols we write $p \vee q$ (p or q or both).

We can now summarize the truth values of each of the two disjunction processes in truth tables.

p	q	$p \veebar q$
T	T	F
T	F	T
F	T	T
F	F	F

p	q	$p \vee q$
T	T	T
T	F	T
F	T	T
F	F	F

Note that the only difference occurs in the first case, where $p \veebar q$ is false when both p is true and q is true. Unless otherwise specified, we will translate "p or q" into the inclusive disjunction form, $p \vee q$.

Many statements that we use in our English language are **conditional** statements. Some examples are:

If today is December 24, then tomorrow is Christmas Day.
If I eat too much, then I'll need a Bromo-Seltzer.
If you're not here after what I'm here after, then you'll be here after I'm gone.
If I get a good summer job, then I'll return to college.

In order to further realize the prevalence of conditional statements in our language we should observe that the "If . . . , then . . ." type of statement is the basic form of mathematical theorems. Again, some examples:

If two angles are right angles, then they are equal.
If $a = b$, then $a^2 = b^2$.
If a function is differentiable at a point, then it is continuous at that point.

Let's take a closer look at a conditional statement. Let p be "I get a good summer job" and let q be "I'll return to college." Suppose Bill makes the following compound statement:

If I get a good summer job, then I'll return to college.

We will symbolize this as $p \to q$, which is often read in either of the two ways:

1. If p, then q
2. p conditional q

Consider that Bill gets a good summer job, and that he also returns to college. Then $p \to q$ will be true (see the first row of the truth table below). Suppose Bill gets a good summer job, but he fails to return to college. We would have to say that when Bill makes the statement $p \to q$, he lies. That is, $p \to q$ is false in the case where p is true but q is false (see the second row of the truth table below).

It is possible that Bill does not get a good summer job, in which case he might still return to college (he inherits $10,000) or he might not (he has no money). In either of these two cases, Bill's original statement $p \to q$ would not be considered untrue. Hence, the last two rows of the truth table below have T's under $p \to q$.

p	q	$p \to q$
T	T	T
T	F	F
F	T	T
F	F	T

As a brief summary, the only time $p \to q$ is false is when p is true and q is false.

The conditional, although used commonly in our language, is often misused and misunderstood. For example, out of the advertising industry often come statements like these:

If you use Crust, then you'll have 47% fewer cavities.
If you use Proteen 27, then your hair will be manageable all day.
If you use Scape in the morning, then your breath will be fresh all day.

Consider the last of these statements and further presume the claim to be true. When we hear such a statement we often equate "using Scape" with "fresh breath" and "not using Scape" with "bad breath." The shock value of this could motivate us to scurry to the store immediately. However, consider the truth table below:

use Scape	fresh breath	use Scape $\to$ fresh breath
T	T	T
T	F	F
F	T	T
F	F	T

It becomes evident that such a reasoning process considers only two (1st and 4th rows) of the three cases (1st, 3rd, and 4th rows) in which "use Scape $\to$ fresh breath" is really true. We should also consider the third row when we hear the original claim: "If you use Scape, then your breath will be fresh all day." Logically, this third row tells us that we can have fresh breath without using Scape at all!

As another example of an anomaly related to the conditional, consider

"2 = 3," which is certainly false, and consider "grass is green," which is true. The compound statement

If 2 = 3, then grass is green

is true. What even sounds more ridiculous,

If 2 = 3, then grass is purple

is also true. (At this point you are probably thinking that logic isn't very logical at all, but don't let these last two compound statements worry you.) "2 = 3" certainly doesn't seem at all related to the color of grass. Don't be bothered by that; there is no reason that it should. The "relatedness" aspect of the conditional is something that has evolved through usage and development of our language. The "relatedness" is not the interesting thing from a mathematical viewpoint; we are only interested in the possible truth values of the conditional.

A connective closely related to the conditional just studied is the **biconditional**. A few examples are:

I will marry you if and only if you love me.
Fill out the information if and only if you have a change of address.
A quadrilateral is a parallelogram if and only if its opposite sides are parallel.

Let's work out the truth table of the biconditional by considering a specific example and using a little common sense. Use p to represent "I will marry you" and q to stand for "you love me." Imagine that Bill tells Addie, "I will marry you if and only if you love me." We will symbolize this statement as $p \leftrightarrow q$ and read it as:

1. p if and only if q
2. p iff q
3. p biconditional q

In any case, what Bill is saying is that Addie's love will lead him to marry her and also that if Addie does not love him, he will not marry her. That is, any biconditional $p \leftrightarrow q$ states that when p is true, q is true, and that when p is false, q is false. For these reasons, we see T's in the first and fourth rows of the truth table along with F's in the second and third rows of the truth table.

p	q	$p \leftrightarrow q$
T	T	T
T	F	F
F	T	F
F	F	T

In brief, $p \leftrightarrow q$ is true when p and q have the same truth values (both true or both false).

Earlier we mentioned that the biconditional is closely related to the conditional. It also is related to the conjunction and disjunction. In fact, there are many interrelationships among all six of the connectives ($\sim$, $\wedge$, $\vee$, $\underline{\vee}$, $\rightarrow$, $\leftrightarrow$) we have studied, but we are going to defer this until section 1.5.

Right now, let's summarize all six of the truth tables studied so far; we will save a little writing by presenting them in slightly compact form.

$\sim$	p
F	T
T	F

p	$\wedge$	q
T	T	T
T	F	F
F	F	T
F	F	F

p	$\underline{\vee}$	q
T	F	T
T	T	F
F	T	T
F	F	F

p	$\vee$	q
T	T	T
T	T	F
F	T	T
F	F	F

p	$\rightarrow$	q
T	T	T
T	F	F
F	T	T
F	T	F

p	$\leftrightarrow$	q
T	T	T
T	F	F
F	F	T
F	T	F

Notice that all we have done here is to write the possible truth values directly underneath the simple statements p and q and to write the truth values of the compound statement directly beneath the symbol for the connective.

EXERCISES ■ SECTIONS 1.1 AND 1.2.

1. Which of the sentences below are statements?

- **a.** Larry, take out the garbage.
- **b.** $2 \neq 3$.
- **c.** If elected president, then I will lower taxes.
- **d.** Where have all the flowers gone?
- **e.** I will buy lunch if and only if I lose the coin toss.

2. Consider the sentence, "This sentence is false." It is not a statement. Why?

3. Assume we know statement p to be true and we know statement q to be false. Determine the truth value of each of the following.

a. $\sim p$ b. $\sim p \vee q$ c. $p \wedge q$ d. $\sim q$
e. $p \wedge \sim q$ f. $\sim(p \wedge q)$ g. $\sim p \wedge \sim q$ h. $p \underline{\vee} q$
i. $p \rightarrow q$ j. $q \rightarrow p$

4. Classify each statement below as either simple, negation, conjunction, disjunction, conditional, or biconditional.

 a. You are not the man you pretend to be.
 b. The sum of nine and four is thirteen.
 c. If you disobey, then you shall be punished.
 d. The Health Food Company tripled its quarterly dividend.
 e. I will attend Brockport or Harvard next fall, but not both.

5. Determine the truth value of each of the following compound statements.

 a. I died yesterday or $1 + 1 = 2$.
 b. If $1 + 1 = 3$, then tomorrow is Sunday.
 c. If tomorrow is Sunday, then $1 + 1 = 2$.
 d. If tomorrow is Sunday, then $1 + 1 = 3$.
 e. $1 + 1 = 2$ or $82 + 5 = 87$, but not both.
 f. $1 + 1 = 2$ and $7 + 11 = 18$.
 g. $7 - 1 = 6$ and $9 - 3 = 12$.
 h. Today is tomorrow's yesterday if and only if $5 + 6 = 10$.

6. Assume we know statement r to be true and statement s to be true. Determine the truth value of each of the following.

 a. $r \vee \sim r$ b. $r \rightarrow s$ c. $r \leftrightarrow s$
 d. $\sim r \rightarrow s$ e. $\sim(r \rightarrow s)$ f. $r \underline{\vee} s$

7. Assume a is true, b is false, and c is true. Determine the truth value of each of the following.

 a. $[(a \wedge b) \rightarrow c] \rightarrow (a \vee c)$
 b. $\sim[(a \rightarrow c) \wedge (b \underline{\vee} c)]$
 c. $a \leftrightarrow \sim b$
 d. $\sim\{[a \wedge (b \vee c)] \rightarrow [(c \underline{\vee} a) \leftrightarrow b]\}$

8. "Can you score with AND and OR?" The game pictured on the next page involves choosing exactly six of the twelve ping pong balls at the top so that you end up with a ping pong ball at the finish (arrows). A ball passes through an AND gate only when both sleeves contain a ball; a ball passes through an OR gate when one or both of the sleeves is filled. Pick six balls so that you will be a winner. (Hint: There is more than one correct answer.)

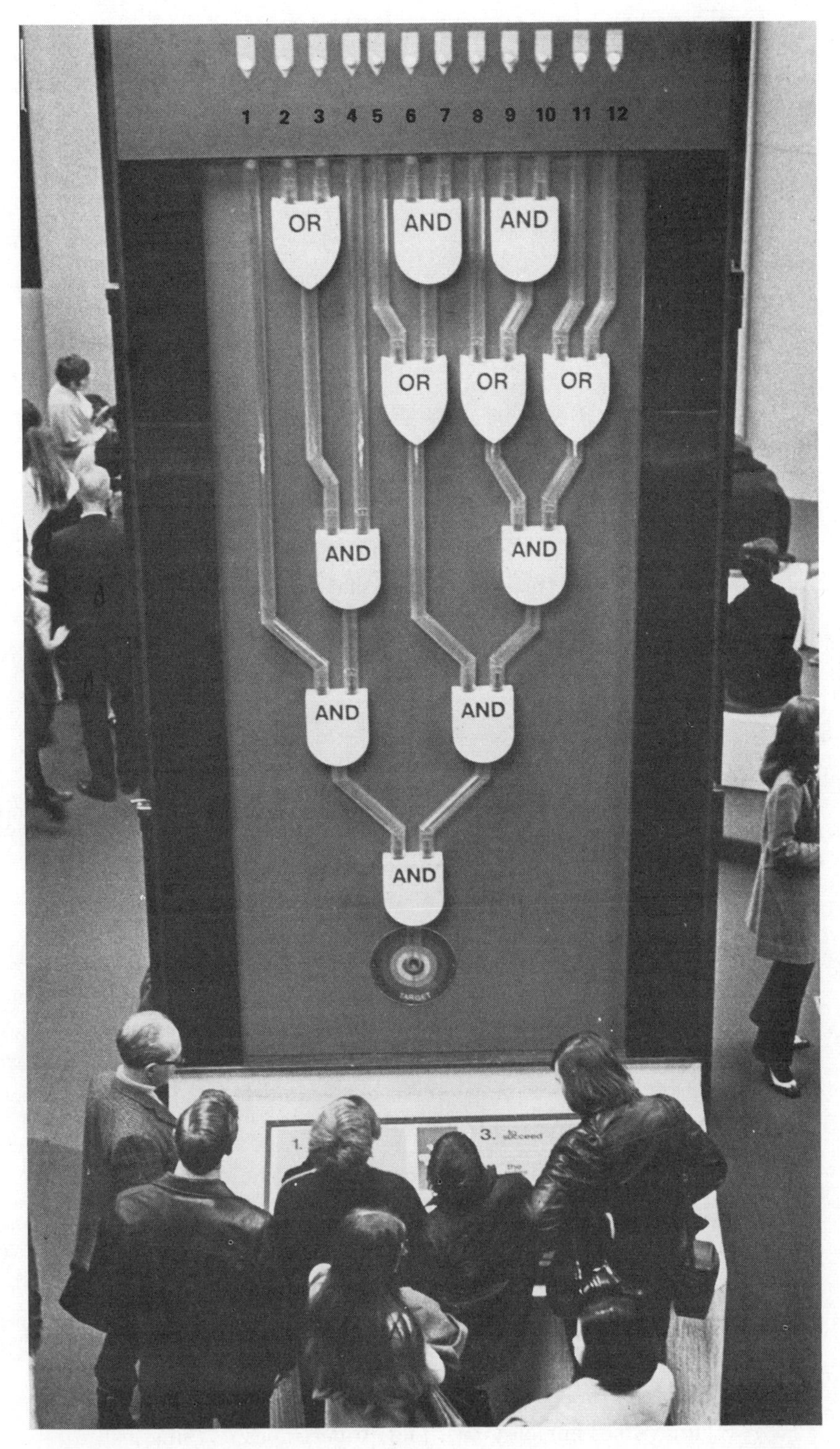

Courtesy of Ontario Science Center, Toronto

1.3 SYMBOLIC REPRESENTATION

We have stated in section 1.2 that lower case letters are used to represent simple statements. Consider the following:

p: Paula sings lead
q: Quincy plays bass
r: Robert sings lead

Our objective now is to represent compound statements symbolically. The statement

Paula does not sing lead

can be represented (as stated in the previous section) by $\sim p$. The statement

It is not the case that Paula sings lead

can also be represented by $\sim p$. The six words "It is not the case that" merely say, "Do the opposite of what follows," and negation is the connective that "does the opposite."

Expanding this notion, let us try to represent symbolically the statement

It is not the case that *Paula sings lead and Quincy plays bass*

We know the words in italic in the statement translate to $p \wedge q$. This conjunction must, however, be negated. So, "It is not the case that Paula sings lead and Quincy plays bass" is symbolized as $\sim(p \wedge q)$. We parenthesize $p \wedge q$ because all of it must be negated. Note:

Paula does not sing lead and Robert does

translates to $\sim p \wedge r$ because here the "not" refers only to statement p.

There are many different connectives used in our language; to describe all of them in this text is not our purpose. We present a few that we will find useful and then move on to bigger and better things in section 1.4. Let us consider the compound statement

Neither Quincy plays bass *nor* Robert sings lead

which is connected, of course, by "neither . . . nor. . . ." How is this written symbolically? To answer this question, we merely inquire: What does it mean?

Neither Quincy plays bass nor Robert sings lead

means

Quincy does not play bass and Robert does not sing lead

which translates to $\sim q \wedge \sim r$. We urge the reader to reread the last sentence to convince himself of its truth.

We now proceed with some examples.

Example 1.3.1. Write "It is not the case that if Paula sings lead, then Quincy plays bass" in symbolic form.

Solution: This statement is the negation of the conditional $p \rightarrow q$. Hence, $\sim(p \rightarrow q)$.

Example 1.3.2. Write "It is not the case that neither Paula doesn't sing lead nor Robert sings lead" symbolically.

Solution: Omitting the first six words we have

Neither Paula doesn't sing lead nor Robert sings lead

and from our previous discussion, this translates to $\sim \sim p \wedge \sim r$ but $\sim \sim p$ is the same as p, as evidenced in the truth table below:

p	$\sim p$	$\sim(\sim p)$
T	F	T
F	T	F

So, "Neither Paula doesn't sing lead nor Robert sings lead" translates to $p \wedge \sim r$ and "It is not the case that neither Paula doesn't sing lead nor Robert sings lead" translates to $\sim(p \wedge \sim r)$.

The connectives "but" and "either . . . or . . ." are introduced in examples 1.3.3 and 1.3.4, respectively.

Example 1.3.3. How would "Paula sings lead, but Quincy does not play bass" translate symbolically?

Solution: Again, we resort to asking ourselves just what the statement means.

Paula sings lead, but Quincy does not play bass

means

Paula sings lead and Quincy does not play bass

which translates to $p \wedge \sim q$. In other words, "but" and "and" are used interchangeably.

Example 1.3.4. "Either Quincy plays bass or Robert sings lead" means the same as "Quincy plays bass or Robert sings lead" and translates to $q \vee r$. Note that the word "either" is superfluous.

There is another situation where a word can be deleted.

Example 1.3.5. "If Paula sings lead, then Quincy plays bass." (This is $p \to q$.) When the word "then" is dropped we have

If Paula sings lead, Quincy plays bass

which certainly has the same meaning. Interchanging the order of the clauses gives

Quincy plays bass, if Paula sings lead

Observe that in this form of the statement, the "if" clause is still "Paula sings lead." Consequently, the statement still translates to $p \to q$.

EXERCISES ■ SECTION 1.3.

1. Use p: I pass the course
q: I pass all the exams
r: I bought the text

and translate each of the following symbolically.

a. I will pass the course but I didn't buy the text.
b. I will pass the course if I pass all the exams.
c. It is not the case that I will pass the course if I pass all the exams.
d. Either I buy the text or I pass all the exams, but not both.
e. I will pass the course if and only if I either buy the text or pass all the exams.

2. Use x: My car started
y: I was late for work
z: I overslept

and translate each of the following into symbolic form.

a. Neither my car started nor was I late for work.
b. I was late for work but my car didn't start.
c. Either my car started or I was late for work, but not both.
d. If I overslept and my car didn't start, then I was late for work.
e. I was late for work if and only if either my car didn't start or I overslept.
f. I was late for work, if I overslept.
g. It is not the case that if my car didn't start and I overslept, then I was late for work.
h. If it is not the case that my car didn't start and I overslept, then I was late for work.

3. Use p: Calvin gets a promotion
q: Bob gets a promotion
r: Jim gets a promotion

and translate each of the following symbolically.

a. It is not the case that if Calvin gets a promotion then Jim gets a promotion.
b. Neither Bob nor Jim get promotions.
c. It is not true that if either Bob or Calvin gets a promotion, then Jim does not get a promotion.
d. Calvin gets a promotion but Jim does not.
e. Calvin doesn't get a promotion if and only if it is not the case that both Jim and Bob get promotions.
f. Calvin gets a promotion if Jim does.

4. Use p, q, r as in exercise 1 and translate each of the following into an English sentence.

a. $p \leftrightarrow \sim q$ **b.** $p \rightarrow (q \wedge r)$ **c.** $(q \underline{\vee} r) \wedge p$
d. $\sim(p \vee r)$ **e.** $\sim(p \rightarrow q)$

1.4 TRUTH TABLES

We are now ready to examine the truth tables of more complex statements. $(p \vee q) \wedge \sim p$ is the statement for which a truth table will be found. Observe that there are five entities to which we must assign truth values: p, q, $p \vee q$, $\sim p$, $(p \vee q) \wedge \sim p$. Begin with p. (We number the columns for reference purposes.)

p	q	$(p$	$\vee$	$q)$	$\wedge$	$\sim p$
T	T	T				T
T	F	T				T
F	T	F				F
F	F	F				F
		(1)				(1)

In step 2, fill in the values for q.

p	q	$(p$	$\vee$	$q)$	$\wedge$	$\sim p$
T	T	T		T		T
T	F	T		F		T
F	T	F		T		F
F	F	F		F		F
		(1)		(2)		(1)

Now we notice that the major connective in $(p \vee q) \wedge \sim p$ is $\wedge$; we must find truth values for each "side" of $\wedge$. They are represented in columns (3) and (4), where (3) represents $p \vee q$, and (4) represents $\sim p$.

p	q	$(p$	$\vee$	$q)$	$\wedge$	$\sim$	p
T	T	T	T	T		F	T
T	F	T	T	F		F	T
F	T	F	T	T		T	F
F	F	F	F	F		T	F
		(1)	(3)	(2)		(4)	(1)

Finally, since the last thing we must do is to take the conjunction of columns (3) and (4) (i.e., $(p \vee q)$ with $\sim p$), the truth table is completed in column (5) below.

p	q	$(p$	$\vee$	$q)$	$\wedge$	$\sim$	p
T	T	T	T	T	F	F	T
T	F	T	T	F	F	F	T
F	T	F	T	T	T	T	F
F	F	F	F	F	F	T	F
		(1)	(3)	(2)	(5)	(4)	(1)

Example 1.4.1. Find the truth table for $(p \rightarrow \sim q) \veebar \sim p$.

Solution: We merely give the truth table here and number the steps.

p	q	$(p$	$\rightarrow$	$\sim$	$q)$	$\veebar$	$\sim$	p
T	T	T	F	F	T	F	F	T
T	F	T	T	T	F	T	F	T
F	T	F	T	F	T	F	T	F
F	F	F	T	T	F	F	T	F
		(1)	(4)	(3)	(2)	(6)	(5)	(1)

In the truth tables, we have used the fact that all possible values of trueness and falseness of p and q must be accounted for. They are the so-called reference columns:

p	q	
T	T	
T	F	
F	T	
F	F	

A truth table involving one simple statement has two rows, corresponding to the two possible truth values, T and F (as in the truth table at the bottom of page two). A truth table involving two simple statements has four rows. For a three-statement (p,q,r) truth table, we again must list all possibilities of trueness and falseness for p, q, r. We do so below.

	p	q	r
case I	T	T	T
case II	T	T	F
case III	T	F	T
case IV	T	F	F
case V	F	T	T
case VI	F	T	F
case VII	F	F	T
case VIII	F	F	F

Now let's find the truth table for $\sim[\sim p \leftrightarrow (q \vee r)]$. Note the use of brackets, [ ], indicating that all of $\sim p \leftrightarrow (q \vee r)$ is to be negated.

p	q	r	$\sim$	$[\sim p$	$\leftrightarrow$	$(q$	$\vee$	$r)]$
T	T	T	T	F	F	T	T	T
T	T	F	T	F	F	T	T	F
T	F	T	T	F	F	F	T	T
T	F	F	F	F	T	F	F	F
F	T	T	F	T	T	T	T	T
F	T	F	F	T	T	T	T	F
F	F	T	F	T	T	F	T	T
F	F	F	T	T	F	F	F	F
			(6)	(1)	(5)	(2)	(4)	(3)

Column (5) is the result of using the definition of biconditional on column (1) with (4). Column (6) (our answer column) is merely the negation of everything within brackets (i.e., the negation of column (5)).

Example 1.4.2. Construct a truth table for $p \vee \sim p$.

Solution: We have the following two-row truth table.

p	p	$\vee$	$\sim p$
T	T	T	F
F	F	T	T
	(1)	(3)	(2)

The truth table for $p \vee \sim p$ is always true. A statement that is always true is called a **tautology.**

Example 1.4.3. Construct a truth table for $\sim(p \rightarrow q) \wedge q$.

Solution:

p	q	$\sim$	$(p \rightarrow q)$	$\wedge$	q
T	T	F	T	F	T
T	F	T	F	F	F
F	T	F	T	F	T
F	F	F	T	F	F

Notice that the answer column is FFFF. A statement with an all-false truth table is called a **self-contradiction.** Hence $\sim(p \rightarrow q) \wedge q$ is a self-contradiction.

Example 1.4.4. Construct a truth table for $\sim[(p \vee \sim r) \leftrightarrow (q \rightarrow r)]$.

Solution:

p	q	r	$\sim$	$[(p$	$\vee$	$\sim r)$	$\leftrightarrow$	$(q \rightarrow r)]$
T	T	T	F	T	T	F	T	T
T	T	F	T	T	T	T	F	F
T	F	T	F	T	T	F	T	T
T	F	F	F	T	T	T	T	T
F	T	T	T	F	F	F	F	T
F	T	F	T	F	T	T	F	F
F	F	T	T	F	F	F	F	T
F	F	F	F	F	T	T	T	T

So the truth table for $\sim[(p \vee \sim r) \leftrightarrow (q \rightarrow r)]$ is F T F F T T T F.

Example 1.4.5. Test each of the statements below to see whether they are tautologies, self-contradictions, or neither.

a. $(p \rightarrow q) \vee (q \rightarrow p)$
b. $\sim(r \rightarrow s)$
c. $(x \underline{\vee} y) \wedge (x \leftrightarrow y)$

Solution:

a. The truth table results in TTTT, so this statement is a tautology.
b. The truth table results in FTFF, so this statement is neither a tautology nor a self-contradiction.
c. This is a self-contradiction because the truth table is FFFF.

EXERCISES ■ SECTION 1.4.

1. Construct truth tables for each of the following compound statements.

   a. $\sim p \rightarrow q$
   b. $\sim(p \rightarrow q)$
   c. $(p \leftrightarrow r) \vee q$
   d. $(p \rightarrow q) \wedge (q \rightarrow p)$
   e. $\sim(p \vee q)$
   f. $\sim p \wedge \sim q$
   g. $\sim[p \rightarrow \sim(q \wedge r)]$
   h. $\sim[(\sim q \vee \sim r) \rightarrow \sim(p \wedge r)]$

2. **a.** We have seen truth tables involving one simple statement (two rows), two simple statements (four rows), and three simple statements (eight rows). Extend this idea and complete the following chart.

Number of simple statements	1	2	3	4	5	10	n
Rows in truth table	2	4	8				

   b. Construct a truth table for $[(p \wedge r) \rightarrow s] \underline{\vee} \sim q$.

3. Test the following compound statements to see whether they are tautologies, self-contradictions, or neither.

   a. $(p \vee q) \rightarrow \sim(p \vee q)$
   b. $p \wedge \sim p$
   c. $p \rightarrow (p \vee q)$
   d. $[(p \wedge q) \vee r] \wedge \sim p$
   e. $p \leftrightarrow p$
   f. $p \leftrightarrow \sim p$
   g. $p \underline{\vee} p$
   h. $p \rightarrow \sim p$
   i. $(p \rightarrow q) \leftrightarrow (\sim q \rightarrow \sim p)$
   j. $[p \wedge (q \vee r)] \leftrightarrow [(p \wedge q) \vee (p \wedge r)]$

4. True or false: The negation of a tautology is a self-contradiction.

1.5 RELATIONS

In section 1.2 we saw that the role played by $\sim$, $\wedge$, $\vee$, $\underline{\vee}$, $\rightarrow$, $\leftrightarrow$ was to connect. In logic, as in other branches of mathematics, we frequently try to compare things. For instance, in the branch of mathematics called arithmetic, we use the symbol $=$ to denote a comparison of two things. For example, $1 + 1 = 2$ compares $1 + 1$ with 2. In the expression $4 < 6$ (read "4 is less than 6"), the symbol $<$ compares 4 with 6. In the branch of mathematics called geometry, comparisons are often made of triangles. When two triangles can be made to coincide, they are said to be congruent. Here congruency is the comparison. Comparisons in logic are called **relations.** We shall study three relations here.

The following statement is called an **implication:**

If today is Thursday, then tomorrow is Friday.

It is a conditional statement that is always true. That is, it is impossible for "today is Thursday" to be true and "tomorrow is Friday" to be false simultaneously. When $p \rightarrow q$ is a tautology, p is said to *imply* q and we write $p \Rightarrow q$. The statement $p \Rightarrow q$ is usually read

1. p implies q
2. p logically implies q
3. q is implied by p

We can then say, "Today being Thursday implies tomorrow being Friday."

Example 1.5.1. Does p imply $(p \vee q)$?

Solution: Yes. We write the truth table of $p \rightarrow (p \vee q)$:

p	q	p	$\rightarrow$	$(p \vee q)$
T	T	T	T	T
T	F	T	T	T
F	T	F	T	T
F	F	F	T	F

Since $p \rightarrow (p \vee q)$ is a tautology, we write $p \Rightarrow (p \vee q)$.

Example 1.5.2. Determine whether $p \vee q$ implies $p \wedge q$, or $p \wedge q$ implies $p \vee q$.

Solution: We check the truth tables:

p	q	$(p \vee q)$	$\rightarrow$	$(p \wedge q)$
T	T	T	T	T
T	F	T	F	F
F	T	T	F	F
F	F	F	T	F

p	q	$(p \wedge q)$	$\rightarrow$	$(p \vee q)$
T	T	T	T	T
T	F	F	T	T
F	T	F	T	T
F	F	F	T	F

We see that $(p \vee q) \rightarrow (p \wedge q)$ is false in cases II and III. So $(p \vee q)$ does not imply $p \wedge q$. Take note, however, that $(p \wedge q) \rightarrow (p \vee q)$ is a tautology. Hence, $(p \wedge q) \Rightarrow (p \vee q)$.

Example 1.5.3. Check the following for implications:

$$a: \sim q \qquad b: \sim(p \vee q) \qquad c: \sim p \wedge \sim q$$

Solution: We leave it to the reader to verify the following truth tables.

p	q	a	b	c
T	T	F	F	F
T	F	T	F	F
F	T	F	F	F
F	F	T	T	T

Does a imply b? No, since $a \to b$ is false in case II. Does a imply c? No, for the same reason. Does b imply a? Yes, $b \to a$ is a tautology. (It does not occur that b is true when a is false.) Does $b \Rightarrow c$? Yes, since $b \to c$ is never false. Also $c \Rightarrow a$ and $c \Rightarrow b$.

Example 1.5.4. Show that $[(p \to q) \wedge (q \to r)] \Rightarrow (p \to r)$.

Solution:

p	q	r	$[(p \to q)$	$\wedge$	$(q \to r)]$	$\to$	$(p \to r)$
T	T	T	T	T	T	T	T
T	T	F	T	F	F	T	F
T	F	T	F	F	T	T	T
T	F	F	F	F	T	T	F
F	T	T	T	T	T	T	T
F	T	F	T	F	F	T	T
F	F	T	T	T	T	T	T
F	F	F	T	T	T	T	T

The fact that $[(p \to q) \wedge (q \to r)]$ implies $(p \to r)$ is called the **transitive property.** We will encounter this again in section 1.7.

Sometimes we can examine an implication without the use of a truth table. In example 1.5.1 we showed that p implies $(p \vee q)$ by demonstrating that $p \to (p \vee q)$ is a tautology. It would suffice to show that $p \to (p \vee q)$ cannot be false. However, $p \to (p \vee q)$ can only be false when p is true and $(p \vee q)$ is false. But p being true means $p \vee q$ must be true (by definition of $\vee$) and hence $p \to (p \vee q)$ can never be false.

In summary, we give three characterizations of $x \Rightarrow y$.

$x \Rightarrow y$ means
- **1.** $x \to y$ is a tautology;
- **2.** whenever x is true, y must be true;
- **3.** it cannot happen that x is true when y is false.

Two statements are said to be **equivalent** when each one implies the other. The symbol for equivalence is $\Leftrightarrow$. (This looks like an old friend, $\leftrightarrow$. The reason for that will become apparent shortly.)

In example 1.5.3 we saw that $\sim(p \vee q) \Rightarrow \sim p \wedge \sim q$ and also $(\sim p \wedge \sim q) \Rightarrow \sim(p \vee q)$. Now this can be simplified to $\sim(p \vee q) \Leftrightarrow$

$\sim p \wedge \sim q$. Observe that the truth tables of $\sim(p \vee q)$ and $\sim p \wedge \sim q$ are identical. That is another test for the equivalence of two statements. It is the same as saying $\sim(p \vee q) \leftrightarrow \sim p \wedge \sim q$ is a tautology. So in brief summary, $x \Leftrightarrow y$ means that $x \leftrightarrow y$ is a tautology. Notice the similarity to the fact that $x \Rightarrow y$ means that $x \rightarrow y$ is a tautology.

We now give some examples.

Example 1.5.5. Show that $\sim(p \wedge q)$ is equivalent to $\sim p \vee \sim q$.

Solution:

p	q	$\sim$	$(p \wedge q)$	$\leftrightarrow$	$(\sim p$	$\vee$	$\sim q)$
T	T	F	T	T	F	F	F
T	F	T	F	T	F	T	T
F	T	T	F	T	T	T	F
F	F	T	F	T	T	T	T

The truth table shows all T's in the "$\leftrightarrow$" column. Hence, $\sim(p \wedge q) \leftrightarrow (\sim p \vee \sim q)$ is a tautology, and therefore $\sim(p \wedge q)$ is equivalent to $(\sim p \vee \sim q)$. We can now write $\sim(p \wedge q) \Leftrightarrow \sim p \vee \sim q$.

The two equivalences we have so far, $\sim(p \vee q) \Leftrightarrow (\sim p \wedge \sim q)$ and $\sim(p \wedge q) \Leftrightarrow (\sim p \vee \sim q)$, are called **DeMorgan's laws.** Recall from section 1.3, the example of writing symbolically

Neither Quincy plays bass nor Robert sings lead

The solution was $\sim q \wedge \sim r$. Now, thanks to the first DeMorgan law, we see that $\sim q \wedge \sim r$ is equivalent to $\sim(q \vee r)$. That is, a "neither . . . nor . . ." statement is the negation of an "either . . . or . . ." statement.

A use of equivalence materializes in the following problem: Does $p \wedge q \wedge r$ have meaning? Remember, in all previous discussions of a compound statement with three components, parentheses were used. In other words, if $(p \wedge q) \wedge r$ is equivalent to $p \wedge (q \wedge r)$, then $p \wedge q \wedge r$ will have meaning. Let's see.

p	q	r	$(p \wedge q)$	$\wedge$	r	p	$\wedge$	$(q \wedge r)$
T	T	T	T	T	T	T	T	T
T	T	F	T	F	F	T	F	F
T	F	T	F	F	T	T	F	F
T	F	F	F	F	F	T	F	F
F	T	T	F	F	T	F	F	T
F	T	F	F	F	F	F	F	F
F	F	T	F	F	T	F	F	F
F	F	F	F	F	F	F	F	F
				↑			↑	

Due to the fact that $(p \wedge q) \wedge r$ and $p \wedge (q \wedge r)$ have the same truth tables, they are equivalent. Beware! In the next example we see parentheses are sometimes unavoidable.

Example 1.5.6. Does $p \wedge q \vee r$ have meaning?

Solution: No, because $(p \wedge q) \vee r$ is not equivalent to $p \wedge (q \vee r)$. You see, here the placement of parentheses is strategic. Notice the difference in the truth tables below.

p	q	r	$(p \wedge q) \vee r$			$p \wedge (q \vee r)$		
T	T	T	T	T	T	T	T	T
T	T	F	T	T	F	T	T	T
T	F	T	F	T	T	T	T	T
T	F	F	F	F	F	T	F	F
F	T	T	F	T	T	F	F	T
F	T	F	F	F	F	F	F	T
F	F	T	F	T	T	F	F	T
F	F	F	F	F	F	F	F	F
				↑			↑	

The last relation we are concerned with is **inconsistency.** Two statements are called inconsistent if they both cannot happen simultaneously. Consider the following statements:

r: The Russians were the first to land on the moon
a: The Americans were the first to land on the moon

These statements are inconsistent and we write $r \mathbf{I} a$, which is read, "r is inconsistent with a."

Example 1.5.7. Show that $p \mathbf{I} \sim p$.

Solution: $p \mathbf{I} \sim p$ because they cannot both happen simultaneously. That is, if p is happening (p is true), then $\sim p$ cannot be happening. If p is not happening (p is false), then it is true that $\sim p$ is happening. This can be analyzed in the table below.

p	$\sim p$
T	F
F	T

Now, when $x \mathbf{I} y$, both x and y can never be true simultaneously. So $x \wedge y$ could never be true; $x \wedge y$ would be a self-contradiction.

Example 1.5.8. Show that $(p \wedge q)$ **I** $(p \wedge \sim q)$.

Solution: The truth table shows that $(p \wedge q) \wedge (p \wedge \sim q)$ is a self-contradiction.

p	q	$(p \wedge q)$	$\wedge$	$(p \wedge \sim q)$
T	T	T	F	F
T	F	F	F	T
F	T	F	F	F
F	F	F	F	F

Consequently, the statements $(p \wedge q)$ and $(p \wedge \sim q)$ are inconsistent.

When two statements are not inconsistent, they are called **consistent** or **compatible.** For instance, the statements p and $p \rightarrow q$ are compatible, since they can both be true, as in case I of the table below.

p	q	p	$\wedge$	$(p \rightarrow q)$
T	T	T	T	T
T	F	T	F	F
F	T	F	F	T
F	F	F	F	T

Summarily, we have seen three relations that may exist between any two statements. The implication $x \Rightarrow y$ means that $x \rightarrow y$ is a tautology. The equivalence $x \Leftrightarrow y$ means that $x \leftrightarrow y$ is a tautology. And the inconsistency x **I** y means that $x \wedge y$ is a self-contradiction.

EXERCISES ■ SECTION 1.5.

1. Check each of the problems below to see if the expression in column I implies the expression in column II, if the expression in column II implies the expression in column I, neither, or both.

	I	*II*
a.	p	q
b.	$p \veebar q$	$p \vee q$
c.	$p \rightarrow (q \vee r)$	$\sim p \vee (\sim q \vee \sim r)$
d.	$p \rightarrow q$	$p \leftrightarrow q$
e.	$(p \wedge q) \vee \sim r$	$(p \wedge q) \vee \sim r$
f.	$\sim p \rightarrow q$	$p \vee q$

2. Show that any statement implies itself.

3. Show that $p \underline{\vee} q$ is equivalent to $(p \vee q) \wedge \sim(p \wedge q)$. (This latter statement is a result of literally translating "p or q, but not both.")
4. Show that when y is implied by x and z is implied by y, then z is implied by x.
5. Which of the following pairs of statements are equivalent?

    a. $\sim(p \leftrightarrow q)$, $p \underline{\vee} q$
    b. $p \wedge (q \vee r)$, $(p \wedge q) \vee (p \wedge r)$
    c. $p \vee (q \wedge r)$, $(p \vee q) \wedge (p \vee r)$
    d. $(p \rightarrow q) \wedge (q \rightarrow p)$, $p \leftrightarrow q$
    e. $p \rightarrow q$, $\sim q \rightarrow \sim p$

6. Of utmost importance in chapter 2 will be the equivalence $(p \rightarrow q) \Leftrightarrow (\sim p \vee q)$. Verify it.
7. Although we constantly use the six basic connectives, they are not all necessary. In fact, we can "get by" merely with $\sim$ and $\vee$. Write a statement equivalent to each of the following using only $\sim$ and $\vee$.

    a. $p \wedge q$
    b. $p \underline{\vee} q$
    c. $p \rightarrow q$
    d. $p \leftrightarrow q$ (Hint: use exercise 5*d*.)

8. Which of the following pairs of statements are compatible?

    a. $p \wedge q \wedge r$, p
    b. $p \wedge q \wedge \sim r$, $p \wedge q \wedge r$
    c. $p \rightarrow q$, $q \rightarrow p$
    d. $p \underline{\vee} q$, $p \leftrightarrow q$
    e. $p \vee q$, $\sim q \rightarrow p$
    f. $p \leftrightarrow (q \vee r)$, $\sim p \wedge \sim q \wedge \sim r$

9. **a.** Show that any statement is implied by a self-contradiction.
    b. Show that any statement implies a tautology.

10. Often times students erroneously invent their own rules of logic.

    a. Show that $\sim(p \rightarrow q)$ is not equivalent to $\sim p \rightarrow q$.
    b. Show that $\sim(p \rightarrow q)$ is not equivalent to $\sim p \rightarrow \sim q$.
    c. Show that $\sim(p \leftrightarrow q)$ is not equivalent to $\sim p \leftrightarrow \sim q$ and that in fact, they are inconsistent.

11. Translate "Neither Picasso nor Der is an impressionist" into symbolic form using just the connectives $\sim$ and $\vee$.

1.6 CONDITIONAL STATEMENTS AND VARIATIONS

In section 1.2 we first introduced the conditional statement. Also in that section we began to weigh the importance of the conditional both in terms of its usage in everyday language as well as in mathematics. In this section we will talk more about the conditional and we will also introduce some variations of it.

We begin our discussion with the conditional statement

If he is a thief, then he is a lawbreaker

Closely related to a conditional statement are its variations, called the **converse, inverse,** and **contrapositive.** In the following table we give the example of the above conditional statement and each of its variations. Let p be "he is a thief," and let q symbolize "he is a lawbreaker."

Statement	Symbolic form	Translation
Original conditional	$p \to q$	If he is a thief, then he is a lawbreaker
Converse of original conditional	$q \to p$	If he is a lawbreaker, then he is a thief
Inverse of original conditional	$\sim p \to \sim q$	If he is not a thief, then he is not a lawbreaker
Contrapositive of original conditional	$\sim q \to \sim p$	If he is not a lawbreaker, then he is not a thief

We can give a brief summary of this table. To form the converse of a conditional, you merely interchange the "if" clause with the "then" clause. The inverse of a conditional is formed by negating both the "if" clause and the "then" clause. To write the contrapositive of a conditional, it is necessary to interchange the two clauses as well as to negate each.

Example 1.6.1. Write the converse, inverse, and contrapositive of "If he burns leaves, then he pollutes the air."

Solution: Converse: "If he pollutes the air, then he burned leaves." Inverse: "If he does not burn leaves, then he does not pollute the air." Contrapositive: "If he does not pollute the air, then he does not burn leaves."

Example 1.6.2. Write the converse, inverse, and contrapositive of $\sim q \to \sim p$.

Solution: Converse: $\sim p \to \sim q$ (by interchanging "if" clause with "then" clause). Inverse: $q \to p$ (by negating "if" clause as well as "then" clause). Contrapositive: $p \to q$ (by interchanging and negating "if," "then" clauses).

Example 1.6.3. Write the converse, inverse, and contrapositive forms of "If he burns the flag, then we will not like him."

Solution: Converse: "If we do not like him, then he burned the flag." Inverse: "If he does not burn the flag, then we will like him." Contrapositive: "If we like him, he did not burn the flag."

Example 1.6.4. What is the contrapositive of the converse of $\sim q \to p$?

Solution: Converse is $p \to \sim q$. Contrapositive of converse is $\sim \sim q \to \sim p$, or simply $q \to \sim p$.

In previous discussions related to compound statements, we have seen that it is often possible to gain information about them from their truth tables. We do that now.

				Original conditional	Converse of original conditional	Inverse of original conditional	Contrapositive of original conditional
p	q	$\sim p$	$\sim q$	$p \to q$	$q \to p$	$\sim p \to \sim q$	$\sim q \to \sim p$
T	T	F	F	T	T	T	T
T	F	F	T	F	T	T	F
F	T	T	F	T	F	F	T
F	F	T	T	T	T	T	T
				(1)	(2)	(3)	(4)

We can immediately note that columns (1) and (4) have the same truth values, where (4) is the contrapositive of (1); also, (1) is the contrapositive of (4). Comparing columns (2) and (3), we see the same truth values there. Note that (3) is the contrapositive of (2), and (2) is the contrapositive of (3). Consequently, we discover that any conditional and its contrapositive are equivalent, i.e.,

$$(p \to q) \Leftrightarrow (\sim q \to \sim p)$$

Also, observe that a conditional is not equivalent to either its converse or inverse.

Example 1.6.5. Write the contrapositive of $\sim q \to p$.

Solution: Interchanging the two clauses and negating each, we have $\sim p \to q$.

Due to the prevalence of the conditional in our language pattern, it is only natural that its wording has evolved into different forms. For example, consider the statement: "If I have \$4, then I have \$3."

Three alternate wordings that have the same meaning are:

1. I have \$4 only if I have \$3
2. Having \$4 is sufficient for having \$3
3. Having \$3 is necessary for having \$4

By letting "I have \$4" be p and "I have \$3" be q, we can summarize the conditional statement and its equivalent wordings.

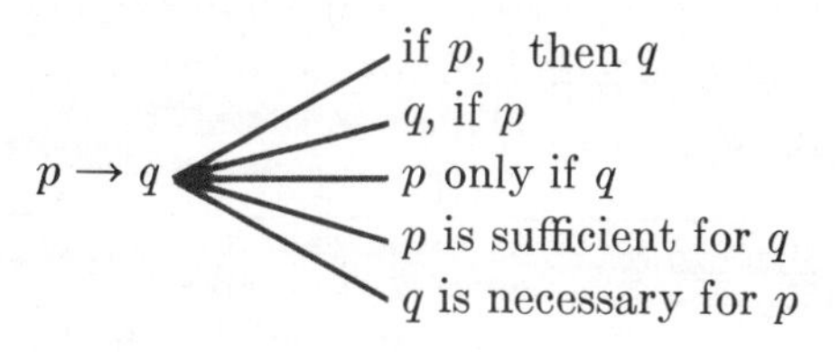

In exercise 5*d* of section 1.5, we saw $(p \leftrightarrow q) \Leftrightarrow (q \to p) \land (p \to q)$. In words, $(q \to p) \land (p \to q)$ can be written "p, if q" and "p only if q," or simply, "p if and only if q." Similarly, $(q \to p) \land (p \to q)$ can be written "p is necessary for q" and "p is sufficient for q," or simply, "p is necessary and sufficient for q." Summarizing, here is a list of verbal translations for the biconditional:

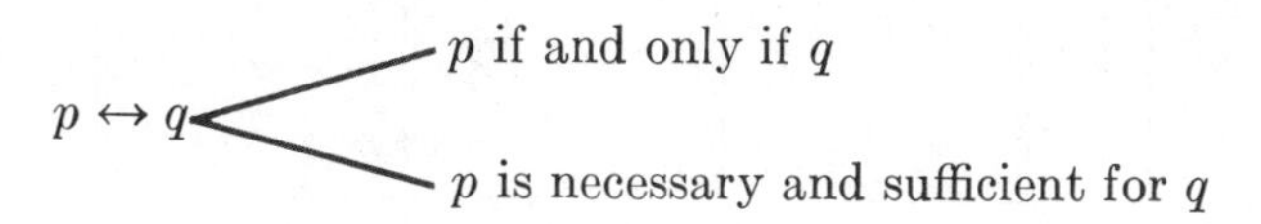

We conclude this section with an example that uses these alternate word forms of the conditional and biconditional statements.

Example 1.6.6. Let p be "profits will increase" and let q be "overhead expenses are decreased." Write a symbolic translation of each of the following:

a. If overhead expenses are decreased, then profits will increase.
b. If overhead expenses are not decreased, then profits will not increase.
c. Overhead expenses decrease, if profits increase.
d. Profits will increase if and only if overhead expenses are decreased.
e. Profits increase only if overhead expenses decrease.
f. Overhead expenses decreasing is sufficient for profits increasing.
g. Overhead expenses decreasing is necessary and sufficient for profits increasing.
h. Overhead expenses decrease only if profits do not increase.
i. It is not the case that profits increasing is sufficient for overhead expenses decreasing.
j. The converse of "Profits increasing is necessary for overhead expenses decreasing."

Solution:

a. $q \to p$
b. $\sim q \to \sim p$
c. $p \to q$
d. $p \leftrightarrow q$
e. This reads p only if q. Hence, $p \to q$.
f. This reads q is sufficient for p. Thus, $q \to p$.
g. $q \leftrightarrow p$.
h. q only if $\sim p$. So we get: $q \to \sim p$.
i. $\sim(p \to q)$
j. "Profits increasing is necessary for overhead expenses decreasing" becomes $q \to p$. So the converse is $p \to q$.

EXERCISES ■ SECTION 1.6.

1. Let p denote "I play the game fairly" and let q denote "I will not lose." Write the converse, inverse, and contrapositive of "If I play the game fairly, I will not lose."
2. Let p be "It snows" and q "we go skiing." Write a symbolic form for each of the following.

   a. If it snows, then we will go skiing.
   b. We will go skiing if it snows.
   c. It snows only if we do not go skiing.
   d. Snow is a sufficient condition for us to go skiing.
   e. We go skiing if and only if it snows.
   f. Snow is necessary for us to go skiing.
   g. Snow is a necessary and sufficient condition for us to go skiing.
   h. The inverse of "We go skiing only if it snows."
   i. The contrapositive of "It does not snow only if we do not go skiing."
   j. The converse of the inverse of "If we go skiing, it snows."

3. Let p be "you will make it" and let q be "we love you." Write a symbolic form for each of the following and then state its equivalent contrapositive.

   a. If we love you, you'll make it.
   b. You'll make it if we love you.
   c. We love you only if you'll make it.
   d. Loving you is a sufficient condition for you making it.
   e. You making it is a necessary condition for us loving you.

   (Note: all answers to the above statements are equivalent.)

1.7 ARGUMENTATION

Our second sentence on page 1 claimed that logic deals with "proving" things. Almost every high school student taking geometry has experienced the unpleasantness sometimes attached to a "proof." Listing statements and supplying reasons is "the way of life" in geometry. In logic, a "proof" is called an **argument.** An argument (just like a proof in geometry) is either valid or invalid. It is made up of two parts: the **premise(s)** (sometimes called **assertions** or **hypotheses**) and the **conclusion.**

When we test the **validity of an argument,** we are testing to see if the premises imply the conclusion. Take for example the following argument:

premise: A token is necessary for a ride on the subway (call it P_1)
premise: I rode the subway (call it P_2)
conclusion: Therefore, I had a token (call it C)

Is it valid? That is, does the conclusion logically come about, knowing only the premises? To answer this question, let's first rewrite the argument in symbolic form:

$$\begin{array}{rl} P_1: & s \to t \\ P_2: & s \\ \hline C: & \therefore t \end{array}$$

($\therefore$ means "therefore.") Examining validity is examining whether or not the premises imply the conclusion. Do both P_1 and P_2 imply C? We use a truth table to test for implication.

s	t	$[(s \to t)$	$\wedge$	$s]$	$\to$	t
T	T	T	T	T	T	T
T	F	F	F	T	T	F
F	T	T	F	F	T	T
F	F	T	F	F	T	F

Yes, the argument is valid. Since $(P_1 \wedge P_2) \to C$ is a tautology, $(P_1 \wedge P_2) \Rightarrow C$.

It is of strategic importance that the reader does not confuse the concept of validity and truth. An argument could be valid and yet it is possible for any of its parts to be false. Consider the argument just discussed.

$$\begin{array}{l} s \to t \\ s \\ \hline \therefore t \end{array}$$

Now assign new statements to the letters:

s: Saturday was yesterday
t: Tomorrow is Tuesday

Our argument now reads:

If Saturday was yesterday, tomorrow is Tuesday
Yesterday was Saturday
Therefore, tomorrow is Tuesday

We are not concerned with the truth values of the individual statements. In argumentation, we are concerned only with whether the conclusion can be deduced from the premises. That is, whenever the premises are assumed, the conclusion must follow.

Example 1.7.1. Test the following argument for validity.

$$\begin{array}{l} p \\ \underline{p \veebar q} \\ \therefore \sim q \end{array}$$

Solution: Does $[p \wedge (p \veebar q)]$ imply $\sim q$? Use a truth table.

p	q	$[p \wedge (p \veebar q)] \rightarrow \sim q$		
T	T	F	T	F
T	F	T	T	T
F	T	F	T	F
F	F	F	T	T

So, $[p \wedge (p \veebar q)] \Rightarrow \sim q$. Consequently, the argument is valid.

When an argument is not valid, we call it a **fallacy.** This occurs, of course, when the premises do not imply the conclusion, i.e., when it is possible for all premises to be true and the conclusion to be false.

Example 1.7.2. Test the following argument for validity:

If I make money, then I will start my own business
I've started my own business
Therefore, I made money

Solution: Symbolically, we have:

$$\begin{array}{l} m \rightarrow b \\ \underline{b \qquad} \\ \therefore m \end{array}$$

The truth table yields

m	b	$[(m \to b)$	$\wedge$	$b]$	$\to$	m
T	T	T	T	T	T	T
T	F	F	F	F	T	T
F	T	T	T	T	F	F
F	F	T	F	F	T	F

which is not a tautology. Hence, the argument is a fallacy.

For arguments whose premises are composed of many simple statements, a truth-table check for validity is obviously a tedious task. For example, consider the argument:

$$\begin{array}{l} P_1\colon p \veebar q \\ P_2\colon p \\ P_3\colon r \to q \\ P_4\colon s \to r \\ \hline C\colon \therefore \sim s \end{array}$$

The required truth table has sixteen rows and it would probably be a while before we can be sure of the argument's validity. So we take another approach. The test for validity does not change: do the premises imply the conclusion? That is, when all the premises are true, is the conclusion true? If the answer is yes, we have validity. Let's see. First, we shall rearrange the premises.

$$\begin{array}{l} P_2\colon p \\ P_1\colon p \veebar q \\ P_3\colon r \to q \\ P_4\colon s \to r \\ \hline C\colon \therefore \sim s \end{array}$$

We stress the fact that we are going to assume the premises to be true and then show that the conclusion must be true. To say P_2 is true merely means p is true. If P_1 is true, $p \veebar q$ must be true. But we already know p is true. These two facts force q to be false. Then since P_3 is true, r must be false. (If r were true and q were false, $r \to q$ would be false.) Now since P_4 is true and r is false, s *must be false*. This is exactly our conclusion: $\sim s$ is true. We have a valid argument.

Example 1.7.3. Test the argument for validity:

$$\begin{array}{l} a \wedge c \\ a \rightarrow b \\ b \underline{\vee} d \\ e \rightarrow d \\ e \leftrightarrow x \\ \hline \therefore c \wedge \sim x \end{array}$$

Solution: Assuming premise one to be true, a is true and c is true, and we write:

$$^{\mathrm{T}}\, a \wedge c\, ^{\mathrm{T}}$$

In premise two we conclude that b must be true (since a is true from premise one).

$$\begin{array}{l} ^{\mathrm{T}}\, a \wedge c\, ^{\mathrm{T}} \\ ^{\mathrm{T}}\, a \rightarrow b\, ^{\mathrm{T}} \end{array}$$

Including premise three and knowing that b is true, we conclude that d must be false.

$$\begin{array}{l} ^{\mathrm{T}}\, a \wedge c\, ^{\mathrm{T}} \\ ^{\mathrm{T}}\, a \rightarrow b\, ^{\mathrm{T}} \\ ^{\mathrm{T}}\, b \underline{\vee} d\, ^{\mathrm{F}} \end{array}$$

With premise four and the fact that d is false, e must be false.

$$\begin{array}{l} ^{\mathrm{T}}\, a \wedge c\, ^{\mathrm{T}} \\ ^{\mathrm{T}}\, a \rightarrow b\, ^{\mathrm{T}} \\ ^{\mathrm{T}}\, b \underline{\vee} d\, ^{\mathrm{F}} \\ ^{\mathrm{F}}\, e \rightarrow d\, ^{\mathrm{F}} \end{array}$$

Since we know that e is false and assume that premise five is true, x must be false. Hence we have:

$$\begin{array}{l} ^{\mathrm{T}}\, a \wedge c\, ^{\mathrm{T}} \\ ^{\mathrm{T}}\, a \rightarrow b\, ^{\mathrm{T}} \\ ^{\mathrm{T}}\, b \underline{\vee} d\, ^{\mathrm{F}} \\ ^{\mathrm{F}}\, e \rightarrow d\, ^{\mathrm{F}} \\ ^{\mathrm{F}}\, e \leftrightarrow x\, ^{\mathrm{F}} \end{array}$$

How about the conclusion? If it is true, we have validity. It is. We have shown c to be true and x to be false when assuming the premises to be true. Therefore $c \wedge \sim x$ is true.

Many times shortcuts can be used in testing for validity. For example, consider:

$$\begin{array}{l} p \rightarrow q \\ q \rightarrow r \\ r \rightarrow s \\ \hline \therefore p \rightarrow s \end{array}$$

Using the transitive property (example 1.5.4), $p \to q$ and $q \to r$ imply $p \to r$. Using it again, $p \to r$ and $r \to s$ imply $p \to s$, and we are finished. In section 1.6 we saw that a conditional statement is equivalent to its contrapositive. This can be used as "auxiliary power," too.

Example 1.7.4. Test the validity of the following argument.

P_1: Ronald's holding a government position (h) is sufficient for him being a member of the group (m).
P_2: Ronald is a member of the group only if he is not a conservative (c).
P_3: Ronald is a conservative if he owns a gun (g).
P_4: Ronald either owns a gun or wears flowered skivvies (s).
C: Therefore, if Ronald holds a government position, he wears flowered skivvies.

Solution: Symbolically we have:

$$\begin{array}{l} P_1\colon h \to m \\ P_2\colon m \to \sim c \\ P_3\colon g \to c \\ P_4\colon g \vee s \\ \hline C\colon \therefore h \to s \end{array}$$

Change P_3 to its contrapositive equivalent, $\sim c \to \sim g$. Change $g \vee s$ to an equivalent, $\sim g \to s$ (see exercise 6, section 1.5). We now have:

$$\begin{array}{l} P_1\colon h \to m \\ P_2\colon m \to \sim c \\ P_3\colon \sim c \to \sim g \\ P_4\colon \sim g \to s \\ \hline C\colon \therefore h \to s \end{array}$$

Now, P_1 and P_2 imply $h \to \sim c$ (by example 1.5.4), and $h \to \sim c$ and P_3 imply $h \to \sim g$. Furthermore, $h \to \sim g$ and P_4 imply $h \to s$. From the premises we conclude $h \to s$. Hence this argument is valid.

EXERCISES ■ SECTION 1.7.

1. Test the following arguments for validity.

a. $$\begin{array}{l} p \leftrightarrow q \\ q \\ \hline \therefore p \end{array}$$

b. $$\begin{array}{l} p \to q \\ \sim q \\ \hline \therefore \sim p \end{array}$$

c. $$\begin{array}{l} p \to q \\ p \vee q \\ \hline \therefore \sim p \end{array}$$

d. $$\begin{array}{l} p \leftrightarrow r \\ r \vee q \\ \sim r \\ \hline \therefore \sim(\sim r \to p) \end{array}$$

e. $p \to q$
$q \land r$
$\therefore p$

f. $\sim p \to q$
$q \to r$
$\sim s \to \sim r$
$s \to u$
$\therefore p \lor u$

g. $p \lor q$
$p \lor \sim q$
$\therefore r$

2. Rewrite the following argument symbolically and test its validity.

If Al gets arrested, Barbara will kill herself.
Either Barbara will kill herself or Al will get arrested, but not both.
Therefore, Barbara will kill herself only if Al doesn't get arrested.

3. Using the method of example 1.7.3, test the following argument to see if it is valid or a fallacy.

Gloria will leave if and only if Michael does.
Gloria left.
Edith will cook if Michael leaves.
Either Edith or Archie will cook, but not both.
Therefore, Archie will not cook.

4. Test the following argument for validity.

If Ralph wins the bet, Sam loses the bet.
Sam lost the bet.
Therefore, Ralph won the bet.

5. Write the following in symbolic form and test for validity.

If George can drive, then Harry will steal a car.
Harry's stealing a car is a necessary and sufficient condition for Isabelle's jilting him.
George cannot drive a car.
Therefore, either Harry stole a car or Isabelle jilted him.

6. Show that whenever the conclusion is a premise, the argument is valid.

7. a. Show that if two premises are inconsistent, anything can be concluded.

b. Show that a tautological conclusion assures a valid argument.

8. The following was adapted from Lewis Carroll's *Alice's Adventures in Wonderland:*

Being a green-eyed kitten is sufficient for not being able to be taught.
If you are a kitten that cannot be taught, you do not love fish.
Kittens love fish if they have whiskers.
Kittens have tails only if they have whiskers.
If you are a kitten without a tail, you do not play with gorillas.

Using all the premises, conclude something that makes the argument valid.

9. Test the following arguments for validity.

a.
$p \wedge q$
$q \rightarrow r$
$p \rightarrow s$
$\therefore r \wedge s$

b.
$p \underline{\vee} q$
$r \leftrightarrow s$
$s \wedge a$
$r \rightarrow \sim q$
$b \rightarrow a$
$\therefore p \wedge b$

c.
$a \rightarrow b$
$b \vee c$
$c \wedge a$
$\therefore c \wedge b$

d.
$r \underline{\vee} s$
$s \rightarrow k$
$k \leftrightarrow t$
t
$\sim s$
$\therefore r$

e.
$a \rightarrow \sim b$
$\sim(b \vee c)$
$c \rightarrow d$
$\therefore a \rightarrow d$

f.
$p \wedge q$
$p \wedge \sim q$
$\therefore r$

1.8 CONSTRUCTING STATEMENTS WITH SPECIFIED TRUTH VALUES

In section 1.4, our objective was to find truth values of given compound statements. In this section we will do just the reverse; we will find a compound statement having a given set of truth values.

To illustrate, consider the truth table below.

p	q	?
T	T	T
T	F	T
F	T	F
F	F	T

Notice that all truth values are already specified. The "?" in the last column indicates that we are looking for some compound statement with truth values TTFT. You may, if lucky, guess a correct answer. However, our objective here is to develop a systematic way of obtaining the answer. We need a compound statement that is true in the first, second, or fourth cases only.

Before explaining how to construct a general compound statement, we will construct statements that are true in just one case. Such statements are called **basic conjunctions.** For the case of two simple statements p, q, there are four basic conjunctions listed below.

p	q	Basic conjunctions
T	T	$p \wedge q$
T	F	$p \wedge \sim q$
F	T	$\sim p \wedge q$
F	F	$\sim p \wedge \sim q$

Recalling that a conjunction of two simple statements is true only when both of the simple statements are true (see the truth table for $\wedge$ in section 1.2), it is easy to see where the above basic conjunctions come from. For example, consider the basic conjunction in case II, where p is true while q is false. In order to make a true conjunction here, we need to use p (which is true) with $\sim q$ (which is true); doing so gives $p \wedge \sim q$. Similar comments follow for the other basic conjunctions.

We now return to the problem of finding a compound statement with the truth table in the beginning of this section. We want to find a compound statement that is true in the first, second, and fourth cases only. Thus we connect the basic conjunctions for each of these cases with $\vee$ to get the answer: $(p \wedge q) \vee (p \wedge \sim q) \vee (\sim p \wedge \sim q)$.

We summarize what we have done in the following table.

p q	$p \wedge q$	$p \wedge \sim q$	$\sim p \wedge \sim q$	$(p \wedge q) \vee (p \wedge \sim q) \vee (\sim p \wedge \sim q)$
T T	T	F	F	T
T F	F	T	F	T
F T	F	F	F	F
F F	F	F	T	T

Observe that the column at the right has the desired truth values, TTFT.

Example 1.8.1. Find a compound statement with the truth values below.

p	q	?
T	T	F
T	F	T
F	T	F
F	F	T

Solution: Here we need the basic conjunctions for cases II and IV:

$$(p \wedge \sim q) \vee (\sim p \wedge \sim q)$$

The same technique just discussed can easily be extended to cover eight-row truth tables as soon as we list the basic conjunctions below:

p	q	r	Basic conjunctions
T	T	T	$p \wedge q \wedge r$
T	T	F	$p \wedge q \wedge \sim r$
T	F	T	$p \wedge \sim q \wedge r$
T	F	F	$p \wedge \sim q \wedge \sim r$
F	T	T	$\sim p \wedge q \wedge r$
F	T	F	$\sim p \wedge q \wedge \sim r$
F	F	T	$\sim p \wedge \sim q \wedge r$
F	F	F	$\sim p \wedge \sim q \wedge \sim r$

Example 1.8.2. Find a compound statement having the truth values indicated in the truth table below.

p	q	r	?
T	T	T	T
T	T	F	T
T	F	T	F
T	F	F	F
F	T	T	F
F	T	F	T
F	F	T	F
F	F	F	F

Solution: We must use the basic conjunctions for cases I, II, and VI:

$$(p \wedge q \wedge r) \vee (p \wedge q \wedge \sim r) \vee (\sim p \wedge q \wedge \sim r)$$

Example 1.8.3. Construct a compound statement having the indicated truth values shown below.

p	q	r	?
T	T	T	T
T	T	F	T
T	F	T	T
T	F	F	T
F	T	T	F
F	T	F	F
F	F	T	T
F	F	F	T

Solution: Here we need the basic conjunctions for all but the fifth and sixth cases. That's a lot of writing. Instead we can use the basic conjunctions for cases V and VI and negate the result:

$$\sim[(\sim p \wedge q \wedge r) \vee (\sim p \wedge q \wedge \sim r)]$$

The reader is asked to verify the technique for this example in exercise 1 at the end of this section.

Example 1.8.4. Write a compound statement that is equivalent to $p \rightarrow q$.

Solution: We know the truth table for $p \rightarrow q$ and we rewrite it below.

p	q	$p \rightarrow q$
T	T	T
T	F	F
F	T	T
F	F	T

Now we reorient our thoughts and ask what compound statement we could construct (using basic conjunctions) having the truth values below.

p	q	?
T	T	T
T	F	F
F	T	T
F	F	T

Method I: Use basic conjunctions for cases I, III, IV. We get

$$(p \wedge q) \vee (\sim p \wedge q) \vee (\sim p \wedge \sim q)$$

Method II: Use the basic conjunction for case II and negate it. We get

$$\sim(p \wedge \sim q)$$

Summarizing for example 1.8.4, we have discussed how to construct a compound statement equivalent to a given compound statement. You may note that the solution in method II involves the negation of a conjunction. This was treated in example 1.5.5 (one of DeMorgan's Laws). Using that law in our solution in method II, $\sim(p \wedge \sim q)$ can be rewritten as $\sim p \vee q$. So it turns out that we have derived the equivalence $(p \rightarrow q) \Leftrightarrow (\sim p \vee q)$.

You may remember that you were asked to verify this equivalence back in the exercises for section 1.5 (see problem 6). Maybe at that time you wondered where it came from—now you know!

EXERCISES ■ SECTION 1.8.

1. The answer to example 1.8.3 was given as $\sim [(\sim p \wedge q \wedge r) \vee (\sim p \wedge q \wedge \sim r)]$. Work out the truth table of this statement to verify that the result really is TTTTFFTT.
2. Using the method derived in this section, find a compound statement with the truth table below.

p	q	?
T	T	T
T	F	T
F	T	T
F	F	F

Observe that the result you get is equivalent to $p \vee q$.

3. Construct a compound statement having truth tables as indicated in the table below.

p	q	r	a	b	c
T	T	T	T	T	F
T	T	F	F	F	T
T	F	T	F	F	T
T	F	F	F	F	T
F	T	T	F	T	F
F	T	F	T	F	T
F	F	T	F	T	F
F	F	F	F	F	T

4. Using the method derived in this section, construct a compound statement having truth values as shown in the table below.

p	q	?
T	T	F
T	F	F
F	T	F
F	F	F

5. Proceeding as in example 1.8.4, construct a statement equivalent to each given statement.

a. $p \leftrightarrow q$ **b.** $\sim(p \rightarrow q)$

c. $p \underline{\vee} q$ **d.** $q \rightarrow \sim p$

1.9 SWITCHING NETWORKS

In this section and the next, we discuss some applications of the principles of logic we have developed so far. This particular section deals with **switching networks.** A switching network is an arrangement of wires and switches connecting two terminals. Before proceeding, however, we will explain a little "basic electricity."

An electric circuit is quite similar to a self-contained water system. Consider the diagram below. The water pump will pump water through

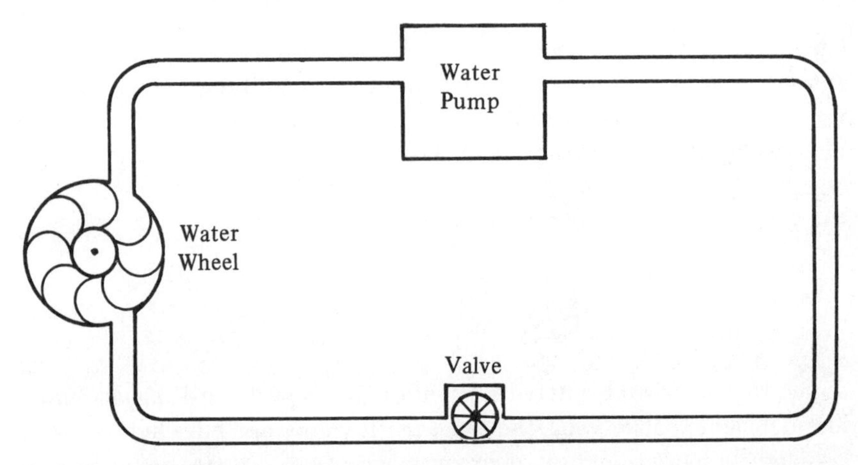

the pipe. The purpose of the system is to turn the water wheel, which we can imagine is connected to some machine that is doing some sort of useful work. The valve controls the whole process: in one position it permits the flow of water (wheel will turn), and in the other position it prohibits the flow of water (wheel will not turn). A simple electric circuit is quite analogous.

The water pump is replaced by an "electricity pump" (a battery or generator), the pipe is replaced by wire, and the valve by a switch. In place of the water wheel we have some piece of electrical apparatus (motor, toaster, bulb, etc.). The water itself is replaced by electric current (actually tiny things called electrons).

So a simple electric circuit looks like this:

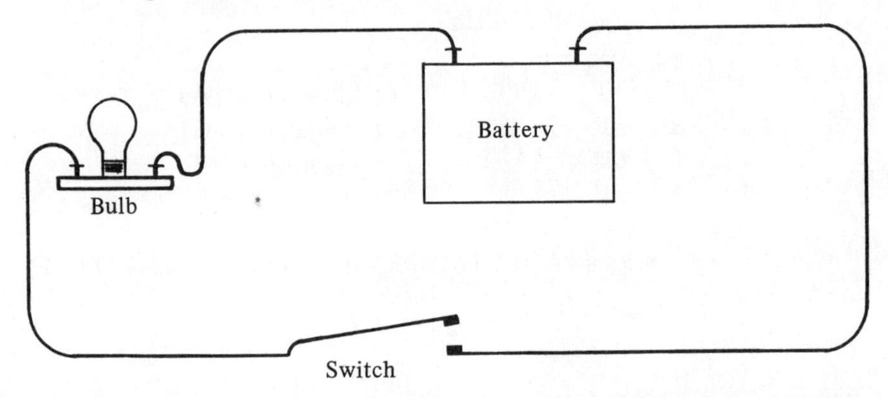

The battery "pumps" electric current through the wire. The purpose of the system is to light the bulb. The switch controls the whole process. An open switch prevents the flow of current, while a closed switch permits current to flow. Since it is the switch that completely controls the process, we depict only that part of the circuit below.

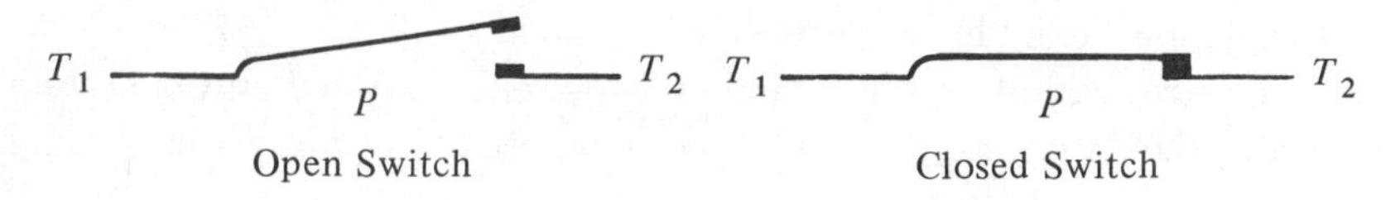

Open Switch Closed Switch

The symbols T_1 and T_2 indicate the switch terminals. Letting p be the statement "switch P is closed," there are two possibilities: either p is true or false. We summarize this, along with the bulb function and corresponding switch positions, in the table below.

p	Bulb lights	
T	T	P closed
F	F	P open

Notice that a T is interpreted as "bulb lights," while an F means "bulb does not light"; also, p has the same truth values as "bulb lights."

A slightly more complicated circuit is when two switches are connected in a **series network,** as shown below. (All four possible switch positions are shown.)

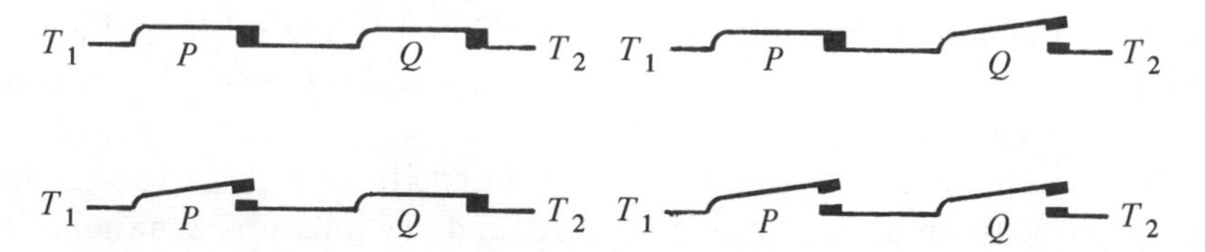

Let p be "switch P is closed" and q be "switch Q is closed." For the bulb to light, current must pass through both switches. We summarize below.

p	q	Bulb lights	
T	T	T	P closed, Q closed
T	F	F	P closed, Q open
F	T	F	P open, Q closed
F	F	F	P open, Q open

Notice that for a series network, the truth values of "bulb lights" (TFFF) are the same as for $p \wedge q$.

Another way to connect switches is in a **parallel network,** shown below. (Again all possible switch positions are pictured.)

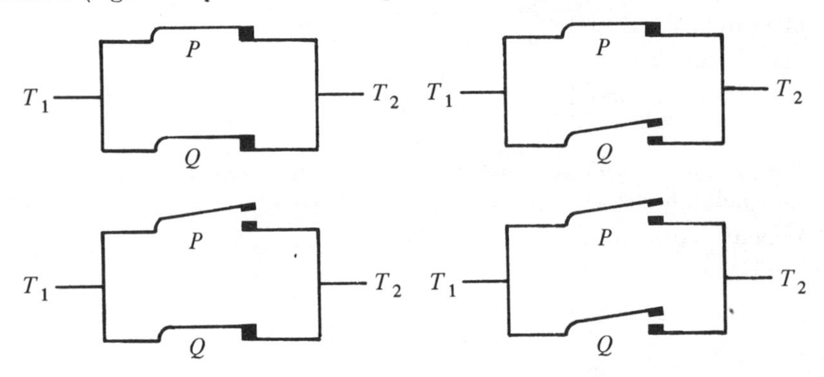

Let p be "switch P is closed," while q denotes "switch Q is closed." Observe that in order for current to pass through the network (thus causing the bulb to light), at least one of the two switches must be closed.

p	q	Bulb lights	
T	T	T	P closed, Q closed
T	F	T	P closed, Q open
F	T	T	P open, Q closed
F	F	F	P open, Q open

Here, "bulb lights" has truth values TTTF. These, of course, are the same as for $p \vee q$.

So far, we have the switch network–logic statement analogies shown below. (We have replaced the switches in the drawings by just a letter to save writing.)

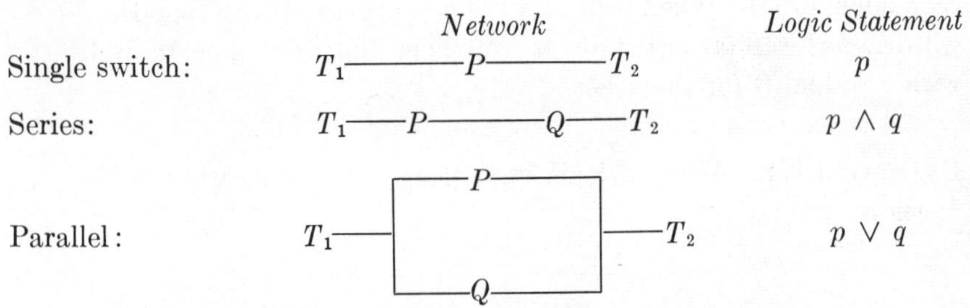

Now we are ready to investigate more interesting and involved networks, which can always be treated as a certain conglomeration of series and parallel parts. Consider the network below.

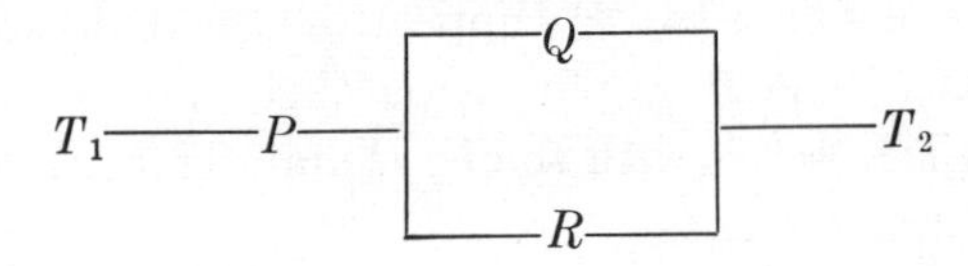

Let the following statements represent the switches:

p: Switch P is closed
q: Switch Q is closed
r: Switch R is closed

The Q, R part of the network is parallel, so it is represented by $q \vee r$. That whole unit is in series with P, so we write $p \wedge (q \vee r)$ as a logic statement representing this network. A truth table gives even more information.

p	q	r	$p \wedge (q \vee r)$	
T	T	T	T	P closed, Q closed, R closed
T	T	F	T	P closed, Q closed, R open
T	F	T	T	P closed, Q open, R closed
T	F	F	F	P closed, Q open, R open
F	T	T	F	P open, Q closed, R closed
F	T	F	F	P open, Q closed, R open
F	F	T	F	P open, Q open, R closed
F	F	F	F	P open, Q open, R open

The T's in the last column indicate the cases in which the bulb will light.

Sometimes in switching networks we use **coupled switches**. This means that two switches are arranged (mechanically) to operate either together (both closed, both open) or oppositely (one open while the other is closed). When switches are coupled together, we use the same capital letter for each such switch. When switches are coupled to operate oppositely, we will use any capital letter for one of them and that same capital letter with a prime (′) for the other.

Example 1.9.1. Write a logic statement that represents the network below.

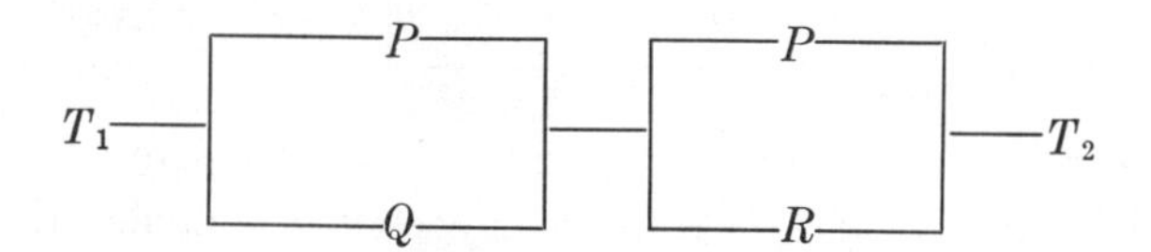

Solution: The fact that two P's appear tells us that those two switches operate together. The P, Q part is parallel, and so is the P, R part. The P, Q unit is in series with the P, R unit. Thus, $(p \vee q) \wedge (p \vee r)$ is the desired statement.

Example 1.9.2. Write a logic statement corresponding to the circuit below.

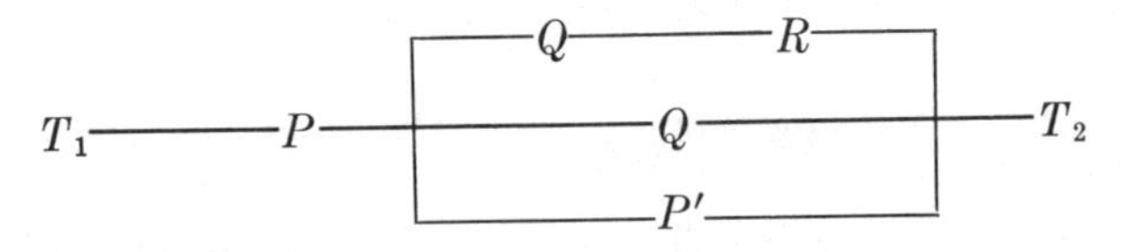

Solution: Consider:

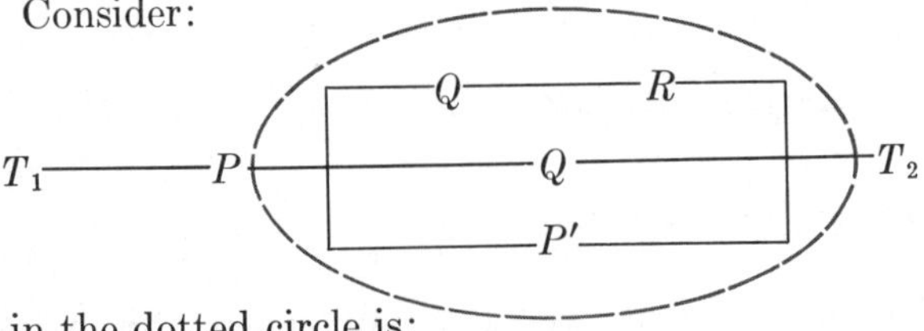

The part in the dotted circle is:

1. itself a parallel unit, and
2. itself in series with the single switch P at the left.

We get $p \wedge [(q \wedge r) \vee q \vee \sim p]$ as our answer. Notice that the logic symbol for P' becomes $\sim p$, assuring us that switches P and P' do operate oppositely, since p and $\sim p$ are, indeed, opposites.

It is also possible to start with a logic statement (involving $\sim$, $\wedge$, $\vee$) and to draw a network corresponding to it.

Example 1.9.3. Starting with $(p \vee q) \wedge (r \vee q \vee \sim q)$, draw the corresponding network.

Solution: Just remember that $\vee$ means parallel, $\wedge$ means series, and $\sim$ becomes $'$.

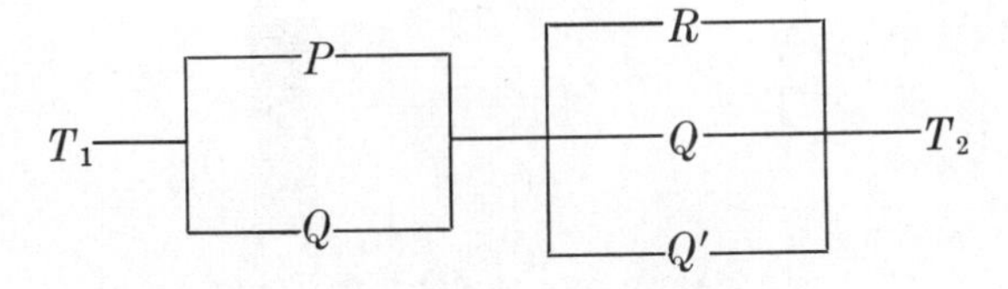

Example 1.9.4. Draw a network corresponding to

$$(p \wedge q) \vee [p \wedge (\sim q \vee \sim p)]$$

Solution:

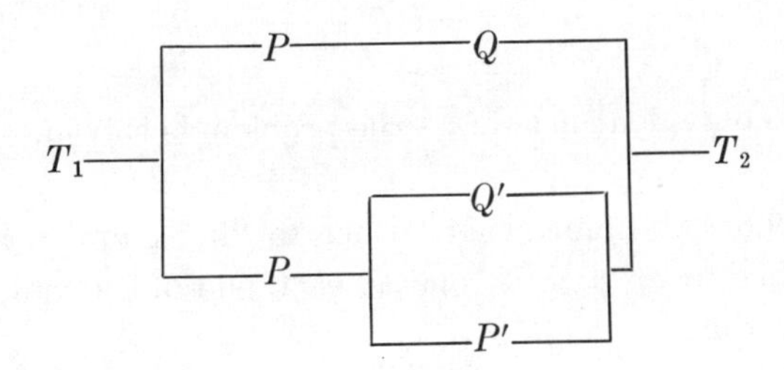

The network theory we have set forth here along with the notion of basic conjunctions (recall section 1.8) can be used to design networks having special properties. The following examples show us how.

Example 1.9.5. An engineer designed the network of switches below.

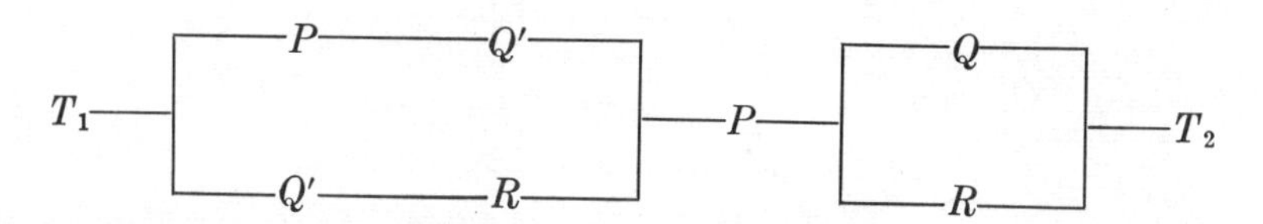

Find a simpler, but equivalent network.

Solution: First we write the logic statement corresponding to this network:

$$[(p \wedge \sim q) \vee (\sim q \wedge r)] \wedge p \wedge (q \vee r)$$

Next we work the truth table of this statement. (We show the final result below.)

p	q	r	$[(p \wedge \sim q) \vee (\sim q \wedge r)] \wedge p \wedge (q \vee r)$
T	T	T	F
T	T	F	F
T	F	T	T
T	F	F	F
F	T	T	F
F	T	F	F
F	F	T	F
F	F	F	F

Now we write a statement equivalent to $[(p \wedge \sim q) \vee (\sim q \wedge r)] \wedge p \wedge (q \vee r)$ using the basic conjunction for case III in the table above. Doing so gives $p \wedge \sim q \wedge r$, which has the network interpretation below.

T_1——P——Q'——R——T_2

This is an equivalent network—the problem is solved.

Example 1.9.6. A homeowner wishes to "hook up" a hall light with two switches to operate it, one at each end of the hall. Design a network to do this.

Solution: Call the switches P, Q. We want the bulb to light when we operate exactly one of the switches. Summarized results appear below, when p is "switch P is closed" and q is "switch Q is closed."

p	q	Bulb lights
T	T	F
T	F	T
F	T	T
F	F	F

Thus the truth table of our network is FTTF. So a logic statement for our network is: $(p \wedge \sim q) \vee (\sim p \wedge q)$. (This last result follows by using the basic conjunctions on cases II and III of the table above.) Hence, a network with the desired property is now known.

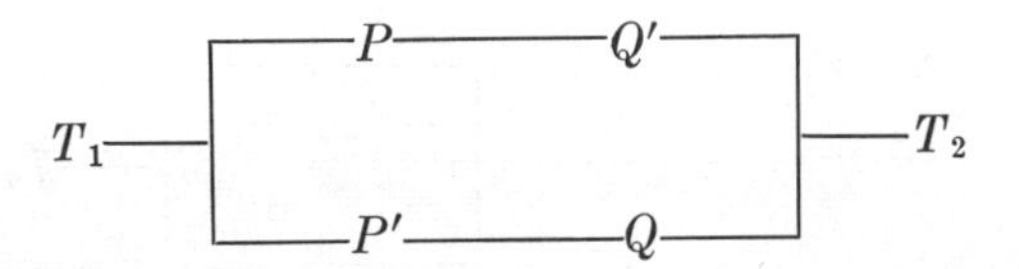

Example 1.9.7. A society of three people wishes to be able to record a majority vote secretly. Design a switching network that will allow the flow of current when a majority votes for an issue.

Solution: Use p for "person 1 closes switch P," thus voting affirmatively. Use q for "person 2 closes switch Q," thus voting affirmatively. Use r for "person 3 closes switch R," thus voting affirmatively.

p	q	r	current flows
T	T	T	T
T	T	F	T
T	F	T	T
T	F	F	F
F	T	T	T
F	T	F	F
F	F	T	F
F	F	F	F

The table tells us that current flows whenever two or more people vote affirmatively. So a logic statement for "current flows" is

$$(p \wedge q \wedge r) \vee (p \wedge q \wedge \sim r) \vee (p \wedge \sim q \wedge r) \vee (\sim p \wedge q \wedge r)$$

We have then, the network shown below.

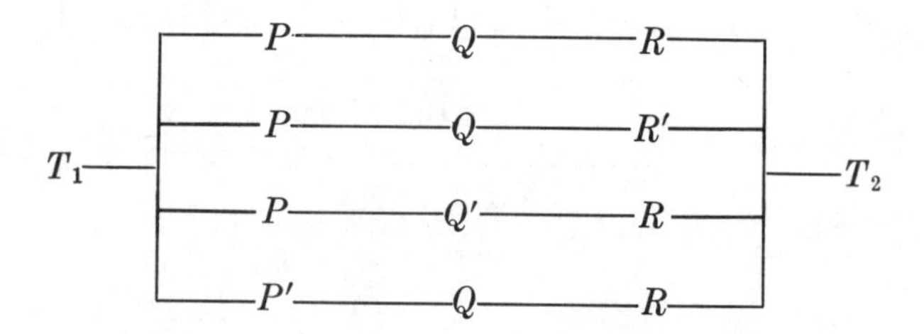

EXERCISES ■ SECTION 1.9.

1. For each of the diagrams below, write the logic statement that corresponds to the network shown.

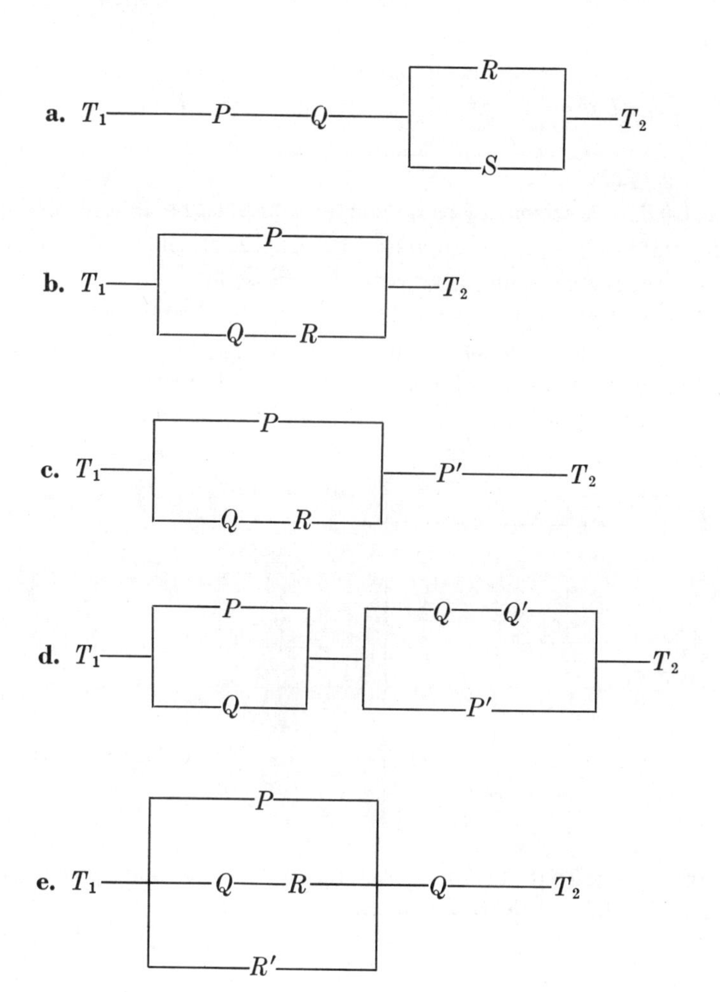

f.

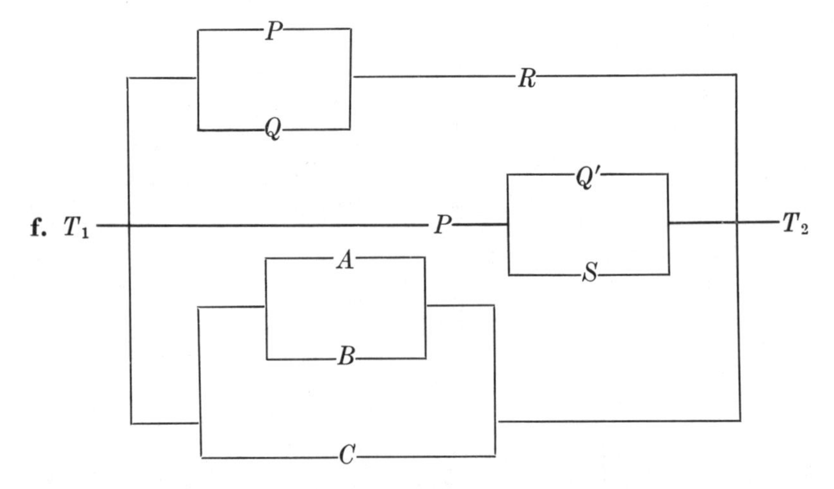

2. For each of the logic statements below, construct the corresponding network.

a. $[p \vee (q \wedge r)] \wedge \sim p$

b. $(p \wedge q) \vee [r \wedge (\sim r \vee \sim q)]$

c. $(p \vee q) \wedge r \wedge (p \vee q)$

d. $p \vee (q \wedge r) \vee [s \wedge (t \vee u)]$

3. Design a network of switches that is simpler than but equivalent to the network shown below.

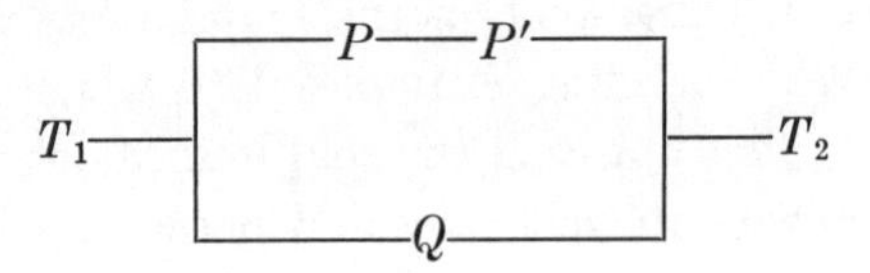

4. Design a network that corresponds to the logic statement $p \rightarrow q$.

5. Four people wish to record a majority vote in favor of a measure by using a switching network whereby an individual votes "yes" by closing a switch. In case of a tie, the outcome is to be the same as the chairman's vote. (The chairman is one of the four people.) Design a network with the above-mentioned properties.

6. Recall that a tautology is a statement that is always true. Show that the network below represents a tautology.

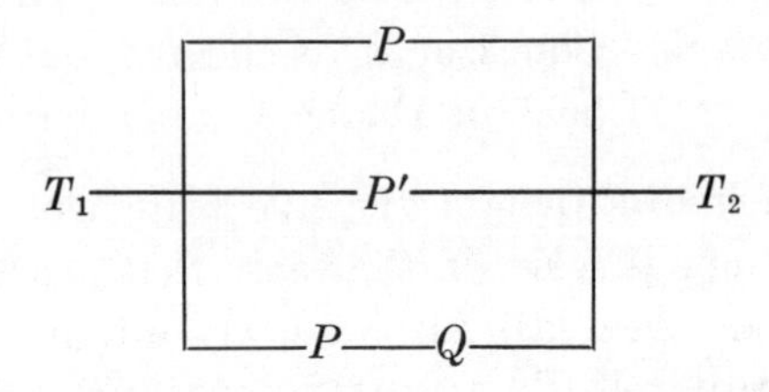

7. Recall that a self-contradiction is a statement that is never true (always false). Show that the network below represents a self-contradiction.

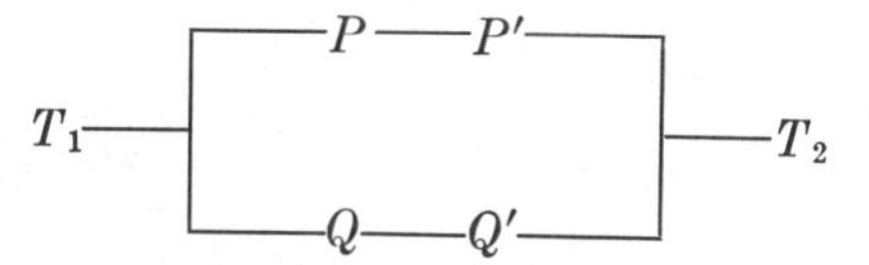

8. Consider the network:

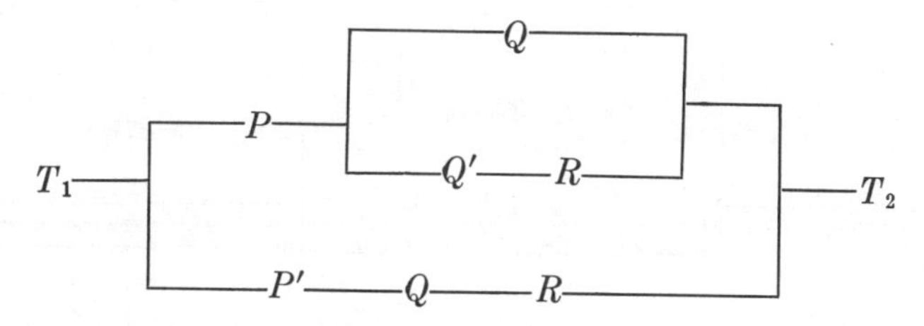

a. Construct a statement for this network.
b. Construct a truth table for your answer to a.
c. Compare your result with example 1.9.7. (The reader should note that the networks are equivalent.)

1.10 LOGIC RECREATIONS

In this section we will use some of the foundations we have developed in this chapter. While reading examples 1.10.1, 1.10.2, and 1.10.3, we suggest that you at first not read the solution. Merely try to think them through. The problems are not easy and perhaps, when you are ready for the solution, more than one reading will be necessary. Don't guess. After all, logic is the art of reasoning; use it!!

Example 1.10.1. Professor Plum knows that exactly two students in his music class cheated on the final exam. He has narrowed his suspects down to Arlo, Bing, Chuck, Duane, and Elvis. He asks the students into his office separately and they make the following statements:

1. Arlo answers, "Bing or Chuck cheated, but not both."
2. Bing answers, "Duane or Arlo cheated, but not both."
3. Chuck answers, "Elvis or Bing cheated, but not both."
4. Duane answers, "Chuck or Elvis cheated, but not both."
5. Elvis answers, "Chuck or Duane cheated, but not both."

Professor Plum knows that of the five statements, exactly four are true. The other one is false (in the sense that neither boy mentioned in that statement cheated). Also, a lowly informer, Stu Lee (whom Plum knows always tells the truth), volunteers:

6. "If Duane cheated, Chuck cheated."

Solution: Let

a: Arlo cheated
b: Bing cheated
c: Chuck cheated
d: Duane cheated
e: Elvis cheated

Since exactly two students cheated, there will have to be exactly two T's in each row of a truth table. All possibilities give:

a	b	c	d	e	Can this be the case?
T	T	F	F	F	NO. This case says (4) and (5) are both false.
T	F	T	F	F	
T	F	F	T	F	NO. This case makes (6) false.
T	F	F	F	T	NO. This case says (1) and (5) are both false.
F	T	T	F	F	NO. This case says (1) and (2) are both false.
F	T	F	T	F	NO. This case makes (6) false.
F	T	F	F	T	NO. (2), (3), (5) would all be false here.
F	F	T	T	F	NO. This case says (3) and (5) are both false.
F	F	T	F	T	NO. This case says (2) and (4) are both false.
F	F	F	T	T	NO. This case makes (6) false.

The only possibility is case II, where a is true and c is true. Hence the cheaters are Arlo and Chuck.

Example 1.10.2. A personnel director, Miss Muff, has accidentally shuffled job applications involving Mr. W, Mr. X, Mr. Y, Mr. Z. She knows one is an artist, one is a bellboy, one is a croupier, and one is a dentist. Trying to cover up her bungle, she asks three of the men two questions and gets the following statements:

Mr. W says, "Mr. Y is an artist." Call it p.
Mr. W says, "Mr. X is a bellboy." Call it q.
Mr. X says, "Mr. Y is a bellboy." Call it r.
Mr. X says, "Mr. Z is a croupier." Call it s.
Mr. Y says, "Mr. W is a bellboy." Call it t.
Mr. Y says, "Mr. Z is a dentist." Call it u.

Miss Muff has determined that each man is telling one truth and one falsehood. Who's who?

Solution: We have the following inconsistencies: $p \mathbf{I} r$, $s \mathbf{I} u$, $r \mathbf{I} t$, $q \mathbf{I} r$, and $q \mathbf{I} t$. From what Miss Muff has determined, we know each of $(p \underline{\vee} q)$, $(r \underline{\vee} s)$, $(t \underline{\vee} u)$ is true. So, the only cases needed for examination are:

p	q	r	s	t	u	Can this be the case?
T	F	T	F	T	F	NO. (p and r cannot both be true.)
T	F	T	F	F	T	NO. (p and r cannot both be true.)
T	F	F	T	T	F	
T	F	F	T	F	T	NO. (s and u cannot both be true.)
F	T	T	F	T	F	NO. (r and t cannot both be true.)
F	T	T	F	F	T	NO. (q and r cannot both be true.)
F	T	F	T	T	F	NO. (q and t cannot both be true.)
F	T	F	T	F	T	NO. (s and u cannot both be true.)

Case III is the only case that satisfies all the conditions of the problem. In case III, p is true, s is true, and t is true. This means:

Mr. Y is the artist
Mr. Z is the croupier
Mr. W is the bellboy

So, Mr. X must be the dentist.

Example 1.10.3. Dick Slooth, super-detective, knows that at 3 a.m., Monday, January 14, someone in the McKinley household murdered Iona Yoo, the maid, in cold blood. He has narrowed down his list of suspects to:

Al, Iona's lover and good friend
Bret, the butler
Carl, the chauffeur and a known communist
Darla, Iona's twin sister

They make the following statements:

1. Al says, "Bret did it."
2. Bret says, "Al's lying."
3. Carl says, "I didn't do it."
4. Darla says, "Al did it."

Detective Slooth knows that exactly one of the statements is true. Who dunnit?

Solution: Let's attack the problem logically. First, using symbols, we have:

a: Al did it
b: Bret did it
c: Carl did it
d: Darla did it

Translating the statements (1)–(4) symbolically, we have:

1. b
2. $\sim b$
3. $\sim c$
4. a

Because there is only one murderer, only four cases are necessary as we look at the table below:

a	b	c	d
T	F	F	F
F	T	F	F
F	F	T	F
F	F	F	T

← Case II, for example, is the case where Bret is the murderer.

Now, we have:

a	b	c	d	(1) b	(2) $\sim b$	(3) $\sim c$	(4) a
T	F	F	F	F	T	T	T
F	T	F	F	T	F	T	F
F	F	T	F	F	T	F	F
F	F	F	T	F	T	T	F

Only one of the statements (1)–(4) is true, so we must examine each of the four rows on the right to determine which has exactly one T. It is the third row, which solves the mystery. Hence, Carl is the ruthless murderer!!

EXERCISE ■ SECTION 1.10.

1. Rework example 1.10.3 except this time assume that Slooth knows that exactly *three* tell the truth.

chapter 1 in review

CHAPTER SUMMARY

In this chapter we have learned how to symbolize and connect statements. Lower case letters are used for simple statements and we have considered six connectives: $\sim$, $\wedge$, $\vee$, $\underline{\vee}$, $\rightarrow$, $\leftrightarrow$. Using one or more of these connectives makes possible the construction of compound statements. In order to understand the meaning of a statement, it becomes necessary to know when it is true and when it is false. A truth table displays that information in concise form and allows us to tell at a glance whether a statement is a tautology, a self-contradiction, or neither. When we compare the truth tables of different statements, we can determine whether those statements are equivalent, whether one implies another, or whether they are inconsistent. We also made applications of our study of logic. Arguments consist of statements organized into premises and a conclusion. Further, an argument is either valid or not valid. Only in a valid argument do the premises imply the conclusion. Switching networks can be represented by statements referring to the position of a switch (open or closed). It is also possible to construct networks with certain specified properties.

VOCABULARY

simple statement
compound statement
negation
truth value
truth table
conjunction
inclusive disjunction
exclusive disjunction
conditional
biconditional
tautology
self-contradiction
implication
transitive property
equivalence
DeMorgan's laws
inconsistency
consistent (compatible)
converse
inverse
contrapositive
valid argument
fallacy
basic conjunctions
switching networks
series network
parallel network

SYMBOLS

$p, q, r, \ldots$
$\sim$
$\wedge$
$\vee$
$\underline{\vee}$
$\rightarrow$
$\leftrightarrow$
$\Rightarrow$
$\Leftrightarrow$
$\mathbf{I}$
$\therefore$

chapter 2
Sets

2.1 INTRODUCTION

A **set,** as a point and a line in geometry, is so fundamental to mathematics that it is left as an undefined concept. It suffices here only to convey an intuitive meaning of set to the reader. Basically, a set is a collection of things. The things in this collection are called **elements** or **members.** In order for sets to have meaning, we must make the assumption that each set is well-defined (we know exactly what is an element of the set and what is not). For example, the set of New Yorkers with bad breath is not well-defined unless a clear-cut definition of bad breath is made, so every New Yorker can be classified.

We shall make other conventions concerning sets, too. Consider the set $S = \{1,3,2,\&\}$. We shall use braces, $\{\quad\}$, and place the elements within them. We say, "1 is an element of S" and write $1 \in S$. Also, we can write $4 \notin S$ to mean "4 is not an element of S." The elements of a set cannot always be listed (as those of S are). For example, consider the set P of all people on Earth. Surely it would be an impossible task to attempt to list the elements of this set. For this reason, it is enough to describe the set as

$$P = \text{the set of people on Earth}$$

Example 2.1.1. Each of the following is true.

a. $7 \in \{1,2,3,7\}$
b. $HT \in \{HH, HT, TH, TT\}$
c. $9 \notin \{1,2,3,7\}$

d. Martha Washington is an element of the set of first ladies of the United States.

e. General Motors is an element of the set of the ten largest U.S. corporations.

The last convention we shall make is the restriction to finite sets. (After all, this is FINITE mathematics!) So, the set of all numbers greater than 2, for example, will be of no interest to us.

The sets that will be of most concern to us will be those whose elements are the result of a certain occurrence, that is, the **set of all logical possibilities** governing a certain situation. For an example, let us consider the following situation. Suppose John and Norbert play a three-game handball tournament in which the victor is he who wins two games first. Then the set of all possible outcomes (the set of all logical possibilities) is

$$\{JJ, JNJ, NJJ, JNN, NJN, NN\}$$

where JNJ, for example, means John wins the first and third games (and hence, the tournament) and Norbert wins the second. Usually the set of all permissible elements in any particular discussion is denoted by $\mathcal{U}$ (called the **universal set**). So, for John and Norbert's tournament,

$$\mathcal{U} = \{JJ, JNJ, NJJ, JNN, NJN, NN\}$$

Example 2.1.2. List the universal set for the experiment of tossing a die once.

Solution: Since a die has six faces, we write:

$$\mathcal{U} = \{1,2,3,4,5,6\}$$

Example 2.1.3. List the set of all logical possibilities, $\mathcal{U}$, for the experiment of tossing a nickel and a dime.

Solution: When you toss, the outcomes could be: (1) a head on the nickel and head on the dime; (2) a head on the nickel and a tail on the dime; (3) a tail on the nickel and a head on the dime; (4) tails on both. We can list set $\mathcal{U}$ as

$$\mathcal{U} = \{HH, HT, TH, TT\}.$$

2.2 OPERATIONS ON SETS

Operations are to sets as connectives are to logic. Consider the following example:

$$\mathcal{U} = \{1,2,3,4,5\}, \qquad A = \{1,3,4\}, \qquad B = \{3,4,5\}$$

In this example $\mathcal{U} = \{1,2,3,4,5\}$, so throughout the example *only these five elements can be discussed.* Now we form the set $\{2,5\}$ consisting of exactly those elements that are NOT elements of set A. We call this set "the **complement** of A" and write $\tilde{A}$. (Note the similarity that exists between the word "not" and the set operation "complement.") Similarly, $\tilde{B} = \{1,2\}$.

Now, by looking at sets A and B, we can form the set $\{3,4\}$ by listing only those elements that are members of set A AND members of set B. This set operation is called **intersection;** we write $A \cap B = \{3,4\}$. (Note the similarity between "and" and "intersection.")

If we take elements that satisfy the condition of being either elements of set A OR elements of set B, we have $\{1,3,4,5\}$. This set is called the **union** of set A with set B; we write $A \cup B = \{1,3,4,5\}$.

Example 2.2.1. Return to the die-tossing experiment of example 2.1.2. Let A = the set of situations where an odd number appears, B = the set of situations where a number greater than 2 appears. Find $\tilde{A}$, $\tilde{B}$, $A \cap B$, $A \cup B$, $A \cap \tilde{B}$, $\tilde{A} \cap \tilde{B}$, $\widetilde{(A \cup B)}$.

Solution: $\mathcal{U} = \{1,2,3,4,5,6\}$, $A = \{1,3,5\}$, $B = \{3,4,5,6\}$

$\tilde{A}$ = the set of elements in $\mathcal{U}$ but not in $A = \{2,4,6\}$

$\tilde{B} = \{1,2\}$

$A \cap B$ = the set of elements that are in A and also in $B = \{3,5\}$

$A \cup B$ = the set of elements that are in either set A or set $B = \{1,3,4,5,6\}$

$A \cap \tilde{B}$ = the set of elements in A but not in $B = \{1\}$

$\tilde{A} \cap \tilde{B} = \{2\}$

$\widetilde{(A \cup B)}$ = the set of elements not in $A \cup B = \{2\}$

Example 2.2.2. Refer to the handball tournament between John and Norbert in the previous section. Let C = the outcomes where John wins the tournament, D = the outcomes where Norbert wins the tournament, E = the set of outcomes where only two games are necessary. Find: $\tilde{C}$, $D \cup E$, $D \cap \tilde{E}$, $\widetilde{(D \cap \tilde{E})}$, $C \cap D$.

Solution: First, $\mathcal{U} = \{JJ,JNJ,NJJ,JNN,NJN,NN\}$, $C = \{JJ,JNJ,NJJ\}$, $D = \{JNN,NJN,NN\}$, $E = \{NN,JJ\}$. Now,

$\tilde{C} = \{JNN,NJN,NN\}$

$D \cup E = \{JJ,JNN,NJN,NN\}$

$D \cap \tilde{E} = \{JNN,NJN\}$

$\widetilde{(D \cap \tilde{E})}$ = everything except JNN and $NJN = \{JJ, JNJ, NJJ, NN\}$

For the intersection of sets C and D, observe that no situation is common to both sets. Hence, the resulting set is the **empty set,** symbolized by $\varnothing$. Whenever we have a set void of elements, we shall denote it by $\varnothing$.

When two sets have no elements in common (that is, their intersection is $\varnothing$) they are said to be **disjoint.** The set of men and the set of women, for example, are disjoint.

EXERCISES ■ SECTIONS 2.1 AND 2.2.

1. In the experiment of tossing a penny and a die simultaneously, we have

$$\mathcal{U} = \{H1,H2,H3,H4,H5,H6,T1,T2,T3,T4,T5,T6\}$$

Let A = the set of situations where an even number appears on the die, B = the set of situations where a head appears on the penny, $C = \{T2,T6\}$. Find:

a. $\tilde{A}$ **b.** $\tilde{B}$ **c.** $\tilde{C}$
d. $A \cap B$ **e.** $A \cup C$ **f.** $\tilde{A} \cap \tilde{B}$
g. $A \cap \tilde{C}$ **h.** $\tilde{A} \cap C$ **i.** $(B \cap A) \cup C$
j. $\widetilde{[(B \cap A) \cup C]}$ **k.** $C \cup (\tilde{A} \cap \tilde{B})$ **l.** $\tilde{A} \cap \tilde{B} \cap \tilde{C}$

2. A set operation frequently mentioned in written works is the **difference** of two sets. The difference of two sets A, B (denoted by $A - B$) is defined to be $A \cap \tilde{B}$. Using A, B, C as in exercise 1 above, find:

a. $A - B$ **b.** $B - C$ **c.** $A - C$ **d.** $B - A$

3. Still another definition sometimes appearing in set theory texts is the concept of **symmetric difference** of two sets. The symmetric difference of A and B (denoted $A \triangle B$) is defined to be:

$$A \triangle B = (A - B) \cup (B - A)$$

Using A, B, C as in exercise 1, find:

a. $A \triangle B$ **b.** $B \triangle C$ **c.** $A \triangle C$

4. The top sales people for the Groin Corporation in 1973 and 1974 were, respectively, A = {Adams, Wells, Young} and B = {Young, Wells, Kaplunkski}. Find:

a. $A \cap B$
b. $A \cup B$
c. $A - B$ (Note: $A - B$ is obtainable without knowing a universal set!!)
d. $B - A$
e. $A \triangle B$

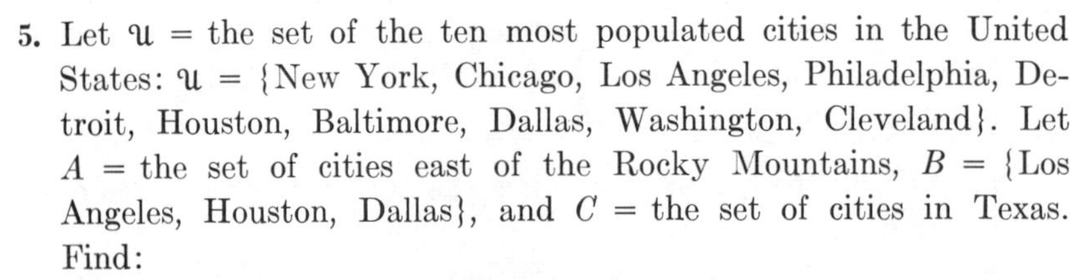

5. Let $\mathcal{U}$ = the set of the ten most populated cities in the United States: $\mathcal{U}$ = {New York, Chicago, Los Angeles, Philadelphia, Detroit, Houston, Baltimore, Dallas, Washington, Cleveland}. Let A = the set of cities east of the Rocky Mountains, B = {Los Angeles, Houston, Dallas}, and C = the set of cities in Texas. Find:

   a. $\tilde{A} \cup B$ **b.** $\widetilde{(A \cap \tilde{B})}$ **c.** $A - B$

   d. $A \triangle B$ **e.** $\widetilde{(A \cup B)}$

6. Let $\mathcal{U}$ = the set of the nine heaviest recorded humans. We list $\mathcal{U}$ in chart form below:

   Robert E. Hughes (1926–1958), U.S., 1069 lbs.
   Mills Darden (1798–1857), U.S., 1020 lbs.
   John H. Craig (1856–1894), U.S., 907 lbs.
   Arthur Knorr (1914–1960), U.S., 900 lbs.
   Mr. T. Valenzuela (1895–1937), Mex., 850 lbs.
   Flora King Jackson (1930–1965), U.S., 840 lbs.
   Ruth Pontico (1904–1941), U.S., 815 lbs.
   David Maguire (1904–1935), U.S., 810 lbs.
   William J. Cobb (1926–   ), U.S., 802 lbs.

   Let A = the set of heaviest human males, B = the set of heaviest human females, C = the set of humans weighing over 1000 lbs., and D = the set of heaviest humans born in the nineteenth century. Find:

   a. $B \cup D$ **b.** $A - C$ **c.** $B \cup C$ **d.** $\tilde{C} \cup D$

   e. $\tilde{C} \cap \tilde{D}$ **f.** $\widetilde{(A - C)}$ **g.** $\tilde{D} \cup B$ **h.** $A \cap C$

7. The ten largest U.S. industrial corporations and their corresponding sales are listed below. (Source: Fortune Magazine, 1972)

Corporation	*Sales (in millions)*
General Motors	\$30,435
Exxon	20,310
Ford Motor	20,194
General Electric	10,240
Chrysler	9,759
International Business Machines	9,533
Mobil Oil	9,166
Texaco	8,693
International Tel. & Tel.	8,557
Western Electric	6,551

Let A be the set of automobile companies. Let B be the set of oil companies. Let C be the set of companies whose sales (in millions) exceeded \$10,000. Find:

a. $\tilde{A} \cap \tilde{B}$ (the set of companies that are neither automobile nor oil companies)

b. $A \cap \tilde{C}$ (the set of companies that are automobile producers with sales less than or equal to 10,000 million dollars)

c. $\widetilde{(A \cup B)} \cap C$ (the set of companies that are neither automobile nor oil companies and whose sales exceed 10,000 million dollars)

2.3 SET RELATIONS

The previous section taught us different operations on sets. In this section we will discuss the **set relations,** which will enable us to compare one set with another. Consider the following sets:

$$\begin{aligned} \mathcal{U} &= \{1,2,3,4,5\} \\ A &= \{1,2,4\} \\ B &= \{3,4,5\} \\ C &= \{3,5\} \\ D &= \{3,5\} \end{aligned}$$

Notice that each element in C is also in set B. More formally, we say that C is a **subset** of B and we write $C \subseteq B$. Keeping this new idea in mind, it is easy to see other examples of the subset relation from the above list. For instance, $B \subseteq \mathcal{U}$, $D \subseteq B$, $D \subseteq \mathcal{U}$.

You may have noticed that sets C and D have a very special property: each is a subset of the other ($C \subseteq D$ and $D \subseteq C$). In this instance we define sets C and D to be **equal** and we write $C = D$.

The third and final set relation we will mention is illustrated by sets A and C, whose intersection is the empty set. As mentioned in the previous section, we call sets A and C **disjoint.**

In summary, then, we have formulated three set relations, namely, the "subset" relation, the "equals" relation, and the "disjoint" relation. Next we look at an example.

EXAMPLE 2.3.1. Given the sets $\mathcal{U} = \{1,2,3,4,5,6\}$, $A = \{1,3,5\}$, $B = \{2,4,6\}$, $C = \{4,2,6\}$, $D = \{3,4,5,6\}$:

a. Is $A \subseteq \mathcal{U}$?

b. Is $\mathcal{U} \subseteq D$?

c. Is $B = C$?

d. Is $\varnothing \subseteq B$?

e. Are A and C disjoint?

Solution:

a. Yes, because every element of A is an element of $\mathcal{U}$.
b. No, because $\mathcal{U}$ has some elements that are not in D (namely, 1 and 2).
c. Yes, because $B \subseteq C$ and $C \subseteq B$. (Note that equal sets need not have their elements listed in the same order.)
d. Yes, because every element of $\varnothing$ is an element of B.
e. Yes, because $A \cap C = \varnothing$.

We have been talking about subsets of a given set and it would seem reasonable to ask, "How many subsets does a given set have?" We will answer this question by taking an easy set to start with—one with just one element, $\{a\}$. The only subsets of this set are itself and $\varnothing$. So, a set with one element has two subsets. The next two examples treat the cases of a set with two and three elements, respectively.

Example 2.3.2. List all the subsets for the set $\{a,b\}$.

Solution: Subsets of this set must have zero, one, or two elements. The subset with zero elements is $\varnothing$. The subsets with one element are $\{a\}$, $\{b\}$. The subset with two elements is $\{a,b\}$. Hence, for a set with two elements we get four subsets.

Example 2.3.3. List all the subsets for the set $\{a,b,c\}$.

Solution: Subsets of this set must have zero, one, two, or three elements each. The subset with zero elements is $\varnothing$. The subsets with one element are $\{a\}$, $\{b\}$, $\{c\}$. The subsets with two elements are $\{a,b\}$, $\{a,c\}$, $\{b,c\}$. The subset with three elements is $\{a,b,c\}$. Thus, for a set with three elements we obtain eight subsets.

In the exercises, the reader is asked to generalize the above results for the number of subsets of any given set.

Frequently we want to know how many elements there are in a set. (This question is dealt with extensively in chapter three.) The set $A = \{1,3, \square, \triangle, \star, 2\}$ is seen to have six elements in it. The **number of elements** in the set A is symbolized $n(A)$, so here it is proper to write $n(A) = 6$.

Example 2.3.4. Consider the experiment mentioned in example 2.1.3 of tossing a nickel and a dime. We found $\mathcal{U} = \{HH, HT, TH, TT\}$. Let A be the subset of $\mathcal{U}$ containing the situations where two heads appear. Let B be the subset of $\mathcal{U}$ containing situations where exactly one head appears. Let C be the subset of $\mathcal{U}$ where zero heads appear. Describe each of A, B, C by listing their members, and compute $n(\mathcal{U})$, $n(A)$, $n(B)$, $n(C)$.

Solution: $A = \{HH\}$, $B = \{HT, TH\}$, $C = \{TT\}$. By counting, $n(\mathcal{U}) = 4$, $n(A) = 1$, $n(B) = 2$, $n(C) = 1$.

The next example will recall the set operations of union and intersection and relate them to logic statements, which were studied in chapter one. Thus, this example will be another link between logic and sets. Later in this chapter (section 2.4), we investigate in detail the many ways in which logic and sets are related to each other.

Example 2.3.5. A survey of prisons in New York State is to be taken where inmates will be classified according to marital status (married, single), where the crime was committed (urban area, rural area), and type of crime (murder, rape, theft). The possible categories are summarized below.

Situation	Marital Status	Area	Type of Crime
1	Married	Urban	Murder
2	Married	Urban	Rape
3	Married	Urban	Theft
4	Married	Rural	Murder
5	Married	Rural	Rape
6	Married	Rural	Theft
7	Single	Urban	Murder
8	Single	Urban	Rape
9	Single	Urban	Theft
10	Single	Rural	Murder
11	Single	Rural	Rape
12	Single	Rural	Theft

Select from this universal set of 12 situations those that satisfy the compound statements below.

a. A person is married and from an urban area.
b. A person is married, from a rural area, and committed theft.
c. A person is single and from an urban area, or is single and committed murder.
d. A person did not commit rape, but is married and from a rural area.
e. A person committed theft and is from an urban area, or is single.

Solution: By selecting the situations that satisfy the compound statements, we find the following results.

a. Situations 1, 2, 3. (We intersect the set of married with the set of urban dwellers.)
b. Situation 6. (We intersect the three sets: married, rural dwellers, and thieves.)

c. Situations 7, 8, 9 represent single urban dwellers. Situations 7 and 10 represent singles who committed murder. The union of these two sets gives 7, 8, 9, 10.
d. Situations 1, 3, 4, 6, 7, 9, 10, 12 represent those not committing rape. Situations 4, 5, 6 represent married rural dwellers. Intersecting these two sets gives 4 and 6.
e. Thieves from urban areas are represented by situations 3 and 9. Singles are represented by 7, 8, 9, 10, 11, 12. Finding the union gives us 3, 7, 8, 9, 10, 11, 12.

Looking back over this section, we have discussed the three set relations "subset," "equals," and "disjoint." We also began to deduce the number of subsets for a set with a given number of elements, and we introduced the $n(A)$ notation to represent the number of elements in a set A. Lastly, in example 2.3.5 we saw an illustration that serves to tie logic and sets together.

EXERCISES ■ SECTION 2.3.

1. Given the sets: $\mathcal{U} = \{1,2,3,4,7,6,5\}$, $A = \{2,4,6\}$, $B = \{1,3,5\}$, $C = \{7\}$, $D = \{4,7,6\}$, $E = \{4,6,7\}$, answer the questions below:

a. Is $A \subseteq D$? **b.** Is $D \subseteq A$?
c. Is $(A \cap D) \subseteq A$? **d.** Is $(D \cup C) \subseteq \mathcal{U}$?
e. Is $(A \cup B \cup C) = \mathcal{U}$? **f.** Is $\varnothing \subseteq \varnothing$?
g. Is there a pair of disjoint sets in the above list? (If so, specify which.)
h. Exhibit a subset of C that is not equal to C.

2. For the list of sets in exercise 1, find the truth value of each of these statements.

a. 4 is an element of D and an element of E.
b. 3 is an element of D or an element of B.
c. If 2 is an element of A, then 7 is an element of C.
d. If 2 is an element of D, then 2 is an element of B.
e. 7 is an element of A, only if 4 is an element of E.
f. 2 being an element of $\mathcal{U}$ is necessary and sufficient for 7 being an element of B.

3. a. We demonstrated in this section that a set with one element has two subsets, a set with two elements has four subsets, and a set with three elements has eight subsets. Generalize this idea by completing the following chart.

Number of elements in set	1	2	3	4	5	10	n
Number of possible subsets							

b. Write the subsets of $\{a,b,c,d\}$.

4. Consider an experiment of tossing a nickel, a dime, and a quarter. The possible situations that can result are listed in $\mathcal{U} = \{HHH, HHT, HTH, THH, HTT, THT, TTH, TTT\}$, where HHT, for instance, means the nickel comes up heads, the dime comes up heads, and the quarter comes up tails. Let A be the subset of $\mathcal{U}$ containing situations with 3 heads. Let B be the subset of $\mathcal{U}$ containing situations with exactly 2 heads. Let C be the subset of $\mathcal{U}$ containing situations with exactly 1 head. Let D be the subset of $\mathcal{U}$ containing situations with 0 heads. Describe each of A, B, C, D by listing their members, and count to determine $n(\mathcal{U})$, $n(A)$, $n(B)$, $n(C)$, $n(D)$, $n(A \cap B)$, $n(B \cup C)$, $n(A \cup \tilde{D})$.

5. A survey is to be taken at recycling centers in Monroe County, and people will be classified according to sex, number of items recycled (glass, metal, paper), and geographical section of county (northwest, northeast, southwest, southeast). The categories are summarized below.

Situation	Sex	Number of items recycled (1, 2, or 3)	Geographical location
1	M	1	NW
2	M	1	NE
3	M	1	SW
4	M	1	SE
5	M	2	NW
6	M	2	NE
7	M	2	SW
8	M	2	SE
9	M	3	NW
10	M	3	NE
11	M	3	SW
12	M	3	SE
13	F	1	NW
14	F	1	NE
15	F	1	SW
16	F	1	SE
17	F	2	NW
18	F	2	NE

19	F	2	SW
20	F	2	SE
21	F	3	NW
22	F	3	NE
23	F	3	SW
24	F	3	SE

Select from this universal set of 24 situations those that satisfy the compound statements below.

a. A person is a male from the northwest or a female who recycles exactly two items.

b. A person is a male who does not recycle two products and is from the northeast, or the person is a female.

c. A person is from the southwest, recycles exactly one item, and is not male.

d. A person is not from the southwest, is not from the northwest, is not a male, and does recycle three items.

e. If a person recycles exactly two items, then that person is from the northwest. (Hint: $(p \rightarrow q) \Leftrightarrow (\sim p \vee q)$.)

2.4 SET DIAGRAMS

In section 2.2, an association between sets and logic began. We established there the links between "and" and "intersection," "or" and "union," "not" and "complement." It is the purpose of this section to study that association further.

In any particular problem concerning sets, we know of the existence of a universal set, $\mathcal{U}$. Let p be a statement about the elements of $\mathcal{U}$. We call the **truth set** of p the set of all situations (in $\mathcal{U}$) that make p true. We denote the truth set of p by P. For example, let $\mathcal{U} = \{1,2,3\}$ and statement p be "the number is odd." Then the truth set of p is $P = \{1,3\}$. Sometimes it will be advantageous to work the other way. That is, given a set, invent a statement that is true for precisely the elements of that set. Such a statement will be called a **corresponding statement** for the given set. With $\mathcal{U} = \{1,2,3\}$ and $Q = \{2,3\}$ a corresponding statement for Q could be q: "the number is greater than 1." Now we look at some examples.

Example 2.4.1. $\mathcal{U} = \{HH,HT,TH,TT\}$ is the set of all logical possibilities when two coins are flipped. Determine the truth set for each of these simple statements:

p: at least one coin is a head
q: the coins match
r: exactly one coin is a head

Solution:

$$P = \{HH,HT,TH\}$$
$$Q = \{HH,TT\}$$
$$R = \{HT,TH\}$$

Example 2.4.2. Letting $\mathcal{U}$, p, q, r, be as in example 2.4.1, find the truth set for each compound statement listed below:

a. $p \wedge q$ **b.** $p \vee q$
c. $\sim p$ **d.** $p \rightarrow r$

Solution:

a. The truth set of $p \wedge q$ is the set of all elements that make $p \wedge q$ true. The statement $p \wedge q$ is "at least one of the coins is a head and they match." The set of elements that satisfy that conjunction is $\{HH\}$, which is none other than $P \cap Q$! The reason for this is clear: an element will be in the truth set of $p \wedge q$ iff it is in the truth set of p and in the truth set of q, that is, iff it is an element of P AND an element of Q (the definition of $P \cap Q$!).
b. The truth set of $p \vee q$ is $P \cup Q = \{HH,HT,TH,TT\}$. The explanation runs parallel to that of part *a*.
c. The truth set of $\sim p$ is the set of all elements which make $\sim p$ true. This is, of course, the set of all elements that make p false, $\tilde{P} = \{TT\}$.
d. Use the fact that $p \rightarrow r$ is equivalent to $\sim p \vee r$ (see exercise 6, section 1.5). From the discussions in parts *a* and *b*, the truth set of a disjunction will merely be the union of the truth sets. So, the set in question is $\tilde{P} \cup R = \{HT,TH,TT\}$.

Example 2.4.3. Consider the experiment where a coin and a die are tossed. Here $\mathcal{U} = \{H1,H2,H3,H4,H5,H6,T1,T2,T3,T4,T5,T6\}$. Suppose $P = \{H2,T2\}$, $Q = \{H3,H6,T3,T6\}$, $R = \{H1,H2,H3,H4,H5,H6,T1\}$. Find the corresponding statements for these sets.

Solution:

p: a two appears on the die
q: a multiple of three appears on the die
r: a head appears or the number one appears

Example 2.4.4. **a.** What is the truth set of a tautology?
b. What is the truth set of a self-contradiction?

Solution:

a. Don't forget that in any particular problem, $\mathcal{U}$ is the set of all elements under consideration. So, when asked for the truth set of a tautology, we must include those elements that make the statement true. Since all elements make a tautology true, the truth set is $\mathcal{U}$. It deems mentioning that a corresponding statement for $\mathcal{U}$ must be a tautology. (Here the quest would be for a statement that is true in all instances.)

b. The truth set for a self-contradiction is the set of elements that make the statement true. Since no element can make a self-contradiction true, the truth set must be $\varnothing$. Also, any corresponding statement for $\varnothing$ would have to be a self-contradiction.

Now that the surface of set theory has been scratched, let's "look" at some developments. We can do this with pictorial representation of sets, called **Venn diagrams.** Basically, Venn diagrams use geometric configurations for sets, with the universal set usually represented as the interior of a rectangle. Subsets of $\mathcal{U}$ frequently take the form of interiors of circles. Consider a pictorial representation of $\mathcal{U}$ with some subset P.

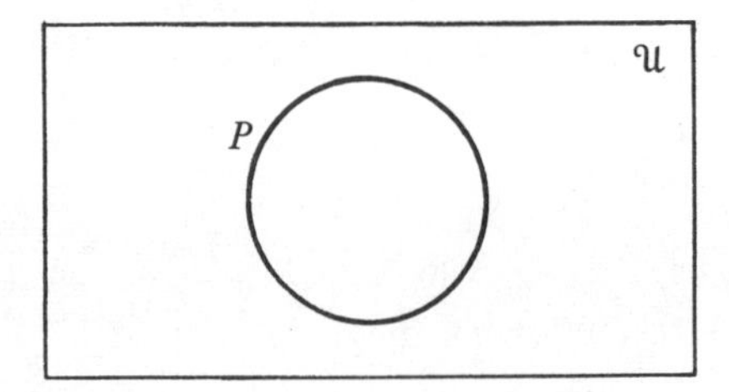

Anything inside the P circle is an element of set P; being outside the circle indicates being a member of set $\tilde{P}$.

When two sets P and Q are introduced, the Venn diagram contains two overlapping circles.

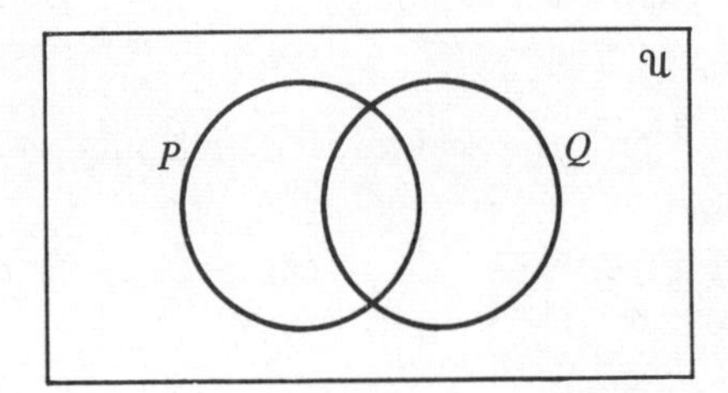

Notice that the overlapping circles divide the universal set into four sections: the intersection of the two sets ($P \cap Q$), that portion that is inside P but outside Q (namely $P \cap \tilde{Q}$), that area that is outside P but inside Q (namely $\tilde{P} \cap Q$), and finally, that that is in neither P nor Q (namely $\tilde{P} \cap \tilde{Q}$). In the following figure, the four sectors are labeled 1, 2, 3, 4, respectively. We urge the reader to compare this with the basic conjunctions for two statements (section 1.8).

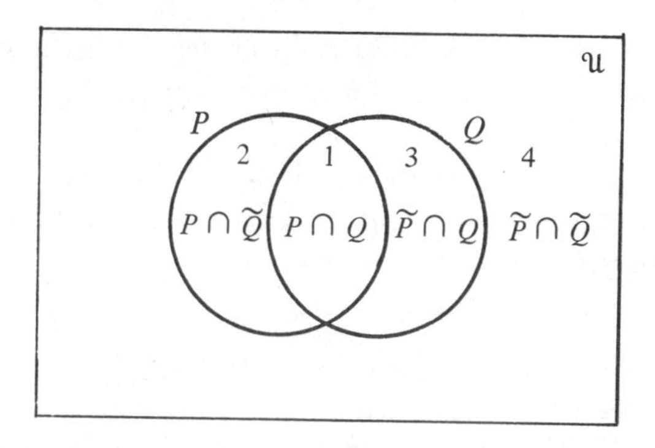

Example 2.4.5. Which of the sectors in the Venn diagram of the preceding figure depicts the truth set for each of the following corresponding statements?

a. $p \vee q$
b. $\sim p$
c. $p \rightarrow q$

Solution:

a. The truth set of $p \vee q$ is $P \cup Q$, which includes sectors 1, 2, 3. Shading the truth set yields the following diagram.

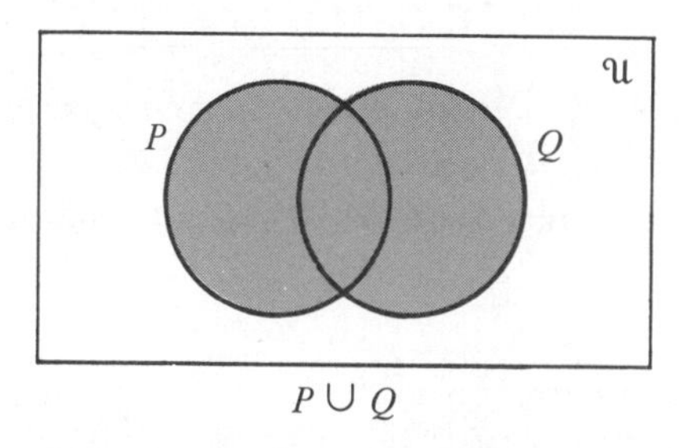

$P \cup Q$

b. The truth set of $\sim p$ is $\widetilde{P}$ and the sectors that fall outside the P circle are 3, 4. We shade the truth set below.

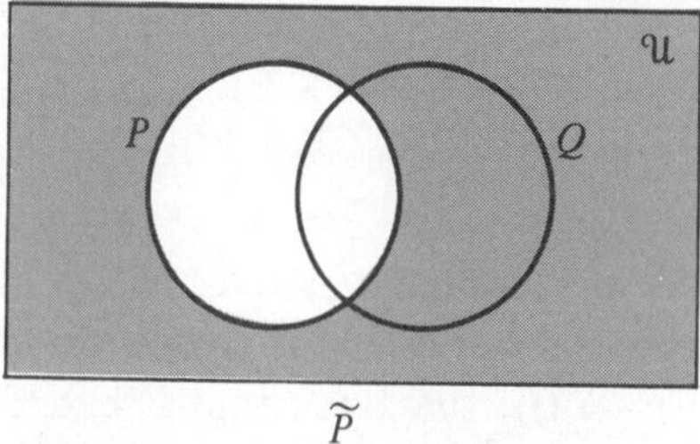

$\widetilde{P}$

c. Recall that the truth set of $p \rightarrow q$ is $\widetilde{P} \cup Q$. Now we must shade everything that is either outside the P circle or inside the Q circle. We shade sectors 1, 3, 4. (This makes sense! After all, $p \rightarrow q$ is true in cases 1, 3, 4 of its truth table.)

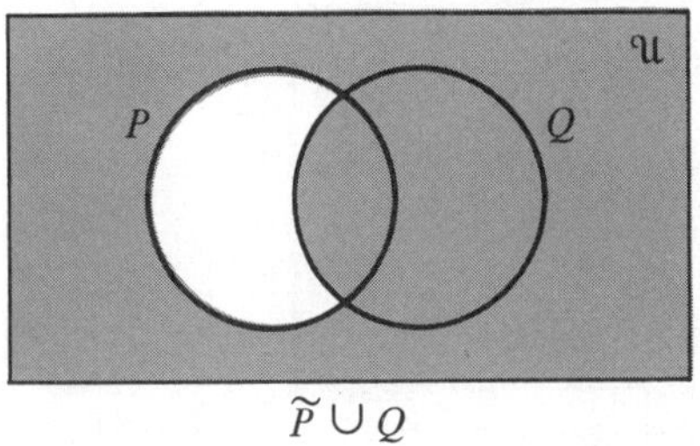

$\tilde{P} \cup Q$

A Venn diagram with three circles representing the sets P, Q, R is easily fabricated. We number the regions below and then analyze them.

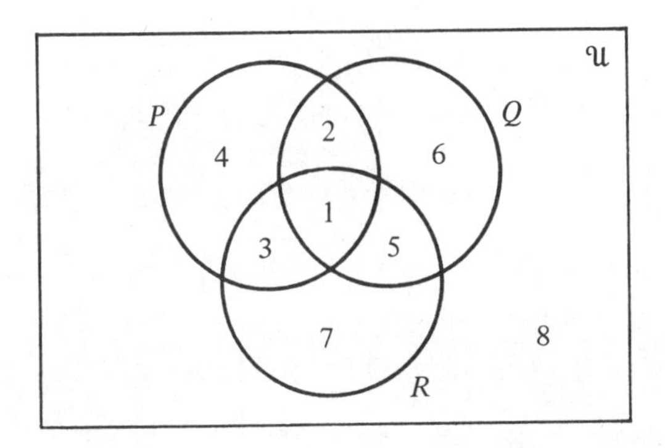

Region number	Truth set represented	Corresponding statement
1	$P \cap Q \cap R$	$p \wedge q \wedge r$
2	$P \cap Q \cap \tilde{R}$	$p \wedge q \wedge \sim r$
3	$P \cap \tilde{Q} \cap R$	$p \wedge \sim q \wedge r$
4	$P \cap \tilde{Q} \cap \tilde{R}$	$p \wedge \sim q \wedge \sim r$
5	$\tilde{P} \cap Q \cap R$	$\sim p \wedge q \wedge r$
6	$\tilde{P} \cap Q \cap \tilde{R}$	$\sim p \wedge q \wedge \sim r$
7	$\tilde{P} \cap \tilde{Q} \cap R$	$\sim p \wedge \sim q \wedge r$
8	$\tilde{P} \cap \tilde{Q} \cap \tilde{R}$	$\sim p \wedge \sim q \wedge \sim r$

Example 2.4.6. Using the preceding Venn diagram, state the regions of the Venn diagram that represent the sets below and shade that area.

a. $Q \cup R$
b. $P \cap (Q \cup R)$
c. $\tilde{P} \cup \tilde{Q}$
d. $\widetilde{(P \cup R)}$

Solution:

a. The regions which represent $Q \cup R$ are 1, 2, 3, 5, 6, 7. We shade the regions as shown on the following page.

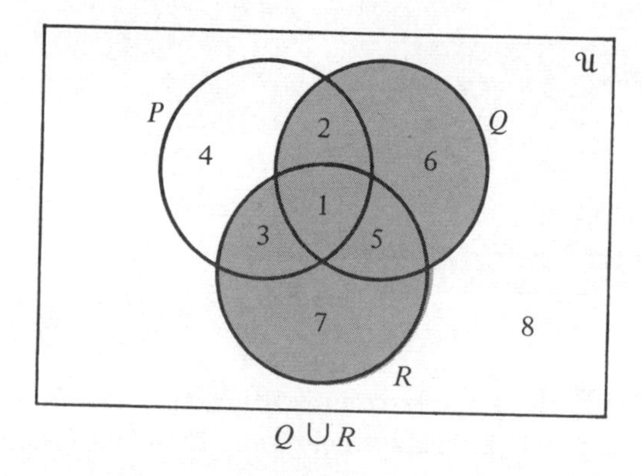

$Q \cup R$

b. The regions that satisfy $P \cap (Q \cup R)$ will be those that are in the P circle and in $Q \cup R$. They are regions 1, 2, 3. These regions are shaded below.

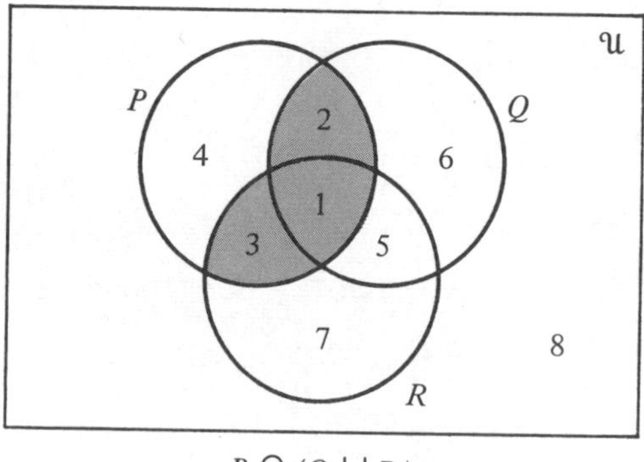

$P \cap (Q \cup R)$

c. $\tilde{P} \cup \tilde{Q}$ is represented by all areas that are either outside P or outside Q. Areas 3, 4, 5, 6, 7, 8 satisfy.

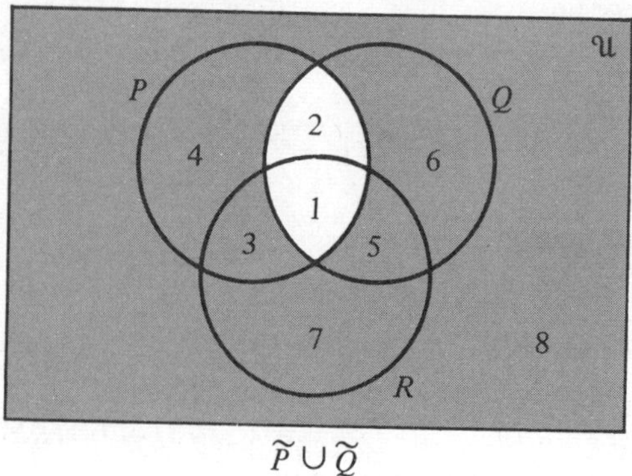

$\tilde{P} \cup \tilde{Q}$

d. $\widetilde{(P \cup R)}$. All regions that are in neither the P circle nor the R circle satisfy. They are 6, 8.

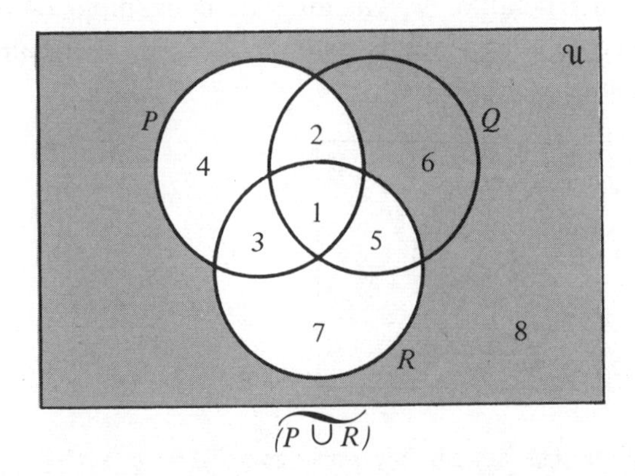

$\widetilde{(P \cup R)}$

Example 2.4.7. Venn diagrams are often useful for examining whether or not two sets are equal. When each is shaded and they have the same representation, they are equal. Show that $\widetilde{(P \cup Q)} = \tilde{P} \cap \tilde{Q}$, which is one of DeMorgan's laws for sets.

Solution: $\widetilde{(P \cup Q)}$ is everything outside of $P \cup Q$:

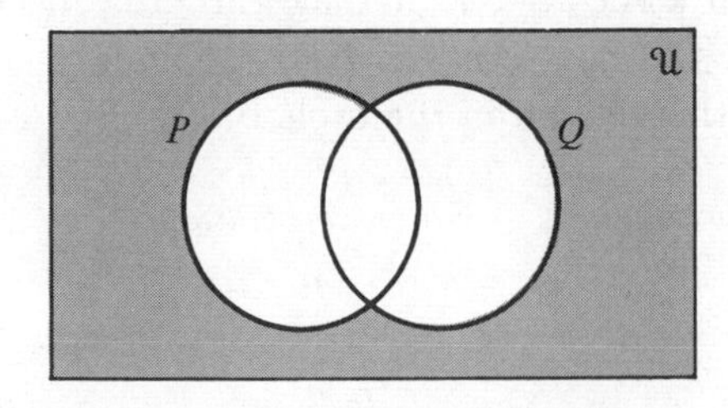

Also $\tilde{P} \cap \tilde{Q}$ is the area that is both outside the P circle and outside the Q circle.

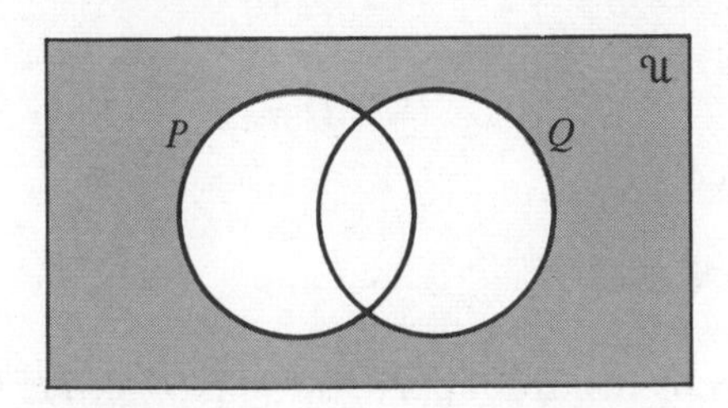

Hence, $\widetilde{(P \cup Q)} = \tilde{P} \cap \tilde{Q}$.

With the aids of truth sets and Venn diagrams, we have seen more vividly the association between logic connectives and set operations. The association must be continued to include the relations. We have studied three logic relations: $p \Rightarrow q$, $p \Leftrightarrow q$, $p \mathbf{I} q$. Recall that $p \Rightarrow q$ when $p \rightarrow q$ is a tautology. To say that $p \rightarrow q$ is a tautology means its truth set, $\tilde{P} \cup Q$, must be all of $\mathcal{U}$. The diagram in example 2.4.5c must, therefore, be adjusted so everything is shaded. This can be accomplished by "shrinking" the unshaded portion.

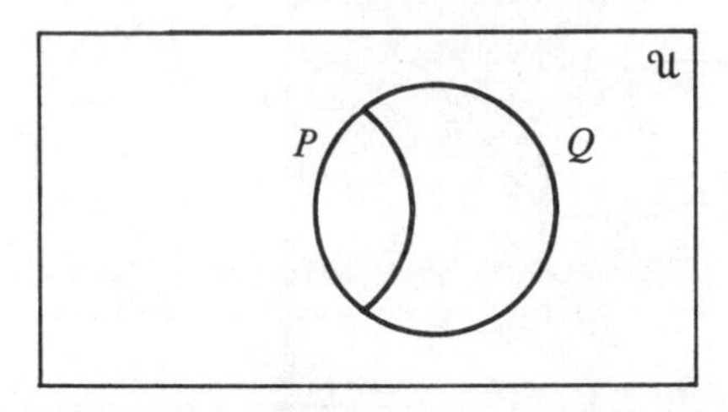

From here it is obvious the set theory analogy of $p \Rightarrow q$ is $P \subseteq Q$.

The second relation, $p \Leftrightarrow q$, is true if and only if $p \Rightarrow q$ and $q \Rightarrow p$. This translates to $P \subseteq Q$ and $Q \subseteq P$ in set language, which is the same as the set relation $P = Q$.

The last logic relation, $p \mathbf{I} q$, means $p \wedge q$ is a self-contradiction. In set theory this translates to $P \cap Q$ is the empty set. This is the relation of disjointness.

The set relations and their associated Venn diagrams are concepts that will be expounded upon in the next section. Here, though, we wish to summarize the material of this section in the table below.

Statement or relation	*Set interpretation*
p	P
$\sim p$	$\tilde{P}$
$p \wedge q$	$P \cap Q$
$p \vee q$	$P \cup Q$
$p \rightarrow q$	$\tilde{P} \cup Q$
$p \Rightarrow q$	$P \subseteq Q$
$p \Leftrightarrow q$	$P = Q$
$p \mathbf{I} q$	$P \cap Q = \varnothing$

EXERCISES ■ SECTION 2.4.

1. Recall that when three coins are flipped, $\mathcal{U} = \{HHH, HHT, HTH, HTT, THH, THT, TTH, TTT\}$. Consider the following simple statements:

   p: all three coins are heads
   q: at least one head appears
   r: at least one tail appears

Find the truth set for the following statements.

a. p **b.** q **c.** r
d. $\sim q$ **e.** $\sim r$ **f.** $q \wedge r$
g. $q \vee r$ **h.** $p \rightarrow q$ **i.** $p \underline{\vee} q$
j. $q \underline{\vee} r$ **k.** $q \leftrightarrow r$ **l.** $\sim(p \vee q)$
m. $\sim(p \rightarrow q)$ **n.** $\sim q \rightarrow \sim r$

2. Construct a Venn diagram for each of the parts below and shade the appropriate area.

a. $\tilde{P} \cup \tilde{Q} \cup \tilde{R}$ **b.** $\tilde{P} \cap \tilde{Q} \cap \tilde{R}$
c. $P - Q$ **d.** the truth set of $\sim p \rightarrow q$
e. the truth set of $(p \rightarrow q) \vee r$

3. Recall that when two sets are equal, their corresponding Venn diagrams are identical. Use this fact to show the following are true.

a. $\widetilde{(X \cap Y)} = \tilde{X} \cup \tilde{Y}$ (DeMorgan's second law)
b. $P \cap (Q \cup R) = (P \cap Q) \cup (P \cap R)$
c. $P \cup (Q \cap R) = (P \cup Q) \cap (P \cup R)$

4. The summary on the previous page is somewhat incomplete.

a. Determine the truth set of $p \underline{\vee} q$. (Hint: use basic conjunctions.)
b. Determine the truth set of $p \leftrightarrow q$. (Hint: use basic conjunctions.)
c. Show the truth set of $p \rightarrow q$ could also be represented as $\widetilde{(P - Q)}$.

5. Create a Venn diagram and then shade the area that represents the given statement.

a. $(p \wedge \sim q) \vee (p \rightarrow q)$ **b.** $(p \wedge q) \vee (p \wedge \sim q)$
c. $(p \wedge q) \wedge (p \wedge \sim q)$ **d.** $\sim p \rightarrow (q \vee r)$
e. $p \wedge q$ when $P \subseteq Q$

6. Show that a corresponding statement for $A \triangle B$ is $a \underline{\vee} b$.

7. Try constructing a Venn diagram with four intersecting subsets of $\mathcal{U}$. Call the sets P, Q, R, S: there should be sixteen sectors. (Hint: don't use four circles.)

2.5 USES OF VENN DIAGRAMS

We begin here by reconsidering the basic notion of a truth set. Since the truth set P of a statement p is composed of elements that make p true, it follows that the truth set of $\sim p$ is composed of elements not in P (i.e., elements of $\tilde{P}$). In other words, when p is true, we represent that in a Venn diagram by the area inside P and when p is false, we represent that in a Venn diagram by the area outside of P (i.e., the area of $\tilde{P}$). Now we will use these ideas to discuss the relationship between tautologies and self-contradictions and sets.

A statement is a tautology when it is always true. So the truth set of a tautology is the universal set $\mathcal{U}$.

Example 2.5.1. Use a Venn diagram to see whether or not $(p \wedge q) \vee (p \vee q) \vee \sim p$ is a tautology.

Solution: First we translate the statement into set language: $(P \cap Q) \cup (P \cup Q) \cup \tilde{P}$. Now consider a Venn diagram.

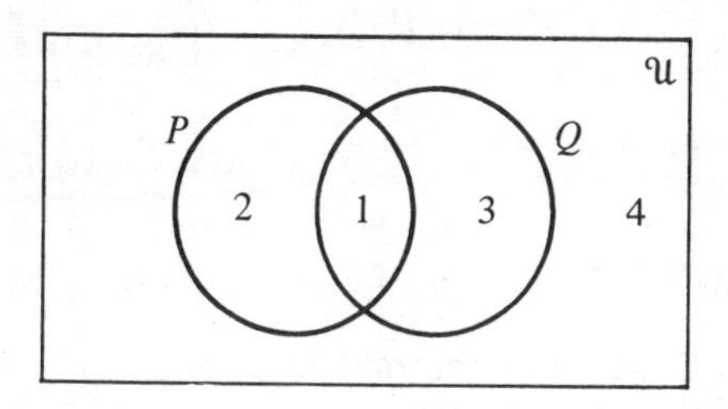

$P \cap Q$ gives region 1.
$P \cup Q$ gives regions 1, 2, 3.
$\tilde{P}$ gives regions 3, 4.
} Taking the union of these gives regions 1, 2, 3, 4.

Since we get the universal set, $(p \wedge q) \vee (p \vee q) \vee \sim p$ is a tautology.

Example 2.5.2. Test $(p \to q) \to q$ to see whether or not it is a tautology.

Solution: Rewriting $(p \to q) \to q$ gives $\sim(\sim p \vee q) \vee q$ as an equivalent statement. So the truth set we must consider is $\widetilde{(\tilde{P} \cup Q)} \cup Q$.

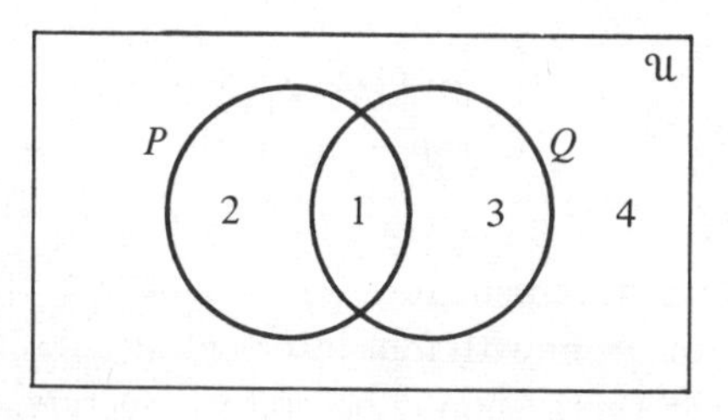

$\tilde{P} \cup Q$ gives regions 1, 3, 4.

$\widetilde{(\tilde{P} \cup Q)}$ gives region 2.
Q gives regions 1, 3.
} Taking the union of these gives regions 1, 2, 3.

Hence, $(p \to q) \to q$ is not a tautology.

It is also possible to test a statement to see if it is a self-contradiction by using Venn diagrams. A self-contradiction is, of course, a statement that is always false. So to test a statement to see if it is a self-contradiction, we can construct a Venn diagram and check to see whether the truth set of the statement is the empty set, $\varnothing$.

Example 2.5.3. It can be shown using a truth table that $(p \leftrightarrow q) \wedge \sim(\sim p \vee q)$ is a self-contradiction. Demonstrate the self-contradiction using a Venn diagram.

Solution: $(p \leftrightarrow q) \wedge \sim(\sim p \vee q)$ can be written $[(p \wedge q) \vee (\sim p \wedge \sim q)] \wedge \sim(\sim p \vee q)$, using the notion of basic conjunctions from section 1.8. So the set statement we need to consider is $[(P \cap Q) \cup$

$(\tilde{P} \cap \tilde{Q})] \cap \widetilde{(\tilde{P} \cup Q)}$.

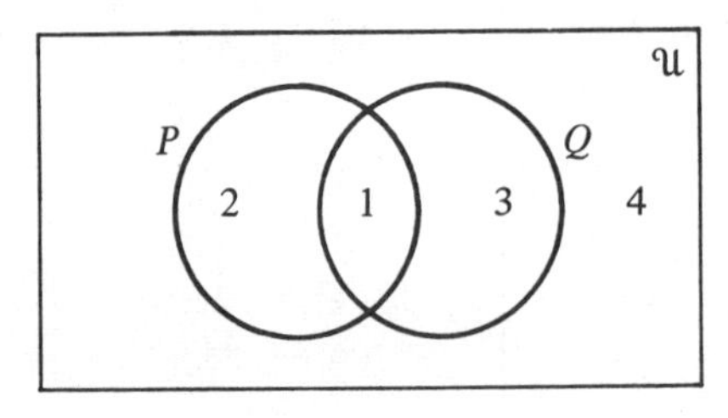

$P \cap Q$ gives region 1.
$\tilde{P} \cap \tilde{Q}$ gives region 4.
$[(P \cap Q) \cup (\tilde{P} \cap \tilde{Q})]$ gives regions 1, 4.
$\widetilde{(\tilde{P} \cup Q)}$ gives region 2.
} Intersecting these gives $\varnothing$.
Thus $(p \leftrightarrow q) \wedge \sim(\sim p \vee q)$ is a self-contradiction.

The last paragraph in section 2.4 pointed out that we would elaborate on how to use Venn diagrams to test statements for equivalence, implication, or disjointness. We now devote our attention to that task. First, recall a partial summary of the table on page 70.

Relation	*Its set interpretation*
$p \Rightarrow q$	$P \subseteq Q$
$p \Leftrightarrow q$	$P = Q$
$p \, \mathbf{I} \, q$	$P \cap Q = \varnothing$

Now suppose we are trying to decide whether $\sim q$ implies $\sim(p \wedge q)$. In set language this would become: Is $\tilde{Q} \subseteq \widetilde{(P \cap Q)}$?

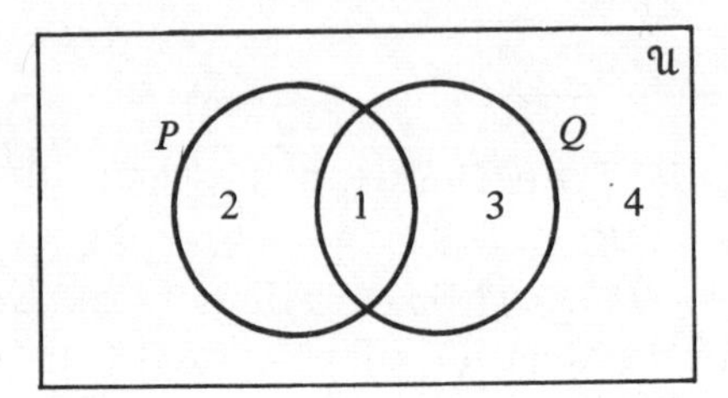

A Venn diagram shows $\tilde{Q}$ is represented by regions 2, 4, while $\widetilde{P \cap Q}$ is represented by regions 2, 3, 4. Hence $\tilde{Q} \subseteq \widetilde{(P \cap Q)}$ and it becomes clear that $\sim q$ does imply $\sim(p \wedge q)$.

Consider the question of whether $\sim(p \vee q)$ is equivalent to $\sim p \wedge \sim q$. In set language we ask: Does $\widetilde{(P \cup Q)} = \tilde{P} \cap \tilde{Q}$? The answer here immediately is "yes." Example 2.4.7 demonstrated this equality of sets.

The statements $\sim p \wedge q$ and $q \rightarrow p$ happen to be inconsistent. (A truth table check of $(\sim p \wedge q) \wedge (q \rightarrow p)$ shows all F's in the $\wedge$ column.) However, this inconsistency can also be demonstrated by showing the truth sets of $\sim p \wedge q$ and $q \rightarrow p$ to be disjoint.

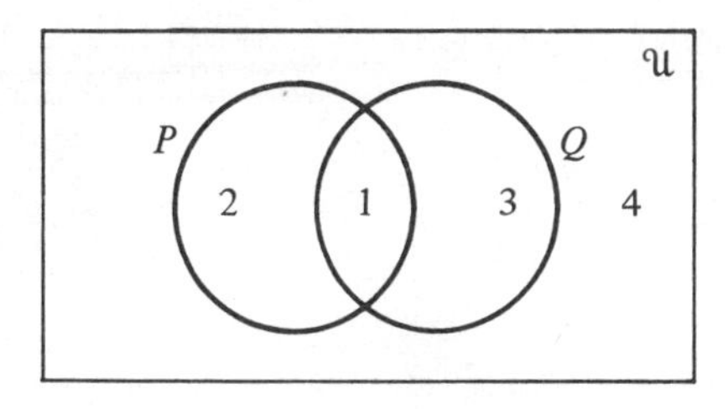

The truth set of $\sim p \wedge q$ is $\tilde{P} \cap Q$. A statement equivalent to $q \rightarrow p$ is $\sim q \vee p$, so the truth set of $q \rightarrow p$ is $\tilde{Q} \cup P$. The truth set of $\sim p \wedge q$ is represented by region 3. The truth set $\tilde{Q} \cup P$ is represented by regions 1, 2, 4. Hence, $(\tilde{P} \cap Q) \cap (\tilde{Q} \cup P) = \varnothing$, and $(\sim p \wedge q)$ **I** $(q \rightarrow p)$.

Example 2.5.4. Consider the pair of statements $\sim p \rightarrow q$ and $p \wedge q$. Test to see if:

a. they are equivalent,
b. either implies the other,
c. they are inconsistent,
d. they are compatible.

Solution:

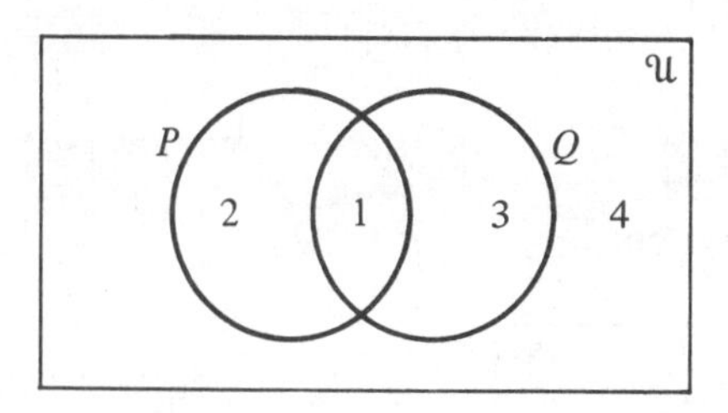

The truth set of $\sim p \rightarrow q$ is $P \cup Q$ (regions 1, 2, 3). The truth set of $p \wedge q$ is $P \cap Q$ (region 1).

a. We ask: Does $(P \cup Q) = (P \cap Q)$? Since regions 1, 2, 3 are not the same as region 1, the given statements are not equivalent.

b. We first ask: Does $(\sim p \rightarrow q) \Rightarrow (p \wedge q)$? In set language the translation becomes: $(P \cup Q) \subseteq (P \cap Q)$. Since regions 1, 2, 3 do not form a subset of region 1, $(\sim p \rightarrow q)$ does not imply $p \wedge q$. However, region 1 is a subset of regions 1, 2, 3. So $(P \cap Q) \subseteq (P \cup Q)$ tells us that $(p \wedge q) \Rightarrow (\sim p \rightarrow q)$.
c. We must ask: "Is $(P \cup Q) \cap (P \cap Q) = \varnothing$? The answer, of course, is no, since $(P \cup Q) \cap (P \cap Q)$ gives region 1. So the statements $\sim p \rightarrow q$ and $p \wedge q$ are not inconsistent.
d. An immediate consequence of part c is that the statements $\sim p \rightarrow q$ and $p \wedge q$ are compatible.

In chapter 1 we discussed the meaning of and how to test for a valid argument. Recall that an argument is said to be valid when the premises imply the conclusion. Consider the argument in symbolic form below.

$$\begin{array}{c} p \rightarrow q \\ \underline{\sim q} \\ \therefore \sim p \end{array}$$

A test for validity could be performed by checking to see whether $[(p \rightarrow q) \wedge \sim q] \Rightarrow \sim p$. Naturally, a truth table could be used here, but let us use a Venn diagram. We must translate the question, "Does $[(p \rightarrow q) \wedge \sim q]$ imply $\sim p$?" into set language. We have then: Is $[(\tilde{P} \cup Q) \cap \tilde{Q}] \subseteq \tilde{P}$?

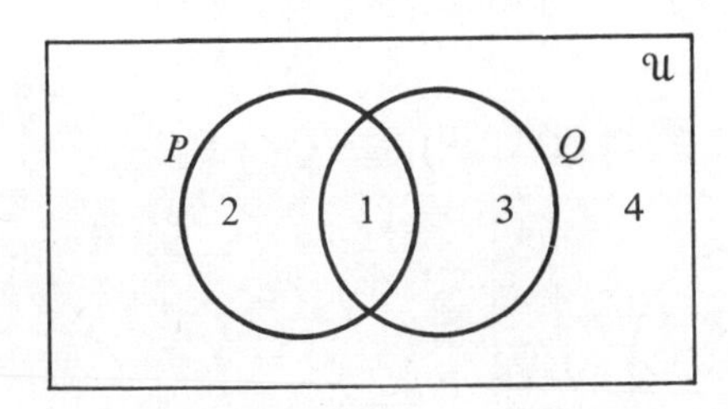

$\tilde{P} \cup Q$ gives regions 1, 3, 4.
$(\tilde{P} \cup Q) \cap Q$ gives region 4. } We see then that
$\tilde{P}$ gives regions 3, 4. } $[(\tilde{P} \cup Q) \cap \tilde{Q}] \subseteq \tilde{P}$.
Hence the original argument is valid.

Example 2.5.5. Use a Venn diagram to check the argument below for validity.

$$\begin{array}{c} p \rightarrow q \\ \underline{q \rightarrow r} \\ \therefore p \rightarrow r \end{array}$$

Solution: We ask: Does $[(p \to q) \wedge (q \to r)]$ imply $p \to r$? (Actually, back in example 1.5.4 we found the answer to this question to be yes.) Translating to set language, we ask: Is $[(\tilde{P} \cup Q) \cap (\tilde{Q} \cup R)] \subseteq (\tilde{P} \cup R)$?

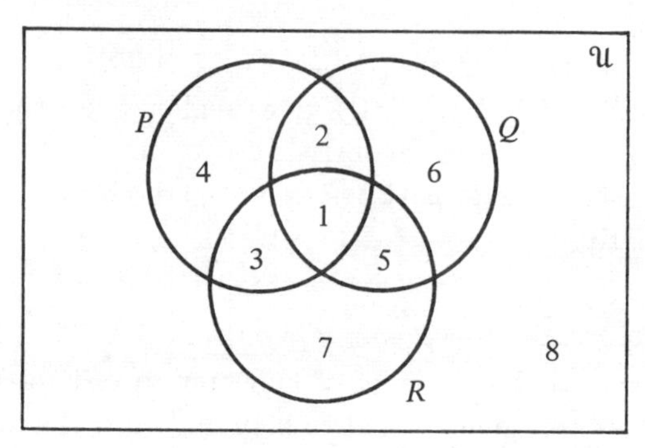

$\tilde{P} \cup Q$ is represented by regions 1, 2, 5, 6, 7, 8.
$\tilde{Q} \cup R$ is represented by regions 1, 3, 4, 5, 7, 8.
$[(\tilde{P} \cup Q) \cap (\tilde{Q} \cup R)]$ is represented by regions 1, 5, 7, 8.
$\tilde{P} \cup R$ is represented by 1, 3, 5, 6, 7, 8.
The last two sentences show us that $[(\tilde{P} \cup Q) \cap (\tilde{Q} \cup R)] \subseteq (\tilde{P} \cup R)$ so the original argument is valid.

Example 2.5.6. Use a Venn diagram to check the validity of the argument:

$$\begin{array}{c} p \vee q \\ \underline{\sim q \to \sim p} \\ \therefore \sim q \wedge \sim p \end{array}$$

Solution: We ask: Is $[(P \cup Q) \cap (Q \cup \tilde{P})] \subseteq (\tilde{Q} \cap \tilde{P})$?

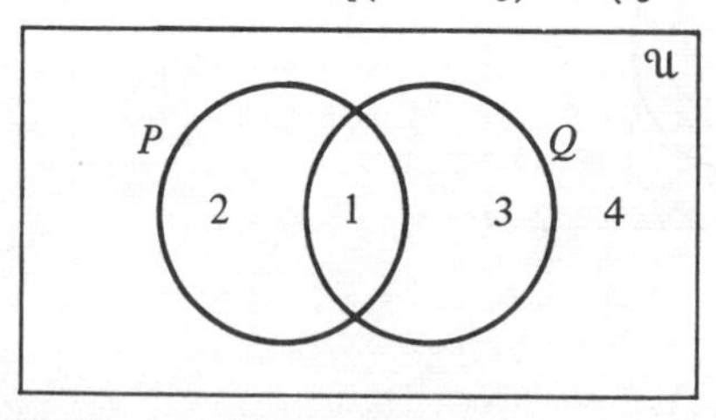

$P \cup Q$ gives regions 1, 2, 3.
$Q \cup \tilde{P}$ gives regions 1, 3, 4.
$[(P \cup Q) \cap (Q \cup \tilde{P})]$ gives regions 1, 3.
$Q \cap \tilde{P}$ gives region 4.
Quite obviously, regions 1, 3 do not form a subset of region 4. The argument is seen to be a fallacy.

Here is an interesting application of Venn diagrams to the typing and donation of blood. Blood contains three different **antigens.** (Antigens are things that are important in the production of antibodies in the blood.) Two of the antigens are symbolized by the letters A, B. Blood is referred to as type AB, type A, type B, or type O, depending on whether it has both A and B antigens, just the A antigen, just the B antigen, or neither the A nor the B. The third antigen is symbolized by Rh. When blood has the Rh antigen, it is referred to as positive (+), and when it is missing the Rh antigen, blood is called negative (−).

A Venn diagram with three mutually overlapping subsets labeled A, B, Rh is shown below.

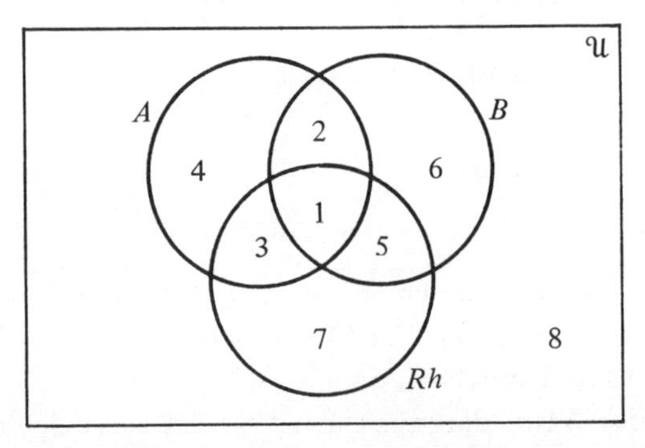

We now list each of the eight regions and note that each corresponds to a certain blood type.

Region	*Subset name*	*Corresponding blood type*
1	$A \cap B \cap Rh$	AB+
2	$A \cap B \cap \widetilde{Rh}$	AB−
3	$A \cap \tilde{B} \cap Rh$	A+
4	$A \cap \tilde{B} \cap \widetilde{Rh}$	A−
5	$\tilde{A} \cap B \cap Rh$	B+
6	$\tilde{A} \cap B \cap \widetilde{Rh}$	B−
7	$\tilde{A} \cap \tilde{B} \cap Rh$	O+
8	$\tilde{A} \cap \tilde{B} \cap \widetilde{Rh}$	O−

We can also show each of the blood types right on the Venn diagram.

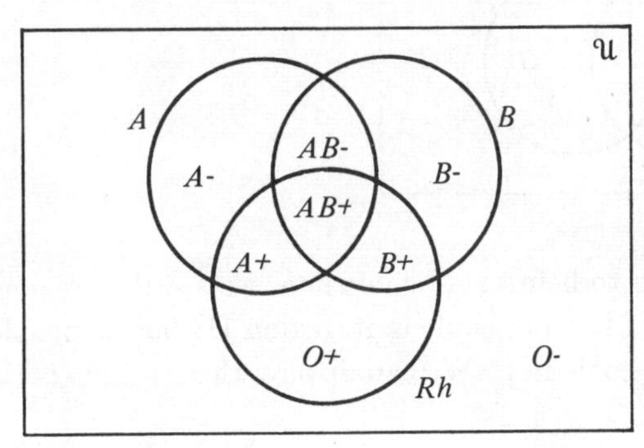

Blood donation is a very important thing on our planet earth. It happens that when a person donates blood, his antigens must be a subset of the receiver's antigens. Using this fact enables us to construct the table below.

Receiver has blood type:	*Can receive blood from donors with:*
AB+	AB+, AB−, A+, A−, B+, B−, O+, O−
AB−	AB−, A−, B−, O−
A+	A+, A−, O+, O−
A−	A−, O−
B+	B+, B−, O+, O−
B−	B−, O−
O+	O+, O−
O−	O−

Notice that no matter what blood a person has, he can receive blood type O− (has none of the three antigens). For this reason, a person with O− blood is often referred to as a universal donor.

Another way in which Venn diagrams can be used is to depict the results from certain numerical survey problems. As an illustration, suppose that **100** students were surveyed to determine their preference for marketing or science courses. The hypothetical results are given below.

15 students were taking just marketing.
45 students were taking both marketing and science.
25 students were taking just science.
15 students were taking neither marketing nor science.

We can display these results in a Venn diagram with two sets, M (for marketing) and S (for science).

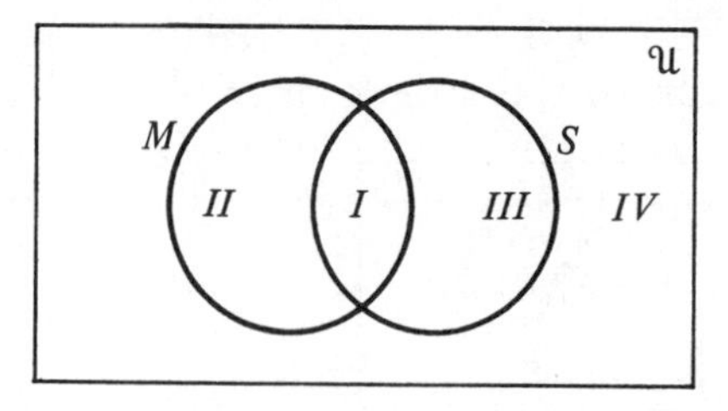

Using Roman numerals to help avoid confusion, we see that region I has 45 people in it, region II has 15 people in it, region III has 25 people in it, and region IV has 15 people in it. We can display these numbers right on the Venn diagram.

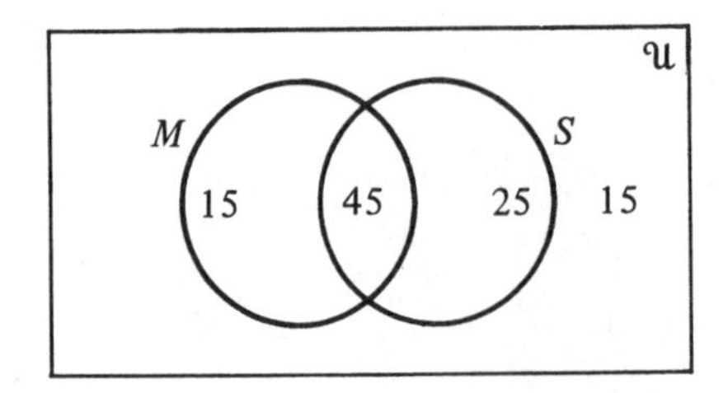

Example 2.5.7. Using the display above, answer:

a. How many students are taking marketing?
b. How many students are taking science?
c. How many students are taking marketing and science?
d. How many students are taking marketing or science?
e. How many students satisfy the statement, "If a student takes marketing, then he takes science"?
f. How many students satisfy the statement, "A student takes marketing if and only if he takes science"?

Solution: By counting with the aid of the Venn diagram,

a. $n(M) = 60$
b. $n(S) = 70$
c. $n(M \cap S) = 45$
d. $n(M \cup S) = 85$
e. Let m be the statement, "A student takes marketing." Let s be the statement, "A student takes science." We need to find the number of students in the truth set of $m \to s$. Hence we want $n(\tilde{M} \cup S)$. By counting, $n(\tilde{M} \cup S) = 85$.
f. Here we need the number of students in the truth set of $m \leftrightarrow s$. Since $(m \leftrightarrow s) \Leftrightarrow [(m \wedge s) \vee (\sim m \wedge \sim s)]$, we want $n((M \cap S) \cup (\tilde{M} \cap \tilde{S}))$. By counting, we get $45 + 15 = 60$ students.

Example 2.5.8. A survey to determine drug usage was conducted among 220 randomly selected people. The results were:

10 people tried both marijuana and hard drugs,
90 people tried marijuana,
20 people tried hard drugs.

Use M for the set of people who tried marijuana. Use H for the set of people who tried hard drugs.

a. Display the survey results on a Venn diagram.
b. Determine the number of people who tried marijuana or hard drugs.
c. Find $n(\tilde{M} \cap \tilde{H})$.
d. Find $n(M \cap \tilde{H})$.

Solution:

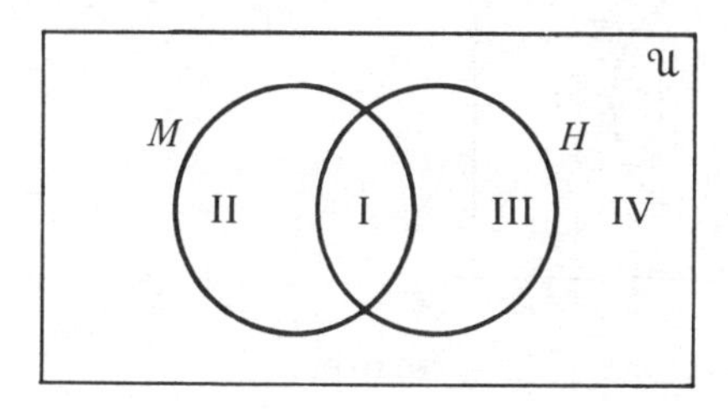

a. In region I, there are 10 people. In all of M there are 90 people, which leaves 80 in region II. Similarly, in H there are 20 people, so region III has 10 of those. Totaling, we see there are 100 people in regions I, II, III, leaving $220 - 100 = 120$ in region IV.

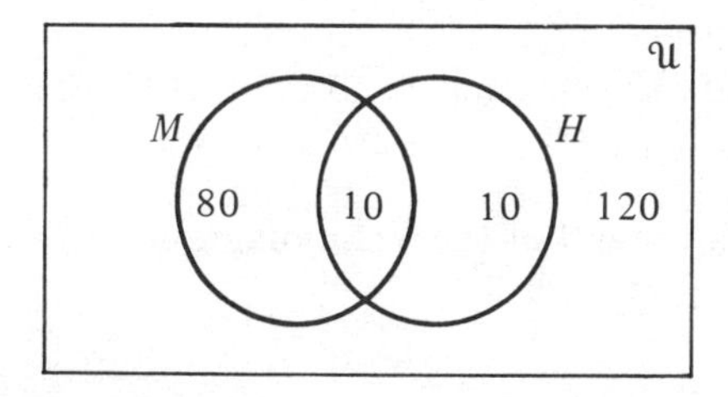

Notice that the sum of all the numbers in the Venn diagram above does total 220 (the number of people surveyed).

b. By counting, $n(M \cup H) = 80 + 10 + 10 = 100$.
c. $n(\tilde{M} \cap \tilde{H}) = 120$
d. $n(M \cap \tilde{H})$ = number who tried just marijuana = 80

Both of the two previous examples investigated dealt with surveys where there were only two sets. We need not always be that restrictive, as is now illustrated.

Example 2.5.9. A survey of 300 people was conducted to determine what items they were recycling (paper, metal, glass). The results are given below.

35 people recycle paper and metal
40 people recycle metal and glass
60 people recycle paper and glass
90 people recycle paper
70 people recycle metal
105 people recycle glass
25 people recycle all three items

a. Display these results on a Venn diagram.
b. Find the number recycling paper only.

c. Find the number recycling paper or glass.
d. Find the number recycling just one item.
e. Find the number recycling at least one item.
f. What fraction of people recycle just one item?
g. What fraction of people recycle no items?
h. What fraction of people recycle something?

Solution:

a. Use P, M, G to stand for paper, metal, glass, respectively.

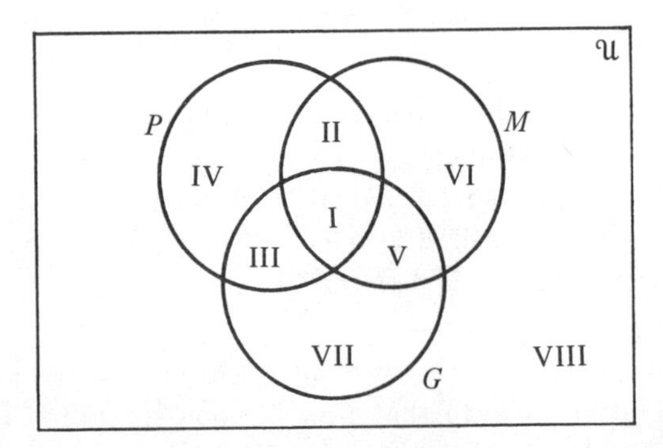

Start with the 25 people who recycle all three items. They go in region I. Now, since 35 people recycle paper and metal, that leaves 10 for region II. Similarly, region III has 35 people. Now regions I, II, III total 70. Since $n(P) = 90$, we get 20 people in region IV. Reasoning similarly we find: region V has 15, region VI has 20, region VII has 30 and region VIII has 145 people.

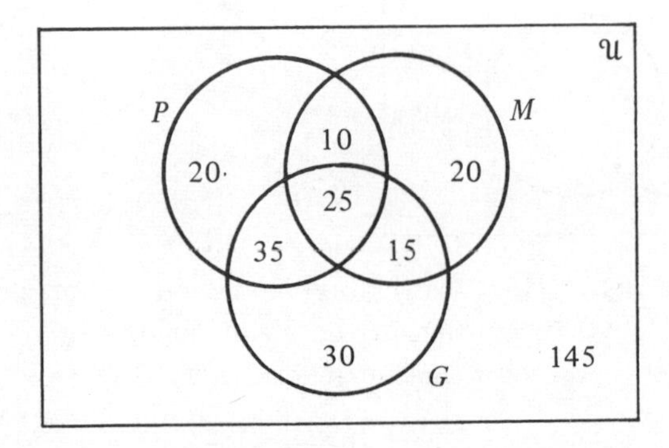

b. The number recycling only paper is the number in region IV, 20.
c. The number recycling paper or glass $= n(P \cup G)$ is found by counting everybody within the set $P \cup G$. We get $20 + 10 + 35 + 25 + 30 + 15 = 135$.
d. The number recycling just one item is the number of people in regions IV, VI, VII, or 70 people.

e. The number recycling at least one item is the total number of people from regions I–VII. That would give us $300 - 145 = 155$ people.

f. Using the result from part d, the fraction is:

$$\frac{70}{300} \text{ or } \frac{7}{30}$$

g. The fraction recycling no item is $\dfrac{145}{300} = \dfrac{29}{60}$.

h. The fraction of people who recycle something is $\dfrac{155}{300} = \dfrac{31}{60}$.

Each of the survey problems that we have just examined, besides being an application of sets and Venn diagrams, has also been illustrative of a *counting problem*. Counting problems are dealt with in depth in the next chapter. Any counting problem will always ask, "How many?" You can imagine that since a whole chapter is devoted to counting problems, they can get rather involved.

For some counting problems, it is convenient to develop formulas that can aid us in their solution. We do that now for one special situation. Consider an experiment where a single card is withdrawn from a deck of 52 cards. Let $\mathcal{U}$ be the set of logical possibilities; then $\mathcal{U}$ is composed of 52 elements (each of the cards that could be drawn). Let p be "the card is an ace." Let q be "the card is a club." Then $n(P) = 4$, $n(Q) = 13$. Note that the statement $p \wedge q$ reads "the card is an ace and a club." Hence, $n(P \cap Q) = 1$. We display these results below.

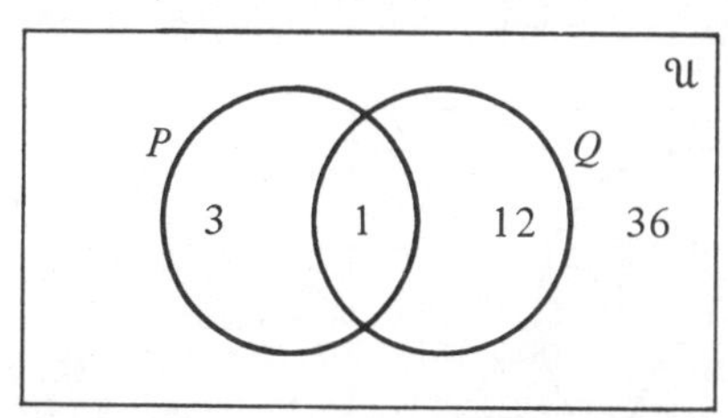

Now we ask the question, "How many ways could an ace or club be drawn?" This is $n(P \cup Q)$. Counting on the Venn diagram gives, correctly, 16. We now propose to show that, in general, $n(P \cup Q) = n(P) + n(Q) - n(P \cap Q)$.

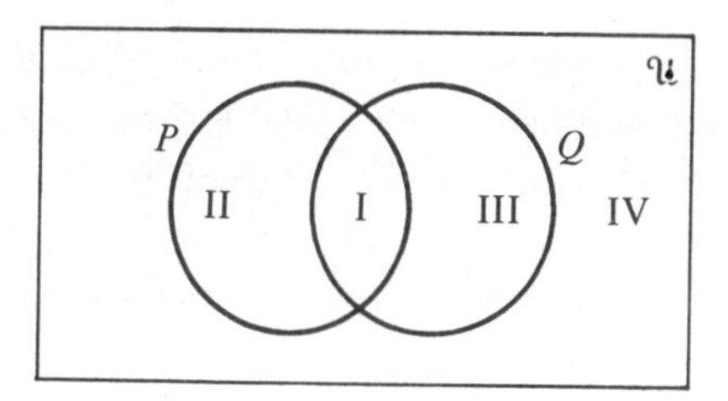

From the Venn diagram, we see that $n(P \cup Q)$ stands for all elements in regions I, II, III; $n(P)$ stands for all elements in regions I, II; $n(Q)$ stands for all elements in regions I, III. Thus $n(P) + n(Q)$ gives elements in regions II and III, but counts the number of elements in region I twice. Hence, the expression $n(P) + n(Q) - n(P \cap Q)$ will "add up" the elements in regions I, II, III once each, which is what we want. Therefore

$$n(P \cup Q) = n(P) + n(Q) - n(P \cap Q)$$

Applying this formula to the card problem above, $n(P \cup Q) = n(P) + n(Q) - n(P \cap Q) = 4 + 13 - 1 = 16$.

EXERCISES ■ SECTION 2.5.

1. Using Venn diagrams, test the compound statements below to see whether they are tautologies, self-contradictions, or neither.

   a. $p \wedge \sim p$
   b. $\sim(p \vee q)$
   c. $(p \vee q) \rightarrow \sim(p \vee q)$
   d. $(p \vee q) \rightarrow (\sim p \vee q)$
   e. $p \leftrightarrow p$
   f. $[(q \wedge p) \vee r] \wedge \sim p$
   g. $(p \rightarrow q) \leftrightarrow (\sim q \rightarrow \sim p)$

2. Using Venn diagrams, check each of the problems below to see if the expression in column I implies the expression in column II, if the expression in column II implies the expression in column I, neither, or both.

	I	II
a.	p	q
b.	$p \rightarrow (q \vee r)$	$\sim p \vee (\sim q \vee \sim r)$
c.	$p \rightarrow q$	$\sim p \vee q$
d.	$p \rightarrow q$	$p \leftrightarrow q$
e.	$p \wedge \sim q$	$\sim(p \rightarrow q)$

3. Using Venn diagrams, check to see which of the following pairs of statements are equivalent.

   a. $p \rightarrow q, \sim q \rightarrow \sim p$
   b. $p \vee q, \sim(p \wedge q)$
   c. $p \leftrightarrow q, p \wedge q$
   d. $\sim p \rightarrow q, p \vee q$.

4. Using Venn diagrams, decide which of the following pairs of statements are compatible.

   a. $p \wedge q \wedge r, p$
   b. $p \rightarrow q, q \rightarrow p$
   c. $p \vee q, \sim q \rightarrow p$
   d. $p \leftrightarrow q, p \wedge q$

5. Test each of the arguments below for validity by using the Venn diagram method developed in this section.

a. p
$\sim p \rightarrow q$
$\therefore \sim q$

b. $\sim p \rightarrow \sim q$
q
$\therefore p$

c. $p \rightarrow q$
$r \vee \sim q$
$\sim r$
$\therefore \sim p$

d. If you make a homemade gun, then you'll be arrested.
If you are a good boy, then you won't be arrested.
Therefore, if you are a good boy, then you won't make a homemade gun.

e. If I like mathematics, then I will study. Either I study or I fail.
Therefore, if I fail, then I do not like mathematics.

6. Refer back to the discussion in the blood-typing problem and fill in the table.

Donor has blood type:	*Can donate to types below:*
AB+	
AB−	
A+	
A−	
B+	
B−	
O+	
O−	

7. In a survey of 31 homes in the Scarsdale area, the following data was compiled:

15 homes had water beds
25 homes had conventional beds
10 homes had both water beds and the conventional type

a. How many of the homes surveyed had neither water beds nor conventional beds?
b. How many homes had water beds only?
c. How many of the homes satisfy the statement, "A home has a water bed if and only if it has a conventional type bed"?

8. A survey by a sociologist of 100 recently married males revealed the following information:

53 married women they had dated in high school

40 married women they had dated in college

10 married women they had dated in both high school and college

68 married women they had dated for a year or more

25 married women they had dated in high school and had dated for a year or more

30 married women they had dated in college and dated for a year or more

3 married women that they had dated both in high school and in college and had dated for a year or more

Let H = set of males who married a woman they dated in high school. Let C = set of males who married a woman they dated in college. Let Y = set of males who married a woman they had dated for a year or more.

a. Display the given information on a Venn diagram with the three mutually overlapping sets H, C, Y.
b. Answer the following questions:

How many males married a woman they had dated only in high school and had dated for less than a year?

How many males married a woman they had dated in high school or college?

How many males married a woman they had dated neither in high school nor in college, but had dated for a year or more?

How many males married a woman they had dated in neither high school nor college?

How many satisfy the statement: "If a male married a woman he had dated in high school, then he dated the woman for a year or more"?

What fraction of males married women they had dated in college and dated for less than a year?

9. A survey of ski areas revealed the information below. The numbers represent number of ski areas in each of the categories.

	Easy access, big vertical drop	Easy access, low vertical drop	Difficult access, big vertical drop	Difficult access, low vertical drop
Vermont	5	2	10	12
New York	4	1	8	15
Quebec	3	6	7	9

Let V = Vermont, Y = New York, and Q = Quebec. Let A = easy access and D = big vertical drop. Find the number of ski areas in the following sets:

a. $V \cap \tilde{A} \cap D$
b. $(Y \cap \tilde{D}) \cup (Q \cap A)$
c. $\widetilde{(V \cup Y)} \cap A$

Find the number of ski areas in the truth sets of the following statements:

d. A ski area is in New York and has neither a big vertical drop nor an easy access.
e. If a ski area is in Vermont, then it has a big vertical drop.

10. In the Clippo Corporation there are 110 employees, including 71 men and 39 women. There are 65 salespersons, 16 secretaries, and 20 management personnel. There are five saleswomen, ten men in management, and one male secretary.

a. Construct a Venn diagram displaying this information. (Be careful: there are four sets involved and not all four overlap.)
b. How many women are neither secretaries nor saleswomen nor in managerial positions?

11. In the Venn diagram of example 2.5.9, verify the following formula:

$$n(P \cup M \cup G) = n(P) + n(M) + n(G) - n(P \cap M) - n(P \cap G) - n(M \cap G) + n(P \cap M \cap G).$$

2.6 TREE DIAGRAMS

Since the beginning of this chapter, many of our examples have contained some universal set $\mathcal{U}$ (the set of logical possibilities). In this section, we will use a technique known as a **tree diagram,** a device that can be used to construct a set of logical possibilities for a given problem.

Look again at example 2.1.3, where two coins were tossed. There we found $\mathcal{U} = \{HH,HT,TH,TT\}$. Realize that it makes no difference whether the coins are tossed together or flipped separately: the same $\mathcal{U}$ results. The reason is quite simple—the coins don't know whether they are being tossed together or separately! Let's view the process as taking place with separate tosses, so we can refer to the "first coin" and the "second coin." Now we show how actually to construct the set $\mathcal{U}$.

Make two columns entitled "first coin" and "second coin," and list all the possibilities for the "first coin" column. Obviously we have just a head or a tail there.

First coin Second coin

H
T

Next take each possibility listed in the "first coin" column (one at a time) and inquire what can happen to the second coin. A sequence of drawings follows.

First coin Second coin

H — *H*
H — *T*
T

First coin Second coin

H — *H*
H — *T*
T — *H*
T — *T*

The diagram immediately above is referred to as a tree diagram (because it sort of "branches out" from left to right). Now we just read all the branches in the tree from left to right and get HH, HT, TH, TT. Those, of course, are exactly the elements of $\mathcal{U}$ for the experiment of tossing two coins.

Let's try the process again for the experiment of tossing three coins that came up in exercise 4 in section 2.3. This time we will need three columns, entitled "first coin," "second coin," and "third coin." A solution follows in a sequence of diagrams.

First coin Second coin Third coin

H
T

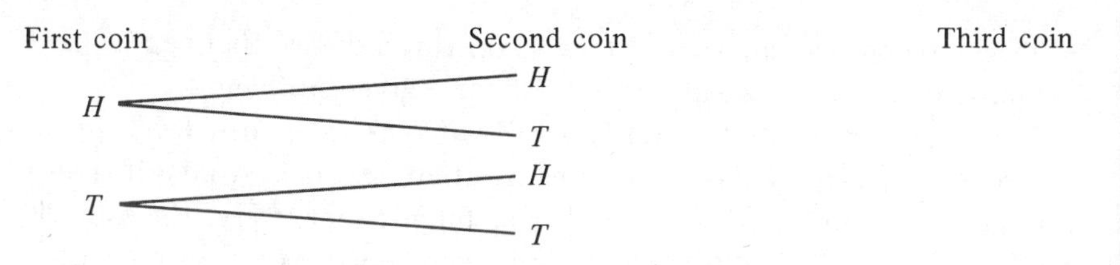

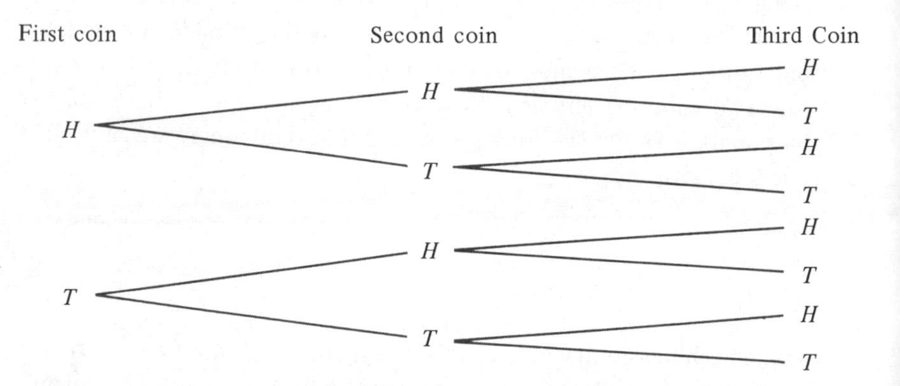

Now if we read this final diagram from left to right along each of its rows, we get HHH, HHT, HTH, HTT, THH, THT, TTH, TTT as the elements of $\mathcal{U}$.

In each of the coin problems just dealt with, we pretended that the coins were separately flipped so that we could refer to "first coin," "second coin," etc. However, many problems have a built-in sequence of events that must, of necessity, be followed.

Example 2.6.1. John and Norbert play a three-game handball tournament in which the victor is he who first wins two games. Use a tree diagram to contruct $\mathcal{U}$.

Solution: Here there can be up to three games, which must be referred to as "first game," "second game," "third game." We follow the sequence of diagrams below.

First game Second game Third game

J

N

First game Second game Third game

J — J, N

N — J, N

At this stage we are not finished with the tree, but we must note that in the first row (JJ), John is the victor and there won't be a third game. In the fourth row (NN), Norbert wins the tournament, thus eliminating any need for a third game there. With these thoughts in mind, we now complete the tree.

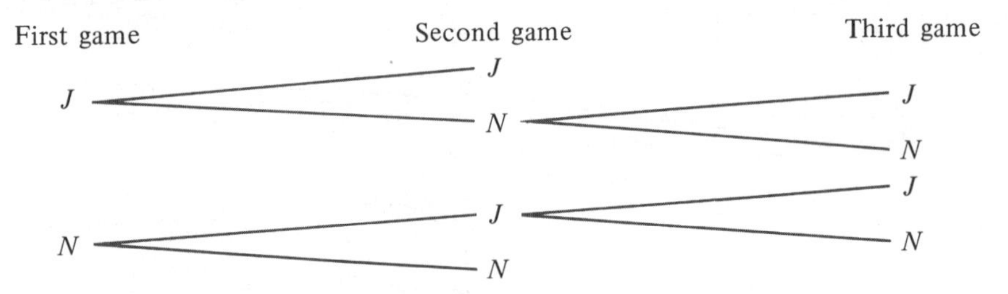

Now, reading all the branches of the tree, we obtain JJ, JNJ, JNN, NJJ, NJN, NN for the elements of $\mathcal{U}$.

In example 2.3.5, we talked about taking a survey of prisons where people were to be classified according to marital status, where the crime was committed, and the type of crime committed. Below is a tree showing the construction of $\mathcal{U}$.

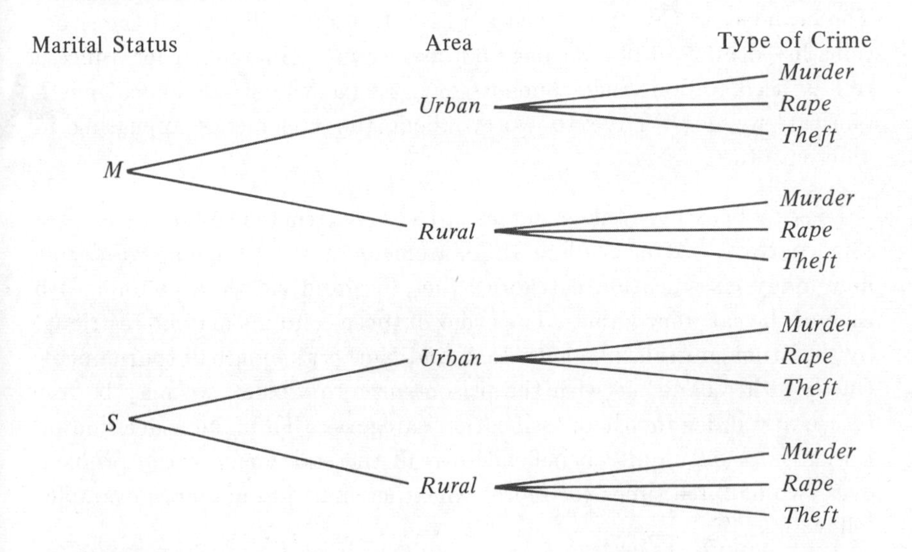

Reading all the branches of this tree gives the 12 situations that appear in the table on page 60.

There is nothing in this problem forcing us to classify first according to "marital status," followed by "area," and "type of crime." We are free to classify people in the order "type of crime," "marital status," and lastly "area." A completed tree is now shown for that order of classification categories.

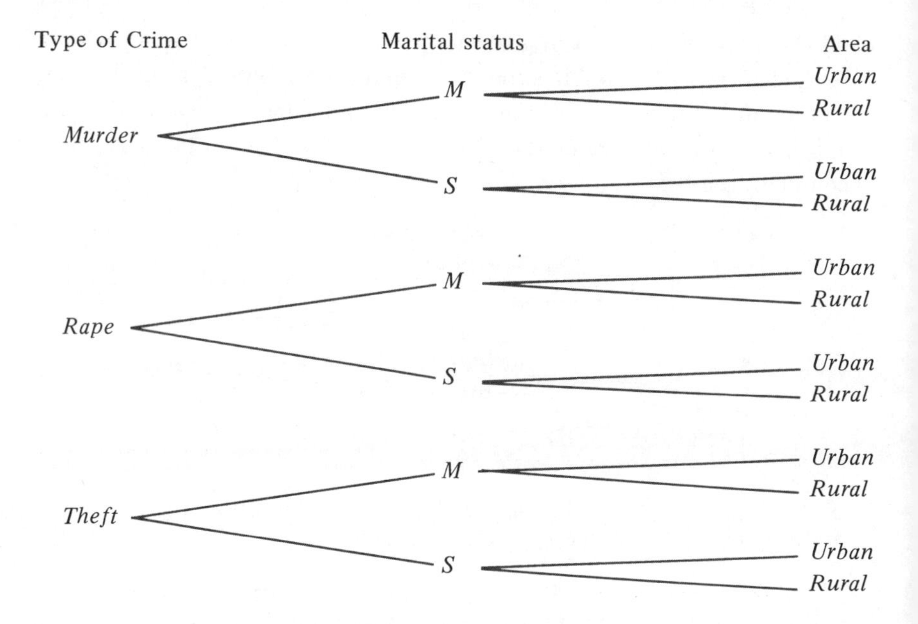

The branches of this tree, when read left to right, still give all the situations listed in the table on page 60, merely in a different order. Since a tree is used to construct elements of a *set* ($\mathcal{U}$, the set of logical possibilities), we don't have to worry about those elements appearing in different order.

Looking back over the examples of trees constructed so far, we can see some patterns. After reading the statement of the problem, we decide how many classification categories there are and entitle a column with each of the category names. The order of those columns may be restricted by the problem itself, like with John and Norbert's handball tournament. On the other hand, as with the prison survey problem, we may be free to use any order for the classification categories. To make a decision on this, it is usually quite beneficial to read the statement of the problem over two or three times (or more) when necessary! Some more examples follow.

Example 2.6.2. The Perkins Corporation has five owners: Meg, Peg, Linda, Tom, and Shelly. Elections will be held to elect officers (president and vice-president). The election for president will be run first, and whoever wins will not be allowed to run in the vice-presidential election. Construct a tree showing the possibilities.

Solution: We must have our first column labelled "president."

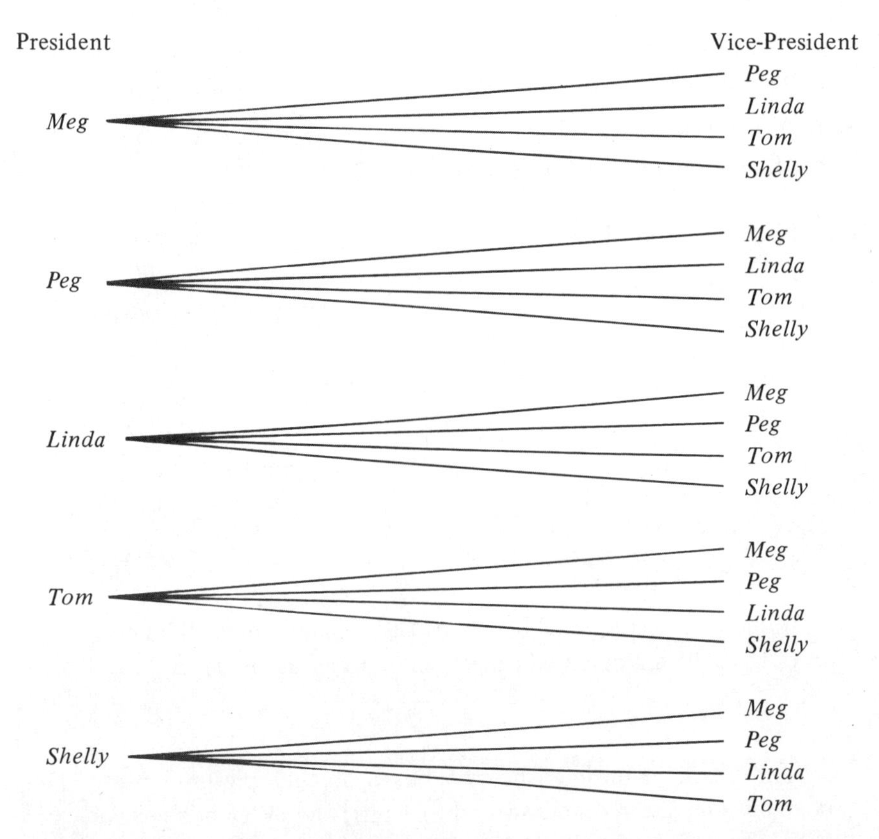

Example 2.6.3. Draw a tree for the above problem with the added restriction that the only nominees for president are Meg and Peg.

Solution:

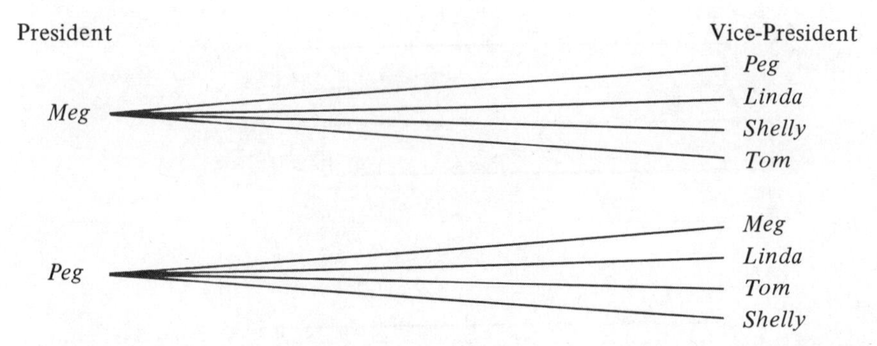

Note that the solution here is just the first eight cases of the tree in the previous example.

Example 2.6.4. A salesperson plans a trip from New York to Miami via Ft. Lauderdale. From New York to Ft. Lauderdale he can travel by bus, train, boat, or plane. From Ft. Lauderdale to Miami he can travel by bus or plane. Construct a tree showing all logical possibilities of transportation. In how many ways can the salesman make the trip?

Solution: There are two classification categories here, New York to Ft. Lauderdale and Ft. Lauderdale to Miami.

New York to Ft. Lauderdale — Ft. Lauderdale to Miami

Bus — *Bus*, *Plane*

Train — *Bus*, *Plane*

Boat — *Bus*, *Plane*

Plane — *Bus*, *Plane*

Reading all the tree branches gives the elements of $\mathcal{U}$. The number of ways that the salesman can make the trip is just $n(\mathcal{U}) = 8$.

Example 2.6.5. Consider the two boxes of balls pictured below. One box contains one red and two white balls; the other box contains two red and one white balls. An experiment consists of choosing a box at random and then choosing two balls from it. (The first ball is not replaced before the second is drawn.)

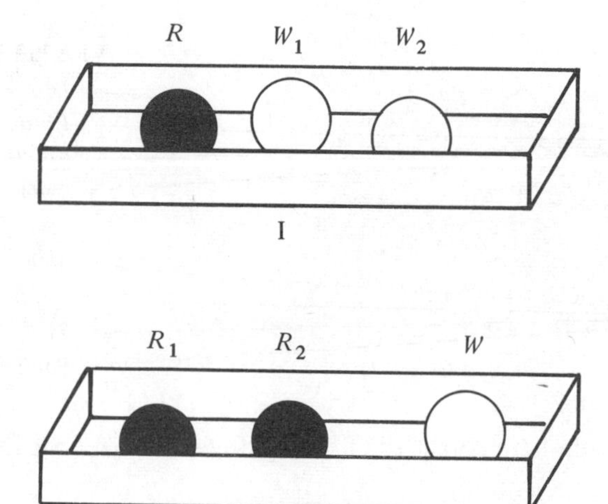

The subscripts are a convenient, notational way to keep track of balls of the same color. Draw a tree showing the logical possibilities.

Solution:

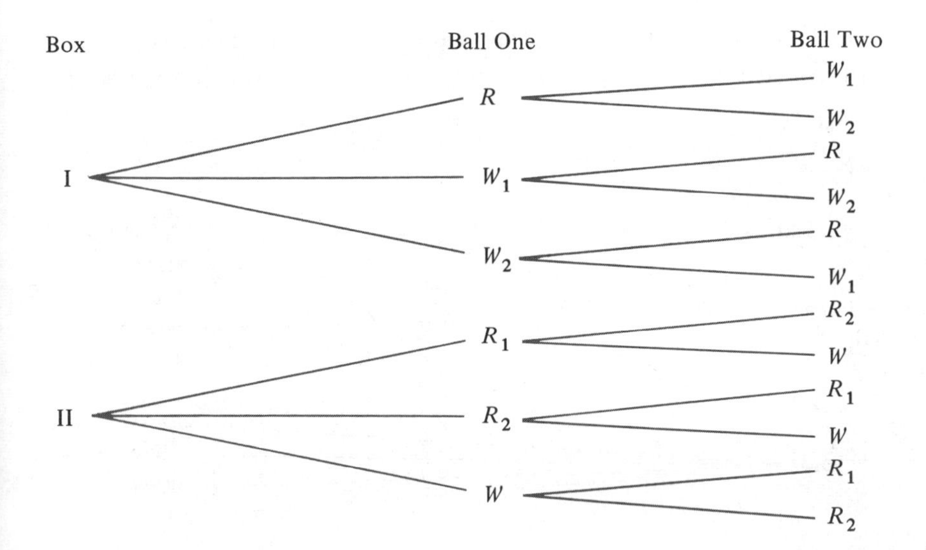

In table form we have:

Situation	*Box*	*Ball one*	*Ball two*
1	I	R	W_1
2	I	R	W_2
3	I	W_1	R
4	I	W_1	W_2
5	I	W_2	R
6	I	W_2	W_1
7	II	R_1	R_2
8	II	R_1	W
9	II	R_2	R_1
10	II	R_2	W
11	II	W	R_1
12	II	W	R_2

Example 2.6.5 involved a certain type of counting situation where replacement (in this case replacement of balls) was not allowed. In the next example, we have a situation where repetition is permitted.

Example 2.6.6. The Frigid Queen Appliance Company manufactures four major appliances (refrigerators, stoves, washers, dryers). They are to advertise one of these appliances on each of two television shows.

a. Draw a tree diagram depicting the choices available to the company. (Note: Here replacement is allowed; that is, the same appliance could be used for both shows.)

b. How many ways can the Frigid Queen Company choose to show their appliances?
c. How many of the ways involve the same appliance being used?
d. What fraction of the ways involve the same appliance being used?

Solution:

a.

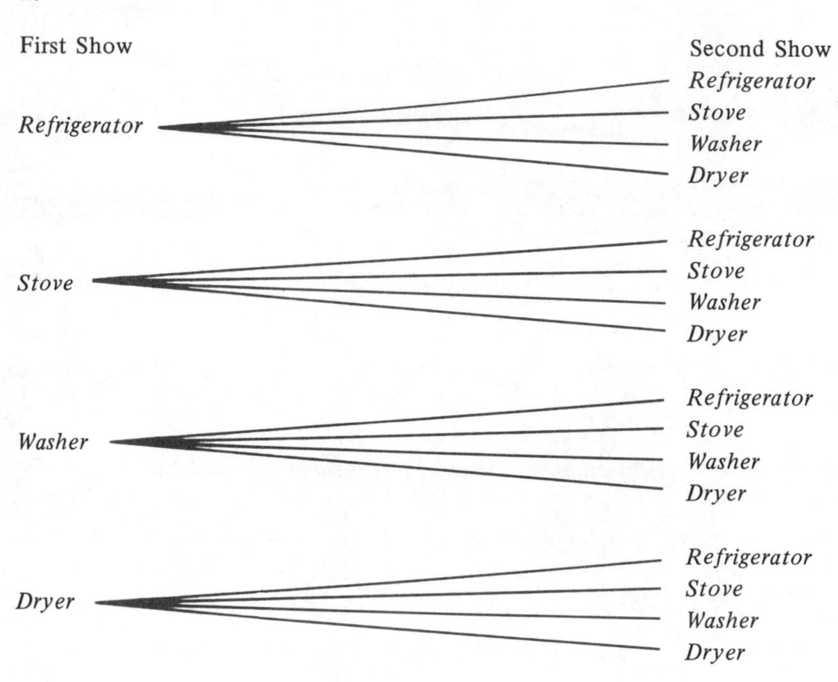

b. There are sixteen ways.
c. The reader should verify that four of the sixteen ways involve pairs of like appliances.
d. $\frac{4}{16} = \frac{1}{4}$

EXERCISES ■ SECTION 2.6.

1. Four coins are tossed (or equivalently, one coin is tossed four times in succession). Construct a tree diagram showing the logical possibilities and list the elements of $\mathcal{U}$.
2. An experiment consists of two parts. First a coin is tossed, then a die is rolled.

   a. Construct a tree showing the logical possibilities.
   b. List the elements of $\mathcal{U}$.
   c. What is $n(\mathcal{U})$?
   d. How many ways are there to get a head and an even number?

e. How many ways are there to get a head or an even number?
f. How many ways satisfy the statement: "If a head appears, then the number on the die is even"?

3. Two dice are tossed.

a. Construct a tree showing the logical possibilities.
b. List the elements of $\mathcal{U}$.
c. Let p be "the sum is seven." Find $n(P)$.
d. Let q be "the sum is eleven." Find $n(Q)$.
e. Let r be "the first die is even." Find $n(R)$.
f. Find $n(P \cap Q)$.
g. Find $n(P \cup Q)$.
h. Find $n(P \cap R)$.
i. Find $n(P \cup R)$.

4. Refer back to exercise 5 on page 62. Draw a tree for that problem showing all 24 logical possibilities.

5. An automobile manufacturer offers three engine sizes (small, medium, large), five body colors (red, orange, yellow, green, blue), and two body styles (sedan, roadster).

a. Draw a tree showing all logical possibilities for a complete car.
b. How many cars will a dealer need for a complete display of all possibilities?

6. Box 1 contains two red and two white balls. Box 2 contains one red and two white balls. An experiment consists of selecting a box and then selecting two balls. Draw a tree showing all logical possibilities when:

a. The first ball is replaced before the second is drawn.
b. The first ball is not replaced before the next is drawn.

7. A ski boot manufacturer offers three styles of fitting (powder injection, foam injection, hot wax injection) and four colors (red, black, yellow, orange). Draw a tree diagram showing all logical possibilities.

8. Sue and Jeanne have a chess tournament in which the victor is she who wins two games in a row or a total of three games. Draw a tree diagram to depict the possible outcomes for the tournament.

9. There are four operations (call them I, II, III, IV) to be done on an assembly line and they can be done in any order. Whenever II follows I, the entire assembly takes 22 minutes; if II precedes I the assembly takes 20 minutes when III follows IV and 15 minutes when III precedes IV. Use a tree diagram to depict all possible outcomes for the assembly and calculate the amount of time it would take an efficiency expert to time all possible assemblies.

2.7 THE CARTESIAN PRODUCT OF SETS

A set operation not mentioned in section 2.4 (because it has no logic analogy) which may be useful is the **Cartesian product** of two sets. The Cartesian product of two sets A, B (denoted $A \times B$ and read "A cross B") is the set of all pairs of elements such that the first element of the pair is an element of A and the second element of the pair is an element of B. For example, consider $A = \{1,2\}$ and $B = \{x,y\}$. Then $A \times B = \{(1,x), (1,y), (2,x), (2,y)\}$. Note that we cannot include the pair $(x,1)$ for example, because to be a pair in $A \times B$ the first of the pair comes from A, the second from B. Hence, the order in which the elements appear is significant. The elements of $A \times B$ are called **ordered pairs.** Also, note in the previous illustration: $n(A \times B) = 4$.

For another illustration consider set $P = \{H,T\}$ and $D = \{H,T\}$. Then $P \times D = \{(H,H), (H,T), (T,H), (T,T)\}$, the set of all occurrences for flipping two coins. This was originally seen in its shortened form, $\{HH,HT,TH,TT\}$.

If $G = \{a,b,c\}$ and $H = \{\&,?\}$, then $G \times H = \{(a,\&), (a,?), (b,\&), (b,?), (c,\&), (c,?)\}$ and we note that $n(G \times H) = 6$.

We shall see, as we work counting problems in chapter three, that the elements in a set are often not as important as the *size* (the number of elements) of the set. Could we have established the size of $G \times H$ without actually forming $G \times H$? The answer is yes. We resort to the following tree diagram.

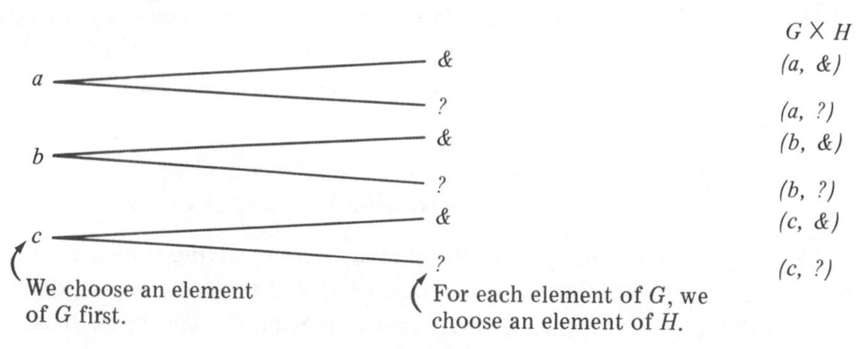

There are six branches in the tree. This is because we can fill the first position in three ways and the second position in two ways. There are 6 ($= 3 \cdot 2$) ways to get all the pairs.

If a set K has five elements and a set L has three elements, then $n(K \times L) = 15$. (This was done without knowing any of the elements!) We can generalize this:

> If set X has r elements and set Y has s elements, there will be $r \cdot s$ elements in $X \times Y$.

We refer to the above as the **r · s rule.**

Example 2.7.1. Let $M = \{\text{Bob, Ted}\}, F = \{\text{Carol, Alice}\}$. Find $M \times F$ and $n(M \times F)$.

Solution: $M \times F$ = the set of all ordered pairs whose first element is Bob or Ted and whose second element is Carol or Alice. Thus $M \times F = \{(\text{Bob, Carol}), (\text{Bob, Alice}), (\text{Ted, Carol}), (\text{Ted, Alice})\}$. Since there are four such ordered pairs, $n(M \times F) = 4$.

Example 2.7.2. Let A be the set of letters in the English alphabet. Find $n(A \times A)$.

Solution: Here it would be very inappropriate to actually find $A \times A$ and count its elements. Instead we use the $r \cdot s$ rule and find

$$n(A \times A) = 26 \cdot 26 = 676$$

The Cartesian product operation can be performed on three sets. That is, we may form $A \times B \times C$. This will be the set of all ordered triplets such that the first element in the triplet comes from A, the second from B, and the third from C. If $A = \{1,2\}$, $B = \{b\}$, and $C = \{5,e\}$, then $A \times B \times C = \{(1,b,5), (1,b,e), (2,b,5), (2,b,e)\}$ and $n(A \times B \times C) = 2 \cdot 1 \cdot 2 = 4$. This notion can be extended so the Cartesian product of any number of sets can be performed.

Example 2.7.3. Let $X = \{1,2,3\}$, $Y = \{4,5\}$, $Z = \{a,b\}$. Find:

a. $X \times Y \times Z$
b. $Y \times Y \times Y$
c. $n(Z \times Z \times Z \times Z \times Z)$

Solution:

a. $X \times Y \times Z = \{(1,4,a), (1,4,b), (1,5,a), (1,5,b), (2,4,a), (2,4,b), (2,5,a), (2,5,b), (3,4,a), (3,4,b), (3,5,a), (3,5,b)\}$
b. $Y \times Y \times Y = \{(4,4,4), (4,4,5), (4,5,4), (4,5,5), (5,4,4), (5,4,5), (5,5,4), (5,5,5)\}$
c. $n(Z \times Z \times Z \times Z \times Z) = 2 \cdot 2 \cdot 2 \cdot 2 \cdot 2 = 32$

EXERCISES ■ SECTION 2.7.

1. Let $\mathcal{U} = \{a,b,c, \ldots ,z\}$, the set of letters in the English alphabet. Let p: the letter is a vowel, q: the letter is a consonant, r: the letter rhymes with "she." Call the truth sets P, Q, R, respectively and find:

a. P
b. Q
c. R
d. $n(P \times \mathcal{U})$
e. $n(P \times \mathcal{U} \times \mathcal{U})$
f. $n(P \times Q \times R)$
g. $n(Q \times \mathcal{U} \times \mathcal{U})$
h. $n(R \times \mathcal{U} \times \mathcal{U})$
i. $n(\mathcal{U} \times \mathcal{U} \times \mathcal{U})$
j. $n(\mathcal{U} \times \mathcal{U} \times P)$

2. Let $V = \{T,F\}$. Find:

a. $V \times V$
b. $V \times V \times V$. Compare this result with the cases in a three-statement truth table.
c. $n(V \times V \times V \times V \times V \times V \times V \times V \times V \times V)$. Compare this result with exercise 2*a*, section 1.4.

3. Let $A = \{1,2\}$, $B = \{2,3\}$, $C = \{1,3,4\}$. Find:

a. $A \times (B \cap C)$
b. $A \times (B \cup C)$
c. $(A \cap B) \times C$
d. $(A \cup B) \times (A \cap C)$
e. $n(A \times B)$
f. $n(A \times B \times C)$

4. Let $C = \{H,T\}$, $D = \{1,2,3,4,5,6\}$. Find:

a. $C \times D$. Compare with exercise two, section 2.6.
b. Let $A = \{h\}$, $B = \{2,4,6\}$. Find $n(A \times B)$ and compare with exercise 2*d*, section 2.6.

5. Let D be as in exercise four above. Find $D \times D$ and compare with exercise 3, section 2.6.

chapter 2 in review

CHAPTER SUMMARY

In this chapter we combined sets by using one or more of the operations $\sim$, $\cap$, $\cup$, $-$, $\triangle$, $\times$. The relationships between these symbols and the symbols of logic were also examined. Every statement has an associated truth set, the set of all things which make it true. Thus the analogy grew between chapter 1 and chapter 2. Two special sets, $\varnothing$ and $\mathcal{U}$, were examined. The relations between sets—equality, the subset relation, and disjointness—were studied and also compared to their logic counterparts. Venn diagrams, a pictorial method of displaying sets and the number of elements in a set, were discussed and used to bridge the way from chapter 1 to chapter 2. Tree diagrams, the $r \cdot s$ rule, and the Cartesian product of sets began a discussion of counting techniques, a concept that will be explored in depth in chapter 3.

VOCABULARY

set	disjoint	truth set
element	difference	corresponding statement
universal set	symmetric difference	Venn diagram

complement
intersection
union
subset
equal sets
tree diagram
Cartesian product
$r \cdot s$ rule

SYMBOLS

$P, Q, R, \ldots$
$\in$
$\mathfrak{U}$
$\tilde{P}$
$\cap$
$\cup$
$-$
$\triangle$
$\subseteq$
$\varnothing$
$n(A)$
$=$
$G \times H$

chapter 3
Counting

3.1 INTRODUCTION

Counting problems are a way of answering the question, "How many?" Calculating the total number of possible outcomes for a given situation without actually listing those outcomes is the concern of this chapter. We have already done some counting! In chapter 2, when asked to find $n(A \cap B)$, for example, we were counting the number of elements in a set. In section 2.7, we used the $r \cdot s$ rule as a counting technique. We shall continue counting in the present chapter.

3.2 ORDERED ARRANGEMENTS (PERMUTATIONS)

There are basically two classifications for counting problems. The first classification includes all situations where **ordered arrangements** (or **permutations**) are counted. By ordered arrangement of elements we mean an array where the order in which an element appears in that array is significant. For example, if we are counting all the possible three-digit numerals that can be formed from the digits 1, 5, 7, 8, we would count 517 and 571 as being different (the order in which the elements appear is different), even though each is composed solely of 1, 5, and 7.

Unordered arrangements (or **combinations**) are those in which the occurrence of an element in an array is all that matters; the order in which elements occur is immaterial. For example, suppose there are 15 applicants for two vacant positions at the Hard-Hat Construction Company. The order in which the applicants are hired is immaterial and the question, "How many choices does the company have for filling the vacancies?" is a question that asks you to count unordered arrangements. We delay discussion of unordered arrangements until the next section and concentrate upon ordered arrangements here.

Now suppose the Attica License Plate Company wanted to know exactly HOW MANY possible different license plates can be made, where the first two positions in the plate are letters and the third, fourth, and fifth are digits. Here, the order in which things appear is significant. (Surely [**YX 628** NEW YORK] is a different plate than [**XY 862** NEW YORK].) We have an ordered arrangement. How many such ordered arrangements are there? One solution could employ a tree diagram. Another (and more practical) approach could be to use our basic counting principle, the $r \cdot s$ rule. (We are not concerned with the license plates themselves, merely with the number of plates.) If A is the set of twenty-six alphabet letters and B is the set of ten digits, we are looking for

$$n(A \times A \times B \times B \times B) = 26 \cdot 26 \cdot 10 \cdot 10 \cdot 10 = 676{,}000.$$

(Now we see the inappropriateness of a tree diagram. At $\frac{1}{4}$ inch per entry, 676,000 entries would produce a tree diagram almost three miles high!)

Let's return to the "three-digit" problem at the beginning of this section. How many three-digit numerals can be formed using the digits 1, 5, 7, 8? If we let $A = \{1,5,7,8\}$, we must find $n(A \times A \times A) = 4 \cdot 4 \cdot 4 = 64$. Or equivalently, we must fill three positions (the three digits):

$$\underline{\qquad}\quad \underline{\qquad}\quad \underline{\qquad}$$

We can fill the first position in any one of four ways (1 or 5 or 7 or 8). After the first position is filled, we can fill the second in any one of four ways and then the last in any one of four ways. So there are:

$$\underline{\quad 4 \quad} \cdot \underline{\quad 4 \quad} \cdot \underline{\quad 4 \quad} = 64$$

ways to "build" the three-digit numeral. This is just the $r \cdot s$ rule revisited.

We have scratched the surface of counting ordered arrangements with a few simple situations. Actually, although the problems get more complex, they don't get more difficult. It is the concept of counting that should be mastered. In the next illustration we look at the two previous illustrations but with certain restrictions. For example, in the "how many digits" problem above, let's count the number of three-digit numerals that can be formed without repeating a digit within any one numeral. Three positions must be filled.

$$\underline{\qquad}\quad \underline{\qquad}\quad \underline{\qquad}$$

There are four possibilities for the first digit.

$$\underline{\ 4\ }\ \underline{\quad}\ \underline{\quad}$$

For the second position there are only three possibilities and then two possibilities for the last digit, since we cannot repeat digits.

$$\underline{\ 4\ } \cdot \underline{\ 3\ } \cdot \underline{\ 2\ } = 24$$

There are twenty-four three-digit numerals whose digits do not repeat made up of 1, 5, 7, 8. (For the nonbeliever they are: 157, 158, 175, 178, 185, 187, 517, 518, 571, 578, 581, 587, 715, 718, 751, 758, 781, 785, 815, 817, 851, 857, 871, 875.) We see that this is *not* $n(A \times A \times A)$ because of the restriction (you cannot repeat) on the second and third digits. We can adjust our $\boldsymbol{r \cdot s}$ **rule** to read:

> If event R can be done r ways and AFTER THAT event S can be done in s ways, then the number of ways R and S can occur is $r \cdot s$.

In the license plate illustration, suppose we slightly alter the problem. How many license plates can be made if the first of the three digits cannot be zero and if the letters involve only vowels? The five positions can be filled as follows.

$$\underline{\ 5\ } \cdot \underline{\ 5\ } \cdot \underline{\ 9\ } \cdot \underline{\ 10\ } \cdot \underline{\ 10\ } = 22{,}500$$

There are five vowels (first two positions); We can't use zero for the first digit (third position); Any of the ten digits can be used here (fourth and fifth positions).

Let's look at some examples.

Example 3.2.1. In the Olympics there are typically three awards for an event: a gold, a silver, and a bronze medal. How many ways exist for awarding the medals if there are twelve entrants for an event?

Solution: There are twelve possibilities for the gold medal, eleven for the silver, and ten for the bronze.

$$\underset{\text{(gold)}}{\underline{\ 12\ }} \cdot \underset{\text{(silver)}}{\underline{\ 11\ }} \cdot \underset{\text{(bronze)}}{\underline{\ 10\ }}$$

Hence, there are 1320 possible ways of awarding the medals.

Example 3.2.2. There are eight horses in the fourth race at Fitzhugh Downs. The "Ultimatafecta" consists of picking the eight horses in their exact order of finish. If someone desired to purchase all possible "Ultimatafecta" tickets, how many would he have to buy?

Solution: There are eight possible horses that could finish first, seven possibilities could finish second, six could finish third, five could finish fourth, four could finish fifth, three could finish sixth, two could finish seventh, and one eighth.

$$\underline{\ 8\ } \cdot \underline{\ 7\ } \cdot \underline{\ 6\ } \cdot \underline{\ 5\ } \cdot \underline{\ 4\ } \cdot \underline{\ 3\ } \cdot \underline{\ 2\ } \cdot \underline{\ 1\ } = 40{,}320$$

Example 3.2.3. The Varsity Club wishes to elect two different members to the posts of president and vice-president. How many possible governing pairs are there if there are five members in the club?

Solution: There are five possibilities for the president, each of which can be paired up with four possibilities (all but the one who's president) for vice-president. Hence, there are $5 \cdot 4 = 20$ possibilities.

Example 3.2.4.

a. How many four-letter "words" (not necessarily from the dictionary) can be formed from the letters in the set $\{h,o,t,s\}$?
b. How many can be formed if the four-letter "words" cannot have repeated letters within a particular arrangement?
c. How many if, in addition to the restriction in (*b*), the "word" must begin with *s* and end in *t*?

Solution:

a. $4 \cdot 4 \cdot 4 \cdot 4 = 256$ possible words.
b. $4 \cdot 3 \cdot 2 \cdot 1 = 24$ possible words.
c. $1 \cdot 2 \cdot 1 \cdot 1 = 2$ possible words.

Traditionally, many questions concerning probability (chapter 4) involve the "balls-in-an-urn" type problem. Suppose we have an urn containing seven differently colored balls. Two balls are selected, one at a time, so that after the first is withdrawn it is recorded and replaced in the urn. In how many ways can the two balls be selected? That's easy, just $7 \cdot 7 = 49$ ways. This differs significantly from the same experiment performed without replacement. That is, a ball is withdrawn, left out, and then another withdrawn (or equivalently, two are withdrawn simultaneously). In how many ways can two balls be withdrawn? There are $7 \cdot 6 = 42$ ways.

Suppose we have an urn containing eight balls and we successively select balls at random without replacement until all balls are selected. In how many ways can this be done? There are $8 \cdot 7 \cdot 6 \cdot 5 \cdot 4 \cdot 3 \cdot 2 \cdot 1$ ways. This is a relatively large number (40,320) and we use a notational shortcut here. By $8!$ we mean $8 \cdot 7 \cdot 6 \cdot 5 \cdot 4 \cdot 3 \cdot 2 \cdot 1$. By $4!$ we mean $4 \cdot 3 \cdot 2 \cdot 1$ and by $n!$ we mean $n(n-1)(n-2) \cdots 3 \cdot 2 \cdot 1$, the product of all whole numbers between n and 1. So $5! = 5 \cdot 4 \cdot 3 \cdot 2 \cdot 1 = 120$; $2! = 2 \cdot 1 = 2$, and $1! = 1$. We shall agree (everyone always has) that $0! = 1$, the only exception to the rule. We read $n!$ as "*n factorial.*" A list of factorials can be found in Table I of the Appendix.

Example 3.2.5. Compute: **a.** $3!$ **b.** $7!$ **c.** $3! \cdot 4!$

d. $\frac{7!}{3!\,4!}$ **e.** $\frac{1000!}{999!}$ **f.** $\frac{4!}{2!}$ **g.** $\frac{5!}{0!\,5!}$

Solution:

a. $3! = 3 \cdot 2 \cdot 1 = 6$

b. $7! = 7 \cdot 6 \cdot 5 \cdot 4 \cdot 3 \cdot 2 \cdot 1 = 5040$

c. $3! = 6$ and $4! = 24$, so $3! \cdot 4! = 6 \cdot 24 = 144$

d. $\frac{7!}{3!\,4!} = \frac{5040}{144} = 35$ or $\frac{7!}{3!\,4!} = \frac{7 \cdot \not{6} \cdot 5 \cdot \not{4} \cdot \not{3} \cdot \not{2} \cdot 1}{\not{3} \cdot \not{2} \cdot 1 \cdot \not{4} \cdot \not{3} \cdot \not{2} \cdot 1} = 35$

e. $\frac{1000!}{999!} = \frac{1000 \cdot 999 \cdot 998 \cdot 997 \cdot 996 \cdots 3 \cdot 2 \cdot 1}{999 \cdot 998 \cdot 997 \cdots 3 \cdot 2 \cdot 1} = 1000$

f. $\frac{4!}{2!} = \frac{4 \cdot 3 \cdot 2 \cdot 1}{2 \cdot 1} = 12$

g. $\frac{5!}{0!\,5!} = \frac{5 \cdot 4 \cdot 3 \cdot 2 \cdot 1}{1 \cdot 5 \cdot 4 \cdot 3 \cdot 2 \cdot 1} = 1$

Let us examine some counting situations with which anyone can identify. Consider the typical juke box. A selection is made by choosing any one of the twenty-six alphabet letters followed by any one of the ten digits {0,1,2,3,4,5,6,7,8,9}. There are $26 \cdot 10 = 260$ different possible selections.

The question, "How many different possible telephone numbers (seven digits) are there?" is one that has perhaps been asked by some telephone executive worrying whether there will be enough numbers to go around. If no restrictions are considered, there are

$$\underline{10} \cdot \underline{10} \cdot \underline{10} \cdot \underline{10} \cdot \underline{10} \cdot \underline{10} \cdot \underline{10} = 10{,}000{,}000$$

numbers, because there are ten choices to fill each digit in the phone number. In real life, however, there are restrictions on the digits. We consider a restriction in the following example.

Example 3.2.6. Suppose the first digit can be neither a zero nor a one. How many phone numbers are there?

Solution: There are eight possibilities for the first digit and ten for each of the others. Now there are

$$\underline{8} \cdot \underline{10} \cdot \underline{10} \cdot \underline{10} \cdot \underline{10} \cdot \underline{10} \cdot \underline{10} = 8{,}000{,}000$$

phone numbers.

There are eight million possible seven-digit phone numbers (where the first digit is neither zero nor one). Obviously there became a need, at some stage in the history of the telephone, for more than 8,000,000 phones. (There are 125,156,700 telephones in this country now.) Thus area codes were invented. If there are 100 different area codes (and there actually are 100 area codes in the continental United States), how many possible phone numbers are there now? That's simply

$$100 \cdot 8{,}000{,}000 = 800{,}000{,}000$$

possible phone numbers.

Another typical counting problem is the following: How many ways can eight people stand in line to buy tickets? Using our "slots," we see that there are eight choices for people to fill the first position. Since we cannot repeat people, the number of ways is

$$\underline{\ 8\ } \cdot \underline{\ 7\ } \cdot \underline{\ 6\ } \cdot \underline{\ 5\ } \cdot \underline{\ 4\ } \cdot \underline{\ 3\ } \cdot \underline{\ 2\ } \cdot \underline{\ 1\ } = 8!$$

Let's change the problem slightly. Suppose there are eight children to line up for a class picture. Suppose five are boys and three are girls. In how many different ways can they line up for the photograph if all the boys want to stand together and all the girls want to stand together? First let's suppose we have the group of boys on the left and the group of girls on the right.

$$\underbrace{(\underline{\quad}\ \underline{\quad}\ \underline{\quad}\ \underline{\quad}\ \underline{\quad})}_{\text{Boys}} \quad \underbrace{(\underline{\quad}\ \underline{\quad}\ \underline{\quad})}_{\text{Girls}}$$

There are five possibilities for the first boy position, four for the second, and so on. There are three possibilities for the first girl position, two for the second, leaving one for the last. This yields:

$$(\underline{\ 5\ } \cdot \underline{\ 4\ } \cdot \underline{\ 3\ } \cdot \underline{\ 2\ } \cdot \underline{\ 1\ }) \cdot (\underline{\ 3\ } \cdot \underline{\ 2\ } \cdot \underline{\ 1\ }) = 720$$

We are not finished yet! We haven't considered the possibilities where the girls are first:

$$\underbrace{(\underline{\quad}\ \underline{\quad}\ \underline{\quad})}_{\text{Girls}} \quad \underbrace{(\underline{\quad}\ \underline{\quad}\ \underline{\quad}\ \underline{\quad}\ \underline{\quad})}_{\text{Boys}}$$

Here, too, there would be 720 ways: $(3 \cdot 2 \cdot 1) \cdot (5 \cdot 4 \cdot 3 \cdot 2 \cdot 1)$. So the final answer is $720 + 720 = 1440$.

We expand upon the above illustration in the three examples that follow.

Example 3.2.7. Ralph is unemployed and has six job interviews tomorrow.

a. In how many ways can he arrange a schedule of interviews?
b. How many interview schedules can he make if he wants four interviews before noon and two after noon?
c. In how many ways can he arrange a schedule of interviews if two of the interviews, those with Company X and Company Y, must be in the afternoon and no others can be in the afternoon?

Solution:

a. $$\underline{\;6\;} \cdot \underline{\;5\;} \cdot \underline{\;4\;} \cdot \underline{\;3\;} \cdot \underline{\;2\;} \cdot \underline{\;1\;} = 720 \text{ ways}$$

b. $$\underbrace{(\underline{\;6\;} \cdot \underline{\;5\;} \cdot \underline{\;4\;} \cdot \underline{\;3\;})}_{\text{The four morning interviews}} \cdot \underbrace{(\underline{\;2\;} \cdot \underline{\;1\;})}_{\text{The two afternoon interviews}} = 720 \text{ ways}$$

c. $$\underbrace{(\underline{\;4\;} \cdot \underline{\;3\;} \cdot \underline{\;2\;} \cdot \underline{\;1\;})}_{\text{The four morning interviews}} \cdot \underbrace{(\underline{\;2\;} \cdot \underline{\;1\;})}_{\text{The two afternoon interviews (Companies } X \text{ and } Y\text{)}} = 48 \text{ ways}$$

Example 3.2.8. A librarian has twelve books to place on a shelf. Five of them are math books, three are history books, and four are biology books. How many different shelf arrangements can be made if all books of the same subject are to stay together?

Solution: Suppose the books are arranged with the biology books first, then history, then math.

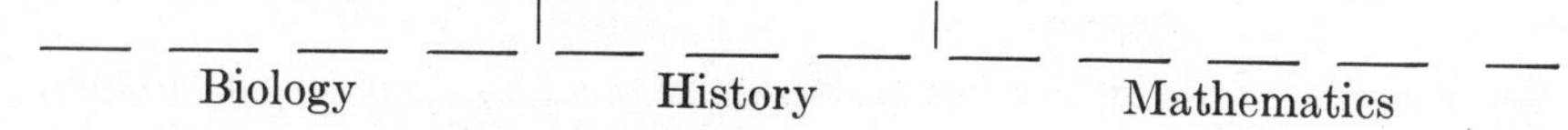

We have four choices for the first biology book, then three, then two, then one. There are three choices for the first history book, and so on. We have:

$$\underline{\;4\;} \cdot \underline{\;3\;} \cdot \underline{\;2\;} \cdot \underline{\;1\;} \mid \underline{\;3\;} \cdot \underline{\;2\;} \cdot \underline{\;1\;} \mid \underline{\;5\;} \cdot \underline{\;4\;} \cdot \underline{\;3\;} \cdot \underline{\;2\;} \cdot \underline{\;1\;}$$

So there are $4! \cdot 3! \cdot 5!$ ways of arranging the books when we have B, H, M (biology, then history, then math). But, the subjects themselves can be arranged in $3! = 6$ ways: BHM, BMH, HBM, HMB, MBH, MHB. Finally, then, the total number of possible arrangements is $3! \cdot 4! \cdot 3! \cdot 5! = 103{,}680$.

Example 3.2.9. Fifty men (twenty Italians, fourteen Poles, five Czechs, and eleven Mexicans) are to be photographed together for a group picture, and all of the same nationality wish to remain together. In how many ways can they arrange themselves for the photograph?

Solution: The four nationalities can be ordered in $4!$ ways. The Italians can arrange themselves in $20!$ ways, the Poles in $14!$ ways, the Czechs in $5!$ ways, and the Mexicans in $11!$ ways. The total number of ways is therefore $4! \cdot 20! \cdot 14! \cdot 5! \cdot 11!$.

We know that if we count the number of four-letter "words" that can be made (without repeating a letter) using the letters of the word "mask," we will find 4! such "words." In fact, they are:

1. m a s k	**7.** a m s k	**13.** s a m k	**19.** k a m s
2. m a k s	**8.** a m k s	**14.** s a k m	**20.** k a s m
3. m s a k	**9.** a s m k	**15.** s m a k	**21.** k s a m
4. m s k a	**10.** a s k m	**16.** s m k a	**22.** k s m a
5. m k a s	**11.** a k s m	**17.** s k a m	**23.** k m a s
6. m k s a	**12.** a k m s	**18.** s k m a	**24.** k m s a

Suppose we wanted to count the number of four-letter "words" that can be made (without repetition of letters) using the letters of the word "sass." Notice that the *m* and *k* in "mask" are each being replaced by an *s*. So, we'll make that replacement in the twenty-four entries:

1. s a s s	**7.** a s s s	**13.** s a s s	**19.** s a s s
2. s a s s	**8.** a s s s	**14.** s a s s	**20.** s a s s
3. s s a s	**9.** a s s s	**15.** s s a s	**21.** s s a s
4. s s s a	**10.** a s s s	**16.** s s s a	**22.** s s s a
5. s s a s	**11.** a s s s	**17.** s s a s	**23.** s s a s
6. s s s a	**12.** a s s s	**18.** s s s a	**24.** s s s a

We no longer have 24 entries. Entries 1, 2, 13, 14, 19, 20 are the same. Entries 3, 5, 15, 17, 21, 23 are the same. Entries 4, 6, 16, 18, 22, 24 are the same, as are 7, 8, 9, 10, 11, 12.

There are four different "words" made up solely of the letters from the word "s a s s":

s a s s, s s a s, s s s a, a s s s

What does our final answer of 4 have to do with the initial 24? Within any one arrangement, the three *s*'s can be interchanged in $3! = 6$ ways, so our answer of 4 is one-sixth of the 24. Hence we have $\frac{4!}{3!}$.

Example 3.2.10. How many distinct ordered arrangements can be formed from all the letters of each word?

a. t h e **b.** s o r r y **c.** m o m m a **d.** k n o c k e r s
e. s p e e d e r **f.** r a d a r **g.** f o o t b a l l
h. s a s s a f r a s s n e s s

Solution:

a. There are 3 different letters, so we have 3!.

b. If all the letters were different, there would be 5! arrangements. However, there are two *r*'s (which can be interchanged in $2! = 2$ ways), hence the answer is $\frac{5!}{2!} = 60$.

c. If all the letters were different, there would be 5! arrangements. We divide them by 3! (the number of ways the *m*'s can be rearranged).

Thus there are $\frac{5!}{3!} = 20$ words.

d. $\frac{8!}{2!}$ **e.** $\frac{7!}{3!}$

f. If all the letters were different, there would be 5! arrangements. Since there are two r's, we must divide by 2!. Since there are two a's we must divide by 2! again. The number of words is $5!/2!2! = 30$.

g. $\frac{8!}{2!\,2!}$

h. $\frac{14!}{7!\,3!}$

The last kind of counting problem we will discuss in this section is the type that is counted in an indirect manner. Suppose we look at the problem of counting the number of three-digit numerals that can be formed (repetitions allowed within any particular arrangement) by selecting digits from the set $\{1,2,3,8,9\}$. There are:

$$\underline{\ 5\ } \cdot \underline{\ 5\ } \cdot \underline{\ 5\ } = 125$$

numerals. How many of those 125 different numerals contain at least one nine? (Just think for a minute and you'll deduce it's not so obvious.) Let's delay that question for awhile and answer: How many of those 125 numerals DO NOT contain a nine? That's not hard:

$$\underline{\ 4\ } \cdot \underline{\ 4\ } \cdot \underline{\ 4\ } = 64 \text{ numerals}$$

Because we are choosing from 1,2,3,8

In other words, of the 125 three-digit numerals, exactly 64 of them do not contain a nine. Therefore, all the rest $(125 - 64)$ contain at least one nine. There are 61 of the 125 that contain at least one nine. It is a good general rule for the reader to keep in mind that whenever you are asked to find "at least one," search for just the opposite ("none") and then subtract that from the total. Actually, if p is "there is at least one" then $\sim p$ is "there is none" and we are using a counting principle:

$$n(P) + n(\tilde{P}) = n(\mathfrak{U})$$

OR

$$n(P) = n(\mathfrak{U}) - n(\tilde{P})$$

Example 3.2.11. Of fifteen musicians (five clarinetists, four pianists, and six guitarists), two are to be selected to appear in a photograph.

a. How many different possible photographs could be made?
b. How many of those contain at least one guitarist?

Solution:

a. $15 \cdot 14 = 210$ possible photographs

b. First count the number of photographs containing no guitarists. There are nine nonguitarists, so we have $9 \cdot 8 = 72$ pairs containing no guitarists. Therefore, the number of photographs containing at least one guitarist is $210 - 72 = 138$.

EXERCISES ■ SECTION 3.2.

1. How many four-letter "words" can be formed from the letters *s, c, r, a, p*, if:
   a. repetitions of a letter within an arrangement are not allowed?
   b. repetitions of a letter within an arrangement are allowed?
2. **a.** How many of the "words" in exercise 1*a* above begin with an *s*?
   b. How many of the "words" in exercise 1*b* above contain at least one *a*?
3. **a.** How many possible social security numbers are there?
   b. How many possible social security numbers are there if 000-00-0000 is not allowed?
4. In how many ways can ten people stand in line for tickets at a box office?
5. Consider the JUMBLE shown at the right.

   a. How many ways are there of arranging the letters in each of the four JUMBLES?
   b. Do each of the four JUMBLES in order to determine the number of ways of arranging the letters of the Surprise Answer.
6. A clerk desires to make a display of curtains. There are eight differently colored curtains. He will pick four of the curtains and arrange them for a sales display. How many ways can he do this?
7. How many four-digit numbers are there if no repetition of digits is allowed within a number and zero cannot be the first digit?
8. A certain matching test for Professor Plum's history course consists of a list of ten historical events and another list of ten people. A student is to match exactly one event with one person. If a

student is totally guessing and must make all ten matches, how many possible sets of answers are there?

9. **a.** How many four-digit numbers can be formed from the digits 0, 1, 2, 3, 4, 5 if no repetitions of digits are allowed within a number? (Zero may not be the first digit.)
   b. How many of those are even? (Hint: count the number of odd four-digit numbers and subtract from your answer to 9*a*.)

10. Evaluate:

a. $\dfrac{5!}{4!}$ **b.** $\dfrac{6!}{2!\,2!\,2!}$ **c.** $2! \cdot 3! \cdot 4!$

d. $\dfrac{8!}{7!}$ **e.** $\dfrac{10!}{8!\,2!}$ **f.** $\dfrac{100!}{98!}$

11. **a.** In a local jail, there are fourteen prospects for a "line-up." (A line-up consists of standing five men before an upstanding citizen.) How many different line-ups are there? (Consider this an ordered arrangement.)
    b. If seven of the men are white, three are Indian, and four are black, how many line-ups contain at least one black man?

12. If six people enter a bus in which there are ten vacant seats, in how many ways can they be seated?

13. **a.** In how many ways can four boys and five girls be seated in a row of nine chairs?
    b. In how many ways can they be seated if boys and girls want to sit alternately?

14. In how many ways can seven boys and seven girls be seated in a row of fourteen chairs, if they must sit in alternate seats?

15. How many different arrangements of flags can be made from four different flags placed one above another on a flagpole?

16. **a.** How many Greek-letter fraternity names are there if each fraternity name contains three letters? (There are twenty-four Greek letters.)
    b. How many of those in 16*a* contain at least one "Σ"?

17. There are seven roads leading to a traffic circle. A traffic engineer must know how many ways there are to enter by one road and leave by another. Do his counting problem.

18. There are 12 flags; three of them are blue, four are red, and five are white. A signal consists of twelve flags, one above another on a flagpole. How many different signals are there? (Hint: Answer the question: "How many distinguishable 'words' can be made using all the letters of the word *BBBRRRRWWWWW*?" Both questions ask the same thing. WHY?)

19. A "Daily Double" involves picking the winning horse in the first

race and the winning horse in the second race. If someone wishes to purchase all possible Daily Double tickets, how many purchases must he make? (Assume eight horses per race.)

20. A purchaser is confronted with the problem of supplying his shoe store with all possible men's shoes in a certain style. The sizes range from six through twelve (including 6, $6\frac{1}{2}$, 7, $7\frac{1}{2}$, . . . , 12), widths *A*, *B*, *C*, *D*, *E*, *EE*, *EEE*, and colors black, brown, tan. What is the minimum number of shoes the store must stock in order to insure itself of having a pair of shoes of each type?
21. A student must take Math 101, English 212, and Biology 102 next semester. Math 101 meets at 8:00, 9:00, 11:00, 12:00, 1:00; English 212 meets at 8:00, 11:00, 12:00; Biology 102 meets at 12:00, 1:00, 2:00. How many possible schedules can the student make? (Hint: use a tree diagram.)
22. The manager of a baseball team has picked the nine men he wants to play ball today.

    a. How many possible batting orders can he make?
    b. How many possible batting orders can he make if the pitcher, Cal Seeyum, must bat ninth?
    c. How many possible batting orders can be made if the pitcher must bat ninth and the shortstop, Speedy Vitalis, must bat first?

23. On a psychology exam, Professor Duerf lists his first ten questions as multiple choice (four possible answers to each question) and the next 15 questions as true–false questions. How many ways are there to answer all of the questions?
24. There exist things called zip codes, which are five-digit numerals.

    a. How many possible zip codes are there?
    b. How many if the numeral 00000 is not included?

25. Morse code consists of a sequence of dots and dashes to represent "things" (usually letters). How many "things" can be represented if there can be at most four dots and/or dashes to a sequence?

3.3 UNORDERED ARRANGEMENTS (COMBINATIONS)

As stated in the last section, unordered arrangements are those in which an element's presence in an array is all that matters; the order in which the elements occur is insignificant. Choosing committees is an illustration of counting unordered arrangements, because the order in which people are chosen is immaterial. Let's attempt to count the number of two-man committees that can be formed from Al, Bob, Chuck, and Dion.

We already know there are $4 \cdot 3 = 12$ ordered arrangements. In fact, they are:

Al, Bob	Bob, Al	Chuck, Al	Dion, Al
Al, Chuck	Bob, Chuck	Chuck, Bob	Dion, Bob
Al, Dion	Bob, Dion	Chuck, Dion	Dion, Chuck

This, however, is not what we want. There is unnecessary duplication. For instance, "Al, Bob" and "Bob, Al" are two arrangements that really represent the same committee. We wish to count only one of them; the order in which they appear doesn't change anything. Eliminating duplication, we have:

Al, Bob	Bob, Chuck
Al, Chuck	Bob, Dion
Al, Dion	Chuck, Dion

That's it! Six committees (or unordered arrangements) can be formed from the four people. Our original 12 was divided by 2.

Let's extend this. Suppose a committee of three is to be chosen from five people: Al, Bob, Chuck, Dion, Efrom. (For brevity's sake, we will call them A, B, C, D, E.) There are $5 \cdot 4 \cdot 3 = 60$ ordered arrangements. They are:

1. $A\ B\ C$	**13.** $B\ A\ C$	**25.** $C\ A\ B$	**37.** $D\ A\ B$	**49.** $E\ A\ B$
2. $A\ B\ D$	**14.** $B\ A\ D$	**26.** $C\ A\ D$	**38.** $D\ A\ C$	**50.** $E\ A\ C$
3. $A\ B\ E$	**15.** $B\ A\ E$	**27.** $C\ A\ E$	**39.** $D\ A\ E$	**51.** $E\ A\ D$
4. $A\ C\ B$	**16.** $B\ C\ A$	**28.** $C\ B\ A$	**40.** $D\ B\ A$	**52.** $E\ B\ A$
5. $A\ C\ D$	**17.** $B\ C\ D$	**29.** $C\ B\ D$	**41.** $D\ B\ C$	**53.** $E\ B\ C$
6. $A\ C\ E$	**18.** $B\ C\ E$	**30.** $C\ B\ E$	**42.** $D\ B\ E$	**54.** $E\ B\ D$
7. $A\ D\ B$	**19.** $B\ D\ A$	**31.** $C\ D\ A$	**43.** $D\ C\ A$	**55.** $E\ C\ A$
8. $A\ D\ C$	**20.** $B\ D\ C$	**32.** $C\ D\ B$	**44.** $D\ C\ B$	**56.** $E\ C\ B$
9. $A\ D\ E$	**21.** $B\ D\ E$	**33.** $C\ D\ E$	**45.** $D\ C\ E$	**57.** $E\ C\ D$
10. $A\ E\ B$	**22.** $B\ E\ A$	**34.** $C\ E\ A$	**46.** $D\ E\ A$	**58.** $E\ D\ A$
11. $A\ E\ C$	**23.** $B\ E\ C$	**35.** $C\ E\ B$	**47.** $D\ E\ B$	**59.** $E\ D\ B$
12. $A\ E\ D$	**24.** $B\ E\ D$	**36.** $C\ E\ D$	**48.** $D\ E\ C$	**60.** $E\ D\ C$

We must eliminate any duplication where the order of the same people is all that is changed. We have:

I.	$A\ B\ C$	(1, 4, 13, 16, 25, 28 above all contain only A, B, C.)
II.	$A\ B\ D$	(2, 7, 14, 19, 37, 40 above all contain only A, B, D.)
III.	$A\ B\ E$	(3, 10, 15, 22, 49, 52 above all contain only A, B, E.)
IV.	$A\ C\ D$	(5, 8, 26, 31, 38, 43 above all contain only A, C, D.)
V.	$A\ C\ E$	(6, 11, 27, 34, 50, 55 above all contain only A, C, E.)
VI.	$A\ D\ E$	(9, 12, 39, 46, 51, 58 above all contain only A, D, E.)
VII.	$B\ C\ D$	(17, 20, 29, 32, 41, 44 above all contain only B, C, D.)

VIII. $B\ C\ E$ (18, 23, 30, 35, 53, 56 above all contain only B, C, E.)
IX. $B\ D\ E$ (21, 24, 42, 47, 54, 59 above all contain only B, D, E.)
X. $C\ D\ E$ (33, 36, 45, 48, 57, 60 above all contain only C, D, E.)

There are ten committees of three that can be chosen from five people, which is 1/6 of our original 60. Why 1/6? Because each group of three can be ordered in $3! = 6$ ways. Then $10 = 60/3!$, and this is the way we shall count unordered arrangements. A few examples follow.

Example 3.3.1. In example 3.2.3 there are five members who wish to elect a governing committee of two. Suppose the restriction as to president, vice-president is to be lifted. How many possible governing committees of two are there?

Solution: If ordered arrangements were counted, there would be $5 \cdot 4 = 20$ ways. In each one of these 20, however, the people can order themselves in 2! ways. So, there are $20/2! = 10$ possible committees.

Example 3.3.2. A foreman has to fire four of ten employees. How many choices has he?

Solution: This is, of course, an unordered arrangement. If we were to count the ordered arrangements, we would have $10 \cdot 9 \cdot 8 \cdot 7$ ways. Each one of those, however, has four men who can order themselves in 4! ways. So, the number of choices the foreman has is

$$\frac{10 \cdot 9 \cdot 8 \cdot 7}{4 \cdot 3 \cdot 2 \cdot 1} = 210$$

To facilitate matters, we shall introduce a notational innovation. Let's return to choosing a committee of three from five people on page 112. We noted there that the ordered arrangements occurred in $5 \cdot 4 \cdot 3 = 60$ ways. It is true (although perhaps not clear why) that $\frac{5!}{2!} = 5 \cdot 4 \cdot 3 = 60$. The number of committees could, therefore, be expressed as $\frac{5!}{2!\,3!} = 10$. In example 3.3.2, the ordered arrangements occurred in $10 \cdot 9 \cdot 8 \cdot 7 \cdot 6 \cdot 5 \cdot 4 \cdot 3$ ways. We can express this number as $10!/2!$. (You see, this "trick" merely employs the fact that in $10!/2!$, the "$2 \cdot 1$" at the end of 10! will cancel with its twin in the denominator.) To obtain the number of unordered arrangements, all we need do is divide that by 8!. Our result is the same:

$$\frac{10!}{2!\,8!} = \frac{10 \cdot 9 \cdot \not{8} \cdot \not{7} \cdot \not{6} \cdot \not{5} \cdot \not{4} \cdot \not{3} \cdot \not{2} \cdot \not{1}}{2 \cdot 1 \cdot \not{8} \cdot \not{7} \cdot \not{6} \cdot \not{5} \cdot \not{4} \cdot \not{3} \cdot \not{2} \cdot \not{1}} = 45$$

The number of unordered arrangements of 7 things chosen from a possible 12 is simply $\frac{12!}{5!\,7!}$. The reason: $\frac{12!}{5!} = 12 \cdot 11 \cdot 10 \cdot 9 \cdot 8 \cdot 7 \cdot 6$ is the number of ordered arrangements of seven things chosen from twelve, and 7! is the number of ways those seven can order themselves. In general, the number of unordered arrangements of j things chosen from n things is $\frac{n!}{(n-j)!\,j!}\cdot$ Don't let the abstractness of n's and j's bother you. When $n = 12$ and $j = 7$, for example, we have $\frac{12!}{5!\,7!}\cdot$ We shall use the notation $\binom{n}{j}$ to mean the number of unordered arrangements of j things chosen from n things. The numbers $\binom{n}{j}$ are called **binomial coefficients,** and $\binom{n}{j}$ is read as "n on j":

$$\binom{n}{j} = \frac{n!}{(n-j)!\,j!}$$

EXAMPLE **3.3.3.** Evaluate **a.** $\binom{6}{2}$ **b.** $\binom{8}{1}$ **c.** $\binom{100}{98}$ **d.** $\binom{5}{5}$

e. The number of committees of six that can be formed from a group of nine people.

Solution:

a. $\binom{6}{2} = \frac{6!}{4!\,2!} = \frac{6\cdot5\cdot\not{4}\cdot\not{3}\cdot\not{2}\cdot\not{1}}{\not{4}\cdot\not{3}\cdot\not{2}\cdot\not{1}\cdot2\cdot1} = 15$

b. $\binom{8}{1} = \frac{8!}{7!\,1!} = \frac{8\cdot\not{7}\cdot\not{6}\cdot\not{5}\cdot\not{4}\cdot\not{3}\cdot\not{2}\cdot\not{1}}{\not{7}\cdot\not{6}\cdot\not{5}\cdot\not{4}\cdot\not{3}\cdot\not{2}\cdot\not{1}\cdot1} = 8$

c. $\binom{100}{98} = \frac{100!}{2!\,98!} = \frac{100\cdot99\cdot\not{98}\cdot\not{97}\cdots\not{3}\cdot\not{2}\cdot\not{1}}{2\cdot1\cdot\not{98}\cdot\not{97}\cdots\not{3}\cdot\not{2}\cdot\not{1}} = \frac{100\cdot99}{2}$

$= 50\cdot99 = 4950$

d. $\binom{5}{5} = \frac{5!}{0!\,5!} = \frac{\not{5}\cdot\not{4}\cdot\not{3}\cdot\not{2}\cdot\not{1}}{1\cdot\not{5}\cdot\not{4}\cdot\not{3}\cdot\not{2}\cdot\not{1}} = 1.$ This makes sense because it answers the question, "How many committees of five can be chosen from five people?" There is only one way!

e. This is $\binom{9}{6} = \frac{9!}{3!\,6!} = \frac{9\cdot8\cdot7\cdot\not{6}\cdot\not{5}\cdot\not{4}\cdot\not{3}\cdot\not{2}\cdot\not{1}}{3\cdot2\cdot1\cdot\not{6}\cdot\not{5}\cdot\not{4}\cdot\not{3}\cdot\not{2}\cdot\not{1}} = 84$

Solutions to parts *a, b, d, e* could have been obtained by the use of Table II in the Appendix. Take a look!

EXAMPLE **3.3.4.** Dom, Gerry, John, Ken, Hank, Sue, Cheryl, Carol, Rose, and Shirley apply for a job. If there are three job vacancies, how many choices has the company?

Solution: Realizing, of course, that the order in which the applicants are chosen is not important, we count the number of three-person "committees" that can be made choosing from ten people:

$$\binom{10}{3} = \frac{10!}{7!\,3!} = \frac{10 \cdot 9 \cdot 8 \cdot 7 \cdot 6 \cdot 5 \cdot 4 \cdot 3 \cdot 2 \cdot 1}{7 \cdot 6 \cdot 5 \cdot 4 \cdot 3 \cdot 2 \cdot 1 \cdot 3 \cdot 2 \cdot 1} = 120$$

(10: the number from which we choose; 3: the number we want)

We present now four more examples on unordered arrangements. Since no two counting problems are alike, it is our purpose merely to inject a bit of insight. Beware! Do not waste your time in search for an underlying technique or "cure-all" to solve all counting problems . . . there is none.

Example 3.3.5. In example 3.3.4, how many choices has the company if Shirley is the daughter of the chairman of the board and must be hired?

Solution: The number of choices has diminished. Since Shirley MUST be hired, we count all two-person committees chosen from nine people:

$$\binom{9}{2} = \frac{9!}{7!\,2!} = \frac{9 \cdot 8 \cdot 7 \cdot 6 \cdot 5 \cdot 4 \cdot 3 \cdot 2 \cdot 1}{7 \cdot 6 \cdot 5 \cdot 4 \cdot 3 \cdot 2 \cdot 1 \cdot 2 \cdot 1} = 36$$

Example 3.3.6. A student is to answer eight of ten questions on an exam. How many choices does he have if he must answer the first two questions?

Solution: This means that of the last eight questions, he must answer six. Thus the number of choices he has is

$$\binom{8}{6} = 28$$

Example 3.3.7. A class contains six boys and two girls.

a. In how many ways can the teacher choose a committee of four?
b. How many of those committees contain at least one girl?

Solution:

a. $\binom{8}{4} = 70$ committees

b. Our old friend, "at least one," makes another appearance. Let's count just the opposite of at least one girl: no girls. There are $\binom{6}{4} =$ 15 committees of all boys (no girls). So there are $70 - 15 = 55$ committees containing at least one girl.

Example 3.3.8. There are twenty members of the ΣΔΣ fraternity. A committee of three is to be chosen to attend the national meeting.

a. How many possible three-man committees are there?
b. How many possible three-man committees can be chosen if two of the twenty extremely hate each other and refuse to go to the meeting together.

Solution:

a. $\binom{20}{3} = 1140$

b. Suppose these two characters are named Ralph and Sam. Let r be the statement, "Ralph goes," s the statement, "Sam goes." From chapter 1 we know all 1140 three-man committees must satisfy exactly one of the following statements:

1. $r \wedge s$ (Ralph and Sam both go.)
2. $r \wedge \sim s$ (Ralph goes but Sam doesn't go.)
3. $\sim r \wedge s$ (Ralph doesn't go and Sam does.)
4. $\sim r \wedge \sim s$ (Neither Ralph nor Sam goes.)

These add up to 1140.

The reader is urged to verify that cases 2 or 3 or 4 satisfy the condition we need; case 1 is the only one that doesn't. There are $\binom{18}{1} =$ 18 committees containing Ralph and Sam (the other vacancy can be filled by any of the 18 other members). So, there are $1140 - 18 = 1122$ committees not containing both Ralph and Sam.

The news article on the facing page came from the Rochester Democrat and Chronicle. It shows that there are more possible "committees" of winners in an election than one might think.

EXERCISES ■ SECTION 3.3.

1. Evaluate the following:

a. $\binom{7}{2}$ **b.** $\binom{7}{5}$ **c.** $\binom{8}{2}$ **d.** $\binom{8}{6}$

e. $\binom{20}{2}$ **f.** $\binom{1000}{0}$ **g.** $\binom{576}{575}$ **h.** $\binom{2}{1}$

i. $\binom{11}{10}$ **j.** $\binom{100,000}{2}$ **k.** $\binom{60}{59}$ **l.** $\binom{4}{1}$

m. $\binom{6}{7}$

2. The town of Eden is about to hold an election for the supervisory board. How many possible six-man boards can be elected from a group of nine nominees?

3. A company has thirty stockholders.

a. How many ways can they elect a chairman of the board, treasurer, and secretary as a governing body?

b. How many ways can they elect a governing body of three?

4. A student is taking a twelve-question essay examination.

a. If he must answer nine of the twelve questions, in how many ways can he take the exam?

b. If he must answer the first six questions and a total of nine, in how many ways can he take the exam?

'QUADRILLIONS' OF WINNERS IN ELECTION

By JIM ROWLEY

There are more than 15.5 quadrillion possible winning combinations in the four major election races in Monroe County. That's 15,561,985,873,674,240.

It's as simple as the number of candidates (n) factorial divided by the number of possible winners (k) factorial times the difference between candidates and winners (n-k) factorial.

That's the formula to calculate possible combinations of elements chosen from a larger field. It was supplied by John H. Hubbard, a mathematics instructor at Harvard University, but it's common knowledge in many 12th grade math classes.

Factorial means multiplying a series of numbers arranged in descending order, such as four, three, two, one. The mathematical symbol for factorial is an exclamation point. So, four factorial (4!) is 4 x 3 x 2 x 1.

In the four major races, there are: Two candidates for sheriff, 14 candidates running for 5 City Council seats, 64 hopefuls seeking 29 seats in the County Legislature and 13 candidates vying for 4 school board seats.

$$C = \frac{n!}{k!\ (n-k)!}$$

C = Total combinations
n = Number of candidates
k = Number of winners
! = Factorial

The possible combinations in each race are multiplied by each other to reach the grand total.

One note of caution—the County Legislature election is not at-large as are the City Council and school board and must be treated as 64 separate races.

Here's an example of how to compute the number of possible combinations of winners in a race:

There are 13 candidates for school board; 4 will be elected. n equals 13 and k equals 4; so it's 13! divided by 4! times 9! You can cancel out 9! from the fraction to get 13 x 12 x 11 x 10 divided by 4!. Using the old math you should get 715, the number of possible winning combinations. If that's too confusing, figure out the sheriff's race.

5. In the ΣΔΣ fraternity problem of example 3.3.8, how many committees of three are there if two of the members, Bruce and Carl, are brothers and refuse to be split for the weekend? (Hint: find the committees that satisfy either $b \wedge c$ or $\sim b \wedge \sim c$.)

6. A euchre deck contains 24 different cards. A euchre hand is made up of five cards. How many possible euchre hands are there?

7. Two cards are withdrawn simultaneously from a regular deck of playing cards.

   a. In how many ways can that be done?
   b. How many of them contain at least one heart?

8. a. A manager must pick nine men to play baseball today. If he has a squad of 21 men, how many choices has he?
   b. How many choices has he if he wants Lefty Ature to pitch today?
9. Ms. Johnson accidentally dropped three vials of pills: one vial contained four headache pills, another vial contained six little liver pills, and the other vial contained two sinus pills. If all the pills look the same and Ms. Johnson took three at random from the floor, in how many ways could she have taken at least one headache pill?
10. There are 75 bicycles produced on an assembly line per day. A sample of five will be selected for testing purposes. How many such samples can be selected?

3.4 MORE COUNTING PROBLEMS

The previous two sections dealt with unordered and ordered arrangement problems. In the present section, we will present some examples of counting problems that are more complex and more interesting.

Example 3.4.1. Find the number of integers of up to four digits each that can be formed from the digits 1, 2, 3, 4 if no digit is repeated within any particular integer.

Solution: The statement of the problem allows integers containing from one to four digits each (i.e., one or two or three or four digits). The number of four-digit integers is $4 \cdot 3 \cdot 2 \cdot 1 = 24$ possibilities. The number of three-digit integers is $4 \cdot 3 \cdot 2 = 24$ possibilities. The number of two-digit integers is $4 \cdot 3 = 12$ possibilities. The number of one-digit integers is $4 = 4$ possibilities. So, in all, there are $24 + 24 + 12 + 4 = 64$ possibilities.

Example 3.4.2. The Attica License Plate Company makes license plates consisting of two letters followed by either two or three or four digits. How many different license plates are possible?

Solution: We break the problem into three parts:

1. plates with two letters followed by two digits,
2. plates with two letters followed by three digits,
3. plates with two letters followed by four digits.

For (1), we have $26 \cdot 26 \cdot 10 \cdot 10 = 67{,}600$ possible plates. For (2), we have $26 \cdot 26 \cdot 10 \cdot 10 \cdot 10 = 676{,}000$ possible plates. For (3), we have $26 \cdot 26 \cdot 10 \cdot 10 \cdot 10 \cdot 10 = 6{,}760{,}000$ possible plates. In all we have $6{,}760{,}000 + 676{,}000 + 67{,}600 = 7{,}503{,}600$ possible plates.

Notice that in order to solve this problem, we had to consider three different cases, because the license plates could have two or three or four digits. The key word in the original statement of the problem is "or." Whenever you see that word, begin to think about what cases are possible. The next two examples further illustrate this same kind of counting technique in more practical situations.

Example 3.4.3. From a group of eight people, how many committees made up of either three or four people can be formed?

Solution: We have two cases. The number of possible three-person committees is $\binom{8}{3} = 56$. The number of possible four-person committees is $\binom{8}{4} = 70$. Hence, in total, there are $56 + 70 = 126$ possible committees.

Example 3.4.4. A master teacher has under his wing a student teacher. There are ten qualities upon which the master teacher grades the student teacher. The master teacher must select the student teacher's best traits for a report. If he must select at least one trait but no more than three, how many possible grade forms can be submitted?

Solution: The answer will be found by adding the number of ways one trait can be selected to the number of ways two traits can be selected to the number of ways three traits can be selected. One trait can be selected in $\binom{10}{1}$ ways. Two traits can be selected in $\binom{10}{2}$ ways, and three traits can be selected in $\binom{10}{3}$ ways. Hence, the number of possible grade forms is

$$\binom{10}{1} + \binom{10}{2} + \binom{10}{3} = 175$$

Some counting problems involve both the ordered arrangement idea (*$r \cdot s$ rule*) as well as the unordered arrangement idea $\left(\textit{binomial coefficients } \binom{n}{j}\right)$. For instance, suppose that from a group of four men and six women we wish to form a committee of five people consisting of two men and three women. A typical committee looks like this:

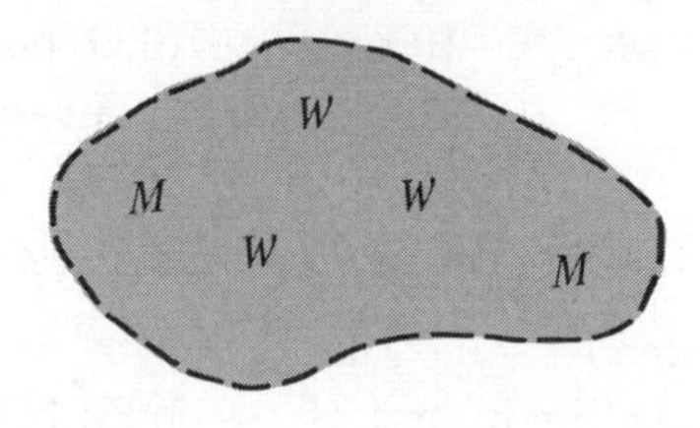

The order in which the people are chosen is immaterial here. All we need to do is to assure ourselves that we have the desired two men and three women. To simplify matters we will choose the men first.

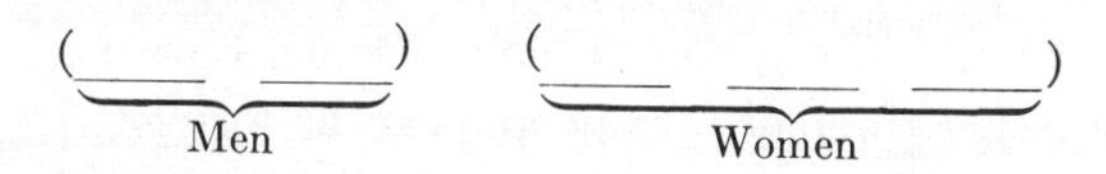

Choosing the men is an unordered arrangement problem where we must pick two men from the four available. So the men can be chosen in $\binom{4}{2} = 6$ ways. Similarly, the women can be chosen in $\binom{6}{3} = 20$ ways. We have here two events: choosing men and choosing women. Choosing men can be done in 6 ways and choosing women can be performed in 20 ways. To form a committee we must do both of these events. That's an $r \cdot s$ rule problem. Hence a committee of two men AND three women can be formed in $\binom{4}{2} \cdot \binom{6}{3} = 6 \cdot 20 = 120$ ways. Some more examples of this type follow.

Example 3.4.5. An employment agency has two offices (A and B). There are a total of eight vacancies, for which office A has 20 applicants and office B has 18 applicants. How many ways are there to fill the eight vacancies if:

a. all are chosen from office A?
b. all are chosen from office B?
c. three are to be chosen from A and five from B?
d. there are no restrictions?

Solution:

a. Here we need to choose eight people from a group of 20. Certainly, the order is immaterial. Thus there are $\binom{20}{8} = 125{,}970$ ways. (Use Table II to figure this out.)

b. For all the people to come from B, we must choose eight from the 18 available. Thus $\binom{18}{8} = 43{,}758$ ways are possible.

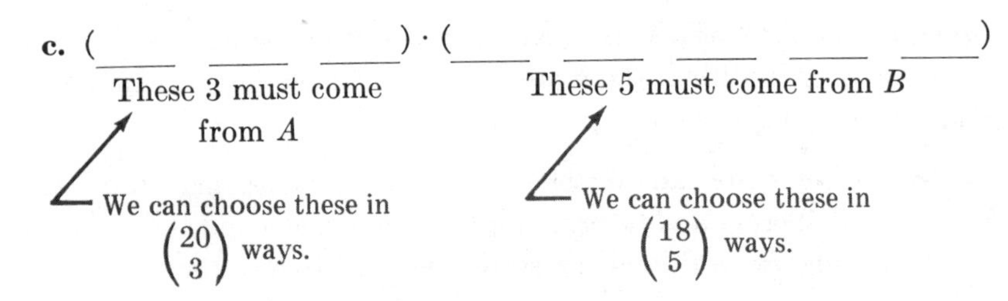

Hence, if three people are to come from A and five from B, there are $\binom{20}{3} \cdot \binom{18}{5} = 1140 \cdot 8568 = 9{,}767{,}520$ possible ways.

d. If there are no restrictions, then we can choose the needed eight from all 38 people available. Thus there are $\binom{38}{8}$ ways.

Example 3.4.6. A football team has thirty members. The coach must choose eleven of the men to play offense and a different eleven men to play defense. How many choices has he? (Assume each player can play either offense or defense.)

Solution: There are thirty men from whom we must choose eleven for offense. This can be done in $\binom{30}{11}$ ways. The eleven defensive men can now be chosen from the remaining 19 men. The total number of team choices is

$$\binom{30}{11} \cdot \binom{19}{11}$$

Example 3.4.7. A baseball team consists of seven outfielders, eight infielders, two catchers, and five pitchers. How many possible starting teams are there if a starting team consists of three outfielders, four infielders, one catcher, and one pitcher?

Solution: The outfielders can be chosen in $\binom{7}{3}$ ways. The infielders can be chosen in $\binom{8}{4}$ ways. The catcher and pitcher can be chosen in $\binom{2}{1}$ and $\binom{5}{1}$ ways, respectively. We need outfielders AND infielders AND a catcher AND a pitcher to make up a team. The number of starting teams is, therefore,

$$\binom{7}{3} \cdot \binom{8}{4} \cdot \binom{2}{1} \cdot \binom{5}{1} = 24{,}500$$

Example 3.4.8. Nine teachers are to leave for a convention in two cars, a Volkswagen and a station wagon. How many ways can they travel if

Werner must drive the Volkswagen and George must drive the station wagon and, in addition, there are to be two people in the Volkswagen and seven in the station wagon?

Solution: There are nine people, but we must choose one from seven (nine minus Werner and George) for the Volkswagen and then six from the remaining six to ride in the station wagon. Hence, there are

$$\binom{7}{1} \cdot \binom{6}{6} = 7 \text{ ways}$$

Example 3.4.9. A dormitory director must assign eight people to three rooms, A, B, C. Room A will have three people, room B will have two people, and room C will have three people. In how many ways can he make the room assignments?

Solution: The room assignments resemble this:

$$\underbrace{(\_\_\_\_ \quad \_\_\_\_ \quad \_\_\_\_)}_{\text{Room } A} \qquad \underbrace{(\_\_\_\_ \quad \_\_\_\_)}_{\text{Room } B} \qquad \underbrace{(\_\_\_\_ \quad \_\_\_\_ \quad \_\_\_\_)}_{\text{Room } C}$$

Since the order of the people within a given room is not important, each "room-filling process" is an unordered arrangement problem. Room A can be filled in $\binom{8}{3}$ ways. After that, since three people have received their room assignment, Room B can be filled in $\binom{5}{2}$ ways. Finally, Room C can be filled in $\binom{3}{3}$ ways. So together, Rooms A and B and C can be filled in $\binom{8}{3} \cdot \binom{5}{2} \cdot \binom{3}{3} = 56 \cdot 10 \cdot 1 = 560$ ways. (Note the use of the $r \cdot s$ rule here in the final step.)

Example 3.4.10. Suppose that in the previous example, there is one pair of students who demand to room together. How many ways can the dormitory director now make the room assignments?

Solution: The particular pair (denote by x's) could live together in any one of the three rooms as illustrated below.

(1) $(\underline{\;x\;}\ \underline{\;x\;}\ \underline{\quad})$ Room A $(\underline{\quad}\ \underline{\quad})$ Room B $(\underline{\quad}\ \underline{\quad}\ \underline{\quad})$ Room C

OR (2) $(\underline{\quad}\ \underline{\quad}\ \underline{\quad})$ Room A $(\underline{\;x\;}\ \underline{\;x\;})$ Room B $(\underline{\quad}\ \underline{\quad}\ \underline{\quad})$ Room C

OR (3) $(\underline{\quad}\ \underline{\quad}\ \underline{\quad})$ Room A $(\underline{\quad}\ \underline{\quad})$ Room B $(\underline{\;x\;}\ \underline{\;x\;}\ \underline{\quad})$ Room C

We will figure the number of ways for (1), then for (2), then for (3), and (due to the presence of the word "or") add these three numbers together. For (1):

A	B	C
(x x ___)	(___ ___)	(___ ___ ___)
can be filled in $\binom{6}{1}$ ways	can be filled in $\binom{5}{2}$ ways	can be filled in $\binom{3}{3}$ ways

With "the pair" in Room A, the assignments can be made in

$$\binom{6}{1} \cdot \binom{5}{2} \cdot \binom{3}{3} = 6 \cdot 10 \cdot 1 = 60 \text{ ways.}$$

For (2):

A	B	C
(___ ___ ___)	(x x)	(___ ___ ___)
can be filled in $\binom{6}{3}$ ways		can be filled in $\binom{3}{3}$ ways

The assignments can be made in $\binom{6}{3} \cdot \binom{2}{2} \cdot \binom{3}{3} = 20 \cdot 1 \cdot 1 = 20$ ways.

For (3):

A	B	C
(___ ___ ___)	(___ ___)	(x x ___)
can be filled in $\binom{6}{3}$ ways	can be filled in $\binom{3}{2}$ ways	can be filled in $\binom{1}{1}$ ways

The assignments can be made in $\binom{6}{3} \cdot \binom{3}{2} \cdot \binom{1}{1} = 20 \cdot 3 \cdot 1 = 60$ ways.

Hence, the number of ways to fill the rooms, being assured that the particular two people are together, is $60 + 20 + 60 = 140$ ways.

Example 3.4.11. Again, let's look at the dormitory room assignment problem of example 3.4.9. Suppose this time, however, that there is one pair of students who refuse to live together. In how many ways can the room assignments be made?

Solution: This question can be answered quickly, because we just figured the number of ways for a pair to live together to be 140 ways. Example 3.4.9 told us that in total (with no restrictions), the assignments could be made in 560 ways. So, with a pair refusing to live together, we can make the assignments in $560 - 140 = 420$ ways. (Problem 19 in the exercises for this section asks the student to figure this counting problem again by actually splitting the two who refuse to live together into separate rooms.)

As a summary of the counting processes discussed, we urge the reader to:

1. look for different cases (your final answer will be the sum of your answers to the cases);
2. look at each separate case and decide whether it is an ordered or unordered arrangement. If it is ordered, use the $r \cdot s$ rule. If it is unordered, use the binomial coefficients. Keep in mind that it may involve both ordered and unordered arrangements.

EXERCISES ■ SECTION 3.4.

1. Find the number of integers of up to five digits each that can be formed from the digits 1, 2, 3, 4, 5, if no digit can be repeated within any particular integer.
2. Find the number of integers of up to five digits each that can be formed from the digits 1, 2, 3, 4, 5, if digits can be repeated within particular integers.
3. In how many different ways can eight boys be seated in a row of eight chairs for a photograph, if two of the boys must sit in adjacent seats?
4. A railway signal has three arms, and each arm can be put into two positions besides its position of rest. If every position of the arms, except that in which they are all at rest, forms a signal, how many different signals can be given?
5. An exam has ten questions. A student must answer at least eight of them. How many ways can he do this?
6. An automobile manufacturer offers the "basic line" car with a choice of any of eleven accessories. Kenny and Becky, newlyweds, need at least two of the accessories for status reasons. However, the size of their welfare check prohibits them from choosing more than five accessories. How many ways can they choose the accessories?
7. At a meeting of 15 people, each shakes hands with all of the others. How many handshakes are there?
8. From a group of five Republicans and four Democrats, find the number of committees of three that can be formed:

a. with no restrictions,
b. with one Republican and two Democrats,
c. with no Republicans and three Democrats.

9. In a league of 10 bocce ball teams, how many games will be played in a season if each team plays two games with each other team?
10. A class consisting of seven boys and three girls is to elect a committee of four. How many committees contain exactly one girl?
11. How many committees of five men and six women can be obtained from a group of seven married couples?
12. From a group of five skiers, six skaters, and ten hockey players, a team of four skiers, three skaters and six hockey players is to be chosen for the U.S. Olympic Team. How many possible teams are there?
13. A box contains 100 light bulbs, ten of which are defective.

a. In how many ways can 20 bulbs be selected?
b. In how many ways can 20 good bulbs be selected?
c. In how many ways can 20 bulbs be selected so exactly two are defective?
d. In how many ways can 20 bulbs be selected so exactly one is defective?
e. In how many ways can 20 bulbs be selected so at least one is defective?

14. Consider the English alphabet, with 26 letters including the five vowels a, e, i, o, u.

a. How many six-letter "words" can be formed containing two different consonants and four different vowels?
b. How many of those in part a contain a z?
c. How many of those in part a begin with a z and end with a q?
d. How many of those in part a begin with a z and contain a q?

15. On a fireman qualification examination, there are five questions on fire-fighting apparatus and five questions on fire-fighting techniques.

a. How many ways are there to choose any eight questions?
b. How many ways are there to choose three on apparatus and four on techniques?
c. How many ways are there to answer eight questions if at least three from each group must be chosen?

16. Ten concerned U.S. Senators are going to investigate foreign aid, drugs on campus, and aid to education. To do this they will form

three committees of four, three, and three each, respectively. How many ways can the ten senators break down into the three committees? (A senator can be on at most one committee.)

17. At the grocery store, Andy must buy milk, eggs, bread, two different kinds of cheese, three different kinds of meat, and a six-pack of beer. How many choices has he if there are three different kinds of milk, two different sizes of eggs, three different brands of bread, seven types of cheese, seven kinds of meat, and eighteen brands of beer?

18. The administration of Attainment Community College recognizes the need for a curriculum committee. The committee is to be made up of one member of the board of trustees, two college administrators, three faculty members and two students. The College population is:

Board of Trustees	5
Administration	20
Faculty	300
Students	5200

a. How many possible curriculum committees can be formed?

b. How many committees can be formed if the Dean of Curriculum (an administrator) must be a member?

19. Consider example 3.4.11. There we had eight people to be assigned to three rooms. Room A will have three people. Room B will have two people. Room C will have three people. There is one pair of students who refuse to live together. Find the number of ways to make the room assignments by splitting the pair. Use x's to denote the pair as we did in example 3.4.10. (Hint: there are six different cases possible for the splitting.)

20. The Mass Transit Bus Company has four busses serving the three suburbs: Arlington, Bilgeville, Cavier City. Assume the busses' traffic pattern within the suburbs is random; at anytime in a given day, how many ways are there for the busses to end up in the suburbs?

21. The Goodyear Publishing Company uses four advertising media to sell books: telephone solicitation (t), personal contact (p), mail brochures (b), and sky-writing (s). From the set $A = \{t,p,b,s\}$, how many choices has Goodyear for promoting this book if:

a. at least one media must be used?

b. at least one but no more than three must be used?

3.5 PASCAL'S TRIANGLE AND THE BINOMIAL THEOREM

In this section we shall see a relationship between the binomial coefficients encountered earlier and another branch of mathematics: algebra.

Among many of the things one does in algebra is the process of raising a binomial like $x + y$ to a power. Direct calculation by multiplication shows:

$$\begin{aligned}
(x+y)^1 &= x + y \\
(x+y)^2 &= x^2 + 2xy + y^2 \\
(x+y)^3 &= x^3 + 3x^2y + 3xy^2 + y^3 \\
(x+y)^4 &= x^4 + 4x^3y + 6x^2y^2 + 4xy^3 + y^4 \\
(x+y)^5 &= x^5 + 5x^4y + 10x^3y^2 + 10x^2y^3 + 5xy^4 + y^5
\end{aligned}$$

We now look at this triangular array of symbols to discover any obvious patterns. In each case, the powers of x decrease starting with the power on the original binomial and descending to zero. The powers of y do just the opposite. The sum of the powers of x and y in any individual term should equal the power to which the binomial is being raised. Knowing that pattern allows us to conjecture what $(x + y)^6$ would be:

$$(x+y)^6 = \_x^6 + \_x^5y + \_x^4y^2 + \_x^3y^3 + \_x^2y^4 + \_xy^5 + \_y^6$$

Here the _ means that there is some coefficient yet to be determined. We'll do that soon.

Now, we look at the triangular array composed of just the coefficients. This pattern of numbers is called **Pascal's triangle.** (Although Pascal (1623–1662) is credited with being the originator of the triangle, the oldest known reference is a work of 1303 by the Chinese algebraist Chu Shï-kié.*)

```
            1   1
          1   2   1
        1   3   3   1
      1   4   6   4   1
    1   5  10  10   5   1
```

Notice that each row of this array begins and ends with a "1." Any other entry may be determined by summing the two numbers that appear in the previous row immediately to the left and right of the number. This is illustrated below by the little triangular outlines, where we see that $3 = 1 + 2$, $5 = 1 + 4$, $10 = 6 + 4$.

* Howard Eves, *An Introduction to the History of Mathematics*, 3rd ed., (New York: Holt, Rinehart, and Winston, 1969), p. 262.

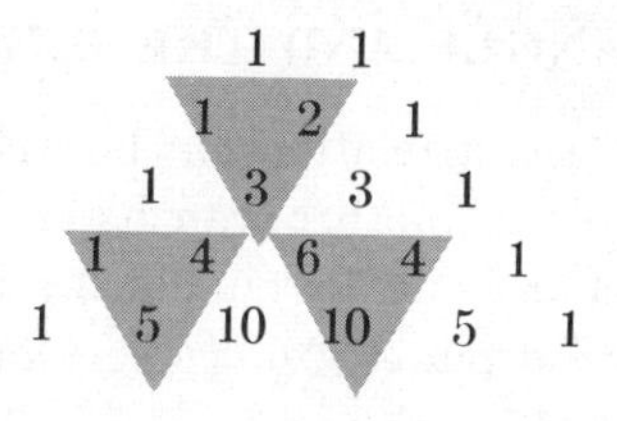

Be aware of the fact that we are proving nothing here; we're only saying that it works!

Now we can write the next (sixth) row of Pascal's triangle

1 1
1 2 1
1 3 3 1
1 4 6 4 1
1 5 10 10 5 1
1 6 15 20 15 6 1

and use it to fill in the coefficients of $(x + y)^6$:

$$(x + y)^6 = \underline{}\,x^6 + \underline{}\,x^5y + \underline{}\,x^4y^2 + \underline{}\,x^3y^3 + \underline{}\,x^2y^4 + \underline{}\,xy^5 + \underline{}\,y^6$$

$$= x^6 + 6x^5y + 15x^4y^2 + 20x^3y^3 + 15x^2y^4 + 6xy^5 + y^6$$

Example 3.5.1. Compute $(x + y)^7$.

Solution: First we write out $(x + y)^7$ without the coefficients:

$$(x + y)^7 = \underline{}\,x^7 + \underline{}\,x^6y + \underline{}\,x^5y^2 + \underline{}\,x^4y^3 + \underline{}\,x^3y^4$$

$$+ \underline{}\,x^2y^5 + \underline{}\,xy^6 + \underline{}\,y^7$$

Now we complete the Pascal's triangle through the seventh row.

1 1
1 2 1
1 3 3 1
1 4 6 4 1
1 5 10 10 5 1
1 6 15 20 15 6 1
1 7 21 35 35 21 7 1

So

$$(x + y)^7 = x^7 + 7x^6y + 21x^5y^2 + 35x^4y^3 + 35x^3y^4 + 21x^2y^5 + 7xy^6 + y^7$$

We can use the ideas presented above to raise other binomials to powers also.

Example 3.5.2. Compute $(2x + 3y)^4$.

Solution: If we let $a = 2x$, $b = 3y$, we merely have $(2x + 3y)^4 = (a + b)^4$. By using the fourth row of Pascal's triangle,

$$(a + b)^4 = a^4 + 4a^3b + 6a^2b^2 + 4ab^3 + b^4$$

Finally, we just substitute $2x$ for a and $3y$ for b:

$$(2x + 3y)^4 = (2x)^4 + 4(2x)^3(3y) + 6(2x)^2(3y)^2 + 4(2x)(3y)^3 + (3y)^4$$

Simplifying:

$$\begin{aligned}(2x + 3y)^4 &= 16x^4 + 4(8x^3)(3y) + 6(4x^2)(9y^2) + 4(2x)(27y^3) + 81y^4 \\ &= 16x^4 + 96x^3y + 216x^2y^2 + 216xy^3 + 81y^4\end{aligned}$$

The shortcut techniques presented above for raising binomials to powers are usable except when the power gets excessively large. For example, computing $(x + y)^{20}$ would require that we work out Pascal's triangle through the twentieth row. That's time-consuming! It just so happens that each of the entries in Pascal's triangle can be written as a binomial coefficient $\binom{n}{j}$. Here n is the number of the row in the triangle, and j takes on the values $0, 1, 2, \ldots, n$. For instance, the first row of Pascal's triangle can be written

$$\binom{1}{0} \quad \binom{1}{1}$$

The second row can be written

$$\binom{2}{0} \quad \binom{2}{1} \quad \binom{2}{2}$$

So the whole triangular array begins to look like this:

$$\begin{array}{ccccccccc} & & & \binom{1}{0} & & \binom{1}{1} & & & \\ & & \binom{2}{0} & & \binom{2}{1} & & \binom{2}{2} & & \\ & \binom{3}{0} & & \binom{3}{1} & & \binom{3}{2} & & \binom{3}{3} & \\ \binom{4}{0} & & \binom{4}{1} & & \binom{4}{2} & & \binom{4}{3} & & \binom{4}{4} \end{array}$$

Again, we have proved nothing—we merely maintain it works! Knowing that Pascal's triangle can be written using the binomial coefficients allows us to write any of its rows without writing all the previous rows. For instance, the fifth row would be

$$\binom{5}{0} \quad \binom{5}{1} \quad \binom{5}{2} \quad \binom{5}{3} \quad \binom{5}{4} \quad \binom{5}{5}$$

Consequently,

$$(x + y)^5 = \binom{5}{0} x^5 + \binom{5}{1} x^4y + \binom{5}{2} x^3y^2 + \binom{5}{3} x^2y^3 + \binom{5}{4} xy^4 + \binom{5}{5} y^5$$

Generalizing these last few developments allows us to display

$$(x + y)^n = \binom{n}{0} x^n + \binom{n}{1} x^{n-1}y + \binom{n}{2} x^{n-2}y^2 + \binom{n}{3} x^{n-3}y^3 + \cdots + \binom{n}{n-1} xy^{n-1} + \binom{n}{n} y^n$$

This is referred to as the **binomial theorem.** It works as long as n is a positive integer. The proof appears in Appendix I. The binomial theorem is a great time-saver. The only thing that could possibly take much time is the actual calculation of the numbers $\binom{n}{j}$. However, as in section 3.3, we can use Table II in the back of the book to aid us here.

Pascal's triangle, besides being used as above, can also be used to count the number of subsets that a given set has. We already investigated this question in section 2.3. Now we look at it in a different light.

Consider a set with three elements, $\{a,b,c\}$. Subsets of this set will have either zero, one, two, or three elements each.

The number of subsets of $\{a,b,c\}$ with zero elements is $\binom{3}{0}$.

The number of subsets of $\{a,b,c\}$ with one element is $\binom{3}{1}$.

The number of subsets of $\{a,b,c\}$ with two elements is $\binom{3}{2}$.

The number of subsets of $\{a,b,c\}$ with three elements is $\binom{3}{3}$.

So, the total number of subsets of a set with three elements is $\binom{3}{0} + \binom{3}{1} + \binom{3}{2} + \binom{3}{3} = 1 + 3 + 3 + 1 = 8$. Of course, exercise 3 on page 61 tells us that a set with three elements has $2^3 = 8$ subsets. A similar discussion for a set with four elements tells us that it has a total of $\binom{4}{0} + \binom{4}{1} + \binom{4}{2} + \binom{4}{3} + \binom{4}{4} = 2^4$ or 16 subsets. Hence Pascal's triangle may also be used to determine the number of subsets of a given set, merely by summing each of the entries in the appropriate row. (A set with n elements has $\binom{n}{0} + \binom{n}{1} + \binom{n}{2} + \cdots + \binom{n}{n} = 2^n$ subsets.)

Looking back over this section, we have seen a tie-in between counting techniques $\left(\text{specifically, } \binom{n}{j}\right)$ and algebra. Also, some of the magic of Pascal's triangle has been presented. (There is more; we just haven't gone into it here.) Many of the ideas presented are summarized in the statement of the binomial theorem, where the numbers $\binom{n}{j}$ turn out to be the coefficients when we raise a binomial to a power. Now it finally makes sense why the numbers $\binom{n}{j}$ have been called binomial coefficients.

EXERCISES ■ SECTION 3.5.

1. Expand each of the binomials below to the indicated power.

   a. $(x + y)^8$ **b.** $(x - 1)^3$ **c.** $(2x - 5y)^3$
   d. $(a - 3b)^4$ **e.** $(-x + y)^3$ **f.** $(-2x - y)^5$

2. Without writing the first thirteen rows, write the members of the fourteenth row of Pascal's triangle.
3. Using Pascal's triangle, figure out how many subsets a set with seven elements has. Include how many contain zero elements, one element, etc., up to seven elements.
4. How many subsets containing four elements each does a set with eight elements have?
5. How many subsets containing nine elements each does a set with 13 elements have?
6. A librarian has six books she has decided she'd like to read.

   a. In how many ways can she select at least one book?
   b. In how many ways can she select at least two books?
   (Hint: selecting books from the set of six available is just like making subsets from a set with six elements.)

7. To improve the efficiency of a certain gasoline, a project engineer has four different additives that will do the job.

   a. If he can use one, two, three, or all four of the additives, how many different mixes of gasoline can he prepare?
   b. Because of financial limitations, it is feasible to use only one or two of the additives. How many mixes of gas can he now prepare?

8. Find the coefficient of the x^7y^3 term in the expansion of $(x + y)^{10}$.
9. Find the coefficient of the x^8y^2 term in the expansion of $(x + y)^{10}$.
10. Find the coefficient of the a^7b^3 term in the expansion of $(a + 2b)^{10}$.
11. Find the sixtieth term of $(x + y)^{100}$.
12. The triangular adding process on page 128 can be expressed by the formula

$$\binom{n}{j} + \binom{n}{j+1} = \binom{n+1}{j+1}$$

Show that the formula works by using the definition of binomial coefficient.

3.6 A SPECIAL APPLICATION TO WAGERING GAMES

The theory of counting is not new to mathematics. It was in about 1140 that the Hebrew writer Rabbi Ben Ezra considered unordered arrangements in the study of astronomy.* Perhaps the first book devoted solely to probability was Liber De Ludo Aleae (Book on Games of Chance) by Girolamo Cardano. The classical story, however, of the birth of probability theory lies in a gambler's dilemma. The French gambler de Méré was losing at a game of dice. He stated that in 24 rolls of a pair of dice, he would obtain a double six at least once. It was our friend Pascal and his colleague Pierre de Fermat who began a mathematical approach to this problem. Today, their efforts are part of the mathematics of probability.

We will consider the problem of de Méré in section 4.9, but in the present section we concern ourselves with a familiarity to dice and cards and the counting situations related to them. We have already discussed dice; they are six-sided cubes with each side depicting a different number of dots, ranging from one to six. We depict the faces below.

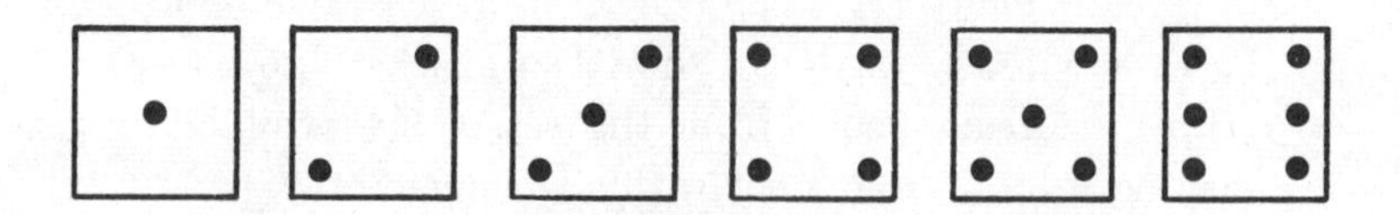

A deck of regular playing cards consists of four suits (spades, hearts, diamonds, clubs), with thirteen denominations in each suit (ace, two, three, . . . , ten, jack, queen, king) for a total of 52 cards. We depict them on the next page.

* D. E. Smith, *History of Mathematics*, vol. 2 (New York: Dover Publications, 1958).

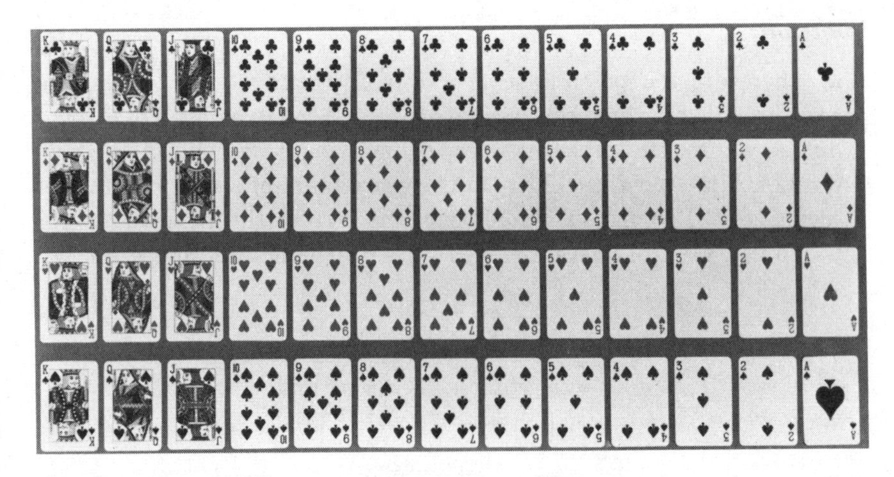

By a **poker** hand is meant any five-card "committee" chosen from the 52 cards. There are, of course, $\binom{52}{5} = 2{,}598{,}960$ possible poker hands. Each of these 2,598,960 hands can be classified into exactly one of the following categories:

Straight flush: five cards in a sequence in a single suit

Four of a kind: four cards of the same denomination plus any fifth card

Full house: three of one denomination with two of another denomination

Flush: five cards in a single suit but not sequentially (i.e. not a straight flush)

Straight: five cards in a row but not all of the same suit

Three of a kind: Three cards of the same denomination, without a matching fourth card or another pair

A pair: two cards of the same denomination, without any other matching card, different pair, or three of a kind.

Two pairs: two cards of the same denomination, two cards of another denomination, and an unmatched fifth card.

A bust: A poker hand that is none of the above.

We use the next example to partition the 2,598,960 poker hands into the nine categories just discussed.

Example 3.6.1. **a.** How many possible straight flushes are there?

b. How many possible four-of-a-kind hands are there?

c. How many possible full houses are there?

d. How many possible flushes are there?

e. How many possible straights are there?

f. How many possible three-of-a-kind hands are there?

g. How many possible two-pair hands are there?

h. How many possible one-pair hands are there?

i. How many possible bust hands are there?

Solution:

a. There are ten possible sequences of five cards (ace through five, two through six, three through seven, four through eight, five through nine, six through ten, seven through jack, eight through queen, nine through king, ten through ace). There are four suits in which the ten sequences can occur. Therefore, there are 40 straight flushes.

b. To get, say, four aces, we need a hand looking like this:

$$\overbrace{(A\ A\ A\ A)}^{\text{aces}} \qquad \overbrace{(N)}^{\text{non-ace}}$$

can get these in $\binom{4}{4}$ ways — can get this in $\binom{48}{1}$ ways

Four aces can thus be dealt in $\binom{4}{4} \cdot \binom{48}{1} = 1 \cdot 48 = 48$ ways. Hence, to get any four of a kind, there are $13 \cdot 48 = 624$ ways.

c. Let's look at a particular full house: $A\ A\ A \quad K\ K$, where A and K denote "ace" and "king," respectively.

$(A\ A\ A)$ can be chosen in $\binom{4}{3}$ ways; $(K\ K)$ can be chosen in $\binom{4}{2}$ ways. This particular full house can be dealt in $\binom{4}{3} \cdot \binom{4}{2} = 4 \cdot 6 = 24$ ways.

However, since the three of a kind can occur in any one of 13 denominations and the pair can occur in any of 12 denominations, the total number of ways to deal a full house is $13 \cdot 12 \cdot 24 = 3{,}744$ ways.

d. A hand of five spades can occur in any of $\binom{13}{5} = 1287$ ways. But there are four suits, so we temporarily have $4 \cdot 1287 = 5148$ ways. We must remember, though, that in those 5148 five-card hands involving one suit, there are 40 straight flushes. We don't want to include those, so we have $5148 - 40 = 5108$ possible flushes.

e. To count the number of straights, consider a five-card sequence starting with an ace:

A	2	3	4	5
4 ways	4 ways	4 ways	4 ways	4 ways

This can be dealt in $4^5 = 1024$ ways. Since the first card can be any one of ten choices (ace, 2,3, . . . ,10), we can deal a five-card sequence in $10 \cdot 1024 = 10{,}240$ ways. Recalling again that there are 40 hands that are both straights and flushes, we deduce there are $10{,}240 - 40 = 10{,}200$ possible straights.

f. To calculate the total number of three-of-a-kind hands, we first look at the number of hands containing three queens.

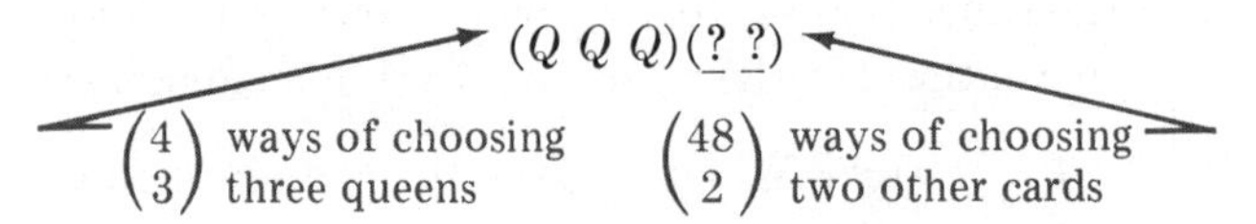

We can choose the denomination of the three-of-a-kind in any of 13 ways. So the number of hands is

$$13 \cdot \binom{4}{3} \cdot \binom{48}{2} = 58{,}656$$

But, included in those arrangements of cards are 3744 full houses. The total number of three-of-a-kind poker hands is thus 58,656 − 3744 = 54,912.

g. To count the number of two-pair hands, we first look at the specific two-pair hand containing a pair of jacks and a pair of queens.

$$(J\ J)(Q\ Q)(\underline{?})$$

can be chosen in $\binom{4}{2}$ ways — can be chosen in $\binom{4}{2}$ ways — can be chosen in $\binom{44}{1}$ ways

This can be done in $\binom{4}{2}\binom{4}{2}\binom{44}{1} = 1584$ ways, and that's just the number of hands containing a pair of jacks and a pair of queens. But we can choose the two denominations in $\binom{13}{2}$ ways. The total number of two-pair hands is, therefore:

$$\binom{13}{2} \cdot 1584 = 123{,}552$$

h. We can count the number of one-pair hands by first counting up all the hands having exactly two kings:

$$(K\ K)\quad (\underline{?}\ \underline{?}\ \underline{?})$$

can be chosen in $\binom{4}{2}$ ways

For the $(\underline{?}\ \underline{?}\ \underline{?})$ part, it would be tempting to say $\binom{48}{3}$, since there are 48 cards (besides the kings) and we must choose three of them. However, that number, $\binom{48}{3}$, would include other pairs and triples, so we must be careful. Let's do this. We know that for

$$(K\ K)\quad (\underline{?}\ \underline{?}\ \underline{?})$$

the kings can be chosen in $\binom{4}{2}$ ways. To be sure of getting no duplicates among the other three cards, we merely choose one of each number, i.e., we choose three of the remaining twelve denominations. That can be done in $\binom{12}{3}$ ways. When we take the suits into consideration, we must multiply by 4^3, since each of the three selections in the $(\underline{?}\ \underline{?}\ \underline{?})$ part can come from any of the four suits. So a hand like

$$(K\ K) \quad (\underline{?}\ \underline{?}\ \underline{?})$$

can be obtained in $\binom{4}{2}\binom{12}{3} 4^3$ ways.

Finally, if we now allow any pair to be present, we have

$$13\binom{4}{2}\binom{12}{3} 4^3 = 1{,}098{,}240$$

ways of getting a poker hand with exactly one pair.

i. Since there are 2,598,960 total possible poker hands, the number of bust hands is

$$2{,}598{,}960 - 40 - 624 - 3744 - 5108$$
$$- 10{,}200 - 54{,}912 - 123{,}552 - 1{,}098{,}240 = 1{,}302{,}540$$

We now summarize the results found in this example.

Type of hand	Number of possibilities
Straight flush	40
Four of a kind	624
Full house	3,744
Flush	5,108
Straight	10,200
Three of a kind	54,912
Two Pairs	123,552
One Pair	1,098,240
Bust	1,302,540
Total	2,598,960

Now it is possible to understand the basic aim in poker—get as rare a hand as possible. The more rare the hand is, the more powerful it is.

Example 3.6.2. There is a game called "draw poker." A player gets five cards and may choose to discard up to three of them in exchange for new cards to again give a total of five, for purposes of improving his hand. (It is all right to assume only one player.) If a player has in his hand the ace of hearts, the ace of clubs, the ace of diamonds, the queen of hearts, and the king of spades, in how many different ways can he improve his hand by discarding the queen and king?

Solution: There are two possible categories into which the player can improve: a full house or four aces. We count each separately. (Keep in mind there are 47 cards remaining in the deck.)

For full houses: $A\ A\ A\ \underline{?}\ \underline{?}$ ← must be some pair

We can make up the pair in the full house by getting 2 kings or 2 queens or 2 twos or 2 threes or 2 fours . . . up to 2 jacks. The two kings can be chosen in $\binom{3}{2}$ ways, as can the 2 queens. (There are only 3 kings and 3 queens left in the deck.) The twos, threes, . . . , jacks can be chosen in $\binom{4}{2}$ ways each. So the number of ways the player can improve his hand by getting a full house is:

$$\binom{3}{2}+\binom{3}{2}+\binom{4}{2}+\binom{4}{2}+\binom{4}{2}+\binom{4}{2}+\binom{4}{2}+\binom{4}{2}+\binom{4}{2} + \binom{4}{2}+\binom{4}{2}+\binom{4}{2} = 66$$

For four aces: To obtain four aces we must get the other ace and any one of the other 46 cards. Thus, there are 46 resulting card hands of four aces here. Therefore the total number of ways he can improve his hand is

$$66 + 46 = 112$$

Let's invent a game. We'll call it WINO. Two dice are tossed. If the dice match, a coin is flipped and when the coin is a head, the player wins. If either a tail is obtained or the dice don't match the player loses and must drink an ounce of wine. In how many ways can the game turn out? Of those, how many are winners and how many are losers?

There are two cases for consideration:

1. when the dice match, and
2. when the dice don't match.

For (1): $\underline{6} \cdot \underline{1} \cdot \underline{2} = 12$ ways when the dice match

choices for first die; choices for matching die; choices for coin

For (2): $\underline{6} \cdot \underline{5} = 30$ ways when the dice do not match

choices for first die; choices for non-matching die

} 42 possible outcomes

In (1) there are 12 ways the game can turn out, of which six are winners and six are losers. In (2) there are 30 ways the game can turn out, of which all are losers. Hence, for the 42 ways the game can turn out, six are winners and 36 are losers.

EXERCISES ■ SECTION 3.6.

1. What fraction of poker hands are:

   a. four-of-a-kind hands? **b.** full houses?
   c. straights? **d.** busts?

2. A euchre deck consists of six denominations (nine, ten, jack, queen, king, ace) in each of the four suits for a total of 24 cards. (See exercise 6, section 3.3.)

   a. How many euchre hands contain exactly one jack?
   b. How many euchre hands contain exactly two jacks of the same color (either heart and diamond or spade and club)?

3. There is a poker game called "seven-card stud," which is just a hand of seven cards from a regular deck of 52 cards.

   a. How many seven-card stud hands are there?
   b. How many seven-card hands contain five or more cards that are of the same suit?
   c. How many seven-card hands contain four cards of the same denomination and three cards that are all different.

4. Andy plays two consecutive games of WINO.

   a. How many ways are there for this to happen?
   b. How many of those ways result in Andy drinking two ounces of wine?

5. In the game of draw poker described in example 3.6.2, a player receives the same hand of ace of hearts, ace of clubs, ace of diamonds, the queen of hearts, and the king of spades. In how many ways can he improve his hand if he chooses to keep all but the queen of hearts?

The reader should compare the answer in exercise 5 with the answer to example 3.6.2 in order to determine which is better strategy for the poker player to follow.

6. In the game of draw poker, a player receives a pair of queens, a pair of aces, and a seven. He discards the seven. How many ways has he of improving his hand?
7. In draw poker, if you have been dealt three spades (ace, nine, king), the seven of hearts, and the six of diamonds, in how many ways can you get a flush in spades?

chapter 3 in review

CHAPTER SUMMARY

An ordered arrangement, or permutation, is an array where the order in which an element appears in that array is significant. We used tree diagrams, the $r \cdot s$ rule, and factorial notation to help count ordered arrangements. The counting principle, $n(P) = n(\mathcal{U}) - n(\tilde{P})$, was also introduced here. If an array does not change when any two of its elements are interchanged, we have an unordered arrangement, or combination. Binomial coefficients were introduced here as $\binom{n}{j}$, the number of combinations of n things choosing j of them at a time. Many counting situations involve both ordered and unordered arrangements, as was the case in section 3.4. There, it was often necessary to examine cases; to obtain the number of entities per case we multiplied and to get the total number of entities we added the results of the cases. Usually, both the $r \cdot s$ rule and binomial coefficients were employed. In section 3.5, binomial coefficients were looked at in a different light, as they pertain to the binomial theorem. Finally, the counting techniques were applied to games of chance in the last section.

VOCABULARY

ordered arrangements (permutations)
unordered arrangements (combinations)
$r \cdot s$ rule
binomial coefficients
Pascal's triangle
binomial theorem
poker

SYMBOLS

$n!$ $\binom{n}{j}$ $(x + y)^n$

chapter 4
Probability

4.1 INTRODUCTION

Almost every human being in the history of man has had cause to concern himself with **probability** sometime in his life. Whenever things happen, probability exists. Actually, the theory of probability concerns itself with the likelihood or the chance of an event occurring—it seeps into our lives daily. Look:

Anyone who has ever listened to the seven o'clock news has heard the expression "PROBABILITY of precipitation."

We all know the important role probability plays in card games and all games of chance.

The clever auto maker estimates the probability that a car will seriously malfunction in the first 12,000 miles (or 12 months, whichever comes first), sees how unlikely it is, and advertises to the poor consumer a good looking guarantee.

Expected life spans and probabilities of living beyond a certain age have all been calculated by insurance companies to assure them of making a profit by selling life insurance.

A man would never have been sent to the moon unless there existed a fairly good chance (probability) he would return.

A decision in business is a decision that weighs the probabilities of different events occurring and chooses the one that is most desirable.

Before we make that trip to the drugstore, we "calculate" the likelihood of getting tied up in traffic.

Physicists use the theory of probability every day in the study of quantum mechanics.
A criminal often tries to design and execute a crime so that the chance of being caught is small.
The politician attempts to word his speeches in order to enhance his chance of securing a person's vote.

So you see, probability is something that is with us in a very real way every day. However, it will be easier to understand the subject if we first consider the less practical situations, like those arising in games of chance.

4.2 SIMPLE EVENTS, COMPOUND EVENTS, AND MUTUALLY EXCLUSIVE EVENTS

As we begin the study of probability, an item of critical importance is the listing of all possible outcomes of a given experiment. Tree diagrams in section 2.6 treated this very problem. The reader should reread that section—right now, before going any further.

When any list of possible outcomes is constructed for an experiment, we obtain some number of situations or **events.** Often times there are different ways of analyzing the same experiment. For instance, suppose we toss a die. We can exhibit two lists of outcomes possible:

1. 1, 2, 3, 4, 5, 6 **2.** odd, even

The first list above is composed of six **simple events,** while the second is composed of two **compound events.** Simple events cannot be "broken down" or decomposed, whereas compound events can be. The distinction between simple and compound events is similar to the distinction between simple and compound statements. A compound event is the disjunction of certain simple events. For instance, the event of an even number occurring is the disjunction of the simple events 2, 4, 6.

For another illustration, look at the experiment of tossing two coins. There are two different lists of possible outcomes:

1. HH, HT, TH, TT **2.** two heads, one head, zero heads

The first list contains four simple events. The second list has one compound event, namely, one head, because "one head" = HT or TH.

Example 4.2.1. Consider the experiment of tossing a coin and rolling a die; this was first looked at in exercise 2 of section 2.6. There are two different lists of possible outcomes:

1. head and even number, head and odd number, tail and even number, tail and odd number
2. $H1$, $H2$, $H3$, $H4$, $H5$, $H6$, $T1$, $T2$, $T3$, $T4$, $T5$, $T6$

a. Which, if either, of the lists contains simple events?
b. Which, if either, of the lists contains compound events?

Solution:

a. The list in (2) has simple events.
b. The list in (1) has all compound events, because
head and even number = $H2$ or $H4$ or $H6$
head and odd number = $H1$ or $H3$ or $H5$
tail and even number = $T2$ or $T4$ or $T6$
tail and odd number = $T1$ or $T3$ or $T5$

Looking back over the example just discussed, as well as the two illustrated before that, we can discover a very important fact. That is, when a tree diagram is used to elaborate all possibilities for an experiment, we automatically get a list just of simple events. In section 4.3 we will prefer to deal with a list of simple events for an experiment.

A special comparison that can be made among events is to see whether or not they are **mutually exclusive events.** That means that the occurrence of any one of the events on a given trial of the experiment prevents the occurrence of all the others. For an illustration, look at the experiment of tossing a die. (Possible outcomes are 1,2,3,4,5,6.) Let there be three events:

event A = outcome is even
event B = outcome is odd
event C = outcome is 1, 2, 4

Events A and B are mutually exclusive, because the occurrence of an even number on a given trial of the experiment prevents the occurrence of the outcome being odd. How about events A and C—are they mutually exclusive? We must ask: "Does the occurrence of an even outcome prevent the occurrence of 1, 2, or 4 on a given trial of tossing a die?" Here, of course, we must correctly answer no. It is proper to call events A and C **non-mutually exclusive.** The reader should also justify that events B and C are non-mutually exclusive. Another way of looking at the notion of mutually exclusive events is to observe that when two events are mutually exclusive, the sets that correspond to the events will be disjoint. When two events are non-mutually exclusive, the sets that correspond to the events will not be disjoint. We now show this.

In the die problem above, we could write $A = \{2,4,6\}$, $B = \{1,3,5\}$, $C = \{1,2,4\}$. Notice that $A \cap B = \varnothing$ ($\therefore A, B$ are mutually exclusive), while $A \cap C \neq \varnothing$ ($\therefore A, C$ are non-mutually exclusive), and $B \cap C \neq \varnothing$ ($\therefore B, C$ are non-mutually exclusive).

Example 4.2.2. A die is tossed. Let

A = a prime number occurs
B = an even number occurs
C = a number greater than five occurs
D = a 1 or 4 occurs

a. Are A and B mutually exclusive?
b. Are A and C mutually exclusive?
c. Are A and D mutually exclusive?
d. Are B and C mutually exclusive?
e. Are B and D mutually exclusive?
f. Are C and D mutually exclusive?

Solution: $A = \{2,3,5\}$, $B = \{2,4,6\}$, $C = \{6\}$, $D = \{1,4\}$.

a. $A \cap B \neq \varnothing$ $\therefore A$ and B are non-mutually exclusive
b. $A \cap C = \varnothing$ $\therefore A$ and C are mutually exclusive
c. $A \cap D = \varnothing$ $\therefore A$ and D are mutually exclusive
d. $B \cap C \neq \varnothing$ $\therefore B$ and C are non-mutually exclusive
e. $B \cap D \neq \varnothing$ $\therefore B$ and D are non-mutually exclusive
f. $C \cap D = \varnothing$ $\therefore C$ and D are mutually exclusive

Example 4.2.3. Consider the problem where a die is tossed. Then $\mathcal{U} = \{1,2,3,4,5,6\}$. Are the six elements (or events) of $\mathcal{U}$ mutually exclusive?

Solution: Yes, because any pair of them, when intersected, gives $\varnothing$.

Example 4.2.4. Consider the problem of tossing two coins in example 2.1.3. There $\mathcal{U} = \{HH, HT, TH, TT\}$. Are the four elements (or events) of $\mathcal{U}$ mutually exclusive?

Solution: Yes, because any pair of them, when intersected, gives $\varnothing$.

Example 4.2.5. Consider John and Norbert's three-game handball tournament first looked at on page 54; $\mathcal{U} = \{JJ, JNJ, NJJ, JNN, NJN, NN\}$. Are these six elements (or events) of $\mathcal{U}$ mutually exclusive?

Solution: Again we must answer yes. Any pair of those events, when intersected, gives $\varnothing$.

The last three examples serve to point out something very important. Namely, when a tree diagram is used to elaborate all possibilities for an experiment, we automatically get a list of mutually exclusive events.

The next section will utilize the concept of a **sample space,** which is simply a list of logical possibilities containing events that are both simple and mutually exclusive. As we have seen in this section, a tree diagram may be used to construct the sample space. Any particular event in a sample space is referred to as a **sample point.** So, for instance, when two coins are flipped, a tree diagram looks like this:

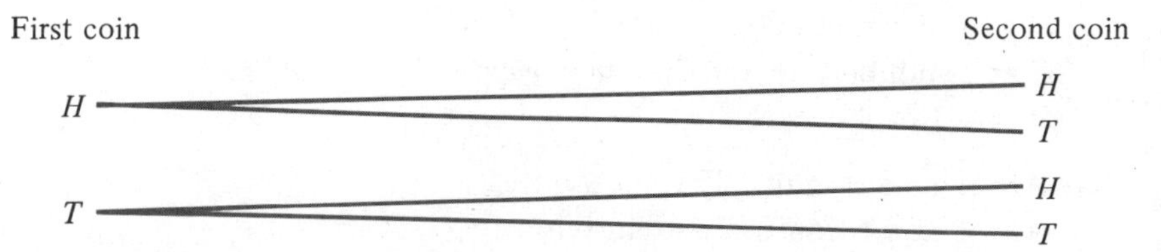

Reading all the branches gives HH, HT, TH, TT, which form the sample space. Also, any one of these four events is properly termed a sample point.

When a tree diagram solution would be lengthy, we can still determine the number of sample points in the sample space by using the counting techniques developed in chapter three.

Example 4.2.6. Three dice are tossed. What is the number of sample points in the sample space?

Solution: The first die can fall 6 ways. The second die can fall 6 ways. The third die can fall 6 ways. Hence, the three dice can fall in $6^3 = 216$ ways. Thus, there are 216 sample points in the sample space.

Example 4.2.7. An experiment consists of dealing three cards from a deck of 52. What is the number of sample points in the sample space?

Solution: A run of the experiment gives three cards. Since this is an unordered arrangement, we can do this in $\binom{52}{3} = 22{,}100$ ways. Hence, there are 22,100 sample points in the sample space.

EXERCISES ■ SECTION 4.2.

1. A die is tossed. Let:

   event A = occurrence of a three
   event B = occurrence of an even number
   event C = occurrence of a number greater than 3
   event D = occurrence of a number less than 2

   a. Is event A simple?
   b. Is event B simple?
   c. Is event C simple?
   d. Is event D simple?
   e. Are events A and B mutually exclusive?
   f. Are events A and C mutually exclusive?

g. Are events A and D mutually exclusive?
h. Are events B and C mutually exclusive?
i. Are events B and D mutually exclusive?
j. Are events C and D mutually exclusive?

2. Three coins are flipped. Construct the sample space.
3. Three coins are flipped. Let:

event A = all three coins are heads
event B = at least one head appears
event C = exactly one tail appears

a. Is event A simple?
b. Is event B simple?
c. Is event C simple?
d. Are events A and B mutually exclusive?
e. Are events A and C mutually exclusive?
f. Are events B and C mutually exclusive?

4. Two dice are tossed. Construct the sample space.
5. Two dice are tossed. Let:

event A = first die is a three
event B = doubles (i.e., both dice the same)
event C = sum is eleven

a. Is event A simple?
b. Is event B simple?
c. Is event C simple?
d. Are events A and B mutually exclusive?
e. Are events A and C mutually exclusive?
f. Are events B and C mutually exclusive?

6. A coin is tossed. If a head results, then the coin is tossed a second time. If a tail is the result on the first toss, then a die is rolled.

a. Construct the sample space for this experiment.
b. How many sample points are there in the sample space?

7. An experiment consists of making a four-digit number by choosing digits from $\{0,1,2,3,4,5,6\}$. If no repetition of digits is allowed and zero may not be the first digit, how many sample points are there in the sample space?
8. An experiment consists of dealing a five-card poker hand. How many sample points are in the sample space?
9. An experiment consists of dealing three jacks and two sevens from a 52-card deck. How many points are there in the sample space?

Construct sample spaces for the following experiments.

10. **a.** Answering a true-false question.
b. Answering a three-question true-false quiz.

11. Determining the day of the week in which a given person dies.

12. The month of someone's birthday.

4.3 THE NATURE OF PROBABILITY

Probability can be called the science of chance. When we are not sure of how likely it is that an event may occur, we can begin to find out by actually performing an experiment. Consider the experiment of tossing a coin. The first thing we want to do is construct the sample space. Naturally, the outcomes that can occur are head (H) or tail (T). The results of 500 tosses are displayed below. The last column, **experimental relative frequency,** is merely the experimental frequency divided by the total number of trials of the experiment.

Possible outcomes (sample points)	Experimental frequency (number of occurrences)	Experimental relative frequency
H	248	$\frac{248}{500} = .496$
T	252	$\frac{252}{500} = .504$

In this experiment, we notice that the experimental relative frequencies are nearly equal. So, we are lead to inquire what would happen to the values of those experimental relative frequencies if we were to continue tossing the coin a large number of times, perhaps even indefinitely. Certainly we could actually make more tosses. However, let's conjecture what would happen. A coin seems like a very symmetrical object and that alone might lead us to suspect that it is just as likely to fall "head" as "tail." Thus, no matter how many tosses we make, one-half of them we would expect to be heads (the other one-half, tails). So, in the long run, we would expect to see each experimental relative frequency approach the theoretical value of 1/2. This theoretical value will be called the **theoretical relative frequency.** When we speak of the *probability* of an event, we mean the theoretical relative frequency of that event. So, when a coin is tossed, the probability of a head, denoted $P(H)$, equals 1/2 and the probability of a tail is $P(T) = 1/2$.

Possible outcomes (sample points)	Experimental frequency	Experimental relative frequency	Theoretical relative frequency or *probability*
H	248	.496	$\frac{1}{2} = .500$
T	252	.504	$\frac{1}{2} = .500$

Consider the experiment of tossing a die 600 times. The sample space with its six sample points appears below.

Sample points	Experimental frequency	Experimental relative frequency
1	97	$\frac{97}{600}$
2	94	$\frac{94}{600}$
3	102	$\frac{102}{600}$
4	99	$\frac{99}{600}$
5	105	$\frac{105}{600}$
6	103	$\frac{103}{600}$

A conjecture as to the theoretical relative frequency of each sample point is now desired. We notice that each of the experimental relative frequencies is "close to" $\frac{100}{600} = \frac{1}{6}$. Also, a die seems rather symmetrically constructed, and the presence of its six faces suggests that we can expect all faces to occur an equal number of times out of the total of 600 tosses. Hence, the table can now be expanded.

Sample points	Experimental frequency	Experimental relative frequency	Theoretical relative frequency or probability
1	97	$\frac{97}{600}$	$\frac{100}{600} = \frac{1}{6}$
2	94	$\frac{94}{600}$	$\frac{100}{600} = \frac{1}{6}$
3	102	$\frac{102}{600}$	$\frac{100}{600} = \frac{1}{6}$
4	99	$\frac{99}{600}$	$\frac{100}{600} = \frac{1}{6}$
5	105	$\frac{105}{600}$	$\frac{100}{600} = \frac{1}{6}$
6	103	$\frac{103}{600}$	$\frac{100}{600} = \frac{1}{6}$

So, we know that $P(1) = 1/6$, $P(2) = 1/6$, $P(3) = 1/6$, $P(4) = 1/6$, $P(5) = 1/6$, $P(6) = 1/6$.

What does $P(1) = 1/6$ mean, anyway? Does it mean that if a die fails to show a 1 in five trials, that on the sixth toss it must come up a 1? Certainly not! Remember, probability is the same as theoretical relative frequency. Thus, PROBABILITY IS A MEASURE OF WHAT TENDS TO HAPPEN IN A VERY LARGE NUMBER OF TRIALS of a given experiment. So, $P(1) = 1/6$ means that when a die is tossed a large number of times, about 1/6 of those tosses are actually 1. Both of the experiments just discussed are such that all the sample points have the same probability. That doesn't always happen, as will now be illustrated.

Consider an "extra thick" coin that can land on its edge, giving us a total of three possibilities (or sample points) when an experiment is defined to be "toss a thick coin once."

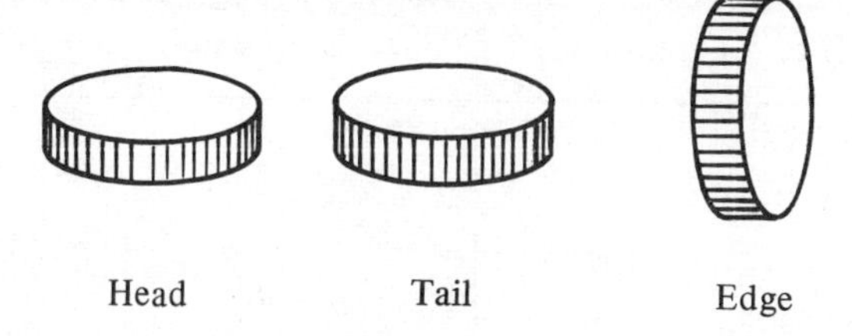

The first two possibilities (head, tail) look the same geometrically—we might expect to see the same probability for each. The third possibility (edge) looks very different from a geometric viewpoint and we might expect it to have a different probability. In order to substantiate our expectations, we now display the results of tossing a thick coin 2000 times.

Sample points	Experimental frequency	Experimental relative frequency	Theoretical relative frequency or probability
H	793	$\frac{793}{2000}$	$\frac{800}{2000} = \frac{2}{5}$
T	802	$\frac{802}{2000}$	$\frac{800}{2000} = \frac{2}{5}$
E	405	$\frac{405}{2000}$	$\frac{400}{2000} = \frac{1}{5}$

We conjecture that the probabilities of head and tail are equal but that the probability of an edge is one-half that of a tail. So, $P(H) = \frac{2}{5}$, $P(T) = \frac{2}{5}$, $P(E) = \frac{1}{5}$. Hence we see that sometimes the sample points in a sample space have equal probabilities and sometimes they do not. However, many times the experiment itself will give us a hint.

Determining the probabilities of the sample points for an experiment is rather time-consuming the way we have done it so far. Let us begin to shorten that time. Notice two things about the probabilities assigned to the sample points (refer to the three previous illustrations):

1. any sample point's probability is a fraction between 0 and 1
2. the sum of the probabilities of all the sample points in any particular sample space is equal to 1.

So, when a die is tossed, we reason to ourselves that since there are six sample points whose equal probabilities must add up to 1, each of the points "gets assigned" a probability of 1/6.

Example 4.3.1. Ten identical cards are numbered 0, 1, 2, . . . , 9 and placed in a hat. An experiment consists of drawing one card.

a. Determine the sample space.
b. Assign probabilities to each of the sample points.

Solution:

a. The sample space consists of 0, 1, 2, . . . , 9.
b. Since there are ten sample points and the ten cards are identical, it seems quite reasonable to assign the probability of 1/10 to each sample point.

Example 4.3.2. Two balanced dice are tossed. Determine the probability of each sample point.

Solution: The sample space is:

1,1	2,1	3,1	4,1	5,1	6,1
1,2	2,2	3,2	4,2	5,2	6,2
1,3	2,3	3,3	4,3	5,3	6,3
1,4	2,4	3,4	4,4	5,4	6,4
1,5	2,5	3,5	4,5	5,5	6,5
1,6	2,6	3,6	4,6	5,6	6,6

There are 36 sample points. Since the dice are balanced, it seems reasonable that all of these 36 points are equally likely. Therefore, we assign the probability of 1/36 to each sample point.

Example 4.3.3. A die has been loaded so that "six" is three times as likely as any of the other possible outcomes.

a. What is the sample space for the experiment of tossing the die once?
b. Assign probabilities to each of the sample points.

Solution:

a. The sample space still consists of the outcomes 1, 2, 3, 4, 5, 6.
b. There are six sample points whose probabilities still must sum up to 1.

Let $P(1) = x$, $P(2) = x$, $P(3) = x$, $P(4) = x$, $P(5) = x$, $P(6) = 3x$. (Since a 2, 3, 4, and 5 are just as likely as a 1 and a 6 is three times as likely as a 1: $P(1) = x$.) Summing up, $x + x + x + x + x + 3x = 1$.

Thus, $8x = 1$, so $x = 1/8$. Hence, $P(1) = 1/8$, $P(2) = 1/8$, $P(3) = 1/8$, $P(4) = 1/8$, $P(5) = 1/8$, $P(6) = 3/8$.

Besides constructing sample spaces and assigning probabilities to the sample points, many times we want to know the probability of some compound event.

EXAMPLE 4.3.4. A die is tossed and then another die tossed. Let:

event A = sum on faces is seven
event B = sum on faces is eleven
event C = first die shows a three

Find:

a. $P(A)$ **b.** $P(B)$ **c.** $P(C)$ **d.** $P(A \text{ and } B)$ **e.** $P(B \text{ and } C)$
f. $P(A \text{ and } C)$ **g.** $P(A \text{ or } B)$ **h.** $P(B \text{ or } C)$ **i.** $P(A \text{ or } C)$

Solution: First we need the sample space and the probability of each sample point. We found those in example 4.3.2.

1,1	2,1	3,1	4,1	5,1	6,1	Each has a probability of $\frac{1}{36}$
1,2	2,2	3,2	4,2	5,2	6,2	
1,3	2,3	3,3	4,3	5,3	6,3	
1,4	2,4	3,4	4,4	5,4	6,4	
1,5	2,5	3,5	4,5	5,5	6,5	
1,6	2,6	3,6	4,6	5,6	6,6	

a. $P(A) = P(\text{sum on faces is seven})$. Sum on faces is seven = 1,6 or 2,5 or 3,4 or 4,3 or 5,2 or 6,1. Each of these has a probability of 1/36. So $P(A) = 6/36$.

b. $P(B) = P(5{,}6 \text{ or } 6{,}5) = 2/36$.

c. $P(C) = P(\text{first die is a three})$. The sample points (3,1), (3,2), (3,3), (3,4), (3,5), (3,6) are those for which the first die is a three. Thus $P(C) = 6/36$.

d. $P(A \text{ and } B)$. A and B means, of course, both A and B. Our experiment consisted of tossing two dice. So, if we ask for the sum to be BOTH seven and eleven, there are no sample points that apply. Therefore $P(A \text{ and } B) = 0$.

e. $P(B \text{ and } C)$. Here again, there is no way to satisfy BOTH B and C. Thus $P(B \text{ and } C) = 0$.

f. $P(A \text{ and } C) = P(\text{sum on faces is seven and first die is a three})$. There is one sample point that applies (3,4). Thus, $P(A \text{ and } C) = 1/36$.

g. $P(A \text{ or } B) = P(\text{sum is seven or sum is eleven})$. There are eight sample points that apply: 1,6 or 2,5 or 3,4 or 4,3 or 5,2 or 6,1 or 5,6 or 6,5. $P(A \text{ or } B) = 8/36$.

h. Similarly, $P(B \text{ or } C) = 8/36$.

i. $P(A \text{ or } C) = P(\text{sum is seven or first die is a three})$. We draw a picture. Note that events A,C are non–mutually exclusive.

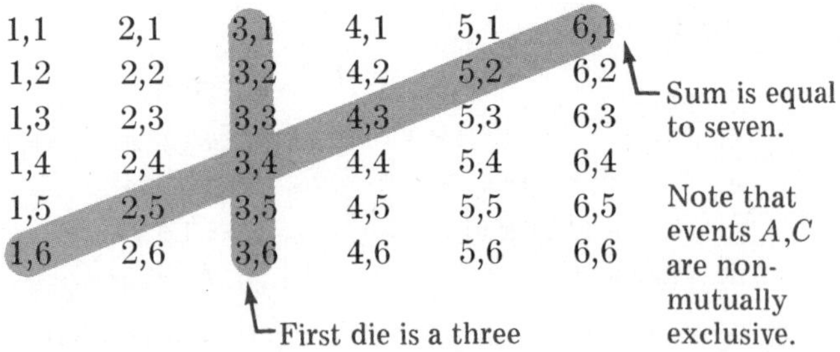

1,1	2,1	3,1	4,1	5,1	6,1	
1,2	2,2	3,2	4,2	5,2	6,2	Sum is equal to seven.
1,3	2,3	3,3	4,3	5,3	6,3	
1,4	2,4	3,4	4,4	5,4	6,4	
1,5	2,5	3,5	4,5	5,5	6,5	Note that events A,C are non-mutually exclusive.
1,6	2,6	3,6	4,6	5,6	6,6	

First die is a three

Notice that "sum is seven" has six sample points, while "first die is a three" has six sample points, too! However, there is one sample point in common, 3,4. (That's because events A,C are non–mutually exclusive.) So, there are a total of 11 sample points whose probabilities we want to count. (We won't count 3,4's probability twice because that would be like saying that 3,4 is twice as likely as any of the others.) Therefore $P(A \text{ or } C) = 11/36$.

We are now in a position to summarize how to compute the probabilities of events.

1. Construct the sample space.
2. Assign probabilities to each of the sample points.
3. Identify (or pick out) the sample points that pertain to the question being asked.
4. Add up the probabilities of those sample points (being careful not to use any point more than once).

EXERCISES ■ SECTION 4.3.

1. A dartboard is numbered 1, 2, 3, . . . , 20. Each of the spaces on such a dartboard is the same size, hence we will presume the numbers to have an equally likely chance of being hit. An experiment consists of throwing one dart.

   a. What is the sample space?
   b. What is the probability of each sample point?
   c. What is the probability of hitting an even number?
   d. What is the probability of hitting a number less than 12?

2. Ten identical cards are numbered 0, 1, 2, . . . , 9 and placed in a hat. An experiment consists of drawing one card. Let:

   event A = an even number appears
   event B = an odd number appears
   event C = a number greater than five appears

Find:

a. $P(A)$
b. $P(B)$
c. $P(C)$
d. $P(A \text{ and } B)$
e. $P(B \text{ and } C)$
f. $P(A \text{ and } C)$
g. $P(A \text{ and } B \text{ and } C)$
h. $P(A \text{ or } B)$
i. $P(B \text{ or } C)$
j. $P(A \text{ or } C)$

3. A clever man has altered a thick coin so that "head" and "tail" are equally likely and each of those is four times as likely as "edge." Assign probabilities to each of the three sample points for the experiment of tossing this coin once.

4. A more clever woman has altered a die so that "2" is twice as likely as "1," "3" is three times as likely as "1," "4" is four times as likely as "1," "5" is five times as likely as "1," and "6" is six times as likely as "1." Assign probabilities to each of the sample points for the experiment of tossing this die once.

5. Small gravestones cost $100. Medium-sized gravestones cost $400. Large gravestones cost $1000. In a moment of great grief, I. M. Knuts randomly selects a gravestone. Presume the choices to be equally likely.

a. What is the probability that he pays $400 (exactly)?
b. What is the probability that he pays at least $400?
c. What is the probability that he pays $50 (exactly)?

6. A local cemetery has a total of 1000 graves. One hundred graves have no gravestone, 200 graves have a small gravestone, 300 graves have a medium-sized gravestone, and 400 graves have a large gravestone. An experiment consists of randomly selecting one gravesite.

a. What is the probability that the gravesite has a small gravestone?
b. What is the probability that the gravesite has a gravestone?

7. Two dice are tossed. Let event A = sum equals five, and event B = first die is odd.

a. Is event A simple?
b. Is event B simple?
c. Are events A and B mutually exclusive?
d. Find $P(A)$.
e. Find $P(B)$.
f. Find $P(A \text{ and } B)$.
g. Find $P(A \text{ or } B)$.

8. Suppose that a fair coin has been tossed 20 times and has come up heads all 20 times. What is the probability that on the next toss we will get a head?

9. A certain board game has a spinner that is composed of the five colors red, orange, yellow, green, and brown, as follows:

After 100 spins the following table was derived.

Color	Times it occurred
Red	29
Orange	32
Yellow	17
Green	4
Brown	18

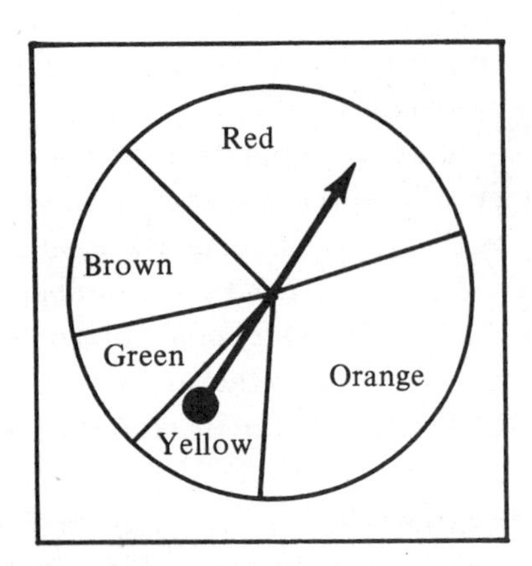

On a single spin,

a. What is the experimental relative frequency of obtaining red?
b. What is the experimental relative frequency of obtaining brown?
c. What is the experimental relative frequency of obtaining red or green?
d. What is the experimental relative frequency of obtaining a color other than red?

10. Suppose that the area of the spinner circle of exercise 9 is 100 square centimeters and the area of the red sector is 30 square centimeters, the area of the orange sector is 30 square centimeters, the area of the yellow sector is 15 square centimeters, the area of the green sector is 5 square centimeters, and the area of the brown sector is 20 square centimeters. Assume the likelihood of a color's occurrence to be proportional to its relative area. Then on a single spin,

a. What is the theoretical relative frequency of obtaining red?
b. What is the theoretical relative frequency of obtaining orange?
c. What is the theoretical relative frequency of obtaining red or green?
d. What is the theoretical relative frequency of obtaining a color other than red?

4.4 SINGLE-ACT EXPERIMENTS

When studying probability, it is helpful to begin with **single-act experiments.** The previous section treated such experiments when we discussed things like "tossing a coin once," "tossing a die once," example 4.3.1, example 4.3.3, and exercises 1–6 at the end of that section. The reader will discover by quickly looking back over each of those experiments that they all involved doing just one thing, that is, performing a single act. So such experiments are not new to us at this point. We will, however, add a few bits of knowledge to what we already know about probability.

When we compute a probability, we construct the sample space, assign probabilities to the sample points, identify those sample points that apply to the question being asked, and finally add up those probabilities.

As already mentioned, probability is a measure of what tends to happen in a large number of trials of the experiment. However, there is another interpretation that we can give to this number in the case of EQUALLY LIKELY sample points. We give an example.

Example 4.4.1. Let a die be tossed. Event A = the number appearing is divisible by three. Find $P(A)$.

Solution: The sample space is 1, 2, 3, 4, 5, 6, with each sample point having a probability of 1/6. Thus $P(A) = P(3 \text{ or } 6) = \frac{1}{6} + \frac{1}{6} = \frac{2}{6}$. Notice that if we term "divisible by three" as success (the other possibilities 1, 2, 4, 5 could be called failure), then

$$P(A) = \frac{\text{number of ways to get a success}}{\text{total number of ways to do experiment}}$$

or equivalently,

$$P(A) = \frac{\text{number of ways to get a success}}{\text{total number of points in sample space}}$$

The result in the above example will happen for any event A as long as the sample points have equally likely probabilities.

Example 4.4.2. A box has in it three red, four yellow, and five blue balls. An experiment is defined as selecting one ball. What is the probability of a yellow?

Solution: The 12 sample points are equally likely. Since there are four yellows, $P(\text{yellow}) = 4/12$.

Probability questions are sometimes stated in terms of an event not happening. We see how to deal with this in the next example.

Example 4.4.3. Let a die be tossed. What is the probability that a 1 does not occur (i.e., that a number greater than one occurs)?

Solution:

Method I

We can attack the problem directly. Let success = 2 or 3 or 4 or 5 or 6.

$$P(\text{a 1 does not occur}) = \tfrac{5}{6}$$

Method II

We can treat the problem indirectly. Let event A = a 1 occurs. We need to find

$$P(A \text{ does not occur}) = P(\tilde{A})$$

Notice that A or $\tilde{A}$ must occur, so $P(A) + P(\tilde{A}) = 1$. Hence $P(\tilde{A}) = 1 - P(A) = 1 - \frac{1}{6} = \frac{5}{6}$.

Although the solution in method II, which uses $P(A) + P(\tilde{A}) = 1$, looks longer, don't hesitate to use it. Often, this method can turn out to be the shortest way of solving a problem—especially in certain problems in section 4.5.

A concept closely related to probability is that of **odds.**

$$\textit{The } \textbf{odds in favor of } \textit{an event A occurring} = \frac{P(A)}{P(\tilde{A})} = \frac{P(A)}{1 - P(A)}$$

$$\textit{The } \textbf{odds against } \textit{event A occurring} = \frac{P(\tilde{A})}{P(A)} = \frac{1 - P(A)}{P(A)}$$

When a die is tossed, the odds in favor of a 2 occurring equal $\frac{1/6}{5/6} = \frac{1}{5}$, commonly written as 1:5 or 1 to 5. Notice that odds in favor of an event A can be interpreted as the ratio

$$\frac{\text{number of successes in sample space}}{\text{number of failures in sample space}}$$

The odds against a two occurring are $\frac{5/6}{1/6} = \frac{5}{1}$ or 5:1 or 5 to 1. Again, we can give another interpretation here. The odds against an event A occurring equal

$$\frac{\text{number of failures in sample space}}{\text{number of successes in sample space}}$$

Odds are a ratio, just as probability is; however, odds can turn out to be greater than one, while probability cannot. Still another interpretation that can be given to odds is the following. In the die illustration, the odds in favor of obtaining a two are 1 to 5. If Al bets Bob $1 that a 2 will occur, Bob should be willing to bet $5 (and hence be "giving" 5 to 1 against obtaining a two).

EXAMPLE 4.4.4. Ten identical cards are numbered 0, 1, 2, . . . , 9 and placed in a hat. An experiment consists of drawing one card. Let event A = an even number appears, event B = a number less than 4 appears. Find:

a. odds for A occurring, **b.** odds for B occurring,
c. odds against B occurring, **d.** odds for A and B occurring,
e. odds for A or B occurring.

Solution: The sample space is 0, 1, 2, . . . , 9.

a. Odds for A occurring $= \dfrac{P(A)}{P(\tilde{A})} = \dfrac{5/10}{5/10} = \dfrac{5}{5}$ or 1 to 1.

b. Odds for B occurring $= \dfrac{P(B)}{P(\tilde{B})} = \dfrac{4/10}{6/10} = \dfrac{4}{6}$ or 2 to 3.

c. Odds against B occurring = 3 to 2.

d. Odds for A and B occurring $= \dfrac{P(A \text{ and } B)}{P(\widetilde{A \text{ and } B})} = \dfrac{2/10}{8/10} = \dfrac{2}{8}$ or 1:4.

e. Odds for A or B occurring $= \dfrac{P(A \text{ or } B)}{P(\widetilde{A \text{ or } B})}$. Now, $P(A \text{ or } B) = \dfrac{7}{10}$.

(Remember not to count sample points common to A and B more than once.) So, odds for A or B occurring $= \dfrac{7/10}{3/10} = \dfrac{7}{3}$ or 7 to 3.

The example just solved points out rather vividly that when you know the probability of an event, you can then compute the odds for (or against) that event. The reverse is also possible: knowing the odds for (or against) an event can lead you to that event's probability.

EXAMPLE 4.4.5. Presume that there is an event A such that the odds in favor of A occurring are 5:3. What is $P(A)$?

Solution: Odds in favor of event A occurring = 5/3. Thus $\dfrac{P(A)}{P(\tilde{A})} = \dfrac{5}{3}$, and $\dfrac{P(A)}{1 - P(A)} = \dfrac{5}{3}$. We now use some algebra.

$$\begin{aligned}3P(A) &= [1 - P(A)]5\\ 3P(A) &= 5 - 5P(A)\\ 3P(A) + 5P(A) &= 5\\ 8P(A) &= 5\\ P(A) &= \tfrac{5}{8}\end{aligned}$$

Example 4.4.6. Presume that there is an event A where the odds against A occurring are 3:2. Find $P(A)$.

Solution: The odds against A are 3:2. Thus odds for A occurring are 2:3.

$$\frac{P(A)}{P(\tilde{A})} = \frac{2}{3}$$

$$\frac{P(A)}{1 - P(A)} = \frac{2}{3}$$

$$3P(A) = 2 - 2P(A)$$

$$5P(A) = 2$$

$$P(A) = \tfrac{2}{5}$$

Often times, the information in a probability problem can be neatly summarized in a Venn diagram. For example, consider example 2.5.8. There, a survey to determine drug usage was conducted among 220 randomly selected people. The results were:

10 people tried both marijuana and hard drugs
90 people tried marijuana
20 people tried hard drugs

Using M for marijuana and H for hard drugs, we see:

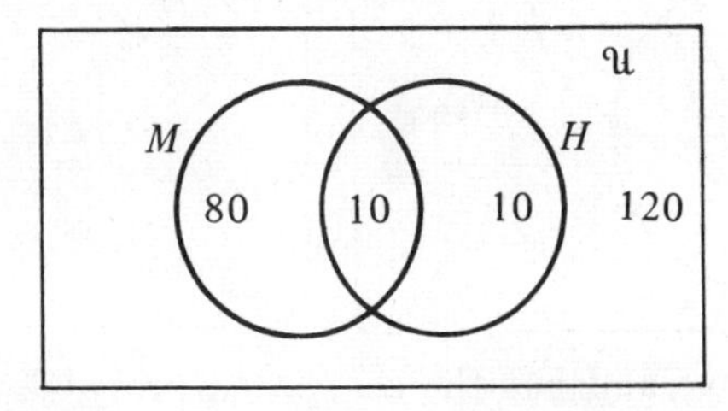

This displays the survey results. We will find it useful to put probabilities on the Venn diagram.

$$P(\text{a person tried just marijuana}) = P(M \cap \tilde{H}) = \tfrac{80}{220} = \tfrac{8}{22}$$

$$P(\text{a person tried both marijuana and hard drugs}) = P(M \cap H) = \tfrac{10}{220} = \tfrac{1}{22}$$

$$P(\text{a person tried just hard drugs}) = P(\tilde{M} \cap H) = \tfrac{10}{220} = \tfrac{1}{22}$$

$$P(\text{a person tried neither marijuana nor hard drugs}) = P(\tilde{M} \cap \tilde{H}) = \tfrac{120}{220} = \tfrac{12}{22}$$

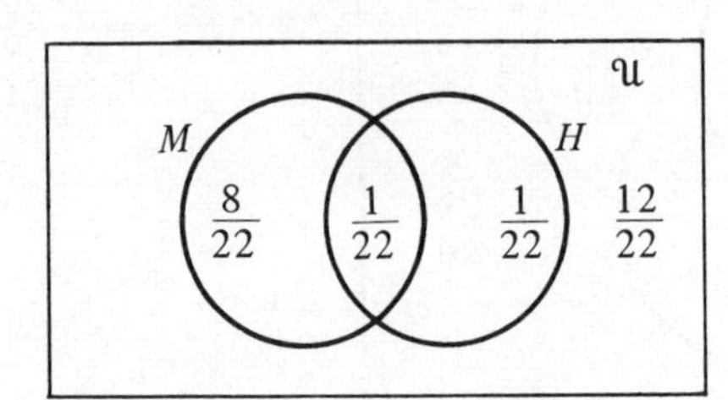

What is the probability that a person tried marijuana? We add up the probabilities inside region M. We get $\frac{8}{22} + \frac{1}{22} = \frac{9}{22}$. What is the probability that a person tried marijuana or hard drugs? That's $P(M \cup H)$. We add all probabilities inside $M \cup H$. So

$$P(M \cup H) = \frac{8}{22} + \frac{1}{22} + \frac{1}{22} = \frac{10}{22} = \frac{5}{11}$$

Example 4.4.7. A survey of 300 people was conducted to determine what items they were recycling (paper, metal, glass).

35 people recycle paper and metal
40 people recycle metal and glass
60 people recycle paper and glass
90 people recycle paper
70 people recycle metal
105 people recycle glass
25 people recycle all three items

In example 2.5.9, where we first dealt with this data, we constructed the Venn diagram below.

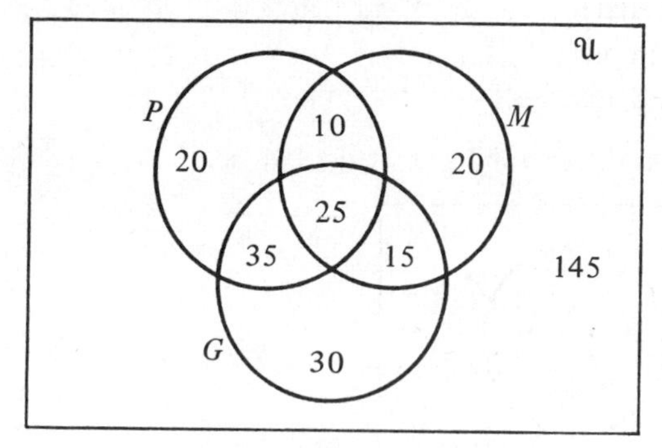

a. Redraw the diagram above and put the associated probabilities in each region.
b. What is the probability that a person recycles just one item?
c. What is the probability that a person recycles no items?
d. What is the probability that a person recycles at least one item?

Solution:

a.

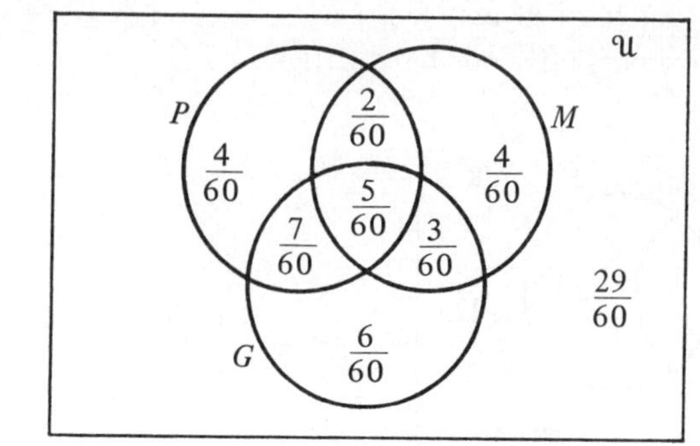

b. $P(\text{a person recycles just one item}) = \frac{4}{60} + \frac{4}{60} + \frac{6}{60} = \frac{14}{60} = \frac{7}{30}$
c. $P(\text{a person recycles no items}) = \frac{29}{60}$
d. Let event A = a person recycles no items. Then,
$P(\text{a person recycles at least one}) = P(\tilde{A}) = 1 - P(A) = 1 - \frac{29}{60} = \frac{31}{60}$.

It is worthy to note that parts *b*, *c*, and *d* of the previous example are essentially the same questions as in parts *f*, *g*, and *h* of example 2.5.9, except then (back in chapter 2) we did not use the word "probability"—we used "fraction of people."

This section dealt with single-act experiments and basically did four things. First, we gave a new interpretation to the probability of an event A: when the sample space has equally likely sample points,

$$P(A) = \frac{\text{number of ways to get a success}}{\text{total number of points in sample space}}$$

Secondly, we used $P(A) + P(\tilde{A}) = 1$ to help us determine the probability of an event A not occurring. Next, we defined the concept of odds. Finally, we used Venn diagrams to aid us in the solution of probability questions. These four types of problems were introduced by using single-act experiments, but in the next section we will see them enter into multi-act experiments in the same way.

EXERCISES ■ SECTION 4.4.

1. In a single selection from a deck of 52 cards, what is the probability of getting:

   a. the ace of spades
   b. an ace
   c. a spade
   d. an ace or a spade
   e. a black jack
   f. a king or queen
   g. a red card
   h. anything but a spade?

2. In a single selection from a deck of 52 cards, what are the odds in favor of:

   a. the ace of spades
   b. an ace
   c. a spade
   d. an ace or a spade?

   What are the odds against:

   e. a black jack
   f. a king or queen
   g. a red card
   h. anything, but a spade?

3. A die is "loaded" so that the probabilities of the numbers 1, 2, 3, 4, 5, 6 are, respectively, $\frac{1}{3}$, $\frac{1}{6}$, $\frac{1}{12}$, $\frac{1}{6}$, $\frac{1}{12}$, $\frac{1}{6}$. An experiment will be to roll a die one time. What is the probability of:

   a. a three or a five
   b. an odd number
   c. an even number
   d. neither a three nor a five?

   What are the odds favoring:

   e. a three or a five
   f. a number greater than 4?

4. Concerning the blood types A, B, AB, O, the percents of the population having each of these types is 41%, 10%, 4%, 45%, respectively. An automobile accident has just occurred and one of the victims needs blood urgently. A person is selected at random from a large crowd of onlookers. What is the probability that the blood type will be:

   a. A
   b. B
   c. AB
   d. O
   e. A or O
   f. B and O
   g. A and O
   h. A or B?

5. The Proton Corporation produces electric pencil sharpeners. Out of every 100,000 pencil sharpeners, it is known that 1000 have faulty wiring, 500 have faulty gears, and 150 have both defects. Let

   p: the sharpener has faulty wiring
   q: the sharpener has faulty gears

   Find:

   a. $P(p \wedge q)$
   b. $P(p \vee q)$
   c. $P(p \wedge \sim q)$
   d. $P(\sim p \wedge \sim q)$
   e. $P(\sim p \vee \sim q)$

6. An elevator at the Mitchell Carob Company has occasional breakdowns:

$$P(\text{exactly zero breakdowns per week}) = .1$$
$$P(\text{exactly one breakdown per week}) = .3$$
$$P(\text{exactly two breakdowns per week}) = .4$$
$$P(\text{exactly three breakdowns per week}) = .1$$
$$P(\text{four or more breakdowns per week}) = .1$$

   What is the probability that (during a given week) there will be at least one breakdown? No more than three breakdowns?

7. The odds for "My Pretty Lady" winning the horse race are 7:5.

   a. What is the probability that "My Pretty Lady" wins?
   b. What is the probability that "My Pretty Lady" loses?

8. The probability of an event A is $2/5$. What are the odds against A's occurrence?

9. A survey of 50 women revealed that: 20 were married, 22 have children, 16 are both married and have children. An experiment consists of selecting a woman in the survey at random.

- **a.** Display the survey results on a Venn diagram, using M = set of women that are married and C = set of women that have children.
- **b.** Redraw the Venn diagram showing probabilities in each of the four regions.
- **c.** Find P(being married).
- **d.** Find P(having children).
- **e.** Find P(being married and having children).
- **f.** Find P(being married or having children).
- **g.** Find P(being married without children).
- **h.** Find P(having children without being married).
- **i.** Find P(neither being married nor having children).

10. Consider the following Venn diagram depicting the results of a survey of lawyers. The associated probabilities are presented for the sets A, B, where A represents the set of lawyers whose fathers were lawyers and B represents the set of lawyers whose mothers were lawyers.

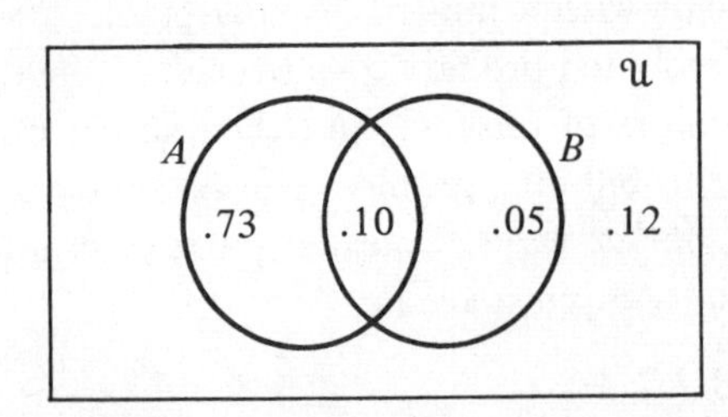

- **a.** Show that $P(A \text{ or } B)$ IS NOT equal to $P(A) + P(B)$.
- **b.** Show $P(A \text{ or } B) = P(A) + P(B) - P(A \text{ and } B)$. Note that this is true all the time. In the special case where $P(A \text{ and } B) = 0$ (i.e., A, B are mutually exclusive events), the equation degenerates to $P(A \text{ or } B) = P(A) + P(B)$.

11. Let $P(x) = .14$, $P(y) = .23$, $P(x \text{ and } y) = .1$. Find:

- **a.** $P(x \text{ or } y)$ (Hint: Use the results in exercise 10*b* above.)
- **b.** $P(\tilde{x} \text{ and } y)$
- **c.** $P(x \text{ and } \tilde{y})$
- **d.** $P(\widetilde{x \text{ and } y})$
- **e.** $P(\widetilde{x \text{ or } y})$

12. Let C and D be mutually exclusive events with $P(C) = .9$ and $P(D) = .05$. Find:

    a. $P(C \text{ and } D)$ b. $P(C \text{ or } D)$

13. Sixty percent of the employees of the Bengil Corporation are college graduates. Of these, ten percent are in sales. Of the non-college graduates, eighty percent are in sales.

    a. What is the probability that an employee selected at random is in sales?

    b. What is the probability that an employee selected at random is neither in sales nor a college graduate?

4.5 MULTI-ACT EXPERIMENTS

A **multi-act experiment** is an experiment that involves doing more than one thing. Tossing two dice, since it is equivalent to tossing one die twice, is a multi-act experiment. Similarly, tossing two coins, tossing three coins, and tossing a coin with a die are also multi-act experiments. There are many other, and more interesting, examples presented in this and future sections. Any multi-act experiment can be solved with the techniques presented so far. In fact, the reader should, at this point, turn back to reread example 4.3.4, which dealt with a multi-act experiment. There we constructed the sample space, assigned probabilities to each of the sample points, identified the sample points of interest, and then added up their associated probabilities. ANY probability problem can be solved in that fashion. Sometimes there are shorter ways. Consider parts g, h, and i of that example (4.3.4) again. The sample space is:

1,1	2,1	3,1	4,1	5,1	6,1	← A
1,2	2,2	3,2	4,2	5,2	6,2	
1,3	2,3	3,3	4,3	5,3	6,3	
1,4	2,4	3,4	4,4	5,4	6,4	
1,5	2,5	3,5	4,5	5,5	6,5	← B
1,6	2,6	3,6	4,6	5,6	6,6	
		↑ C				

A = sum on faces is seven
B = sum on faces is eleven
C = first die is a three

Now we ask for $P(A \text{ or } B)$, $P(B \text{ or } C)$, and $P(A \text{ or } C)$. Exercise 10 at the end of the previous section gave us $P(X \text{ or } Y) = P(X) + P(Y) - P(X \text{ and } Y)$. Let us use that here.

$$P(A \text{ or } B) = P(A) + P(B) - P(A \text{ and } B) = \tfrac{6}{36} + \tfrac{2}{36} - 0 = \tfrac{8}{36}$$

(This checks with the previous solution.) Note that since events A, B are mutually exclusive, $P(A \text{ and } B) = 0$.

$$P(B \text{ or } C) = P(B) + P(C) - P(B \text{ and } C) = \tfrac{2}{36} + \tfrac{6}{36} - 0 = \tfrac{8}{36}$$

(This checks with the previous solution.) Note again that with events B, C being mutually exclusive, $P(B \text{ and } C) = 0$.

$$P(A \text{ or } C) = P(A) + P(C) - P(A \text{ and } C) = \tfrac{6}{36} + \tfrac{6}{36} - \tfrac{1}{36} = \tfrac{11}{36}$$

(This checks with the previous solution.) Note here that events A, C are non-mutually exclusive.

The last three probabilities computed point out that:

$$P(X \text{ or } Y) = P(X) + P(Y) - P(X \text{ and } Y)$$

when X, Y are non-mutually exclusive events;

$$P(X \text{ or } Y) = P(X) + P(Y)$$

when X, Y are mutually exclusive events.

Let's look at another problem that is similar to the one just done.

Example 4.5.1. A coin is tossed and a die is rolled. Let:

event A = a head appears
event B = an even number appears
event C = a two or a tail appears (but not both)

Find:

a. $P(A \text{ or } B)$
b. $P(B \text{ or } C)$
c. $P(A \text{ or } C)$

Solution: The sample space is:

$$\left.\begin{matrix} H1 & T1 \\ H2 & T2 \\ H3 & T3 \\ H4 & T4 \\ H5 & T5 \\ H6 & T6 \end{matrix}\right\} \text{Each has a probability of } \tfrac{1}{12}$$

$$\left.\begin{aligned} A &= \{H1,H2,H3,H4,H5,H6\} \\ B &= \{H2,H4,H6,T2,T4,T6\} \\ C &= \{H2,T1,T3,T4,T5,T6\} \end{aligned}\right\} \text{Here we choose to write } A, B, C \text{ as sets}$$

a. $P(A \text{ or } B) = P(A) + P(B) - P(A \text{ and } B)$
$= \frac{6}{12} + \frac{6}{12} - \frac{3}{12}$
$= \frac{9}{12}$

b. $P(B \text{ or } C) = P(B) + P(C) - P(B \text{ and } C)$
$= \frac{6}{12} + \frac{6}{12} - \frac{3}{12}$
$= \frac{9}{12}$

c. $P(A \text{ or } C) = P(A) + P(C) - P(A \text{ and } C)$
$= \frac{6}{12} + \frac{6}{12} - \frac{1}{12}$
$= \frac{11}{12}$

When dealing with single-act experiments in the previous section, we considered the topic of odds and also we used the relationship $P(A) + P(\tilde{A}) = 1$. In the next example we point out that both of these ideas are handled in the same way when applied to a multi-act experiment.

Example 4.5.2. A coin is tossed and a die is rolled. Let:

A = a number less than six appears
B = a three appears

Find: **a.** $P(A)$ **b.** $P(A \text{ or } B)$ **c.** Odds favoring A or B

Solution: The sample space is, of course:

$H1$	$T1$
$H2$	$T2$
$H3$	$T3$
$H4$	$T4$
$H5$	$T5$
$H6$	$T6$

$A = \{H1, H2, H3, H4, H5, T1, T2, T3, T4, T5\}$
$B = \{H3, T3\}$

a. We want $P(A)$.

Method I

By counting sample points directly, $P(A) = \frac{10}{12}$.

Method II

$P(\tilde{A}) = P(\text{a six appears}) = \frac{2}{12}$
$P(A) + P(\tilde{A}) = 1$
$P(A) = 1 - P(\tilde{A}) = 1 - \frac{2}{12} = \frac{10}{12}$

b. $P(A \text{ or } B) = P(A) + P(B) - P(A \text{ and } B)$
$= \frac{10}{12} + \frac{2}{12} - \frac{2}{12}$
$= \frac{10}{12}$

c. $$\text{Odds favoring } A \text{ or } B = \frac{P(A \text{ or } B)}{P\widetilde{(A \text{ or } B)}} = \frac{10/12}{2/12} = \frac{10}{2}$$

$\therefore$ Odds favoring A or B = 5 to 1.

EXAMPLE 4.5.3. Two dice are tossed. Find the probability that the sum is three or more.

Solution: Possible sums are 2, 3, 4, . . . , 12. We need $P(3 \text{ or } 4 \text{ or } \cdots \text{ or } 12)$. That's lengthy. Let A = sum is three or more. Then $\tilde{A}$ = sum is less than three (i.e. sum is 2). Hence $P(A) = 1 - P(\tilde{A}) = 1 - \frac{1}{36} = \frac{35}{36}$. (That's quick!)

EXAMPLE 4.5.4. Consider the set of integers $\{1,2,3,4,5\}$. An experiment consists of making two selections (i.e., 2-digit numbers) from the given set, where the first integer is replaced before the second is drawn.

a. Construct the sample space.
b. Assign probabilities to each of the sample points.
c. Find P(both integers are odd).
d. Find P(neither integer is odd).
e. Find P(exactly one integer is odd).

Solution:

a. A tree diagram (try it) will yield:

1,1	2,1	3,1	4,1	5,1
1,2	2,2	3,2	4,2	5,2
1,3	2,3	3,3	4,3	5,3
1,4	2,4	3,4	4,4	5,4
1,5	2,5	3,5	4,5	5,5

b. The probability of each sample point equals 1/25.
c. The simple events (1,1), (1,3), (1,5), (3,1), (3,3), (3,5), (5,1), (5,3), (5,5) pertain here. So, P(both integers are odd) = 9/25.
d. The simple events (2,2), (2,4), (4,2), (4,4) pertain here. So, P(neither integer is odd) = 4/25.
e. The simple events (1,2), (1,4), (2,1), (2,3), (2,5), (3,2), (3,4), (4,1), (4,3), (4,5), (5,2), (5,4) pertain here. So, P(exactly one integer is odd) = 12/25.

Consider an alternate approach to part *c* of the previous example:

$$P(\text{both integers are odd}) = P(\underbrace{\text{first integer is odd}}_{A} \text{ and } \underbrace{\text{second integer is odd}}_{B})$$

So, P(both integers are odd) = $P(A \text{ and } B)$. We now want to observe that

$$P(A \text{ and } B) = P(A) \cdot P(B)$$

Look: $P(A \text{ and } B) = 9/25$ (from part *c*, example 4.5.4). Furthermore, $P(A) = 3/5$ (because, for selecting the first integer, there are 3 odd numbers available in {1,2,3,4,5}), and $P(B) = 3/5$ (because, for selecting the second integer, there are still 3 odd numbers in {1,2,3,4,5}).

$$P(A) \cdot P(B) = \tfrac{3}{5} \cdot \tfrac{3}{5} = \tfrac{9}{25}$$

Hence, $P(A \text{ and } B) = P(A) \cdot P(B)$.

We now take a very close look at these two events A and B. The original experiment consisted of drawing two integers (first one replaced before the second is drawn). What we actually did was to decompose the experiment into two distinct events. You can think of them as happening consecutively (i.e., one after the other). These two events A, B are called **independent events,** which means that the occurrence of either one of them does not affect the probability of occurrence of the other. What we have found is that for two independent events A, B:

$$P(A \text{ and } B) = P(A) \cdot P(B)$$

When two events are not independent, they are said to be **dependent.** An obvious question arises. Can we multiply probabilities of dependent events when we are looking for the probability of the conjunction of two events? Let's see. Consider the set of integers {1,2,3,4,5} and define an experiment to be selecting two integers (the first not replaced before the second is drawn). Ask: what is P(both odd)? We can easily make this into an "and" type of problem by writing:

$$P(\text{both odd}) = P(\underbrace{\text{odd on first}}_{A} \text{ and } \underbrace{\text{odd on second}}_{B}) = P(A \text{ and } B)$$

Here A, B are dependent events. Hence, we don't know whether or not $P(A \text{ and } B)$ equals $P(A) \cdot P(B)$ now. Let's "play it safe." We'll construct the sample space, assign probabilities to each of the sample points, identify the sample points of interest, and then finally add up their associated probabilities. Drawing a tree diagram shows the sample space:

1,2	2,1	(3,1)	4,1	(5,1)	Each has a probability of 1/20
(1,3)	2,3	3,2	4,2	5,2	
1,4	2,4	3,4	4,3	(5,3)	
(1,5)	2,5	(3,5)	4,5	5,4	

The ● indicates "both odd." So, $P(\text{both odd}) = 6/20$.

Now, let's see what happens if we try computing $P(A) \cdot P(B)$. We already calculated that $P(A) = 3/5$, and $P(B) = ?$, depending upon whether A actually occurred or not. Assuming A actually did occur, $P(B) = 2/4$. Hence $P(A) \cdot P(B) = \tfrac{3}{5} \cdot \tfrac{2}{4} = \tfrac{6}{20}$. (This checks with the solution above.) So, for dependent events A, B we can write:

$$P(A \text{ and } B) = P(A) \cdot \underbrace{P(B)}_{\text{Computed by assuming } A \text{ does occur first}}$$

Example 4.5.5. A pouch contains four black and three white marbles. Two marbles are to be drawn. If the first marble is replaced before the second is drawn, find:

a. P(both black)
b. P(both white)
c. P(exactly one black)

If the first is not replaced before the second is drawn, find:

d. P(both black)
e. P(both white)
f. P(exactly one black)

Solution:

a. P(both black) = P(first black and second black)

Independent because of replacement of first ball

$\therefore$ P(both black) = P(first black) · P(second black)

So,

$$P(\text{both black}) = \tfrac{4}{7} \cdot \tfrac{4}{7} = \tfrac{16}{49}$$

b. P(both white) = P(first white and second white)

Independent because of replacement of first ball

$\therefore$ P(both white) = P(first white) · P(second white)

So,

$$P(\text{both white}) = \tfrac{3}{7} \cdot \tfrac{3}{7} = \tfrac{9}{49}$$

c. We need P(exactly one black). The event "exactly one black" is a little tricky to decompose. We can do it, though:

P(exactly one black)
= P(first black and second white or first white and second black)

Independent — Independent — Mutually exclusive

"And" tells us to multiply. "Or" is an indicator to add.

$$P(\text{exactly one black}) = [P(\text{first black}) \cdot P(\text{second white})] + [P(\text{first white}) \cdot P(\text{second black})]$$

So,

$$P(\text{exactly one black}) = (\tfrac{4}{7} \cdot \tfrac{3}{7}) + (\tfrac{3}{7} \cdot \tfrac{4}{7}) = \tfrac{12}{49} + \tfrac{12}{49} = \tfrac{24}{49}$$

d. P(both black) = P(first black and second black)

Dependent, because first is not replaced

$$P(\text{both black}) = P(\text{first black}) \cdot \underbrace{P(\text{second black})}_{\substack{\text{Assuming first} \\ \text{was black}}}$$

$$P(\text{both black}) = \tfrac{4}{7} \cdot \tfrac{3}{6} = \tfrac{12}{42}$$

e. Decomposing as in part *b*,

$$P(\text{both white}) = P(\text{first white}) \cdot \underbrace{P(\text{second white})}_{\substack{\text{Assuming first} \\ \text{was white}}}$$

$$P(\text{both white}) = \tfrac{3}{7} \cdot \tfrac{2}{6} = \tfrac{6}{42}$$

f. Decomposing as in part *c*,

$$P(\text{exactly one black}) = P(\text{1st black}) \cdot P(\text{2nd white}) + P(\text{1st white}) \cdot P(\text{2nd black})$$

$$P(\text{exactly one black}) = \tfrac{4}{7} \cdot \tfrac{3}{6} + \tfrac{3}{7} \cdot \tfrac{4}{6} = \tfrac{24}{42}$$

Looking back over this last example, we can observe that the probabilities of compound events can be computed by decomposing that event into "or" problems (where adding is the key) or "and" problems (where multiplying is the key). The next four examples further illustrate this technique.

Example 4.5.6. The probability that a man will live ten more years is 1/3. The probability that a woman will live ten more years is 1/2. Presume these events to be independent. Find:

a. P(both will live for ten more years)
b. P(at least one will die during next ten years)

Solution:

a. P(both will live ten more years)

$= P$(husband will live ten more years and wife will live ten more years)
$= P$(husband will live ten years) $\cdot$ P(wife will live ten more years)
$= \frac{1}{3} \cdot \frac{1}{2} = \frac{1}{6}$

b. Let A = both will live ten more years; then $\tilde{A}$ = it is not the case that both will live ten more years (i.e., at least one will die during next ten years). Then

P(at least one will die during next ten years) $= P(\tilde{A}) = 1 - P(A)$

So,

$$P(\text{at least one will die during next ten years}) = 1 - \tfrac{1}{6} = \tfrac{5}{6}$$

Example 4.5.7. A TV station calls phone numbers that are chosen at random from a list of 100,000 numbers. The recipients of the phone calls are asked to name the amount in the grand jackpot. It is his if he guesses correctly.

a. What is the probability that a particular phone number on the station's list will be called?
b. Presume that the probability of knowing the jackpot is 1/250. What is the probability that a person whose phone number is on the list will win the jackpot?
c. What are the odds against winning?

Solution:

a. Since the numbers are randomly selected, they are equally likely. Thus, each has a probability of 1/100,000.
b. In order to win the jackpot, two things must happen: a number must be selected AND the recipient must guess the jackpot. (These two events certainly seem to be independent—we'll presume that to be the case.)

$$\begin{aligned} P(\text{winning jackpot}) &= P(\text{number is selected and jackpot is guessed}) \\ &= P(\text{number is selected}) \cdot P(\text{jackpot is guessed}) \\ &= \frac{1}{100{,}000} \cdot \frac{1}{250} = \frac{1}{25{,}000{,}000} \end{aligned}$$

c. $\text{Odds against winning} = \dfrac{P(\text{not winning})}{P(\text{winning})} = \dfrac{\dfrac{24{,}999{,}999}{25{,}000{,}000}}{\dfrac{1}{25{,}000{,}000}}$

$$\text{Odds against winning} = 24{,}999{,}999 \text{ to } 1$$

Example 4.5.8. Three persons work independently at deciphering a secret message in code. The respective probabilities that they will decipher it are $\frac{1}{6}$, $\frac{1}{5}$, $\frac{1}{3}$. What is the probability that the message will be deciphered?

Solution: $P(\text{deciphering})$ must be calculated. Let event A = no one deciphers the code. Then $\tilde{A}$ = at least one person will decipher the code. Hence $P(\text{deciphering}) = P(\tilde{A}) = 1 - P(A)$.

$$P(A) = P(\text{person "1" does not decipher and person "2" does not decipher and person "3" does not decipher})$$

$$P(A) = \tfrac{5}{6} \cdot \tfrac{4}{5} \cdot \tfrac{2}{3} = \tfrac{40}{90} = \tfrac{4}{9}$$

$$\therefore P(\text{deciphering}) = 1 - \tfrac{4}{9} = \tfrac{5}{9}$$

Example 4.5.9. Three dice are tossed.

a. What is the probability that all three dice show a 6?
b. What is the probability of exactly one 6 showing?

Solution:

a. $P(\text{all show a 6}) = P(\text{6 on first and 6 on second and 6 on third})$

$$P(\text{all show a 6}) = P(\text{6 on first}) \cdot P(\text{6 on second}) \cdot P(\text{6 on third})$$
$$= \tfrac{1}{6} \cdot \tfrac{1}{6} \cdot \tfrac{1}{6} = \tfrac{1}{216}$$

b. $P(\text{exactly one 6}) = P(\underbrace{6,\tilde{6},\tilde{6}} \text{ or } \underbrace{\tilde{6},6,\tilde{6}} \text{ or } \underbrace{\tilde{6},\tilde{6},6})$, where, for instance,

These are mutually exclusive events.

$6, \tilde{6}, \tilde{6}$ means six on the first die and something else on the other two.

$$P(\text{exactly one 6}) = P(6,\tilde{6},\tilde{6}) + P(\tilde{6},6,\tilde{6}) + P(\tilde{6},\tilde{6},6)$$
$$= P(6) \cdot P(\tilde{6}) \cdot P(\tilde{6}) + P(\tilde{6}) \cdot P(6) \cdot P(\tilde{6}) + P(\tilde{6}) \cdot P(\tilde{6}) \cdot P(6)$$
$$= \tfrac{1}{6} \cdot \tfrac{5}{6} \cdot \tfrac{5}{6} + \tfrac{5}{6} \cdot \tfrac{1}{6} \cdot \tfrac{5}{6} + \tfrac{5}{6} \cdot \tfrac{5}{6} \cdot \tfrac{1}{6}$$
$$= 3(\tfrac{25}{216}) = \tfrac{75}{216}$$

Many probability questions can be answered directly by using some of the counting techniques developed in chapter 3. Consider the set of digits $\{0,1,2,3,4,5,6\}$. Suppose an experiment involves making a four-digit number. What is the probability that such a number will be even? The sample space consists of all possible ways to do the experiment. That's $6 \cdot 7 \cdot 7 \cdot 7 = 2{,}058$ sample points that are equally likely. In the beginning of section 4.4, we pointed out that when sample points are equally likely,

$$P(A) = \frac{\text{number of ways to get a success}}{\text{total number of points in sample space}}$$

Here, success is getting an even four-digit number. We can do that in $6 \cdot 7 \cdot 7 \cdot 4 = 1{,}176$ ways. So $P(\text{a four-digit number being even}) = \tfrac{1176}{2058}$.

Example 4.5.10. The Venal Corporation has an East Coast plant and a West Coast plant. The East Coast plant has 100 employees, the West Coast plant has 150 employees. Twenty persons are to be laid off by a random drawing.

a. What is the probability they will all be from the same plant?
b. What is the probability that there will be ten from each plant chosen?

Solution:

a. The total number of points in the sample space is $\binom{250}{20}$. The twenty to be laid off must all come from the East Coast plant OR the

West Coast plant. That can happen in $\binom{100}{20} + \binom{150}{20}$ ways. Hence the desired probability is:

$$\frac{\binom{100}{20} + \binom{150}{20}}{\binom{250}{20}}$$

b. We need ten from the East Coast plant AND ten from the West Coast plant. That can be done in $\binom{100}{10} \cdot \binom{150}{10}$ ways. So the desired probability is

$$\frac{\binom{100}{10} \cdot \binom{150}{10}}{\binom{250}{20}}$$

Example 4.5.11. During an emergency operation, Marcus Willoby (who has no medical or surgical experience) cuts three nerves. He has forgotten which nerves pair up. Presuming all nerve hook-ups are equally likely, what is the probability that he gets the nerves hooked up correctly?

Solution: There are six nerve endings and Marcus must correctly pick three pairs. This is similar to assigning six people to three dormitory rooms, two per room, which can be done in $\binom{6}{2} \cdot \binom{4}{2} \cdot \binom{2}{2} = 90$ ways (see example 3.4.9). When applied to pairing nerves, we must divide the 90 by 3! because we are not concerned with the order of the three pairs. There is only one correct pairing, so the desired probability is $\frac{1}{15}$.

EXERCISES ■ SECTION 4.5.

1. An experiment involves tossing two dice. Let:

   event A = sum on faces is nine
   event B = exactly one of the dice is a two
   event C = doubles (both dice the same)

   Find:

   a. $P(A)$ **b.** $P(B)$ **c.** $P(C)$ **d.** $P(A \text{ and } B)$

e. $P(A \text{ and } C)$ **f.** $P(B \text{ and } C)$

g. $P(A \text{ or } B)$
h. $P(A \text{ or } C)$
i. $P(B \text{ or } C)$
(g–i) Do these by using the formula in exercise 10 of section 4.4.

j. Odds in favor of C

2. An experiment involves tossing two dice. Possible sums that can occur are 2, 3, 4, . . . , 12. Find the probability of each sum.

3. Hank, Gordon, and Kevin are policemen. At the police academy's target range, Hank hits the target 4 out of every 5 times, Gordon hits it 2 out of every 5 times, and Kevin hits it 3 out of every 4 times. They all fire one shot at a target simultaneously.

a. What is the probability the target was hit by all three policemen?
b. What is the probability the target was missed?
c. What is the probability the target was hit (at least once)?

4. Two integers are selected from the set $\{1,2,3,4,5\}$. Find the probability that at least one of the integers is even. (Presume that the second selection can be a repeat of the first.)

5. a. A coin is tossed and then it is tossed a second time. Are these events independent?
b. A card is selected from a regular deck of 52 cards. It is not replaced and a second card is selected. Let event A be the occurrence of an ace on the first card and event B the occurrence of a king on the second card. Are events A and B independent?

6. A bag contains 4 black and 5 blue marbles. A marble is drawn and then replaced, after which a second marble is drawn. What is the probability that the first is black and the second blue?

7. If, in exercise 6, the first marble drawn is not replaced before the second is drawn, what is the probability that the first is black and the second is blue?

8. Alice and Sean are salespersons at two different companies. The probability of Alice winning the monthly sales contest at her company is $\frac{1}{3}$. The probability of Sean winning the contest in his company is $\frac{1}{5}$. What is the probability that:

a. both will win the contests?
b. neither will win the contest?
c. exactly one will win the contest?

9. Of the 100 salespersons of Goodyear Publishing Company, 80 are rated as excellent, 15 as good, four as fair, and one as poor. Five salespeople are randomly selected to attend the annual convention. What is the probability that:

a. all five are excellent?
b. none of the five are excellent?
c. of the five, none are good or excellent?
d. all are good?

10. Three coins are tossed. What is the probability that:

a. all fall tails?
b. there will be exactly two heads?

11. A library has three copies of a certain book. If five people use copies of the book at different times, what is the probability that they all use the same copy?
12. A small insurance company has written theft insurance for two different businesses. In any one year, the probability that business A is burglarized is .01. In any one year, the probability that business B is burglarized is .15. (Assume these are independent events.) Find the probability that:

a. both will be burglarized this year.
b. neither will be burglarized this year.
c. exactly one will be burglarized this year.

13. An airplane has two engines. Each engine has a probability of failure of .02. The plane will crash when one or both engines fail. What is the probability of crashing? What is the probability of not crashing?
14. An urn contains 7 red marbles and 3 white marbles. Three marbles are drawn from the urn, one after the other without replacement. Find the probability that the first two are red and the third is white.
15. A box contains 4 red and 3 blue poker chips.

a. Three are selected randomly (with replacement). Find the probability that all three are red.
b. Three are selected randomly (without replacement). Find the probability that all three are red.

16. A coin is "fixed" so that on a single toss, $P(H) = \frac{3}{5}$ and $P(T) = \frac{2}{5}$.

a. List the sample space for tossing the coin twice.
b. Assign probabilities to each of the four sample points.
c. Find $P(2H)$.
d. Find $P(1H)$.
e. Find $P(0H)$.

17. The probability that a certain door is locked is 1/2. The key to the door is one of nine keys hanging on a key rack. Two keys are

selected at random before approaching the door. What is the probability that the door can be opened without returning for another key? (Hint: let event A = door is locked and you do not select correct key.)

18. A box of ten stopwatches contains two that are defective. You select three stopwatches at random. What is the probability that all three of the stopwatches work properly?

19. A committee of five is selected from six lawyers, seven engineers, and four doctors. What is the probability that all on the committee are of the same profession?

20. Six married couples are playing in a room. If two people are chosen at random, find the probability that:

a. they are married to each other.
b. one is male and one is female.

Problems 21–25 refer to the game of poker previously seen in example 3.6.1.

21. What is the probability of being dealt a straight flush?
22. What is the probability of being dealt a full house?
23. What is the probability of being dealt a straight?
24. What is the probability of a "bust"?
25. What is the probability of at least one pair?
26. When a "fixed" die is tossed, assume that the likelihood of a side's occurrence is proportional to its surface area. If this die (unlike most, which are cubes) is a rectangular solid with dimensions 2 cm by 3 cm by 4 cm, use the table below to calculate the probabilities when a pair of these dice are tossed:

No. of dots	Dimensions of side	Theoretical relative frequency or probability
1	2 cm × 4 cm	$\frac{8}{52}$
2	2 cm × 3 cm	$\frac{6}{52}$
3	3 cm × 4 cm	$\frac{12}{52}$
4	3 cm × 4 cm	$\frac{12}{52}$
5	2 cm × 3 cm	$\frac{6}{52}$
6	2 cm × 4 cm	$\frac{8}{52}$

Find the probability that when a pair of these dice is tossed you will obtain:

a. a sum of twelve
b. a sum of seven
c. doubles

d. either doubles or a sum of seven
e. either doubles or a sum of eight, but not both

Exercise 27 is an illustration of the "sometimes not-so-obvious" nature of probability.

27. The **classical birthday problem**
In order to illustrate that probability is not always a self-evident phenomenon, the *classical birthday problem* is often examined. Suppose there are twenty-five people in a room. What is the probability B that at least one pair will have the same birthday? (Hint: Look, of course, at the complement event: $\tilde{B}$ = all have different birthdays. Then calculate $1 - P(\tilde{B})$. You'll find that this happens greater than 50% of the time—something that's not, perhaps, too obvious!)

4.6 CONDITIONAL PROBABILITY

The sample space for the experiment of tossing two coins is $\{HH,HT,TH,TT\}$ and the probability that both are heads is 1/4. When asked, "What is the probability of obtaining two heads KNOWING THAT one of the coins is a head?" we are asked to find the probability of an event that is affected by an added stipulation or condition. Such a query leads to the study of **conditional probabilities** and involves the notion of information being used (perhaps) to alter a sample space.

For example, in the illustration of the coins, above, we must first realize that the following two questions certainly ask different things:

1. What is the probability of obtaining two heads?
2. What is the probability that two heads occur, knowing that one of the coins is a head?

In fact, the sample spaces are different. Because, in question (2), we KNOW that one of the coins must be a head, our sample space becomes $\{HH, HT, TH\}$ and TT is no longer a logical possibility. Hence, the probability of obtaining two heads under the condition that one is a head, is 1/3.

Let us look at another illustration. The following Venn diagram was obtained on page 79. Here M = the set of students taking marketing,

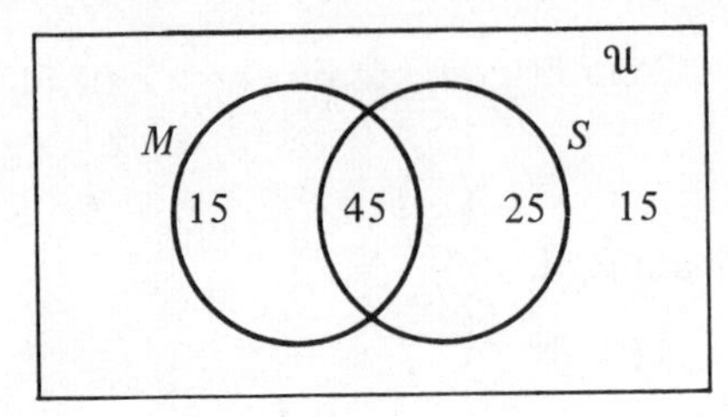

and S = the set of students taking science. The probability that a student selected at random is taking both marketing and science is, of course, 45/100. Suppose we know something about that student. Suppose we know that he is taking science. Under that condition, what is the probability he is taking both marketing and science? Since we know he is taking science, he must be one of the 70 (= 45 + 25) students pictured below:

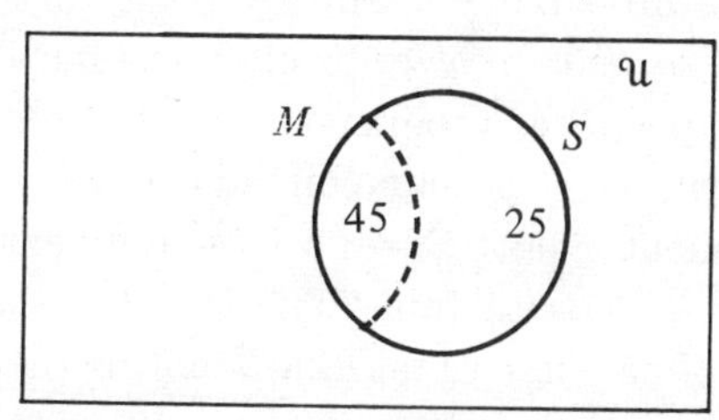

Hence, knowing that a student selected at random is taking science, the probability that he is taking both marketing and science is 45/70.

Example 4.6.1. Two dice are tossed.

a. What is the probability that the sum of the dice is 8, knowing that doubles has occurred?
b. What is the probability a 2 appears, knowing the sum of the dice is 7?
c. What is the probability a 5 occurs, knowing both dice are 4s?

Solution:

a. The altered sample space is $\{(1,1), (2,2), (3,3), (4,4), (5,5), (6,6)\}$. One of those six pairs has the sum of 8. Hence, P(sum is 8 knowing doubles has occurred) = 1/6.
b. The altered sample space is $\{(1,6), (2,5), (3,4), (4,3), (5,2), (6,1)\}$, so P(a 2 appears knowing the sum is 7) = 2/6.
c. The sample space becomes $\{(4,4)\}$, so P(a 5 occurs knowing both dice are 4s) = 0.

Example 4.6.2. The Students for the Environment Club has seven members: Bess, Eleanor, Harry, Charlotte, Laura, Edna, and Tom. A committee of three is to be elected. What is the probability the committee consists of three girls, knowing that Bess is on the committee?

Solution: Without the use of the sample space, it is a straightforward counting problem to see that there are $\binom{6}{2} = 15$ committees containing Bess. However, only $\binom{4}{2} = 6$ committees contain three girls including Bess. Hence, the desired probability is $\dfrac{\binom{4}{2}}{\binom{6}{2}} = \dfrac{6}{15}$.

To save space, we use a notational abbreviation. The **probability of event A knowing event B has occurred** shall be denoted $P(A \mid B)$. With equally likely sample points, it is true that $P(A \mid B) = \dfrac{n(A \cap B)}{n(B)}$, for this is just a symbolic way of expressing conditional probability as calculated in the two previous examples. It makes sense, because the denominator is the total number of ways in which B can occur, while the numerator is the number of ways A AND B occur. The formula can be adjusted to read:

$$P(A \mid B) = \frac{P(A \cap B)}{P(B)}$$

because of the following algebraic manipulation.

Let N = the total number of ways in which the experiment can be performed (i.e., the total number of sample points in the sample space). Then

$$P(A \cap B) = \frac{n(A \cap B)}{N}$$

$$P(B) = \frac{n(B)}{N}$$

which leads to:

$$P(A \mid B) = \frac{n(A \cap B)}{n(B)} = \frac{\dfrac{n(A \cap B)}{N}}{\dfrac{n(B)}{N}} = \frac{P(A \cap B)}{P(B)}$$

The handiness of the formula will become apparent in the examples to follow.

Example 4.6.3. Let M = the set of males, C = the set of colorblind people. Suppose $P(M \cap C) = 7/100$, $P(\tilde{M} \cap C) = 1/1000$, $P(M) = \frac{1}{2}$.

a. A colorblind person is chosen at random; what is the probability that person is a female?

b. A female is chosen at random; what is the probability she is colorblind?

Solution: A Venn diagram proves helpful but not essential. The reader should verify the numbers.

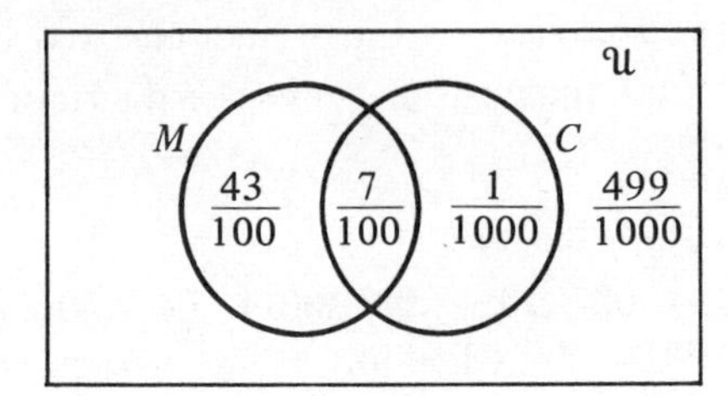

$$\text{a. } P(\tilde{M} \mid C) = \frac{P(\tilde{M} \cap C)}{P(C)} = \frac{1/1000}{71/1000} = \frac{1}{71}$$

$$\text{b. } P(C \mid \tilde{M}) = \frac{P(C \cap \tilde{M})}{P(\tilde{M})} = \frac{1/1000}{1/2} = \frac{1}{500}$$

Example 4.6.4. Let $P(P) = \frac{8}{16}$, $P(Q) = \frac{9}{16}$, and $P(P \cup Q) = \frac{11}{16}$. Find:

a. $P(P \mid Q)$
b. $P(Q \mid P)$
c. $P(P \mid \tilde{Q})$

Solution: A Venn diagram depicting the associated probabilities is appropriate.

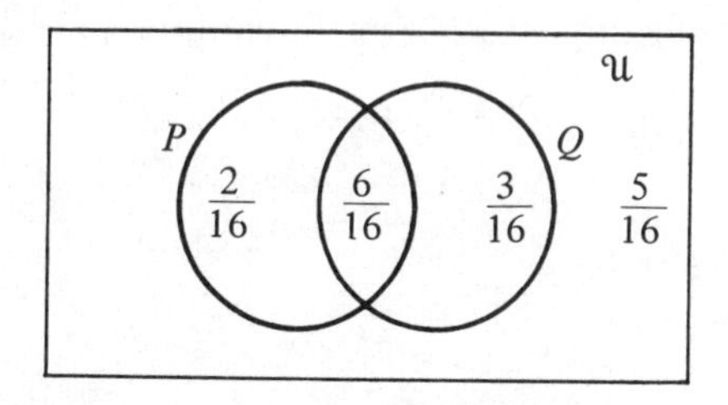

$$\text{a. } P(P \mid Q) = \frac{P(P \cap Q)}{P(Q)} = \frac{\frac{6}{16}}{\frac{9}{16}} = \frac{6}{9}$$

$$\text{b. } P(Q \mid P) = \frac{P(Q \cap P)}{P(P)} = \frac{\frac{6}{16}}{\frac{8}{16}} = \frac{6}{8}$$

$$\text{c. } P(P \mid \tilde{Q}) = \frac{P(P \cap \tilde{Q})}{P(\tilde{Q})} = \frac{\frac{2}{16}}{\frac{7}{16}} = \frac{2}{7}$$

Example 4.6.5. At the PDQ Super Market, there are three checkout counters serviced by Al, Barbara, and Carl. They handle 20%, 30%, and 50% of all customers, respectively. It is known that 2% of Al's customers, 3% of Barbara's customers, and 4% of Carl's customers have purchased sirloin steak. A customer is chosen at random after being checked out. What is the probability the customer was checked out by Carl, if you know the customer has sirloin steak?

Solution: Let C be the event that the customer was checked out by Carl, and let S be the event that the customer purchased sirloin steak. Then

$$P(C) = .50$$

$$P(S) = .033 \qquad (.033 = .02 \cdot .20 + .03 \cdot .30 + .04 \cdot .50)$$

$$P(C \cap S) = .5 \cdot .04 = .02$$

We now must find $P(C \mid S)$.

$$P(C \mid S) = \frac{P(C \cap S)}{P(S)} = \frac{.020}{.033} = \frac{20}{33}$$

EXERCISES ■ SECTION 4.6.

1. A survey of 25 people in an urban area reveals that 10 of them have televisions (T), 18 of them have automobiles (A), and 7 of them have both.

   a. Make a Venn diagram depicting the various regions and numbers of people in each.
   b. Make a Venn diagram depicting the associated probabilities in each of the regions.

   A person is selected at random.

   c. Find $P(A \mid T)$.
   d. Find $P(T \mid A)$.

2. At the Sam Houston Institute of Physiology, there are 1000 students, 800 of whom are males. Furthermore, 250 males and 85 females are studying mathematics. A student is selected at random. What is the probability the student is male, knowing that the student is *not* studying mathematics?

3. In a survey of 60 college students suffering from illnesses, data was compiled and summarized in the Venn diagram below. Here, R = the set of students with ringworm, S = the set of students with syphilis, and T = the set of students with tuberculosis.

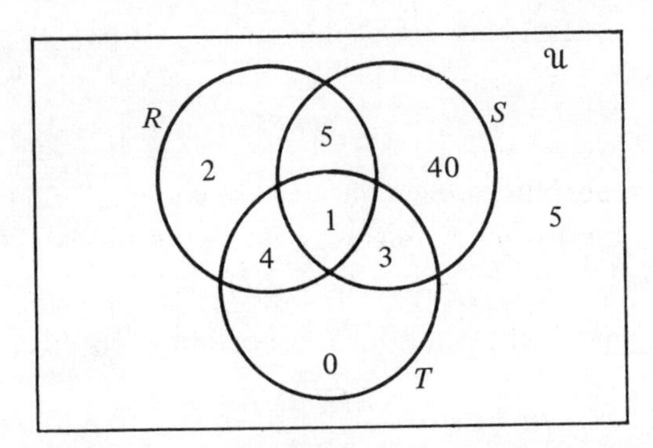

   a. What is the probability a student chosen at random has syphilis, knowing he has tuberculosis?
   b. Find $P(\tilde{T} \mid R)$.
   c. Find $P(T \mid R \cup S)$.

4. A letter is chosen at random from the word "sassafrassness."

a. What is the probability it is an s, given it is a consonant?
b. What is the probability it is an s, given it is a vowel?

5. The Johnson family has three children, ages 2, 3, and 19. (Boys and girls are equally likely.)

a. Given that the two-year-old child is a boy, what is the probability all are boys?
b. Given that at least one is a boy, what is the probability all are boys?

6. Let $P(A \cap B) = 1/4$, $P(A) = 7/12$, $P(B) = 5/12$.

a. Find $P(A \mid B)$.
b. Find $P(B \mid A)$.
c. Find $P(\tilde{A} \mid \tilde{B})$.
d. Find $P(A \cup B \mid \tilde{A} \cup \tilde{B})$.

7. Find the probability that the next person you meet was born on a Saturday if you know that he was born on either Friday, Saturday, or Tuesday.

8. Three cards are selected from a regular deck of playing cards.

a. What is the probability of obtaining three clubs, knowing that one of the three is the king of clubs?
b. What is the probability of obtaining exactly two queens, knowing one of the cards is the queen of hearts?

9. Suppose the 100 U.S. senators are projected to be categorized as follows in the year 2001:

	Republican	Democrat	Other
Men	31	29	2
Women	2	34	2

a. What is the probability a Senator is a woman?
b. Knowing the Senator is a woman, what is the probability she is a Republican?

10. Show what happens to the conditional probability formula

$$P(A \mid B) = \frac{P(A \cap B)}{P(B)}$$

when A and B are independent events. Doesn't it make sense, considering what the word independent means?

11. An advertising agency conducted a poll prior to selecting a 9:00

television program to sponsor. The results of the poll are depicted below.

	Watched channel 8 at 9:00	Watched channel 10 at 9:00	Didn't watch TV at 9:00
City I	600	750	150
City II	500	750	250
City III	10,000	21,000	3,500

Assume there are only these two channels in each of the three cities.

a. Knowing that a person watched TV at 9:00, what is the probability he watched channel 8?
b. Knowing that a person watched TV at 9:00 and lives in city III, what is the probability he watched channel 8?
c. Knowing that a person watched channel 8, what is the probability he lives in city III?
d. Knowing that a person didn't watch TV at 9:00, what is the probability he lives in city III?
e. Knowing that a person didn't watch channel 8, what is the probability he watched channel 10?

12. What is the probability that the next woman you meet was born in September,

a. knowing that her astrological sign is Virgo?
b. knowing that her astrological sign is Gemini?

13. At Brockport State College, 70% of the student population are female, 18% of the males are black, and 21% of the females are black. Suppose a black student is chosen at random. What is the probability that person is female?

14. Two dice are tossed. Let event A = the sum equals seven. Let event B = the sum equals eight. Let event C = the first die is a three.

a. Find $P(A \mid B)$. **b.** Find $P(A \mid C)$.

15. Lola, Nick, and Monica work on a brassiere production line. They work on 30%, 25%, and 45%, of all brassieres, respectively. Of their output, 4%, 2%, and 3% are defective, respectively. A brassiere is chosen at random and found to be defective.

a. What is the probability it was on Lola's line?
b. What is the probability it was on Nick's line?
c. What is the probability it was on Monica's line?

4.7 AN APPLICATION TO GENETICS

It is the purpose of this section to study a marriage of two sciences: biology and mathematics. That branch of biology called **genetics** concerns itself with heredity. It also provides a model for the mathematician studying probability. Keep in mind, please, that we present simplifications here.

We begin by making some general assumptions. First, any inherited trait is determined by genes. The genes are carried by both parents (we will call them mother and father). The offspring possess two sets of genes, one from each parent, constituting the offspring's **genotype** for that trait. For example, suppose Hank and Carol are married and Carol is pregnant. We wish to examine the inherited trait of eye color. As soon as the sperm and egg united, the genotype of the baby was determined. If the sperm carried a brown-eyed gene (B) and the egg possessed a blue-eyed gene (b), then the baby's genotype shall consist of one brown-eyed gene and one blue-eyed gene (Bb). In the case of eye color, we say that brown is the **dominant characteristic** and blue is the **recessive characteristic,** meaning that when both genes are present, the baby's eye color will be the dominant one, brown. (Think, in the Bb genotype, as though the B "overpowers" the b.) The physical appearance of an individual trait is referred to as the **phenotype.** For our eye-color simplification, three possible genotypes and two possible phenotypes are as witnessed below.

Genotype	Phenotype	Biological term
BB	Brown	pure dominant (or *homozygous* brown)
Bb	Brown (because of the dominant B)	hybrid (or *heterozygous*)
bb	Blue (note absence of brown gene)	pure recessive (or *homozygous* blue)

Every expectant mother thinks about what her child will look like when it's born. Will it have blue eyes and dark hair or brown eyes and light hair? This is a question in probability and we slightly simplify and rephrase it to read: "What is the probability my baby will have blue eyes?" To answer it, we must know the parents' genotypes.

Suppose the father has blue eyes; his genotype is bb. Suppose the mother is hybrid (Bb). To determine the offspring's possible genotypes, we look at the following table:

		Mother	
		B	b
Father	b		
	b		

Since one of the father's genes with one of the mother's genes determines the offspring's genotype, we fill in the chart (known as a **Punnett square**) as follows:

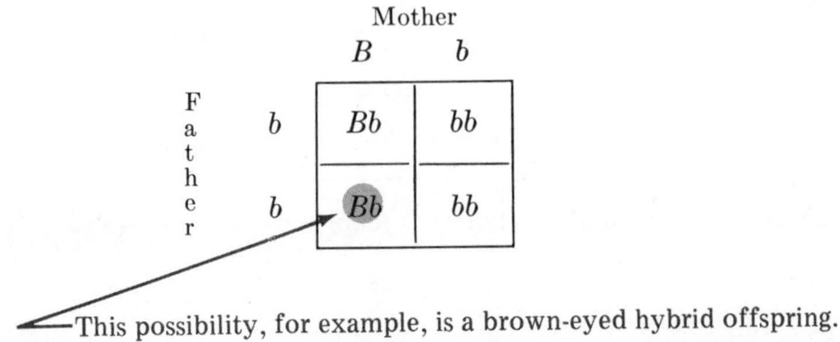

There are four possibilities for the offspring's genotype, two of which are bb. Hence, the offspring (of Bb and bb parents) has a $\frac{2}{4} = \frac{1}{2}$ probability of having blue eyes.

Note, there are six possible matings that can occur among the three genotypes (Bb with bb, Bb with Bb, bb with bb, BB with Bb, BB with BB, BB with bb). One of those six, Bb with bb, has been examined. We examine the other five in the following example.

Example 4.7.1. What is the probability an offspring has brown eyes if:

a. the parents are Bb and Bb?
b. the parents are bb and bb?
c. the parents are BB and Bb?
d. the parents are BB and BB?
e. the parents are BB and bb?

Solution:

a. We fill in the chart:

	B	b
B	BB	Bb
b	Bb	bb

Hence, the probability of the offspring having brown eyes is $3/4$.

b.

	b	b
b	bb	bb
b	bb	bb

Here, the probability is $\frac{0}{4} = 0$. Two blue-eyed parents cannot have a brown-eyed offspring.

c.

	B	B
B	BB	BB
b	Bb	Bb

Here, the probability that the offspring has brown eyes is $\frac{4}{4} = 1$.

d.

	B	B
B	BB	BB
B	BB	BB

Again, P(offspring having brown eyes) $= 1$.

e.

	B	B
b	Bb	Bb
b	Bb	Bb

Here, the probability of having a brown-eyed child when the parents are BB and bb is 1.

We summarize this example in the table on the facing page.

The sex of an individual is determined in a way very similar to eye color. Call the sex-determining genes x for the female characteristic and y for the male characteristic. Every father is xy and every mother is xx, so we can see the 50:50 ratio of men to women in the Punnett square below.

	x	x
x	xx	xx
y	xy	xy

Parents	Genotype of offspring	Probability of offspring genotype	Phenotype of offspring	Probability of offspring phenotype
BB with BB	BB Bb bb	1 0 0	brown blue	1 0
BB with Bb	BB Bb bb	$\frac{2}{4}$ $\frac{2}{4}$ 0	brown blue	1 0
BB with bb	BB Bb bb	0 1 0	brown blue	1 0
Bb with Bb	BB Bb bb	$\frac{1}{4}$ $\frac{2}{4}$ $\frac{1}{4}$	brown blue	$\frac{3}{4}$ $\frac{1}{4}$
Bb with bb	BB Bb bb	0 $\frac{2}{4}$ $\frac{2}{4}$	brown blue	$\frac{1}{2}$ $\frac{1}{2}$
bb with bb	BB Bb bb	0 0 1	brown blue	0 1

Thus, there is a $\frac{1}{2}$ probability of a male offspring.

In genetics, there is the law of **independent assortment,** which states that genes representing different traits unite independently. So the probability of two independent traits occurring together is found by multiplying the probabilities of their occurring separately. We use this law in the next example.

Example 4.7.2. What is the probability of a blue-eyed father (bb) and a brown-eyed mother (Bb) having a blue-eyed son?

Solution: We have $P(\text{blue-eyed son}) = P(\text{blue-eyed male}) = P(\text{blue-eyed}) \cdot P(\text{male})$, because of independence. Thus

$$P(\text{blue-eyed son}) = \tfrac{1}{2} \cdot \tfrac{1}{2} = \tfrac{1}{4}$$

Example 4.7.3. Suppose two parents, both *Bb*, plan to have three children. What is the probability that all will be blue-eyed males?

Solution: P(all are blue-eyed males) $= P$(1st is blue-eyed male AND 2nd is blue-eyed male AND 3rd is blue-eyed male) $= P$(1st is blue-eyed male) $\cdot$ P(2nd is blue-eyed male) $\cdot$ P(3rd is blue-eyed male). Now, P(1st is blue-eyed male) $= P$(2nd is blue-eyed male) $= P$(3rd is blue-eyed male) $= P$(blue-eyed) $\cdot$ P(male) $= \frac{1}{4} \cdot \frac{1}{2} = \frac{1}{8}$.

So, P(all are blue-eyed males) $= \frac{1}{8} \cdot \frac{1}{8} \cdot \frac{1}{8} = \frac{1}{512}$

The concepts we've used so far for human beings can be extended, of course, to most life forms in the animal kingdom, and even to plants. In fact, Gregor Mendel, the father of genetics, did his original experiments with garden peas. We discuss the genetics of the garden pea in the following example.

Example 4.7.4. Let's examine three traits: plant height, seed texture, and seed color. We let:

$$\begin{aligned} T &= \text{tall plant} \\ t &= \text{short plant} \\ R &= \text{round seed} \\ r &= \text{wrinkled seed} \\ Y &= \text{yellow seed color} \\ y &= \text{green seed color} \end{aligned}$$

Here a capital letter denotes dominance. When a ($TtRryy$) is crossed with a ($TtRRYy$), what is the probability of obtaining a plant with tall, round, green traits?

Solution: Because of the law of independent assortment in biology, and the nature of independent events in probability, P(tall and round and green) $= P$(tall) $\cdot$ P(round) $\cdot$ P(green), and we examine each of P(tall), P(round), P(green) separately.

To find P(tall), we check the Punnett square for $Tt \times Tt$, where the multiplication sign "$\times$" denotes "crossed with."

	T	t
T	TT tall	Tt tall
t	Tt tall	tt short

So, $P(\text{tall}) = 3/4$. To find $P(\text{round})$, we check the Punnett square for $Rr \times RR$.

	R	r
R	RR round	Rr round
R	RR round	Rr round

So, $P(\text{round}) = 1$. To find $P(\text{green})$, we check the Punnett square for $yy \times Yy$.

	y	y
Y	Yy yellow	Yy yellow
y	yy green	yy green

So, $P(\text{green}) = \frac{2}{4} = \frac{1}{2}$. Finally,

$$P(\text{tall and round and green}) = \tfrac{3}{4} \cdot 1 \cdot \tfrac{2}{4} = \tfrac{6}{16} = \tfrac{3}{8}.$$

Speaking of plants, one of our favorites is the four o'clock flower. Consider the trait of color, where:

$$R = \text{Red flower}$$
$$r = \text{white flower}$$

In the four o'clock flower, **codominance** (sometimes called **incomplete dominance**) occurs. By codominance, we mean the hybrid Rr will not be red but will be pink. (Think of codominance as a "blending" of the phenotypes.) We discuss this in the example following.

Example 4.7.5. Suppose a white (rr) and a red (RR) four o'clock flower are crossed. Determine the probabilities of all possible phenotypes of such a cross. (The results of a cross are sometimes referred to as the **F_1 generation** or **first filial generation.**) Then cross two of the F_1 generation and determine the probabilities of all possible phenotypes. (The results of a cross between two F_1 generations is referred to as the F_2 generation.)

Solution: We use a Punnett square to determine the probabilities of RR crossed with rr.

	R	R
r	Rr pink	Rr pink
r	Rr pink	Rr pink

Hence, the only occurring phenotype is pink, and it occurs with probability 1. To determine the phenotypic probabilities of the F_2 generation, we must cross Rr with Rr:

	R	r
R	RR red	Rr pink
r	Rr pink	rr white

Hence, the probability of red in the second generation = 1/4. The probability of a pink flower in the second generation is 1/2 and the probability of white in the second generation is 1/4.

We conclude this section with another example on codominance.

Example 4.7.6. A classical example of codominance occurs in a certain breed of cattle. Their coat color can be red (RR), white (rr), or roan (Rr), which is a red-white mixture. Suppose two cattle are crossed, a roan and a white, to produce an F_1 generation, and then two of the F_1 generation are crossed to produce an F_2 generation. What is the probability of an F_2 generation offspring being roan?

Solution: First we must produce the F_1 generation by crossing Rr with rr.

	R	r
r	Rr roan	rr white
r	Rr roan	rr white

We see that 1/2 of the F_1 generation is roan; 1/2 is white. To create the F_2 generation, we must take all possible matings in F_1. There are four cases:

1. mate a roan male with a roan female,
2. mate a roan male with a white female,
3. mate a white male with a roan female,
4. mate a white male with a white female.

} The probability of any one case is 1/4.

1. We mate a roan male with a roan female.

	R	r
R	RR red	Rr roan
r	Rr roan	rr white

So the probability we get roan is 1/2 in this case. Since this case occurs only 1/4 of the time, the probability of getting the roan from the beginning is $\frac{1}{4} \cdot \frac{1}{2} = \frac{1}{8}$.

2. We mate a roan male with a white female.

	R	r
r	Rr roan	rr white
r	Rr roan	rr white

One-half of these are roan, but this case occurs only 1/4 of the time, so the probability of obtaining a roan from the beginning is $\frac{1}{4} \cdot \frac{1}{2} = \frac{1}{8}$.

3. We mate a white male and a roan female. The argument is the same as case 2 above. The probability of roan $= 1/8$.

4. We mate a white male with a white female.

	r	r
r	rr white	rr white
r	rr white	rr white

The probability of roan in this case is 0.

Finally, the probability of an F_2 roan occurrence is $\frac{1}{8} + \frac{1}{8} + \frac{1}{8} + 0 = \frac{3}{8}$.

For the reader who is interested in knowing more about the mathematics of genetics, we cite the reference: *Biosphere: A Study of Life* by N. M. Jessop (Englewood Cliffs, N.J.: Prentice-Hall, 1970).

EXERCISES ■ SECTION 4.7.

In problems 1 and 2 let:

D = dark hair (dominant)
d = blond hair
B = brown eyes (dominant)
b = blue eyes

1. If the mother is $DdBb$ and the father is $ddbb$, what is the probability that:

a. an offspring will have blue eyes?
b. their next child will be a blond-haired, blue-eyed girl?
c. if they have seven children, all will have blond hair and blue eyes?
d. if they have seven children, all will be blond-haired, blue-eyed girls?

2. a. If the mother, Mrs. Jones, is $ddbb$ and the father, Mr. Jones, is $DDBB$, what is the probability that their child has brown hair?
b. Suppose their child marries a $Ddbb$. What is the probability the Jones' grandchild will have blue eyes?

3. In example 4.7.6, suppose a roan is crossed with a roan. Calculate phenotypic probabilities in the F_1 generation. Calculate phenotypic probabilities in F_2. (Form F_2 by mating all possible F_1's with each other.)

4.8 EXPECTATION

Arlo and Bob are playing a game of coin flipping. It goes like this: each flips a coin and if both turn up heads, Arlo gives Bob $1.00. Otherwise, Bob gives Arlo 30¢. A question that can quite naturally arise is, "Who has the advantage?" or "How much will the person with the advantage make per trial (on the average)?" To study the answers to these questions, the concept of **expectation** (or **expected value**) is introduced. Expectation is the value of a game or experiment averaged over a large number of trials. We now illustrate.

Presume that Arlo and Bob play the coin flipping game 200 times. For any one of these games, the possible outcomes are HH, HT, TH, TT. Since each of these is equally likely, each would tend to appear 50 times (in 200 games). The table below summarizes the money transactions.

Possibility	Arlo (per game)	Arlo (in total)	Bob (per game)	Bob (in total)
HH appears 50 times	loses $1.00 each time	loses $50 in all	wins $1.00 each time	wins $50 in all
HT appears 50 times	wins 30¢ each time	wins $15 in all	loses 30¢ each time	loses $15 in all
TH appears 50 times	wins 30¢ each time	wins $15 in all	loses 30¢ each time	loses $15 in all
TT appears 50 times	wins 30¢ each time	wins $15 in all	loses 30¢ each time	loses $15 in all
This is what would tend to happen		Arlo's net losses are $5		Bob's net winnings are $5

Since Arlo's net losses are $5 (for 200 games), then ON THE AVERAGE he loses $2\frac{1}{2}$¢ per game ($5 ÷ 200). So Arlo's expectation is $-2\frac{1}{2}$¢. (The minus sign symbolizes the loss.) In actuality, of course, Arlo never loses $2\frac{1}{2}$¢. He either wins $1.00 or loses 30¢, but ON THE AVERAGE he loses $2\frac{1}{2}$¢ per game. Further, Bob's net winnings are $5 (for 200 games). So, ON THE AVERAGE he wins $2\frac{1}{2}$¢ per game ($5 ÷ 200). Thus, Bob's expectation is $+2\frac{1}{2}$¢. (The plus sign signifies the gain of money.) We can see, then, that for this coin flipping game, Bob has the advantage (his expectation is positive) while Arlo has the disadvantage (his expectation is negative).

Let's work another illustration with a gambling setting. Suppose Peggy approaches Maureen's Money Maker (a die game at a local carnival), where for 75¢ a roll she takes her chances at winning $1.00 (over and above the 75¢ bet) when she rolls either a 5 or a 6, or losing her 75¢ when she rolls a 1 or a 2 or a 3 or a 4. Again, let's presume the game is played many times (say 600). A table follows.

Possibility	Maureen (per game)	Maureen (in total)	Peggy (per game)	Peggy (in total)
A 1 appears 100 times	wins 75¢	wins $75	loses 75¢	loses $75
A 2 appears 100 times	wins 75¢	wins $75	loses 75¢	loses $75
A 3 appears 100 times	wins 75¢	wins $75	loses 75¢	loses $75
A 4 appears 100 times	wins 75¢	wins $75	loses 75¢	loses $75
A 5 appears 100 times	loses $1.00	loses $100	wins $1.00	wins $100
A 6 appears 100 times	loses $1.00	loses $100	wins $1.00	wins $100
This is what would tend to happen		Maureen's net winnings are $100		Peggy's net losses are $100

Here, Maureen's expectation is $\$100 \div 600 = \$\frac{1}{6} = 16\frac{2}{3}$¢. Thus, Peggy's expectation is $-16\frac{2}{3}$¢. So, as Maureen offers this game to Peggy and other customers, she will tend to make $16\frac{2}{3}$¢ per game (on the average from each customer).

It is worthwhile to realize that an expectation, once calculated, can be used to make a prediction. For instance, in Arlo and Bob's coin flipping game, we found Arlo's expectation equal to $-2\frac{1}{2}$¢. That meant a loss of $2\frac{1}{2}$¢ per game in the long run. This $2\frac{1}{2}$¢-per-game loss is a prediction because we are never assured that each of the possibilities HH, HT, TH, TT will occur exactly 50 times when 200 games are played. So when an expectation is calculated, just realize that you cannot use it to tell what will definitely happen in one trial of a game. It is a concept that can be used to predict when a large number of games are played. Notice also that when your expectation is positive, you will tend to win in the long run, and when your expectation is negative, you will tend to lose in the long run. An expectation can be zero (see example 4.8.3), meaning that in the long run you tend neither to win nor lose (the game is said to be **fair** to both parties).

Summarily, it is seen that we have computed expectations by first listing all possible outcomes for the game being played. Then we considered a hypothetical number of games to be played. (This does not affect the numerical value of the expectation: see exercise 1 at the end of this section.) Lastly, we listed all the money transactions in order to determine net wins and losses and then divided each by the number of games played to get the expectation. This process can be time-consuming, so we now present a formula that summarizes the above procedure and shortens the computation time. Expectation (or expected value) E is given by:

$$E = p_1m_1 + p_2m_2 + p_3m_3 + \cdots + p_km_k$$

where each p represents the PROBABILITY of an event and each m represents the NET AMOUNT of money (won or lost) when that event occurs. There are k terms in this formula; k may be different from one problem to the next (read on to see this).

Now we illustrate the use of this formula for Arlo and Bob's coin flipping game. Recall that for HH, Arlo loses \$1.00 (Bob wins the \$1.00) and for any of HT, TH, TT, Arlo wins 30¢ (Bob loses the 30¢). Now we compute Arlo's expectation, E_A. There are just two terms ($k = 2$) in this computation because there is a total of two ways to win or lose money.

$$E_A = \underbrace{\underbrace{\tfrac{1}{4}}_{\text{Probability of } HH}\underbrace{(-100¢)}_{\text{Money lost by Arlo on } HH}}_{\text{This is } p_1m_1} + \underbrace{\underbrace{\tfrac{3}{4}}_{\text{Probability of } HT,\ TH, \text{ or } TT}\underbrace{(+30¢)}_{\text{Money won by Arlo on } HT,\ TH, \text{ or } TT}}_{\text{This is } p_2m_2} = -25¢ + 22\tfrac{1}{2}¢ = -2\tfrac{1}{2}¢$$

Next we compute Bob's expectation, E_B.

$$E_B = \underbrace{\tfrac{1}{4}}_{\text{Probability of } HH}\underbrace{(+100¢)}_{\text{Money won by Bob on } HH} + \underbrace{\tfrac{3}{4}}_{\text{Probability of } HT,\ TH, \text{ or } TT}\underbrace{(-30¢)}_{\text{Money lost by Bob on } HT,\ TH, \text{ or } TT} = 25¢ - 22\tfrac{1}{2}¢ = +2\tfrac{1}{2}¢$$

Notice carefully that when money is won, we use a plus sign and when money is lost we use a minus sign. Some examples follow.

Example 4.8.1. Compute (using the expectation formula) both Maureen's and Peggy's expectation on Maureen's Money Maker game.

Solution: Let E_M be Maureen's expectation and E_P be Peggy's:

$$E_M = \tfrac{4}{6}(+75¢) + \tfrac{2}{6}(-100¢)$$
$$= 50¢ - 33\tfrac{1}{3}¢ = +16\tfrac{2}{3}¢ \qquad \text{(Checks with previous solution.)}$$
$$E_P = \tfrac{4}{6}(-75¢) + \tfrac{2}{6}(+100¢)$$
$$= -50¢ + 33\tfrac{1}{3}¢ = -16\tfrac{2}{3}¢ \qquad \text{(Checks with previous solution.)}$$

Example 4.8.2. There are 1000 tickets sold on a raffle. First prize is \$200, second prize is \$150, and each of the 2 third prizes are for \$100 each. Tickets cost \$1 apiece.

a. Suppose you buy a ticket. What is your expectation?
b. Suppose you buy two tickets. What is your expectation?
c. Do you have the advantage in this game?
d. What should be the price of a ticket in order for this game to be "fair?"

Solution: The probability of winning first prize is 1/1000, the probability of winning second prize is 1/1000, and the probability of winning third prize is 2/1000. Let E denote expectation.

a. $$E = \tfrac{1}{1000}(\$199) + \tfrac{1}{1000}(\$149) + \tfrac{1}{1000}(\$99) + \tfrac{1}{1000}(\$99) + \tfrac{996}{1000}(-\$1)$$

In this step there are two important things to notice:

1. each of the prizes is reduced by the cost of the ticket (\$1);
2. there are five terms ($k = 5$) going into the computation of E (four prizes and the concept of a loss).

Continuing,

$$E = \frac{\$199}{1000} + \frac{\$149}{1000} + \frac{\$99}{1000} + \frac{\$99}{1000} - \frac{\$996}{1000}$$
$$= -\frac{\$450}{1000} = -45¢$$

On the average, you can expect to lose 45¢ per game.

b. When you buy two tickets, you can expect to lose 90¢, so $E_2 = -90¢$. (E_2 denotes the expectation for two tickets.)
c. No; since your expectation is negative, you have the disadvantage.
d. Since your expectation is $-45¢$, you tend to lose 45¢ each time you play this raffle (on the average). Hence the proprietor of the raffle tends to win 45¢ on each ticket. Since you pay \$1.00 per ticket and 45¢ of that is clear profit to the proprietor, a price for a ticket that would make the game fair is 55¢. (See exercise 3 of this section.)

Example 4.8.3. A lottery is set up with 1500 tickets to be sold and a first prize of \$100 and a second prize of \$50. What should be the price of a ticket in order for this lottery to be fair?

Solution: We present two solutions; read both.

Method I

Proprietor pays out:
$100 for first prize
50 for second prize
$150 total
Proprietor takes in:
$1500(P)$, where P represents the price of a ticket.
For a fair game,
$1500P$ must equal $150

$$1500P = \$150$$

$$P = \frac{\$150}{1500} = 10¢$$

Method II

Let P be the price of a ticket.

$$E = \frac{1}{1500}(100 - P) + \frac{1}{1500}(50 - P) + \frac{1498}{1500}(-P)$$

But E must be zero for the lottery to be fair, so

$$\frac{1}{1500}(100 - P) + \frac{1}{1500}(50 - P) - \frac{1498P}{1500} = 0$$

$$\frac{100 - P}{1500} + \frac{50 - P}{1500} - \frac{1498P}{1500} = 0$$

$$\frac{100 - P + 50 - P - 1498P}{1500} = 0$$

$$100 - P + 50 - P - 1498P = 0$$

$$150 - 1500P = 0$$

$$-1500P = -150$$

$$P = \frac{1}{10} \text{ or } 10¢$$

The gambling situations that we have been discussing so far have ramifications not yet pointed out. For instance, at the beginning of the section we discussed Maureen's Money Maker game. Recall that she will pay out $1.00 when a player rolls a 5 or 6 and she will collect 75¢ when a player rolls 1 or 2 or 3 or 4. We calculated $E_M = 16\frac{2}{3}¢$, which told us that on the average, Maureen makes $16\frac{2}{3}¢$ per game from each player. Now let's look at the business aspect of this situation. Suppose Maureen has found that on a typical weekend, about 1200 customers play her game. How much profit can she expect to make? That's just the profit per game multiplied by the number of games played, or $(16\frac{2}{3}¢) \cdot 1200 = \$\frac{1}{6} \cdot 1200$ or $200. Here, then, is another outlook on matters. Probability and its "cousin," expectation, can be used to calculate expected profits in a business venture. The next three examples illustrate this.

Example 4.8.4. A plumbing company is bidding on a school contract that will yield a profit of $30,000 or a loss (incurred due to strikes, unavailability of supplies, etc.) of $12,000. The company's past experience indicates that the probability of making the above profit is 5/6, whereas the probability of suffering the loss is 1/6. What is the company's expectation?

Solution: $E = \frac{5}{6}(\$30{,}000) + \frac{1}{6}(-\$12{,}000) = \$25{,}000 - \$2{,}000$

So, $E = \$23{,}000$. There are some interesting interpretations. First, the company is gambling here and has a positive expectation. Hence, the company is operating from an advantage. Further, the \$23,000 can be interpreted as follows. On plumbing jobs similar to the one above, sometimes the company will make a profit of \$30,000 (about 5/6 of the time). Sometimes the company will lose \$12,000 (about 1/6 of the time). So, on the average, the company will make \$23,000 per job.

EXAMPLE 4.8.5. A mutual-fund company can sell a certain lot of stocks at a profit of \$5000 with probability of 1/5. They have a probability equal to 2/5 for selling at a loss of \$4000. The probability that they break even (no profit or loss) is 2/5.

a. What is the expectation for the company on this sale?
b. Is it to the company's advantage (mathematically) to sell?

Solution:

a. $E = \frac{1}{5}(\$5000) + \frac{2}{5}(-\$4000) + \frac{2}{5}(\$0)$
$= \$1000 - \$1600 + 0 = -\$600$

b. No, because their expectation is negative.

EXAMPLE 4.8.6. Suppose the mutual-fund company in the example above has its stock analysts go to work on this problem. They feel that in one month, there will be a probability of 1/10 for selling the lot of stocks at a profit of \$1000, a probability of 3/10 for selling the stock at a profit of \$500, a probability of 4/10 for breaking even, and a probability of 2/10 for losing \$3000. Should they wait for a month to sell (on the basis on this information)?

Solution:

$$\begin{aligned} E &= \tfrac{1}{10}(\$1000) + \tfrac{3}{10}(\$500) + \tfrac{4}{10}(\$0) + \tfrac{2}{10}(-\$3000) \\ &= \$100 + \$150 + \$0 - \$600 \\ &= \$250 - \$600 = -\$350. \end{aligned}$$

The expectation here is still negative, but numerically it is not as large as the expectation in example 4.8.5. Hence, it would seem that the company should wait for a month to sell the lot of stocks.

Another application of expected value is to the calculation of yearly premiums (cost per year) that an insurance company charges its customers. Suppose that for a certain event you have purchased an insurance policy that will pay \$2000 when the event occurs and will pay nothing if the event fails to occur. Further, suppose that the insurance company has determined that the probability of this event occurring is 7/100. (This information could result from keeping records over a long period of time.) We now treat this situation as a "bet" between you and the company. Let C represent the yearly premium you pay. Then,

$$E = \tfrac{7}{100}(2000 - C) + \tfrac{93}{100}(-C)$$

Now, we first presume that this betting situation is a fair one, so that $E = 0$. Then

$$\frac{7}{100}(2000 - C) + \frac{93}{100}(-C) = 0$$

$$\frac{14000 - 7C}{100} + \frac{-93C}{100} = 0$$

$$\frac{14000 - 7C - 93C}{100} = 0$$

$$14000 - 7C - 93C = 0$$

$$-100C = -14000$$

$$C = \frac{-14000}{-100} = \$140$$

So \$140 is the yearly premium you should pay for this bet to be fair. However, because insurance companies have many expenses (overhead, employees' salaries, etc.), this figure of \$140 must be increased. The amount of the increase is, of course, an accounting problem and will not be covered here.

Example 4.8.7. A contact-lens wearer wishes to purchase a \$200 breakage policy on her contact lenses (of that value). From past insurance company experience, it has been found that the probability of breaking two lenses (during a year) is 1/20, the probability of breaking one lens (during a year) is 3/20, and the probability of no breaks is 16/20. What yearly premium should she pay so that the company can break even?

Solution: Let C = cost of yearly premium. Then

$$E = \frac{1}{20}(200 - C) + \frac{3}{20}(100 - C) + \frac{16}{20}(-C)$$

This must equal zero:

$$\frac{1}{20}(200 - C) + \frac{3}{20}(100 - C) + \frac{16}{20}(-C) = 0$$

$$\frac{200 - C}{20} + \frac{300 - 3C}{20} + \frac{-16C}{20} = 0$$

$$200 - C + 300 - 3C - 16C = 0$$

$$500 - 20C = 0$$

$$C = \frac{500}{20} = \$25$$

Thus far in this section, we have dealt with expectation as it applies to monetary situations. In the formula

$$E = p_1m_1 + p_2m_2 + \cdots + p_km_k$$

the m's represent a certain amount of money. However, we can also reorient our thinking so that instead of representing a number of dollars, the m's represent many other things. The following example points this out.

Example 4.8.8. Consider a box containing six light bulbs, two of which are defective. Successive bulbs are withdrawn and tested until a good one is obtained. How many drawings can be expected?

Solution: In this case, $E = p_1m_1 + p_2m_2 + \cdots + p_km_k$, where each p represents the probability of an event and each m represents the number of draws needed for that event. Note that with a box having four good bulbs and two defective ones, we would need at most three draws before a good bulb is obtained. The probability of getting a good bulb with one draw is 4/6. The probability of getting a good bulb with exactly two draws is $\frac{2}{6} \cdot \frac{4}{5} = \frac{8}{30}$. The probability of getting a good bulb with exactly three draws is $\frac{2}{6} \cdot \frac{1}{5} \cdot \frac{4}{4} = \frac{2}{30}$. Hence,

$$\begin{aligned} E &= \tfrac{4}{6}(1) + \tfrac{8}{30}(2) + \tfrac{2}{30}(3) \\ &= \tfrac{4}{6} + \tfrac{16}{30} + \tfrac{6}{30} = \tfrac{7}{5} = 1\tfrac{2}{5} \end{aligned}$$

That is, on the average, $1\frac{2}{5}$ draws are made before a good bulb is obtained.

This type of problem can be useful for a person checking the quality of items coming off an assembly line. Suppose that a machine produces, within a three-hour period, a certain number of items with the percent of defectives known (from past experience). Thus, we could compute the expected number of selections needed to get a nondefective item (as in example 4.8.8). If we then actually do select items from the lot produced and find that it takes many more selections than expected to obtain a nondefective item, it would be sensible to infer that the machining process was out of control.

EXERCISES ■ SECTION 4.8.

1. Refer to the coin flipping game played by Arlo and Bob at the beginning of this section.
   - **a.** Construct a table like the one on page 191 for a series of 1000 such games.
   - **b.** What would Arlo's net losses be?
   - **c.** What would Bob's net winnings be?

d. What would Arlo's loss per game be (on the average)?
e. What would Bob's win per game be (on the average)?

2. Compute the amount of money that Bob should give Arlo (when Arlo wins) in order to make their coin flipping game a fair one.
3. Compute the expectation for example 4.8.2 when the price of a ticket is 55¢.
Note: The solution to part *d* of example 4.8.2 showed us that the game was fair when a ticket costs 55¢. Thus, for a fair game, $E = 0$.
4. A carnival game involves betting certain amounts of money when dice are tossed. Suppose that one of the offers is to bet $1 on the sum equaling 7 for one toss of a pair of dice. When a 7 occurs you get $5 (this includes the original $1 bet). For any other sum that occurs, you lose your dollar.

a. Compute your expectation.
b. Is it wise (mathematically) to play this game?
c. If you played this game 100 times, how much would you stand to win (lose)?

5. A national magazine offers a free lottery to 2,000,000 U.S. residents. To enter, all that need be done is to fill out a card and send it in. (You must buy an 8¢ stamp yourself.) First prize is $50,000. Second prize is $25,000. (Assume that only 1,000,000 people decide to enter the lottery.)

a. Compute your expectation.
b. Is it wise (mathematically) to buy an 8¢ stamp to enter this contest?

6. Jack and Jill went up the hill to play a game of cards. One card was chosen from the deck to see if it was a picture card (a jack, queen, or king). If it was, Jack gave Jill $1.00. If not, Jill gave Jack a quarter.

a. To whose advantage was the game?
b. What is that person's expectation?

7. It is known that whenever salesman Tim Hardsel approaches a customer, the probability he can sell the customer three or more items is .05, the probability he can sell exactly two items is .2, the probability he can sell exactly one item is .6, and the probability he sells nothing is .15. Now, Tim's commission is $50 if he can sell three or more items and $20 if he sells one or two items. Find Tim's expected commission.

8. Rebecca studies the stock market regularly. Before buying 200 shares of stock, each costing $18, she estimates that she could sell each share of the stock within six months for \$15, \$18, \$20, \$40 with probabilities .2, .5, .2, .1, respectively. If Rebecca decides to buy the stock and then resell it within six months, what is her expected profit?

9. The probability that a woman 20 years old will live for one year is .9921. She purchases a one-year life insurance policy that will pay \$10,000 upon her death during that year.

a. What is the probability that she will die during the year in question?

b. How much should she pay (neglecting company expenses) for this policy?

10. Albert is a terrible geometry student. His father plans to award him for good grades. He promises Albert \$10 for an A, \$5 for a B. On the other hand, Albert must pay his father \$4 for a grade lower than a C. Assume that $P(A) = P(B) = \frac{1}{10}$, $P(C) = \frac{2}{10}$, $P(D) = \frac{3}{10}$, $P(F) = \frac{3}{10}$. What is Albert's expectation for this wager?

11. The probability that Mr. Setek will sell his house and make a profit of \$4000 is $\frac{2}{10}$. The probability that he will sell at a profit of \$2000 is $\frac{3}{10}$. The probability that he will break even is $\frac{4}{10}$ and the probability of selling at a loss of \$1000 is $\frac{1}{10}$. What is his expected profit?

12. A building contractor is considering bidding on one of two contracts for new downtown buildings (A and B). It has been estimated that a profit of \$200,000 would be made on building A. Bidding costs for the contractor on building A would be \$10,000. On building B, the estimated profit is \$500,000 and bidding costs would be \$20,000. The probability of being awarded the contract is $\frac{2}{5}$ on building A and $\frac{1}{5}$ on building B. (Assume bidding costs are incurred only in the case the contract is *not* obtained.)

a. What is the contractor's expectation for building A?

b. What is the contractor's expectation for building B?

c. For which job should the contractor bid?

13. A box has one red, two yellow, and three green gumballs. Gumballs are selected (without replacement) until a yellow one is obtained. How many selections can be expected?

14. A truck is carrying eight cartons of meat, two of which are spoiled. A carton is selected and examined until a spoiled one is obtained. How many selections can be expected?

15. An inebriated gentlemen goes home and finds his front door locked. On his key chain he has six keys but cannot even begin to think about which one opens the door. Assuming he will not try the same key twice, what is the expected number of tries needed to find the right key?

4.9 BINOMIAL EXPERIMENTS AND BINOMIAL PROBABILITIES

This chapter has, so far, developed methods of computing probabilities for many different kinds of situations. Often times we have considered the probabilities associated with tossing coins or dice, or withdrawing certain colored balls from a box. These experiments might not really seem to be related to real life situations, but in many respects they are. Coin tossing type experiments, for instance, are conducted daily in social science, behavioral science, physical science, and industry.

To illustrate this, consider the familiar polls that are taken to determine people's preference for a certain political candidate. When asked the question: "do you prefer candidate A?" the possible response is "yes" or "no" (that's like a head or tail). The physicist may be interested in the question, "Do the X rays penetrate the unknown substance?" The answer is "yes" or "no." The civil engineer is worried about the steel I-beam construction. "Is it sufficient to support the weight of the bridge plus vehicles?" The answer here is "yes" or "no" also. The automobile manufacturer wants to determine "Has our recent advertising campaign caused people to buy more of our autos?" Again the answer is "yes" or "no."

The psychologist tests an animal to see if a learned behavior pattern can be altered or not. A new drug will prove effective or ineffective. A rocket engine will either work or not. A manufactured item selected from the assembly line will be defective or nondefective. A person insured by a life insurance company on a 20-year term policy will live or die during those 20 years.

All of these situations, in some respects, are similar. In one sense they all involve the occurrence of only one of two possible outcomes. One outcome is usually favorable in some way, while the other is not. The favorable outcome we call **success** and the other we label **failure.** In another sense, each of the hypothetical experiments mentioned must be repeated some number of times to gain information that is as concrete as possible. (For example, the rocket engine we talked about as either working or not working would be tested many times before we drew any firm conclusions about it.)

These concepts are just part of a **binomial experiment,** for which we now list all the criteria.

1. Each trial of the experiment results in one of two possible outcomes (called success or failure).

2. There are n identical trials of the experiment.
3. The trials must be independent. (We can check this by seeing if probabilities remain the same from trial to trial.)
4. We let p = probability of success on any particular trial. We let q = probability of failure on any particular trial. (Note: $p + q = 1$)
5. We are interested in x, which is the number of successes during the n trials.
6. We are interested in $P(x)$, which is the probability of obtaining x successes in the total of n trials.

The three basic properties that a binomial experiment must exhibit are properties 1, 2, 3 above. Items 4, 5, 6 are really just notational matters for the variables involved. We will now present (without proof for the time being) the **binomial probability distribution formula** that will be used in this section:

$$P(x) = \binom{n}{x} p^x q^{n-x}$$

where:

p = probability of success on a single trial of the binomial experiment
q = probability of failure on a single trial of the binomial experiment
n = total number of trials
x = the number of successes that can occur
$P(x)$ = the probability of x successes occurring out of all n trials

(Note: x can assume any of the integral values $0, 1, 2, \ldots, n$.)

Consider a first example. Suppose a coin is tossed three times (or three coins are tossed at once). This is a binomial experiment because:

1. each trial occurs as just head or tail (one of two ways),
2. each of the three tosses is identical (the coin is merely tossed three times),
3. choosing to call a head success, p = probability of a head = $\frac{1}{2}$ on each separate toss.

The number of heads (successes), x, can be 0, 1, 2, or 3 here, since $n = 3$. We can now compute:

$$P(0) = \binom{3}{0} (\tfrac{1}{2})^0 (\tfrac{1}{2})^3 = \tfrac{1}{8}$$

$$P(1) = \binom{3}{1} (\tfrac{1}{2})^1 (\tfrac{1}{2})^2 = \tfrac{3}{8}$$

$$P(2) = \binom{3}{2} (\tfrac{1}{2})^2 (\tfrac{1}{2})^1 = \tfrac{3}{8}$$

$$P(3) = \binom{3}{3} (\tfrac{1}{2})^3 (\tfrac{1}{2})^0 = \tfrac{1}{8}$$

The tossing of a die does not, at first, appear to be a binomial experiment because there are six outcomes possible (instead of two). However, we can define success as the occurrence of a 6 and failure as anything else. Hence, $p = \frac{1}{6}$ and $q = \frac{5}{6}$. Suppose an experiment is to toss a die and we choose to do it a total of two times ($n = 2$). This is a binomial experiment because:

1. there are only two possible outcomes (success = 6 and failure = non-6),
2. there are two identical trials ($n = 2$),
3. the trials are independent because $p = \frac{1}{6}$ on each trial.

The number of sixes (successes), x, can be 0, 1, or 2 here, since $n = 2$. We now compute:

$$P(0) = \binom{2}{0}\left(\tfrac{1}{6}\right)^0\left(\tfrac{5}{6}\right)^2 = \tfrac{25}{36}$$

$$P(1) = \binom{2}{1}\left(\tfrac{1}{6}\right)^1\left(\tfrac{5}{6}\right)^1 = \tfrac{10}{36}$$

$$P(2) = \binom{2}{2}\left(\tfrac{1}{6}\right)^2\left(\tfrac{5}{6}\right)^0 = \tfrac{1}{36}$$

A quick check on these computations is possible. Observe the sample space below.

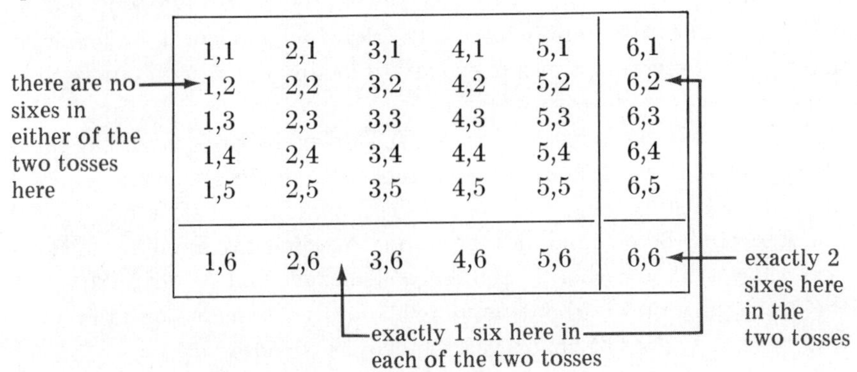

$$P(\text{no sixes}) = P(0) = \tfrac{25}{36}$$
$$P(\text{exactly one six}) = P(1) = \tfrac{10}{36}$$
$$P(\text{exactly two sixes}) = P(2) = \tfrac{1}{36}.$$

Example 4.9.1. Let a pouch contain 4 black and 3 white marbles. Two marbles are to be drawn. If the first marble is replaced before the second is drawn, find:

a. P(both black)
b. P(both white)
c. P(exactly one black)

Solution: First observe that this is a binomial experiment, because with replacement of the first marble, the two trials will be identical. Further, if we define success = a black marble, $p = \frac{4}{7}$ on each of the two trials and also x = number of black marbles.

a. P(both black) means we should let $x = 2$.

$$\therefore P(2) = \binom{2}{2}(\tfrac{4}{7})^2(\tfrac{3}{7})^0 = \tfrac{16}{49}$$

b. P(both white) means we should let $x = 0$.

$$\therefore P(0) = \binom{2}{0}(\tfrac{4}{7})^0(\tfrac{3}{7})^2 = \tfrac{9}{49}$$

c. P(exactly one black) means $x = 1$, so

$$P(1) = \binom{2}{1}(\tfrac{4}{7})^1(\tfrac{3}{7})^1 = \tfrac{24}{49}$$

The reader is now directed to check this answer and compare the ease of solution with example 4.5.5, parts *a, b, c*, on page 167, which is the same problem. Further note that parts *d, e, f* could NOT be solved as a binomial experiment because there, the first marble was not replaced before the second was drawn, and this causes p to change value on the second trial to $\frac{3}{7}$. So when any trial after the first involves the "no replacement" idea, we cannot correctly treat the problem as binomial in nature. However, sometimes it is okay to be approximate in our computations, as is now illustrated.

Suppose that in a population of 10,000 voters, 50% of them favor candidate A. If we withdraw two voters (without replacement, which is the usual way in polling) and sample their opinion, P(first favors A) $= \frac{1}{2}$ (because 50% favor candidate A), and P(second favors A) $= \frac{4999}{9999}$ or $\frac{5000}{9999}$ (depending on whether the first person favors A or not). Although the probability of favoring candidate A does not remain the same from trial to trial, it doesn't change very much, as we observe:

$$\begin{aligned} \tfrac{1}{2} &= .50000 \cdots \\ \left.\begin{aligned} \tfrac{4999}{9999} &= .499949994 \cdots \\ \tfrac{5000}{9999} &= .500050005 \cdots \end{aligned}\right\} & \text{ each very nearly equals } \tfrac{1}{2} \end{aligned}$$

This happens solely because the size of the population of voters (10,000) is relatively large in comparison with the size of the sample (2).

In the following five examples, we will see how the binomial probability distribution formula can be useful to us in many different settings.

Example 4.9.2. In the U.S. about 20% of the population is left-handed. If a group of ten people is randomly selected, what is the probability that exactly four people will be left-handed?

Solution: Call success = being left-handed. Here $n = 10$, $p = .2$, $q = .8$, and the value of x that we need is 4.

$$\therefore P(4) = \binom{10}{4}(.2)^4(.8)^6$$

There is a lot of multiplying here:

$$\binom{10}{4} = 210$$

$$(.2)^4 = (.2)(.2)(.2)(.2) = .0016$$

$$(.8)^6 = (.8)(.8)(.8)(.8)(.8)(.8) = .262144$$

So finally, $P(4) = 210(.0016)(.262144) = .0880803840$. Table III in Appendix III gives binomial probabilities correct to three places. To use the table in this case, observe that part of the table where you see this:

n	x	. . .	.20	p . . . . .
10	0			
	1			
	2			
	3			
	4		088	
	5			
	.			
	.			
	.			

Hence, $P(4) = .088$. (You must insert the decimal point—they are omitted in the table.)

Example 4.9.3. A basketball player makes $\frac{1}{3}$ of his shots.

a. What is the probability that he makes all of his next three shots?

b. What is the probability that he makes exactly two of his next three shots?

Solution: Past experience indicates that the player makes $\frac{1}{3}$ of his shots. Thus on any particular shot, there is a probability of $\frac{1}{3}$ of making it. We'll assume that the shots are independent. Taking a total of three shots gives us $n = 3$. Success = makes shot.

a. $P(3) = \binom{3}{3}(\frac{1}{3})^3(\frac{2}{3})^0 = \frac{1}{27}$

b. $P(2) = \binom{3}{2} (\frac{1}{3})^2 (\frac{2}{3})^1 = \frac{6}{27} = \frac{2}{9}$

Example 4.9.4. Suppose that 50% of a certain type of seed germinate. If ten such seeds are planted, what is the probability that at least two seeds will germinate?

Solution: Here we need $P(2) + P(3) + P(4) + \cdots + P(10)$. That's a lot of busy work. Instead we find from our table III,

$$P(0) = .001$$
$$P(1) = .010$$

and then, $P(\text{at least two}) = 1 - [P(0) + P(1)] = .989$.

Example 4.9.5. Let us look again at exercise 13 at the end of section 4.5. There, each of two engines has a probability of not working equal to .02. Also the plane will crash if one or more engines fail. What is the probability of crashing?

Solution: Call success = engine does not work. We treat this as a binomial experiment with $n = 2$ (there are two engines—each gets its chance to not work or work). Then $p = .02$, $q = .98$, and $x = 1$ or 2. So

$$\begin{aligned} P(\text{crashing}) &= P(1) + P(2) \\ &= \binom{2}{1} (.02)^1 (.98)^1 + \binom{2}{2} (.02)^2 (.98)^0 = .0396 \end{aligned}$$

Example 4.9.6. A new serum was tested to ascertain its effectiveness in preventing the common cold. Ten people were inoculated with the new drug and observed for a period of one year. Nine survived the year without a cold. It is further known that when the serum is not used, the probability of surviving the year without a cold is only .3. If the serum really happens to be ineffective, what is the probability that nine will actually survive a year without a cold?

Solution: Starting with the assumption that the vaccine is ineffective and letting success = surviving the year without a cold, $p = .3$, $q = .7$, also $n = 10$, and $x = 9$. Then

$$P(9) = \binom{10}{9} (.3)^9 (.7)^1 = .000138 \qquad \text{(rounded off)}$$

At this stage the reader is urged to reread each of the examples 4.9.2 through 4.9.6 with the intent of checking each against the defining criteria discussed at the beginning of this section.

We now briefly turn to the question of the origin of the binomial probability distribution formula. So let us imagine that we have a binomial experiment being run a total of n times. In those n trials, let there be x successes and, hence, $n - x$ failures. One sequence of outcomes of all the trials would be:

$$\underbrace{(S\ S\ S\ S\ \cdots\ S)}_{x \text{ of these}}\underbrace{(F\ F\ F\ F\ \cdots\ F)}_{n - x \text{ of these}}$$

Because of the independence of trials for a binomial experiment, the probabilities of this sequence would be:

$$\underbrace{(p\ p\ p\ p\ \cdots\ p)}_{x \text{ of these}}\underbrace{(q\ q\ q\ q\ \cdots\ q)}_{n - x \text{ of these}} = p^x q^{n-x}$$

However, $p^x q^{n-x}$ is only the probability of one particular possible sequence of successes and failures. We must now ask how many different sequences are possible. That is, with a total of n things (trials), how many ways are there to choose x of them (successes)? We draw a picture for this counting problem below.

$$\underbrace{(\_\ \_\ \_\ \cdots\ \_)}_{x \text{ slots here to be filled with } S\text{'s}}\underbrace{(\_\ \_\ \_\ \cdots\ \_)}_{n - x \text{ slots here to be filled with } F\text{'s}}$$

order of S's *not* important → x slots here to be filled with S's: $\binom{n}{x}$ ways

order of F's *not* important → $n - x$ slots here to be filled with F's: $\binom{n-x}{n-x}$ ways

Now, $\binom{n}{x}\binom{n-x}{n-x} = \binom{n}{x} \cdot 1 = \binom{n}{x}$. So, $p^x q^{n-x}$ must be multiplied by $\binom{n}{x}$. Hence,

$$P(x) = \binom{n}{x} p^x q^{n-x}$$

EXERCISES ■ SECTION 4.9.

1. Do example 4.5.10 by treating it as a binomial experiment where $n = 3$, success $= 6$ occurs on a trial, $p = 1/6$, $q = 5/6$ and $x =$ the appropriate value.
2. Do exercise 16 (c,d,e) on page 173 by treating it as a binomial experiment with $n = 2$, success $=$ head, $p = 3/5$, $x =$ appropriate value.
3. An ordinary die is thrown four times. What is the probability that exactly two sixes will occur?

4. A ten-question true-false exam is given. A student guesses on every question. (Assume he is just as likely to guess correctly as incorrectly.) He hopes to get at least a 70% grade on the test (seven or more questions correct). What is the probability of getting at least a 70%?
5. A coin is thrown 10 times. What is the probability of:
   a. exactly five heads occurring?
   b. between three and seven heads (inclusive) occurring?

   In this problem we notice that the probability of the most likely situation (five heads) is fairly low (.246). However, if we allow for normal fluctuation in the tosses, we see a high probability for between three and seven heads (.890).
6. A baseball player's batting average is 300. (This means that .300 or 30% of the time he gets a hit.) If he comes to bat five times, what is the probability that he gets:
   a. exactly one hit?
   b. at least one hit?
7. An antiaircraft gun will hit a plane with a probability of .4.
   a. If there are four guns firing at a particular plane, what is the probability that the plane gets hit at least once?
   b. How many guns must be fired if it is desired that the probability for at least one hit should be greater than .9?
8. In a certain country, 30% of all losses due to fraudulent, dishonest, or criminal acts are covered by insurance. What is the probability that among eight cases (randomly selected from court files), only one was covered by insurance?
9. A thick coin, when tossed, will land heads with a probability of 3/7, tails with a probability of 3/7, and will land on edge with a probability of 1/7. If this coin is tossed a total of six times, what is the probability that it lands on edge exactly two times?
10. In its training program, a company has a drop-out rate of .20. If eight trainees start the program, what is the probability that six or more finish?
11. Records show that 1% of the trucks in a trucking firm have accidents each day. If ten trucks are operating on a given day, what is the probability of exactly one accident?
12. It is known that 40% of the rats used in an experiment will die after being administered a given dose of an experimental drug. Should the experimenter be very surprised if among 10 rats, none died after having received the given dose of the drug?
13. On a certain assembly line, it is known that five out of every 100 items is unacceptable.

a. What is the probability that in a sample of ten items, exactly one is unacceptable?
b. What is the probability that in a sample of ten items, none is unacceptable?
c. What is the probability that in a sample of ten items, exactly two items are unacceptable?

14. In this problem we return to the problem of de Méré that was mentioned at the beginning of section 3.6. So we now ask the question, "What is the probability of obtaining at least one double 6 in 24 rolls of a pair of dice?" (Hint: first compute the probability of no double sixes and subtract this from 1.)

chapter 4 in review

CHAPTER SUMMARY

In this chapter we learned that probabilities pertain to events, which are outcomes when an experiment is performed. These events can be either simple or compound, either mutually exclusive or non–mutually exclusive. The formula $P(A \text{ or } B) = P(A) + P(B) - P(A \text{ and } B)$ was used. We also learned ways of constructing a sample space and, when this was inappropriate, we used the counting techniques of chapter 3 to determine the size of a sample space. The nature of an event's probability from the viewpoint of experimental frequencies was also discussed. In the section on single-act experiments, the distinction between equally likely events and non–equally likely events was made, odds were introduced as an interpretation of probability, and the formula $P(E) = 1 - P(\tilde{E})$ was employed. In the multi-act experiments section, the distinction between independence and dependence was investigated. The concept of independence was further examined in section 4.6 (conditional probability), as was the formula $P(A \mid B) = P(A \cap B) \div P(B)$. In section 4.7 probability theory was applied to genetics. Expectation, the value of a game averaged over a large number of trials, was the topic of section 4.8. Finally, when an experiment with two possible outcomes is performed n times and each trial of the experiment is independent of the others, we have a binomial experiment. The formula used to obtain the probability of exactly x successes in n trials of a binomial experiment is given by $P(x) = \binom{n}{x} p^x q^{n-x}$, where p is the probability of success in any trial and $q = 1 - p$.

VOCABULARY

probability
simple event
compound event
mutually exclusive events
sample space
sample point
experimental relative frequency
theoretical relative frequency
single-act experiments
odds in favor of
odds against
multi-act experiments
independent events
conditional probability
genetics
genotype
dominant characteristic
recessive characteristic
Punnett square
independent assortment
codominance
expectation
fair game
binomial experiment
success
failure
binomial distribution formula

SYMBOLS

$P(E)$
$2:3$
$P(\tilde{E})$
$P(A \mid B)$
$P(x) = \binom{n}{x} p^x q^{n-x}$

chapter 5

Vectors and Matrices

5.1 INTRODUCTION

In this chapter we will develop some skills needed for the applications to probability and algebra appearing in sections 5.5 and 5.6. These skills will also be employed in chapters 6 and 7. We will be working with objects called **matrices.** A **matrix** is merely a rectangular array of numbers (called **components** or **elements**) arranged into rows and columns. For example,

$$\mathbf{A} = \begin{pmatrix} 1 & 5 & 6 & 3 \\ -11 & 15 & -16 & 41 \\ 9 & \frac{1}{2} & 0 & 8.3 \end{pmatrix}$$

is a twelve-component matrix with three rows and four columns. The component in the second row and third column of $\mathbf{A}$ is -16, the component in the third row and second column is $\frac{1}{2}$, and so on.

A matrix has dimensions just as a rectangle does. The rectangle's dimensions are usually given as length $\times$ width. Similarly, the matrix's **dimensions** are

$$(\text{number of rows}) \times (\text{number of columns})$$

We say $\mathbf{A}$, therefore, is a 3×4 (read "3 by 4") matrix.

Example 5.1.1. What are the dimensions of the following matrices?

$$\mathbf{B} = \begin{pmatrix} 1 & 0 \\ 0 & 0 \end{pmatrix} \qquad \mathbf{C} = \begin{pmatrix} 5 & -4 & 5 \\ -11 & 0 & -1 \\ -17 & 0 & -1 \end{pmatrix} \qquad \mathbf{D} = (1 \quad 9 \quad 0 \quad 0 \quad 6)$$

$$\mathbf{E} = \begin{pmatrix} 5 \\ 9 \\ -11 \\ 1.6 \\ 1.7 \end{pmatrix} \qquad \mathbf{F} = \begin{pmatrix} 2 & 2 & 2 \\ 2 & 2 & 2 \end{pmatrix}$$

Solution: **B** is a 2×2 matrix
C is a 3×3 matrix
D is a 1×5 matrix
E is a 5×1 matrix
F is a 2×3 matrix

Matrices that have the same "length" as "width" (like **B** and **C** above) are called **square** matrices. Matrices with exactly one row (like **D** above) go by the special name **row vectors,** and matrices with exactly one column (like **E** above) are called **column vectors.**

We shall spend the remainder of this chapter discussing the properties, operations, and applications of matrices.

5.2 VECTOR OPERATIONS

The operations of addition, subtraction, and multiplication will now be looked at for vectors.

Suppose the Nelorich Ice Cream Company has two ice cream factories. Factory No. 1 produces 30 gallons of vanilla, 40 gallons of chocolate, 12 gallons of strawberry, and 2 gallons of black raspberry per day. Factory No. 2 produces 50 gallons of vanilla, 20 gallons of chocolate, 2 gallons of strawberry, and no black raspberry per day. This data can be represented by row vectors. We let **A** denote the row vector representing the output of Factory No. 1 and **B** denote the row vector representing the output of Factory No. 2:

$$\begin{array}{lcccc} \mathbf{A} = & (30 & 40 & 12 & 2) \\ \mathbf{B} = & (50 & 20 & 2 & 0) \\ & \uparrow & \uparrow & \uparrow & \uparrow \\ & \text{vanilla} & \text{chocolate} & \text{strawberry} & \text{black raspberry} \end{array}$$

(We note that in order for this notation to be of any use, we list the flavor amounts in the same order for both vectors.)

Obviously, the Nelorich Company is producing a daily total of 80 gallons of vanilla, 60 gallons of chocolate, 14 gallons of strawberry, and 2 gallons of black raspberry. We can represent the company's total ice cream output (that is, the sum of the two factories' outputs) in terms of a vector that is the sum **A** + **B** as follows:

$$\mathbf{A} + \mathbf{B} = (30 \quad 40 \quad 12 \quad 2) + (50 \quad 20 \quad 2 \quad 0) = (80 \quad 60 \quad 14 \quad 2)$$

In other words, to add row vectors, we merely add the components, position by position. Column-vector addition is performed similarly:

$$\begin{pmatrix} 1 \\ 5 \\ 7 \end{pmatrix} + \begin{pmatrix} 6 \\ 4 \\ 0 \end{pmatrix} + \begin{pmatrix} 20 \\ 30 \\ -40 \end{pmatrix} = \begin{pmatrix} 27 \\ 39 \\ -33 \end{pmatrix}$$

Here we note that vector addition is meaningful only for vectors of the same dimensions. Subtraction of vectors is performed in a similar manner:

$$\begin{pmatrix} 9 \\ 5 \\ 4 \end{pmatrix} - \begin{pmatrix} 1 \\ 2 \\ 3 \end{pmatrix} = \begin{pmatrix} 8 \\ 3 \\ 1 \end{pmatrix}$$

An example follows.

EXAMPLE 5.2.1. Let $\mathbf{R} = (5 \quad 6 \quad -11 \quad 4.1)$, $\mathbf{S} = (0 \quad 0 \quad 0 \quad 9)$, $\mathbf{T} = (1 \quad 3 \quad 5 \quad 6)$,

$$\mathbf{U} = \begin{pmatrix} 9 \\ 3 \\ 0.5 \\ 8.6 \end{pmatrix} \qquad \mathbf{V} = \begin{pmatrix} 5 \\ 7 \\ 9 \\ 11 \end{pmatrix} \qquad \mathbf{W} = \begin{pmatrix} 1 \\ 2 \\ 3 \end{pmatrix}$$

Find:

a. $\mathbf{R} + \mathbf{S}$ **b.** $\mathbf{R} - \mathbf{T}$ **c.** $\mathbf{R} + \mathbf{V}$ **d.** $\mathbf{U} + \mathbf{V}$
e. $\mathbf{V} + \mathbf{W}$ **f.** $\mathbf{T} - \mathbf{R}$ **g.** $\mathbf{V} - \mathbf{U}$

Solution:

a. $(5 \quad 6 \quad -11 \quad 4.1) + (0 \quad 0 \quad 0 \quad 9) = (5 \quad 6 \quad -11 \quad 13.1)$
b. $(5 \quad 6 \quad -11 \quad 4.1) - (1 \quad 3 \quad 5 \quad 6) = (4 \quad 3 \quad -16 \quad -1.9)$
c. **R** + **V** has no meaning because **R** and **V** have different dimensions.

d. $$\begin{pmatrix} 9 \\ 3 \\ 0.5 \\ 8.6 \end{pmatrix} + \begin{pmatrix} 5 \\ 7 \\ 9 \\ 11 \end{pmatrix} = \begin{pmatrix} 14 \\ 10 \\ 9.5 \\ 19.6 \end{pmatrix}$$

e. **V** + **W** has no meaning because **V** and **W** have different dimensions.
f. $(1 \quad 3 \quad 5 \quad 6) - (5 \quad 6 \quad -11 \quad 4.1) = (-4 \quad -3 \quad 16 \quad 1.9)$

g. $\begin{pmatrix} 5 \\ 7 \\ 9 \\ 11 \end{pmatrix} - \begin{pmatrix} 9 \\ 3 \\ 0.5 \\ 8.6 \end{pmatrix} = \begin{pmatrix} -4 \\ 4 \\ 8.5 \\ 2.4 \end{pmatrix}$

For an illustration of vector multiplication, we look at diving competition. Jacque Strappé has chosen six optional dives he will perform. Each dive has a degree of difficulty as listed below:

Dive	Degree of difficulty
1. Forward dive from the layout position	1.4
2. Backward dive from the tuck position	1.6
3. Reverse dive with $1\frac{1}{2}$ somersaults from pike position	2.5
4. Inward dive from the layout position	1.7
5. Inward double somersault from the tuck position	2.3
6. Backward dive with $\frac{1}{2}$ twist from the pike position	1.9

We represent this information in a row vector **A**:

$$\mathbf{A} = (1.4 \quad 1.6 \quad 2.5 \quad 1.7 \quad 2.3 \quad 1.9)$$

Now a team of judges awards Jacque the following scores:

Dive	*Score*
No. 1	28.5
No. 2	20.0
No. 3	17.5
No. 4	26.0
No. 5	21.5
No. 6	27.0

This we represent in a column vector **B**:

$$\mathbf{B} = \begin{pmatrix} 28.5 \\ 20.0 \\ 17.5 \\ 26.0 \\ 21.5 \\ 27.0 \end{pmatrix}$$

Jacque's total score in the diving competition is obtained by multiplying each score by its degree of difficulty and adding the four results. That is:

$$\underbrace{1.4 \cdot 28.5}_{\text{Dive No. 1}} + \underbrace{1.6 \cdot 20.0}_{\text{Dive No. 2}} + \underbrace{2.5 \cdot 17.5 + 1.7 \cdot 26.0 + 2.3 \cdot 21.5 + 1.9 \cdot 27.0}_{\text{etc.}}$$

which is precisely the product of the vectors **A** and **B**:

$$\mathbf{A} \cdot \mathbf{B} = (1.4 \quad 1.6 \quad 2.5 \quad 1.7 \quad 2.3 \quad 1.9) \cdot \begin{pmatrix} 28.5 \\ 20.0 \\ 17.5 \\ 26.0 \\ 21.5 \\ 27.0 \end{pmatrix} = 260.6$$

Jacque's total score in the diving competition was 260.6 points. (Notice that we use a "$\cdot$" to represent this multiplication.) So, when we multiply a row vector by a column vector (notice each has the same number of components), the result is the product of the first components, added to the product of the second components, added to the product of the third components, and so on. Observe that the result of vector multiplication is a number, not a vector.

Example 5.2.2. Suppose John is on a diet that restricts him to certain foods. The chart below depicts that food with its caloric content and the number of servings of that food John had last Tuesday.

Food	*Calories per serving*	*Tuesday's food consumption (in servings)*
apple	70	2
lima beans	260	1
steak	230	4
whole wheat bread	55	2
cabbage	25	2
grapefruit	55	3
beef liver	60	0
water	0	10

Solution: Depict John's consumption last Tuesday by a row vector, the calories of each food by a column vector, and use vector multiplication to compute John's total caloric intake for last Tuesday. Let $\mathbf{T} = (2 \quad 1 \quad 4 \quad 2 \quad 2 \quad 3 \quad 0 \quad 10)$,

$$\mathbf{C} = \begin{pmatrix} 70 \\ 260 \\ 230 \\ 55 \\ 25 \\ 55 \\ 60 \\ 0 \end{pmatrix}$$

Then $\mathbf{T} \cdot \mathbf{C}$
$= 2 \cdot 70 + 1 \cdot 260 + 4 \cdot 230 + 2 \cdot 55 + 2 \cdot 25 + 3 \cdot 55 + 0 \cdot 60 + 10 \cdot 0$
$= 140 + 260 + 920 + 110 + 50 + 165 + 0 + 0 = 1645$. That is, 1645 was John's caloric intake for last Tuesday.

Example 5.2.3. Let

$$\mathbf{A} = (1 \quad 5 \quad -3 \quad 7), \qquad \mathbf{B} = \begin{pmatrix} 2 \\ 4 \\ 6 \\ 8 \end{pmatrix}, \qquad \mathbf{C} = \begin{pmatrix} 1 \\ 3 \\ 5 \\ 7 \end{pmatrix}, \qquad \mathbf{D} = \begin{pmatrix} 2 \\ 6 \\ 5 \end{pmatrix}.$$

Find: **a. A · B** **b. B · A** **c. A · C** **d. B · C** **e. A · D**

Solution:

a. $\mathbf{A} \cdot \mathbf{B} = (1 \quad 5 \quad -3 \quad 7) \cdot \begin{pmatrix} 2 \\ 4 \\ 6 \\ 8 \end{pmatrix}$

$$= 1 \cdot 2 + 5 \cdot 4 + (-3) \cdot 6 + 7 \cdot 8 = 60$$

b. B · A has no meaning. In order to multiply a column vector and a row vector, we must write the row vector first. The reason will become clearer in section 5.3.

c. $\mathbf{A} \cdot \mathbf{C} = (1 \quad 5 \quad -3 \quad 7) \cdot \begin{pmatrix} 1 \\ 3 \\ 5 \\ 7 \end{pmatrix}$

$$= 1 \cdot 1 + 5 \cdot 3 + (-3) \cdot 5 + 7 \cdot 7 = 50$$

d. B · C has no meaning because vector multiplication can be performed only for the product of a row vector and column vector.

e. A · D has no meaning because vector multiplication can be performed only on a row vector and column vector of the same length (dimensions).

One additional operation that can be performed is the product of a vector with a number. For example, if we have a list of linear measurements in inches (say 2″, 10″, 12″, 36″, 100″) and wish to convert them to centimeters, we must multiply each measurement in inches by 2.54, because 1″ = 2.54 cm. The vector representation looks like this:

$$2.54 \cdot (2 \quad 10 \quad 12 \quad 36 \quad 100) = (5.08 \quad 25.4 \quad 30.48 \quad 91.44 \quad 254)$$

That is, 2″ = 5.08 cm, 10″ = 25.4 cm, 12″ = 30.48 cm, and so on. The same operation can occur on column vectors:

$$7 \cdot \begin{pmatrix} 1 \\ 2 \\ 3 \\ 4 \\ 5 \end{pmatrix} = \begin{pmatrix} 7 \\ 14 \\ 21 \\ 28 \\ 35 \end{pmatrix}$$

EXERCISES ■ SECTIONS 5.1 AND 5.2.

1. State the dimensions of each of the following matrices:

$$\mathbf{A} = \begin{pmatrix} 1 & 2 & -10 \\ 0 & 3 & 5 \\ 6 & 6 & 6 \end{pmatrix} \qquad \mathbf{C} = \begin{pmatrix} -5 & 7 & 9 & 11 \\ 6 & 8 & 10 & 12 \end{pmatrix}$$

$$\mathbf{B} = \begin{pmatrix} 1 & 2 & 3 & 0 \\ 4 & 5 & 6 & 0 \\ 7 & 8 & 9 & 0 \end{pmatrix} \qquad \mathbf{D} = \begin{pmatrix} 1 & 3 & 4 \\ 3 & 8 & 3 \\ 3 & 5 & 6 \\ 1 & 2 & 7 \\ 4 & 0 & 5 \\ 0 & 7 & 2 \end{pmatrix}$$

$$\mathbf{E} = \begin{pmatrix} a_{11} & a_{12} & a_{13} & a_{14} & \cdots & a_{1n} \\ a_{21} & a_{22} & a_{23} & a_{24} & \cdots & a_{2n} \\ & & & \cdot & & \\ & & & \cdot & & \\ & & & \cdot & & \\ a_{m1} & a_{m2} & a_{m3} & a_{m4} & \cdots & a_{mn} \end{pmatrix}$$

Note: **E** is a matrix that will be referred to in the next section. The component a_{ij} is found in the ith row, jth column.

2. Let $\mathbf{X} = \begin{pmatrix} 1 \\ -3 \\ 6 \end{pmatrix}$, $\quad \mathbf{Y} = \begin{pmatrix} 0 \\ 2 \\ 6 \end{pmatrix}$, $\quad \mathbf{Z} = (2 \quad 1 \quad 3)$.

Find:

a. $\mathbf{X} + \mathbf{Y}$ **b.** $2\mathbf{Y}$ **c.** $\mathbf{X} - \mathbf{X}$
d. $2\mathbf{X} + \mathbf{Y}$ **e.** $\frac{1}{2}(\mathbf{X} + \mathbf{Y})$ **f.** $\frac{1}{2}\mathbf{X} + \frac{1}{2}\mathbf{Y}$
g. $\mathbf{X} \cdot \mathbf{Z}$ **h.** $\mathbf{Z} \cdot \mathbf{X}$ **i.** $\mathbf{Z} \cdot 2\mathbf{X}$
j. $\mathbf{Z} \cdot \mathbf{Y}$ **k.** $\mathbf{Z} \cdot (\mathbf{X} + \mathbf{Y})$ **l.** $\mathbf{Z} \cdot (\mathbf{Y} - \mathbf{X})$

3. Vector **A** below represents the daily high temperatures in Rochester, New York, for the first week of July, 1974. Vector **B** represents the daily low temperatures for the same period.

	M	T	W	Th	F	S	S
A =	(91	78	84	89	90	91	81)
B =	(65	61	63	70	72	74	69)

Find the daily range of temperatures by calculating **A** − **B**.

4. Friendly Felix Auto Sales specializes in luxury cars and sells the following during the month of April, 1975: 5 Rolls Royces, 10

Cadillacs, 12 Lincoln Continentals, 15 Mercedes, and 1 Edsel. The average selling price for each car is as follows: \$30,000 for Rolls Royces, \$10,000 for Cadillacs, \$10,000 for Lincoln Continentals, \$15,000 for Mercedes, \$850 for Edsels. Calculate Felix's total monetary intake, $\mathbf{A} \cdot \mathbf{B}$, by representing the numbers of cars by row vector **A** and the average prices of each by column vector **B**.

5. Last semester Cheryl received two A's, one B, one C, one D, and one F. If an A is worth four points, a B three points, a C two points, a D one point, and an F zero points, find Cheryl's total points by vector multiplication.
6. Repeat exercise five for Hank, who received four B's and one D. Hint: the vectors to be multiplied here are:

$$\begin{pmatrix} 0 & 4 & 0 & 1 & 0 \end{pmatrix} \quad \text{and} \quad \begin{pmatrix} 4 \\ 3 \\ 2 \\ 1 \\ 0 \end{pmatrix}$$

7. M. L. Pierce is about to purchase some stock. He will buy 100 shares of AT&T, 50 shares of Eastman Kodak, and 50 shares of United Park Mines. They sell for \$53.80, \$134.20, and \$1.75 per share, respectively. Represent the numbers of shares in a row vector, the prices per share in a column vector, and use vector multiplication to find M. L. Pierce's total investment.
8. Two vectors are **equal** if they have the same dimensions and the corresponding components are equal. We can use this fact to solve vector equations. If $\begin{pmatrix} x \\ y \end{pmatrix} = \begin{pmatrix} 5 \\ -7 \end{pmatrix}$, for example, $x = 5$ and $y = -7$. Solve the following vector equations:

   a. $\begin{pmatrix} x \\ y \end{pmatrix} = \begin{pmatrix} 1 \\ 4 \end{pmatrix} + \begin{pmatrix} 7 \\ -11 \end{pmatrix}$

   b. $(x \quad y) = (9 \quad 23)$

   c. $5 \cdot \begin{pmatrix} w \\ x \\ y \\ z \end{pmatrix} = \begin{pmatrix} 20 \\ 40 \\ 60 \\ 80 \end{pmatrix}$

   d. $(a \quad b \quad c \quad d \quad e) + (1 \quad 2 \quad 3 \quad 4 \quad 5) = (19 \quad 0 \quad 4 \quad 0 \quad 6)$

   e. $\begin{pmatrix} x \\ y \end{pmatrix} + \begin{pmatrix} 11 \\ -10 \end{pmatrix} = \begin{pmatrix} -14 \\ -15 \end{pmatrix}$

9. A druggist has labeled six jars of a particular drug in ounces as follows: (5 15 25 30 40 50). A nurse must convert each of these ounce measurements to cubic centimeter (cc) measurements. If 1 ounce = 30 cc, do the nurse's calculation.

5.3 OPERATIONS WITH MATRICES

Just as we added (and subtracted) vectors in the previous section, we can add (and subtract) matrices. We obey the same two rules we obeyed then:

1. The matrices must be of the same dimensions.
2. The corresponding components are added (or subtracted).

For example, if $\mathbf{A} = \begin{pmatrix} 1 & 0 & 4 \\ 5 & -1 & 3 \end{pmatrix}$ and $\mathbf{B} = \begin{pmatrix} 5 & 2 & 6 \\ 7 & -2 & -3 \end{pmatrix}$,

$$\mathbf{A} + \mathbf{B} = \begin{pmatrix} 1 & 0 & 4 \\ 5 & -1 & 3 \end{pmatrix} + \begin{pmatrix} 5 & 2 & 6 \\ 7 & -2 & -3 \end{pmatrix} = \begin{pmatrix} 6 & 2 & 10 \\ 12 & -3 & 0 \end{pmatrix}$$

and

$$\mathbf{A} - \mathbf{B} = \begin{pmatrix} -4 & -2 & -2 \\ -2 & 1 & 6 \end{pmatrix}$$

It shall be convenient in this section to refer to the general $m \times n$ matrix:

$$\mathbf{A} = \begin{pmatrix} a_{11} & a_{12} & a_{13} & \cdots & a_{1n} \\ a_{21} & a_{22} & a_{23} & \cdots & a_{2n} \\ & & \vdots & & \\ a_{m1} & a_{m2} & a_{m3} & \cdots & a_{mn} \end{pmatrix}$$

The subscripts automatically locate a given component. For example, the component a_{23} is in the second row, third column. The component a_{mn} is in the mth row, nth column. We incorporate the matrix **A** into the next example.

Example 5.3.1. Let

$$\mathbf{A} = \begin{pmatrix} a_{11} & a_{12} & a_{13} & \cdots & a_{1n} \\ a_{21} & a_{22} & a_{23} & \cdots & a_{2n} \\ & & \vdots & & \\ a_{m1} & a_{m2} & a_{m3} & \cdots & a_{mn} \end{pmatrix}$$

$$\mathbf{B} = \begin{pmatrix} b_{11} & b_{12} & b_{13} & \cdots & b_{1n} \\ b_{21} & b_{22} & b_{23} & \cdots & b_{2n} \\ & & \vdots & & \\ b_{m1} & b_{m2} & b_{m3} & \cdots & b_{mn} \end{pmatrix}$$

$$\mathbf{C} = \begin{pmatrix} 10 & 9 & 8 & -1 \\ 7 & 6 & 5 & -2 \\ 4 & 3 & 2 & -3 \end{pmatrix} \qquad \mathbf{D} = \begin{pmatrix} 3 & -1 & 4 & 1 \\ 5 & 9 & 2 & 6 \\ 3 & 5 & 3 & 1 \end{pmatrix}$$

Find:

a. A + B **b. C + D** **c. C − D**

Solution:

$$\textbf{a. } \mathbf{A} + \mathbf{B} = \begin{pmatrix} a_{11} + b_{11} & a_{12} + b_{12} & a_{13} + b_{13} & \cdots & a_{1n} + b_{1n} \\ a_{21} + b_{21} & a_{22} + b_{22} & a_{23} + b_{23} & \cdots & a_{2n} + b_{2n} \\ & & \cdot & & \\ & & \cdot & & \\ & & \cdot & & \\ a_{m1} + b_{m1} & a_{m2} + b_{m2} & a_{m3} + b_{m3} & \cdots & a_{mn} + b_{mn} \end{pmatrix}$$

Here, we can see component-by-component addition taking place.

$$\textbf{b. } \mathbf{C} + \mathbf{D} = \begin{pmatrix} 10+3 & 9-1 & 8+4 & -1+1 \\ 7+5 & 6+9 & 5+2 & -2+6 \\ 4+3 & 3+5 & 2+3 & -3+1 \end{pmatrix}$$

$$= \begin{pmatrix} 13 & 8 & 12 & 0 \\ 12 & 15 & 7 & 4 \\ 7 & 8 & 5 & -2 \end{pmatrix}$$

$$\textbf{c. } \mathbf{C} - \mathbf{D} = \begin{pmatrix} 10-3 & 9+1 & 8-4 & -1-1 \\ 7-5 & 6-9 & 5-2 & -2-6 \\ 4-3 & 3-5 & 2-3 & -3-1 \end{pmatrix}$$

$$= \begin{pmatrix} 7 & 10 & 4 & -2 \\ 2 & -3 & 3 & -8 \\ 1 & -2 & -1 & -4 \end{pmatrix}$$

We note here that when two $m \times n$ matrices are added (or subtracted) the result is an $m \times n$ matrix.

One advantage of a matrix is its compact representation of data. Suppose that in 1972, the following statistics applied:

	Sales (in millions)		
1972	Sedans	Station wagons	Convertibles
Ford Motor Co.	4	2.8	0.0
General Motors	5	1.6	0.0
Volkswagen	3.2	1.7	0.6

We represent the data in the convenient arrangement, call it matrix **A**, and write

$$\mathbf{A} = \begin{array}{c} \\ \\ \\ \\ \end{array}\begin{pmatrix} 4 & 2.8 & 0.0 \\ 5 & 1.6 & 0.0 \\ 3.2 & 1.7 & 0.6 \end{pmatrix} \begin{array}{l} \text{FMC} \\ \text{GM} \\ \text{VW} \end{array}$$

(columns: sedans, wagons, conv.)

Suppose for 1973's data, we use matrix **B**:

$$\mathbf{B} = \begin{pmatrix} 3.8 & 3.1 & 0.0 \\ 4.0 & 2.2 & 0.0 \\ 3.6 & 2.8 & 0.2 \end{pmatrix} \begin{array}{l} \text{FMC} \\ \text{GM} \\ \text{VW} \end{array}$$

(columns: sedans, wagons, conv.)

The sum **A** + **B** represents the total output for 1972–1973.

$$\mathbf{A} + \mathbf{B} = \begin{pmatrix} 7.8 & 5.9 & 0.0 \\ 9.0 & 3.8 & 0.0 \\ 6.8 & 4.5 & 0.8 \end{pmatrix} \begin{array}{l} \text{FMC} \\ \text{GM} \\ \text{VW} \end{array}$$

(columns: sedans, wagons, conv.)

Matrix **A** + **B** shows, for example, that GM produced 3.8 million station wagons in the years 1972 and 1973.

A **matrix** can be **multiplied** by a number just as vectors were in section 5.2. If

$$\mathbf{C} = \begin{pmatrix} 2 & 7 & 1 \\ 8 & -2 & 8 \end{pmatrix}$$

Then

$$\tfrac{1}{2}\mathbf{C} = \begin{pmatrix} 1 & \frac{7}{2} & \frac{1}{2} \\ 4 & -1 & 4 \end{pmatrix}$$

and

$$10\mathbf{C} = \begin{pmatrix} 20 & 70 & 10 \\ 80 & -20 & 80 \end{pmatrix}$$

The multiplication of two matrices is a somewhat more complex task and special rules must be obeyed. Let

$$\mathbf{X} = \begin{pmatrix} 7 & 4 \\ 3 & 2 \\ 0 & 1 \end{pmatrix} \quad \text{and} \quad \mathbf{Y} = \begin{pmatrix} 5 & 9 & 20 & 40 \\ 6 & 10 & 30 & 50 \end{pmatrix}$$

We wish to find the product of **X** with **Y**, denoted **X** · **Y**. First, note that the number of columns of **X** equals the numbers of rows of **Y**. This is essential. Also, **X** is a 3 × 2 matrix and **Y** is a 2 × 4 matrix. The product **X** · **Y** will be a 3 × 4 matrix; **X** · **Y** will have the number of rows of **X** and the number of columns of **Y**:

$$\begin{pmatrix} 7 & 4 \\ 3 & 2 \\ 0 & 1 \end{pmatrix} \cdot \begin{pmatrix} 5 & 9 & 20 & 40 \\ 6 & 10 & 30 & 50 \end{pmatrix} = \begin{pmatrix} \_ & \_ & \_ & \_ \\ \_ & \_ & \_ & \_ \\ \_ & \_ & \_ & \_ \end{pmatrix}$$

Now, to obtain the component a_{ij} of $\mathbf{X} \cdot \mathbf{Y}$, we multiply the ith row vector of $\mathbf{X}$ by the jth column vector of $\mathbf{Y}$. That is, to find the component in the second row, fourth column of $\mathbf{X} \cdot \mathbf{Y}$, we take the second row of $\mathbf{X}$ and multiply it by the fourth column of $\mathbf{Y}$:

$$\begin{pmatrix} 7 & 4 \\ 3 & 2 \\ 0 & 1 \end{pmatrix} \cdot \begin{pmatrix} 5 & 9 & 20 & 40 \\ 6 & 10 & 30 & 50 \end{pmatrix} = \begin{pmatrix} - & - & - & - \\ - & - & - & 220 \\ - & - & - & - \end{pmatrix}$$

$$(3 \quad 2) \cdot \begin{pmatrix} 40 \\ 50 \end{pmatrix} = 3 \cdot 40 + 2 \cdot 50 = 220$$

This process is continued until twelve vector multiplications have been performed to give the twelve components of the 3×4 matrix $\mathbf{X} \cdot \mathbf{Y}$.

$$\mathbf{X} \cdot \mathbf{Y} = \begin{pmatrix} 7 & 4 \\ 3 & 2 \\ 0 & 1 \end{pmatrix} \cdot \begin{pmatrix} 5 & 9 & 20 & 40 \\ 6 & 10 & 30 & 50 \end{pmatrix}$$

$$= \begin{pmatrix} 7 \cdot 5 + 4 \cdot 6 & 7 \cdot 9 + 4 \cdot 10 & 7 \cdot 20 + 4 \cdot 30 & 7 \cdot 40 + 4 \cdot 50 \\ 3 \cdot 5 + 2 \cdot 6 & 3 \cdot 9 + 2 \cdot 10 & 3 \cdot 20 + 2 \cdot 30 & 3 \cdot 40 + 2 \cdot 50 \\ 0 \cdot 5 + 1 \cdot 6 & 0 \cdot 9 + 1 \cdot 10 & 0 \cdot 20 + 1 \cdot 30 & 0 \cdot 40 + 1 \cdot 50 \end{pmatrix}$$

$$= \begin{pmatrix} 59 & 103 & 260 & 480 \\ 27 & 47 & 120 & 220 \\ 6 & 10 & 30 & 50 \end{pmatrix}$$

The reader is urged to verify the calculations.

We now summarize the rules for matrix multiplication. To calculate $\mathbf{A} \cdot \mathbf{B}$:

1. The number of columns of $\mathbf{A}$ must equal the number of rows of $\mathbf{B}$.
2. If $\mathbf{A}$ is an $m \times p$ matrix and $\mathbf{B}$ is a $p \times n$ matrix, $\mathbf{A} \cdot \mathbf{B}$ is an $m \times n$ matrix.
3. A component in the ith row, jth column of $\mathbf{A} \cdot \mathbf{B}$ is obtained by multiplying the ith row of $\mathbf{A}$ by the jth column of $\mathbf{B}$.

Some examples follow.

Example 5.3.2. Let

$$\mathbf{A} = \begin{pmatrix} 0 & -1 & -2 \\ 5 & 6 & 2 \end{pmatrix} \quad \mathbf{B} = \begin{pmatrix} 1 & 2 & 3 \\ -3 & -2 & -1 \\ -2 & 0 & 2 \end{pmatrix} \quad \mathbf{C} = \begin{pmatrix} 1 & 0 \\ 0 & 1 \end{pmatrix}$$

$$\mathbf{D} = \begin{pmatrix} 5 & -1 \\ 6 & 2 \end{pmatrix} \quad \mathbf{E} = \begin{pmatrix} 6 & -2 \\ 0 & 1 \end{pmatrix}$$

Find:

a. $\mathbf{A} \cdot \mathbf{B}$ **b.** $\mathbf{A} \cdot \mathbf{C}$ **c.** $\mathbf{C} \cdot \mathbf{A}$ **d.** $\mathbf{C} \cdot \mathbf{D}$
e. $\mathbf{D} \cdot \mathbf{E}$ **f.** $\mathbf{E} \cdot \mathbf{D}$ **g.** $(3\mathbf{D}) \cdot \mathbf{E}$

Solution:

a. Since **A** is a 2×3 and **B** is a 3×3 matrix, $\mathbf{A} \cdot \mathbf{B}$ is a 2×3 matrix:

$$\mathbf{A} \cdot \mathbf{B} = \begin{pmatrix} 0 & -1 & -2 \\ 5 & 6 & 2 \end{pmatrix} \cdot \begin{pmatrix} 1 & 2 & 3 \\ -3 & -2 & -1 \\ -2 & 0 & 2 \end{pmatrix}$$

$$= \begin{pmatrix} 0 \cdot 1 + (-1) \cdot (-3) + (-2) \cdot (-2) & 0 \cdot 2 + (-1) \cdot (-2) + (-2) \cdot 0 & 0 \cdot 3 + (-1) \cdot (-1) + (-2) \cdot 2 \\ 5 \cdot 1 + 6 \cdot (-3) + 2 \cdot (-2) & 5 \cdot 2 + 6 \cdot (-2) + 2 \cdot 0 & 5 \cdot 3 + 6 \cdot (-1) + 2 \cdot 2 \end{pmatrix}$$

$$= \begin{pmatrix} 7 & 2 & -3 \\ -17 & -2 & 13 \end{pmatrix}$$

b. $\mathbf{A} \cdot \mathbf{C}$ cannot be found because **A** has three columns and **C** has two rows.

c. $\mathbf{C} \cdot \mathbf{A}$ is a 2×3 matrix.

$$\mathbf{C} \cdot \mathbf{A} = \begin{pmatrix} 1 & 0 \\ 0 & 1 \end{pmatrix} \cdot \begin{pmatrix} 0 & -1 & -2 \\ 5 & 6 & 2 \end{pmatrix}$$

$$= \begin{pmatrix} 1 \cdot 0 + 0 \cdot 5 & 1 \cdot (-1) + 0 \cdot 6 & 1 \cdot (-2) + 0 \cdot 2 \\ 0 \cdot 0 + 1 \cdot 5 & 0 \cdot (-1) + 1 \cdot 6 & 0 \cdot (-2) + 1 \cdot 2 \end{pmatrix}$$

$$= \begin{pmatrix} 0 & -1 & -2 \\ 5 & 6 & 2 \end{pmatrix}$$

d. $\mathbf{C} \cdot \mathbf{D}$ is a 2×2 matrix.

$$\mathbf{C} \cdot \mathbf{D} = \begin{pmatrix} 1 & 0 \\ 0 & 1 \end{pmatrix} \cdot \begin{pmatrix} 5 & -1 \\ 6 & 2 \end{pmatrix} = \begin{pmatrix} 5 & -1 \\ 6 & 2 \end{pmatrix}$$

e. $$\mathbf{D} \cdot \mathbf{E} = \begin{pmatrix} 5 & -1 \\ 6 & 2 \end{pmatrix} \cdot \begin{pmatrix} 6 & -2 \\ 0 & 1 \end{pmatrix} = \begin{pmatrix} 30 & -11 \\ 36 & -10 \end{pmatrix}$$

f. $$\mathbf{E} \cdot \mathbf{D} = \begin{pmatrix} 6 & -2 \\ 0 & 1 \end{pmatrix} \cdot \begin{pmatrix} 5 & -1 \\ 6 & 2 \end{pmatrix} = \begin{pmatrix} 18 & -10 \\ 6 & 2 \end{pmatrix}$$

g. $$3\mathbf{D} = \begin{pmatrix} 15 & -3 \\ 18 & 6 \end{pmatrix}$$

so $$(3\mathbf{D}) \cdot \mathbf{E} = \begin{pmatrix} 15 & -3 \\ 18 & 6 \end{pmatrix} \cdot \begin{pmatrix} 6 & -2 \\ 0 & 1 \end{pmatrix} = \begin{pmatrix} 90 & -33 \\ 108 & -30 \end{pmatrix}$$

Note that some basic rules pertaining to the multiplication of real numbers do not apply to matrix multiplication. We can see in the previous example that **A** · **C** is not equal to **C** · **A**; and **D** · **E** is not equal to **E** · **D**.

Example 5.3.3. Suppose the E-Z Screw Company produces three kinds of screw drivers: 3-inch, 7-inch, and 12-inch. Factory No. 1 produces 400, 500, 100, respectively, of each type. Factory No. 2 produces 100, 200, 700, respectively, of each type. In addition, it is known that each 3-inch screw driver costs $.12 to produce and $.03 to ship; each 7-inch screw driver costs $.15 to produce and $.03 to ship; each 12-inch screw driver costs $.20 to produce and $.05 to ship. Let **A** be the 2×3 matrix representing factory output, and let **B** represent the 3×2 matrix showing costs for each screw driver. Find the costs per factory, matrix **A** · **B**.

Solution:

$$\mathbf{A} = \begin{matrix} \\ \text{Factory 1} \\ \text{Factory 2} \end{matrix} \begin{matrix} 3'' & 7'' & 12'' \\ \left(\begin{matrix} 400 \\ 100 \end{matrix}\right. & \begin{matrix} 500 \\ 200 \end{matrix} & \left.\begin{matrix} 100 \\ 700 \end{matrix}\right) \end{matrix}$$

$$\mathbf{B} = \begin{matrix} \\ 3'' \\ 7'' \\ 12'' \end{matrix} \begin{matrix} \text{production costs} & \text{transportation costs} \\ \left(\begin{matrix} .12 \\ .15 \\ .20 \end{matrix}\right. & \left.\begin{matrix} .03 \\ .03 \\ .05 \end{matrix}\right) \end{matrix}$$

$$\mathbf{A} \cdot \mathbf{B} = \begin{matrix} \\ \text{Factory 1} \\ \text{Factory 2} \end{matrix} \begin{matrix} \text{production costs} & \text{transportation costs} \\ \left(\begin{matrix} 400 \cdot .12 + 500 \cdot .15 + 100 \cdot .20 \\ 100 \cdot .12 + 200 \cdot .15 + 700 \cdot .20 \end{matrix}\right. & \left.\begin{matrix} 400 \cdot .03 + 500 \cdot .03 + 100 \cdot .05 \\ 100 \cdot .03 + 200 \cdot .03 + 700 \cdot .05 \end{matrix}\right) \end{matrix}$$

$$\mathbf{A} \cdot \mathbf{B} = \begin{matrix} \\ \text{Factory 1} \\ \text{Factory 2} \end{matrix} \begin{matrix} \text{production costs} & \text{transportation costs} \\ \left(\begin{matrix} 143 \\ 182 \end{matrix}\right. & \left.\begin{matrix} 32 \\ 44 \end{matrix}\right) \end{matrix}$$

This can be interpreted as follows:

total production cost for Factory No. 1 is $143,
total production cost for Factory No. 2 is $182,
total transportation cost for Factory No. 1 is $32, and
total transportation cost for Factory No. 2 is $44.

We will now look at the previous example to substantiate the way matrices are multiplied. We have already seen that in order to formulate the matrix product **A** · **B**, the matrix **A** must have the same number of columns as **B** has rows. Furthermore, the result **A · B** will have the same number of rows as **A** and the same number of columns as **B**. That is,

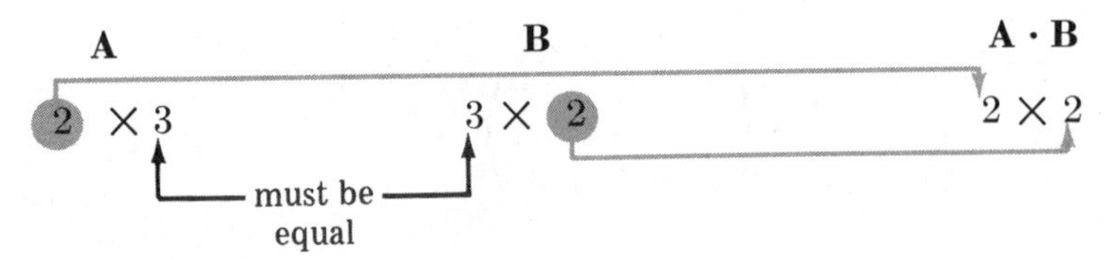

The previous example illustrates the rationale for this. In order for the matrix product **A · B** to be meaningful, the columns of **A** must refer to the same thing as the rows of **B**. (In our example, the columns of **A** and rows of **B** refer to types of screwdrivers.) Notice that the rows of **A · B** refer to factories, as do the rows of **A**; the columns of **A · B** refer to costs, as do the columns of **B**. The following summary unfolds:

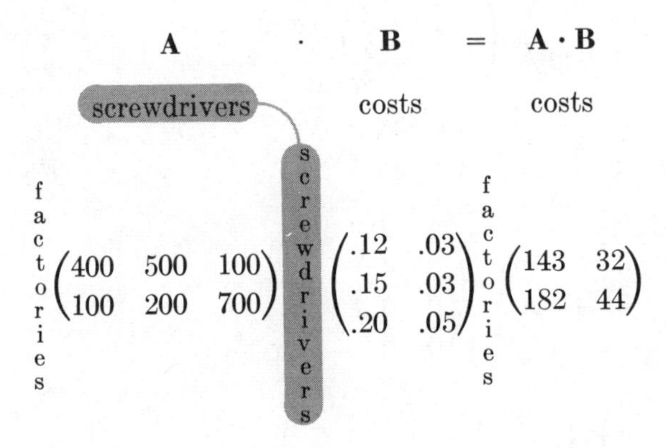

EXERCISES ■ SECTION 5.3.

1. See the remark following example 5.3.2 and investigate which of the following properties hold for matrices.

   a. **A + B = B + A.**

   b. **A + O = A** ("**O**" more generally is referred to as the **additive identity.** That is, it doesn't "change" things upon addition. Can you exhibit a matrix **X** such that **A + X = A**?)

   c. **A · I = A** ("**I**" more generally is referred to as the **multiplicative identity.** Can you exhibit a matrix such that when multiplied by any other matrix **A**, the matrix **A** remains the same? Hint: look at **C** in example 5.3.2 and generalize this.)

   d. If **A · B = O**, then **A = O** or **B = O**. (Hint: see exercise 6 on following page.)

In exercises 2–18, perform the indicated matrix operation(s) when possible.

2. $\begin{pmatrix} 1 & 7 & -1 \\ 5 & 0 & 0 \end{pmatrix} \cdot \begin{pmatrix} 5 \\ 7 \\ -2 \end{pmatrix}$

3. $\begin{pmatrix} 5 & -6 & 0 \\ \frac{1}{2} & 7 & 3 \\ 9 & 7 & \frac{4}{5} \end{pmatrix} - \begin{pmatrix} 5 & -6 & 0 \\ 3 & 7 & \frac{1}{2} \\ 9 & 7 & \frac{4}{5} \end{pmatrix}$

4. $5\begin{pmatrix} 2 & 1 \\ 3 & 7 \\ -1 & -2 \end{pmatrix} + 6\,(1 \quad 9 \quad 8)$

5. $5\begin{pmatrix} 2 & 1 \\ 3 & 7 \\ -1 & -2 \end{pmatrix} + 6\begin{pmatrix} 1 & 9 \\ 3 & 10 \\ 20 & 0 \end{pmatrix}$

6. $\begin{pmatrix} 1 & -5 & 4 \\ 6 & 2 & 0 \end{pmatrix} \cdot \begin{pmatrix} -2 & -1 \\ 6 & 3 \\ 8 & 4 \end{pmatrix}$

7. $\begin{pmatrix} -1 & 0 & 5 & 10 \\ 2 & 3 & 4 & 20 \end{pmatrix} \cdot \begin{pmatrix} 1 & 2 \\ 3 & 4 \end{pmatrix}$

8. $\begin{pmatrix} -1 & 0 & 5 & 10 \\ 2 & 3 & 4 & 20 \end{pmatrix} \cdot \begin{pmatrix} 1 & 2 \\ 3 & 4 \\ 0 & 0 \\ 0 & 0 \end{pmatrix}$

9. $\begin{pmatrix} 1 & -5 \\ 5 & 2 \\ 7 & 3 \end{pmatrix} \cdot \begin{pmatrix} 7 & 6 \\ -6 & 4 \end{pmatrix}$

10. $\begin{pmatrix} 1 & 2 & 3 \\ 4 & 5 & 6 \\ 7 & 8 & 9 \end{pmatrix} \cdot \begin{pmatrix} 1 & 0 & 0 \\ 0 & 1 & 0 \\ 0 & 0 & 1 \end{pmatrix}$ (Compare with exercise 1*c*, above.)

In exercises 11–18, let

$$\mathbf{X} = \begin{pmatrix} 7 & 3 & 2 \\ 11 & 15 & \frac{1}{8} \end{pmatrix} \qquad \mathbf{Y} = \begin{pmatrix} 1 & 1 & 1 \\ 0 & 1 & 1 \\ 0 & 0 & 1 \end{pmatrix} \qquad \mathbf{Z} = \begin{pmatrix} 1 & 2 & 3 \\ 3 & 2 & \frac{7}{8} \end{pmatrix}$$

11. Find $\mathbf{X} \cdot \mathbf{Y}$.
12. Find $\mathbf{Y} \cdot \mathbf{X}$.
13. Find $(3 \cdot \mathbf{X}) + \mathbf{Z}$.
14. Find $3(\mathbf{X} + \mathbf{Z})$.
15. Find $(\mathbf{X} + \mathbf{Z}) \cdot \mathbf{Y}$.
16. Find $\mathbf{Y} \cdot (\mathbf{X} + \mathbf{Z})$.
17. Find $(\mathbf{X} - \mathbf{Z}) \cdot \mathbf{Y}$.
18. Find $(\mathbf{X} \cdot \mathbf{Y}) \cdot \mathbf{Y}$.

19. The convenience of matrices with respect to expectation (see section 4.8) is seen in the following sociological survey. Ten thousand people were surveyed to determine a relationship between an individual's educational status (according to the highest degree earned) and the number of times the individual married. The totals are depicted in the table below.

	Never married	Married exactly once	Married exactly twice	Married more than twice
No high school diploma	25	1500	150	50
High school diploma	400	1050	325	120
At most a bachelor's degree	650	2500	1125	300
At most a master's degree	125	1175	435	15
Doctor's degree	35	5	5	10

A 5×4 matrix, call it **X**, can easily depict the above information. Furthermore, $\frac{1}{10{,}000} \cdot \mathbf{X}$ would be the division of each component of **X** by 10,000, the total in the survey. So $\frac{1}{10{,}000} \cdot \mathbf{X}$ is, in effect, a **probability matrix**.

a. Find $\frac{1}{10{,}000} \cdot \mathbf{X}$.

b. Suppose, in addition, a sociologist has compiled the following information.

	Students and faculty at Monroe Comm. College	Residents of Los Angeles	The people on a given block in Penfield, N.Y.
No high school diploma	125	2,150,000	31
High school diploma	5400	2,900,000	28
At most a bachelor's degree	100	2,600,000	5
At most a master's degree	230	1,350,000	4
Doctor's degree	50	700	0

Depict the above in a 3×5 (be careful) matrix **Y**.

c. Calculate the expected number of people you would find in the various locations by computing $\mathbf{Y} \cdot \left(\frac{1}{10{,}000} \cdot \mathbf{X}\right)$. (We are assuming here that the data in part *a* can be generalized to any of the given locations.)

20. A textbook publisher categorizes costs into: production costs, advertising expense, payroll costs, and miscellaneous expenses. For a certain biology text they are 40%, 10%, 5%, 5% of the wholesale price, respectively. For a literature text, they are 20%, 15%, 5%, 5%, respectively, and for a mathematics text they are 40%, 15%, 10%, 5%, respectively.

a. Depict the above cost percentage as a 4×3 matrix $\mathbf{C}$.

b. The wholesale price of the biology text is $8; the wholesale price of the literature text is $6; the wholesale price of the mathematics text is $10. Using suitable dimensions, depict a wholesale price matrix $\mathbf{P}$, and find the publisher's categorized cost by multiplying $\mathbf{C}$ and $\mathbf{P}$ in the correct order.

c. Multiply the 4×1 matrix solution of *b*, above, by the matrix $\mathbf{S} = (1 \quad 1 \quad 1 \quad 1)$. That is, find $\mathbf{S} \cdot (\mathbf{C} \cdot \mathbf{P})$. What does the result represent?

21. A psychiatrist, doing research on children with exactly two siblings, has found a relation between the age of an individual (with respect to his siblings) and certain physiopsychologic traits (call them T_1, T_2, T_3). They are depicted in the matrix $\mathbf{B}$ below:

$$\mathbf{B} = \begin{matrix} \text{oldest child} \\ \text{middle child} \\ \text{youngest child} \end{matrix} \overset{\begin{matrix} T_1 & T_2 & T_3 \end{matrix}}{\begin{pmatrix} 0.6 & 0.3 & 0.0 \\ 0.3 & 0.5 & 0.2 \\ 0.0 & 0.2 & 0.7 \end{pmatrix}}$$

where 0.6, for example, means 60% of oldest children have trait T_1. If two different samples S_1, S_2 have the following distributions of children:

S_1:	200 oldest	200 middle	200 youngest
S_2:	500 oldest	100 middle	100 youngest

find the expected number of children with traits T_1, T_2, T_3 for each sample.

22. Jennifer is about to panel her house. She is interested in four different types, *A*, *B*, *C*, *D*. There are three lumber yards which carry the panelling, Yard 1, Yard 2, Yard 3. Yard 1 regularly offers the panelling at \$4, \$3, \$2, \$1, respectively. Yard 2 regularly offers the panelling at \$4.50, \$3.50, \$2.50, \$1.50, respectively. Yard 3 regularly offers the panelling at \$5, \$3, \$1, \$.75, respectively. Jennifer has decided she needs either five A panels, three B panels, four C panels, and one D panel, or four A panels, four B panels, four C panels, and one D panel. Use matrix multiplication to find the cheapest choice and lumber yard for Jennifer.

23. To multiply the second column (only) of a 3×3 matrix by a constant k, we use the following "device":

$$\begin{pmatrix} a_{11} & a_{12} & a_{13} \\ a_{21} & a_{22} & a_{23} \\ a_{31} & a_{32} & a_{33} \end{pmatrix} \cdot \begin{pmatrix} 1 & 0 & 0 \\ 0 & k & 0 \\ 0 & 0 & 1 \end{pmatrix}$$

a. Verify that the desired product is this matrix:

$$\begin{pmatrix} a_{11} & ka_{12} & a_{13} \\ a_{21} & ka_{22} & a_{23} \\ a_{31} & ka_{32} & a_{33} \end{pmatrix}$$

b. Extend this idea to multiply the nth column of an $m \times m$ matrix by a constant.

24. Redo problem 22 when lumber yard 2 offers a "25% off" sale.

25. Suppose matrix $\mathbf{A}$ is an $m \times n$ matrix. By interchanging rows for columns, we form a new matrix $\mathbf{A}^T$, called the **transpose** of $\mathbf{A}$. ($\mathbf{A}^T$ is, of course, an $n \times m$ matrix.) For example, if

$$\mathbf{A} = \begin{pmatrix} 1 & -1 & 6 \\ -3 & 2 & 4 \end{pmatrix} \quad \text{then} \quad \mathbf{A}^T = \begin{pmatrix} 1 & -3 \\ -1 & 2 \\ 6 & 4 \end{pmatrix}$$

Find the transpose of each matrix below.

a. $\begin{pmatrix} 1 & 2 & 3 \\ 4 & 5 & 6 \end{pmatrix}$ **b.** $(5 \quad 7 \quad -6)$ **c.** $\begin{pmatrix} 1 \\ 2 \\ \frac{1}{2} \end{pmatrix}$

d. $\begin{pmatrix} 1 & 0 & 0 & 0 \\ 0 & 1 & 0 & 0 \\ 0 & 0 & 1 & 0 \\ 0 & 0 & 0 & 1 \end{pmatrix}$ **e.** $\begin{pmatrix} \frac{1}{2} & \frac{2}{5} & \frac{1}{10} \\ 1 & 0 & 0 \\ \frac{1}{2} & \frac{1}{3} & \frac{1}{6} \end{pmatrix}$

26. Determine which of the following statements (if any) are not true.

a. $(\mathbf{A} + \mathbf{B})^T = \mathbf{A}^T + \mathbf{B}^T$ **b.** $(\mathbf{A} \cdot \mathbf{B})^T = \mathbf{A}^T \cdot \mathbf{B}^T$

c. $(\mathbf{A} \cdot \mathbf{B})^T = \mathbf{B}^T \cdot \mathbf{A}^T$ **d.** $(\mathbf{A}^T)^T = \mathbf{A}$

5.4 THE INVERSE OF A MATRIX

When the number 5 is multiplied by 1, the product is, of course, still 5. This property is unique for 1 and 1 is called the **identity** for multiplication. Zero is the identity for addition because nothing changes when zero is added to anything. In the previous section (exercise 1*c*), we took one look at the matrix multiplicative identity. In general, if **A** is an $m \times n$ matrix, and we wish to exhibit an identity matrix **I** such that $\mathbf{A} \cdot \mathbf{I} = \mathbf{A}$, then **I** must be an $\mathbf{n} \times \mathbf{n}$ matrix. (Why?) It will look like this:

$$\mathbf{I} = \begin{pmatrix} 1 & 0 & 0 & 0 & 0 & \cdots & 0 \\ 0 & 1 & 0 & 0 & 0 & \cdots & 0 \\ 0 & 0 & 1 & 0 & 0 & \cdots & 0 \\ \cdot & & & & & & \cdot \\ \cdot & & & & & & \cdot \\ \cdot & & & & & & \cdot \\ 0 & 0 & 0 & 0 & 0 & \cdots & 1 \end{pmatrix}$$

For example, if **A** is the 5×3 matrix

$$\mathbf{A} = \begin{pmatrix} 1 & -1 & 10 \\ 2 & 6 & 20 \\ 3 & -11 & 30 \\ 4 & 9 & 40 \\ 7 & 8 & 0 \end{pmatrix}$$

then to multiply **A** by

$$\mathbf{I} = \begin{pmatrix} 1 & 0 & 0 \\ 0 & 1 & 0 \\ 0 & 0 & 1 \end{pmatrix}$$

yields

$$\mathbf{A} \cdot \mathbf{I} = \begin{pmatrix} 1 & -1 & 10 \\ 2 & 6 & 20 \\ 3 & -11 & 30 \\ 4 & 9 & 40 \\ 7 & 8 & 0 \end{pmatrix} \cdot \begin{pmatrix} 1 & 0 & 0 \\ 0 & 1 & 0 \\ 0 & 0 & 1 \end{pmatrix} = \begin{pmatrix} 1 & -1 & 10 \\ 2 & 6 & 20 \\ 3 & -11 & 30 \\ 4 & 9 & 40 \\ 7 & 8 & 0 \end{pmatrix} = \mathbf{A}$$

That is, $\mathbf{A} \cdot \mathbf{I} = \mathbf{A}$.

We will confine our discussion throughout the remainder of this chapter to square matrices.

We say that when a matrix **A** can be multiplied by a matrix **B** to yield the identity matrix, then **B** is **the inverse of A**. That is, whenever $\mathbf{A} \cdot \mathbf{B} = \mathbf{I}$, then **B** is the inverse of **A** (or equivalently, **A** and **B** are inverses of each other). For example, if

$$\mathbf{A} = \begin{pmatrix} 1 & 2 \\ 3 & 4 \end{pmatrix} \quad \text{and} \quad \mathbf{B} = \begin{pmatrix} -2 & 1 \\ \frac{3}{2} & -\frac{1}{2} \end{pmatrix}$$

then

$$\mathbf{A} \cdot \mathbf{B} = \begin{pmatrix} 1 \cdot (-2) + 2 \cdot \frac{3}{2} & 1 \cdot 1 + 2 \cdot (-\frac{1}{2}) \\ 3 \cdot (-2) + 4 \cdot \frac{3}{2} & 3 \cdot 1 + 4 \cdot (-\frac{1}{2}) \end{pmatrix} = \begin{pmatrix} 1 & 0 \\ 0 & 1 \end{pmatrix}$$

We denote the inverse of **A** by $\mathbf{A}^{-1}$. So,

$$\mathbf{A} \cdot \mathbf{A}^{-1} = \mathbf{I}$$

To determine whether two matrices are inverses of each other, just multiply and see if the identity matrix is obtained.

The goal of the remainder of this section is to develop a method for FINDING the inverse of a given matrix. The method we shall employ involves changing the rows of the given matrix by means of **row transformations.** There are three types of row transformations:

1. Multiply a row of the matrix by any number except zero.
2. Add a multiple of any row to another row. (Subtract a multiple of any row from another row.)
3. Interchange any two rows.

Given a matrix **A**, we shall perform the row transformations to change matrix **A** to the identity matrix. WHEN THESE SAME TRANSFORMATIONS ARE PERFORMED ON THE IDENTITY MATRIX, THE RESULT IS THE INVERSE.

For example, consider $\mathbf{A} = \begin{pmatrix} 2 & 4 \\ 3 & -1 \end{pmatrix}$. We perform the transformations in the following manner:

Step 1. Multiply the top row by $\frac{1}{2}$. We do this keeping in mind that the identity matrix has a 1 in the a_{11} position. So,

$$\begin{pmatrix} 2 & 4 \\ 3 & -1 \end{pmatrix} \xrightarrow{\text{step 1}} \begin{pmatrix} 1 & 2 \\ 3 & -1 \end{pmatrix}$$

Step 2. Subtract 3 times the new first row from the second row. We do this because a_{21} is a zero in the identity matrix. So,

$$\begin{pmatrix} 1 & 2 \\ 3 & -1 \end{pmatrix} \xrightarrow{\text{step 2}} \begin{pmatrix} 1 & 2 \\ 0 & -7 \end{pmatrix}$$

Step 3. Multiply the new second row by $-\frac{1}{7}$, because the identity matrix has a "1" in the a_{22} position. So,

$$\begin{pmatrix} 1 & 2 \\ 0 & -7 \end{pmatrix} \xrightarrow{\text{step 3}} \begin{pmatrix} 1 & 2 \\ 0 & 1 \end{pmatrix}$$

Step 4. Finally, subtract twice the bottom row from the top row.

$$\begin{pmatrix} 1 & 2 \\ 0 & 1 \end{pmatrix} \xrightarrow{\text{step 4}} \begin{pmatrix} 1 & 0 \\ 0 & 1 \end{pmatrix}$$

We have in four steps transformed the given matrix $\mathbf{A}$ into the identity matrix. The inverse $\mathbf{A}^{-1}$ will now be obtained by applying these exact operations on $\mathbf{I}$. We obtain

$$\begin{pmatrix} 1 & 0 \\ 0 & 1 \end{pmatrix} \xrightarrow{\text{step 1}} \begin{pmatrix} \frac{1}{2} & 0 \\ 0 & 1 \end{pmatrix}$$

$$\begin{pmatrix} \frac{1}{2} & 0 \\ 0 & 1 \end{pmatrix} \xrightarrow{\text{step 2}} \begin{pmatrix} \frac{1}{2} & 0 \\ -\frac{3}{2} & 1 \end{pmatrix}$$

$$\begin{pmatrix} \frac{1}{2} & 0 \\ -\frac{3}{2} & 1 \end{pmatrix} \xrightarrow{\text{step 3}} \begin{pmatrix} \frac{1}{2} & 0 \\ \frac{3}{14} & -\frac{1}{7} \end{pmatrix}$$

$$\begin{pmatrix} \frac{1}{2} & 0 \\ \frac{3}{14} & -\frac{1}{7} \end{pmatrix} \xrightarrow{\text{step 4}} \begin{pmatrix} \frac{1}{14} & \frac{2}{7} \\ \frac{3}{14} & -\frac{1}{7} \end{pmatrix}$$

So,

$$\begin{pmatrix} \frac{1}{14} & \frac{2}{7} \\ \frac{3}{14} & -\frac{1}{7} \end{pmatrix} = \mathbf{A}^{-1}$$

The reader is urged to verify that $\mathbf{A} \cdot \mathbf{A}^{-1} = \mathbf{I}$.

We can use a convenient notation for working on both the given matrix and the identity matrix simultaneously. The two matrices, separated by a vertical line, are both placed between parentheses. If $\mathbf{C} = \begin{pmatrix} 10 & -2 \\ 6 & 3 \end{pmatrix}$ we have $\left(\begin{array}{cc|cc} 10 & -2 & 1 & 0 \\ 6 & 3 & 0 & 1 \end{array}\right)$. To find $\mathbf{C}^{-1}$ we must perform the row transformations simultaneously on both sides of the separated matrix.

First multiply row one by $\frac{1}{10}$:

$$\left(\begin{array}{cc|cc} 1 & -\frac{1}{5} & \frac{1}{10} & 0 \\ 6 & 3 & 0 & 1 \end{array}\right)$$

Now subtract six times the first row from the second row:

$$\left(\begin{array}{cc|cc} 1 & -\frac{1}{5} & \frac{1}{10} & 0 \\ 0 & \frac{21}{5} & -\frac{3}{5} & 1 \end{array}\right)$$

Multiply the second row by $\frac{5}{21}$:

$$\left(\begin{array}{cc|cc} 1 & -\frac{1}{5} & \frac{1}{10} & 0 \\ 0 & 1 & -\frac{1}{7} & \frac{5}{21} \end{array}\right)$$

Add $\frac{1}{5}$ times the second row to the first row:

$$\left(\begin{array}{cc|cc} 1 & 0 & \frac{1}{14} & \frac{1}{21} \\ 0 & 1 & -\frac{1}{7} & \frac{5}{21} \end{array}\right)$$

So,

$$\mathbf{C}^{-1} = \begin{pmatrix} \frac{1}{14} & \frac{1}{21} \\ -\frac{1}{7} & \frac{5}{21} \end{pmatrix}$$

A check reveals:

$$\mathbf{C}\cdot\mathbf{C}^{-1} = \begin{pmatrix} 10 & -2 \\ 6 & 3 \end{pmatrix}\cdot\begin{pmatrix} \frac{1}{14} & \frac{1}{21} \\ -\frac{1}{7} & \frac{5}{21} \end{pmatrix} = \begin{pmatrix} \frac{10}{14}+\frac{2}{7} & \frac{10}{21}-\frac{10}{21} \\ \frac{6}{14}-\frac{3}{7} & \frac{6}{21}+\frac{15}{21} \end{pmatrix}$$

$$= \begin{pmatrix} 1 & 0 \\ 0 & 1 \end{pmatrix} = \mathbf{I}$$

An example follows.

Example 5.4.1. Find the inverse of each of the following matrices and use matrix multiplication to check for the identity matrix.

a. $\mathbf{P} = \begin{pmatrix} 0 & 5 \\ 1 & 0 \end{pmatrix}$ **b.** $\mathbf{Q} = \begin{pmatrix} 1 & 2 & 3 \\ -2 & 4 & 0 \\ 0 & 8 & 10 \end{pmatrix}$

Solution:

a. $\left(\begin{array}{cc|cc} 0 & 5 & 1 & 0 \\ 1 & 0 & 0 & 1 \end{array}\right)$

Here it would be most convenient to use row transformation 3 (interchange the two rows):

$$\left(\begin{array}{cc|cc} 1 & 0 & 0 & 1 \\ 0 & 5 & 1 & 0 \end{array}\right)$$

Now, $\frac{1}{5}$ times the second row should do it!

$$\left(\begin{array}{cc|cc} 1 & 0 & 0 & 1 \\ 0 & 1 & \frac{1}{5} & 0 \end{array}\right)$$

So, $$\mathbf{P}^{-1} = \begin{pmatrix} 0 & 1 \\ \frac{1}{5} & 0 \end{pmatrix}$$

Check: $$\mathbf{P}\cdot\mathbf{P}^{-1} = \begin{pmatrix} 0 & 5 \\ 1 & 0 \end{pmatrix}\cdot\begin{pmatrix} 0 & 1 \\ \frac{1}{5} & 0 \end{pmatrix} = \begin{pmatrix} 1 & 0 \\ 0 & 1 \end{pmatrix}$$

b. The inverse of a 3×3 matrix will, of course, take more steps.

$$\left(\begin{array}{ccc|ccc} 1 & 2 & 3 & 1 & 0 & 0 \\ -2 & 4 & 0 & 0 & 1 & 0 \\ 0 & 8 & 10 & 0 & 0 & 1 \end{array}\right)$$

Adding twice row one to row two yields:

$$\left(\begin{array}{ccc|ccc} 1 & 2 & 3 & 1 & 0 & 0 \\ 0 & 8 & 6 & 2 & 1 & 0 \\ 0 & 8 & 10 & 0 & 0 & 1 \end{array}\right)$$

The motivation here is to get a zero in the second row

Subtracting $\frac{1}{4}$ times row two from row one yields:

$$\left(\begin{array}{ccc|ccc} 1 & 0 & \frac{3}{2} & \frac{1}{2} & -\frac{1}{4} & 0 \\ 0 & 8 & 6 & 2 & 1 & 0 \\ 0 & 8 & 10 & 0 & 0 & 1 \end{array}\right)$$

The motivation here is to get the zero in the first row

Multiplying row two by $\frac{1}{8}$ yields:

$$\left(\begin{array}{ccc|ccc} 1 & 0 & \frac{3}{2} & \frac{1}{2} & -\frac{1}{4} & 0 \\ 0 & 1 & \frac{3}{4} & \frac{1}{4} & \frac{1}{8} & 0 \\ 0 & 8 & 10 & 0 & 0 & 1 \end{array}\right)$$

Here, we're trying for the one in the second row

Subtracting eight times row two from row three yields:

$$\left(\begin{array}{ccc|ccc} 1 & 0 & \frac{3}{2} & \frac{1}{2} & -\frac{1}{4} & 0 \\ 0 & 1 & \frac{3}{4} & \frac{1}{4} & \frac{1}{8} & 0 \\ 0 & 0 & 4 & -2 & -1 & 1 \end{array}\right)$$

We want a zero here

Multiplying row three by $\frac{1}{4}$ yields:

$$\left(\begin{array}{ccc|ccc} 1 & 0 & \frac{3}{2} & \frac{1}{2} & -\frac{1}{4} & 0 \\ 0 & 1 & \frac{3}{4} & \frac{1}{4} & \frac{1}{8} & 0 \\ 0 & 0 & 1 & -\frac{1}{2} & -\frac{1}{4} & \frac{1}{4} \end{array}\right)$$

We want a one here

Subtracting $\frac{3}{4}$ times the third row from the second row yields:

$$\left(\begin{array}{ccc|ccc} 1 & 0 & \frac{3}{2} & \frac{1}{2} & -\frac{1}{4} & 0 \\ 0 & 1 & 0 & \frac{5}{8} & \frac{5}{16} & -\frac{3}{16} \\ 0 & 0 & 1 & -\frac{1}{2} & -\frac{1}{4} & \frac{1}{4} \end{array}\right)$$

We want a zero

Finally, subtracting $\frac{3}{2}$ of row three from row one should do it!

$$\left(\begin{array}{ccc|ccc} 1 & 0 & 0 & \frac{5}{4} & \frac{1}{8} & -\frac{3}{8} \\ 0 & 1 & 0 & \frac{5}{8} & \frac{5}{16} & -\frac{3}{16} \\ 0 & 0 & 1 & -\frac{1}{2} & -\frac{1}{4} & \frac{1}{4} \end{array}\right)$$

Finally, this zero

So, $$\mathbf{Q}^{-1} = \begin{pmatrix} \frac{5}{4} & \frac{1}{8} & -\frac{3}{8} \\ \frac{5}{8} & \frac{5}{16} & -\frac{3}{16} \\ -\frac{1}{2} & -\frac{1}{4} & \frac{1}{4} \end{pmatrix}$$

Check:

$$\mathbf{Q} \cdot \mathbf{Q}^{-1} = \begin{pmatrix} 1 & 2 & 3 \\ -2 & 4 & 0 \\ 0 & 8 & 10 \end{pmatrix} \cdot \begin{pmatrix} \frac{5}{4} & \frac{1}{8} & -\frac{3}{8} \\ \frac{5}{8} & \frac{5}{16} & -\frac{3}{16} \\ -\frac{1}{2} & -\frac{1}{4} & \frac{1}{4} \end{pmatrix} = \begin{pmatrix} 1 & 0 & 0 \\ 0 & 1 & 0 \\ 0 & 0 & 1 \end{pmatrix}$$

So, we have seen that when finding an inverse, we transform the given matrix into the identity matrix. However, can that always be done? Suppose $\mathbf{S} = \begin{pmatrix} 9 & 12 \\ 3 & 4 \end{pmatrix}$. No matter how hard you try, you won't be able to convert $\mathbf{S}$ to $\begin{pmatrix} 1 & 0 \\ 0 & 1 \end{pmatrix}$. Watch:

$$\left(\begin{array}{cc|cc} 9 & 12 & 1 & 0 \\ 3 & 4 & 0 & 1 \end{array}\right)$$

$$\left(\begin{array}{cc|cc} 1 & \frac{4}{3} & \frac{1}{9} & 0 \\ 3 & 4 & 0 & 1 \end{array}\right)$$

$$\left(\begin{array}{cc|cc} 1 & \frac{4}{3} & \frac{1}{9} & 0 \\ 0 & 0 & -\frac{1}{3} & 1 \end{array}\right)$$

None of the three permissible row transformations will convert that second component in the top row to zero while leaving the 1 in the first component. When this happens we say, very simply, that **S** has no inverse. An indication that you should stop trying (i.e., that the inverse doesn't exist) is a row of zeros. During the transformation process, whenever a row of zeros occurs, the original matrix has no inverse.

Now that we have investigated the process of finding the inverse of a square matrix, we will apply the procedure in the next section.

EXERCISES ■ SECTION 5.4.

1. Determine whether or not the following pairs of matrices are inverses of one another by seeing if the product under matrix multiplication is the identity.

   a. $\begin{pmatrix} 1 & 2 \\ 3 & 4 \end{pmatrix} \quad \begin{pmatrix} -2 & 1 \\ \frac{3}{2} & -\frac{1}{2} \end{pmatrix}$

   b. $\begin{pmatrix} 1 & 0 & 8 \\ 5 & 10 & -20 \\ 0 & 0 & 2 \end{pmatrix} \quad \begin{pmatrix} 1 & 0 & 4 \\ -\frac{1}{2} & \frac{1}{10} & 3 \\ 0 & 0 & \frac{1}{2} \end{pmatrix}$

c. $\begin{pmatrix} 0 & 0 & 0 & 1 \\ 1 & 0 & 0 & 0 \\ 0 & 1 & 0 & 0 \\ 0 & 0 & 1 & 0 \end{pmatrix}$ $\begin{pmatrix} 0 & 1 & 0 & 0 \\ 0 & 0 & 1 & 0 \\ 0 & 0 & 0 & 1 \\ 1 & 0 & 0 & 0 \end{pmatrix}$

Find the inverse of each of the following matrices, if possible. Check your answer using matrix multiplication.

2. $\begin{pmatrix} 6 & 12 \\ 0 & 4 \end{pmatrix}$ **3.** $\begin{pmatrix} 5 & 7 \\ 3 & 4 \end{pmatrix}$ **4.** $\begin{pmatrix} \frac{1}{2} & \frac{1}{2} \\ \frac{1}{2} & 1 \end{pmatrix}$

5. $\begin{pmatrix} 1 & 0 & 0 \\ -11 & 7 & 3 \\ 6 & 5 & 2 \end{pmatrix}$ **6.** $\begin{pmatrix} 1 & 2 & 4 \\ 8 & 11 & 9 \\ 6 & 0 & 9 \end{pmatrix}$ **7.** $\begin{pmatrix} 8 & 2 \\ 16 & 4 \end{pmatrix}$

8. $\begin{pmatrix} 1 & 10 & 0 \\ 3 & 5 & 1 \\ 6 & 2 & 2 \end{pmatrix}$ **9.** $\begin{pmatrix} 0 & 0 & 0 & 5 \\ 5 & 0 & 0 & 0 \\ 0 & 5 & 0 & 0 \\ 0 & 0 & 5 & 0 \end{pmatrix}$ **10.** $\begin{pmatrix} 1 & 2 & 3 & 4 \\ 4 & 3 & 1 & 0 \\ 0 & 5 & 6 & 1 \\ 7 & 2 & 5 & 1 \end{pmatrix}$

11. $\begin{pmatrix} 6 & 5 & 4 \\ 3 & 2 & -1 \\ 1 & 2 & 3 \end{pmatrix}$ **12.** $\begin{pmatrix} 4 & 2 & 10 \\ 6 & 1 & 3 \\ 5 & 5 & 5 \end{pmatrix}$

5.5 APPLICATIONS

In this section we will examine two important applications of matrices, both incorporating the use of the matrix inverse.

The first way in which matrices will be "put to work" is in solving a system of linear equations. Such a system may look like this:

$$\begin{aligned} x + 2y &= 1 \\ 3x + 4y &= -1 \end{aligned}$$

The object is to find numbers that, when substituted for x and y, will satisfy both equations simultaneously. The above is called **a system of two equations in two unknowns.** Similarly,

$$\begin{aligned} x + y + z &= 7 \\ 2x + 3y - z &= 2 \\ x - 11y + 6z &= 1 \end{aligned}$$

is a system of three equations in three unknowns, and

$$\begin{aligned} x + y + z + w &= 8 \\ y + w &= 11 \end{aligned}$$

is a system of two equations in four unknowns.

To solve a system of equations, we employ matrices. More specifically, we first form the **matrix of coefficients** of the system. For

$$\begin{aligned} x + 2y &= 1 \\ 3x + 4y &= -1 \end{aligned}$$

the matrix of coefficients is:

$$\begin{pmatrix} 1 & 2 \\ 3 & 4 \end{pmatrix}$$

Next, there are two unknowns (x and y). We form a column vector of the unknowns

$$\begin{pmatrix} x \\ y \end{pmatrix}$$

and a column vector of constants (the right hand side of both equations):

$$\begin{pmatrix} 1 \\ -1 \end{pmatrix}$$

We then write:

$$\begin{pmatrix} 1 & 2 \\ 3 & 4 \end{pmatrix} \cdot \begin{pmatrix} x \\ y \end{pmatrix} = \begin{pmatrix} 1 \\ -1 \end{pmatrix}$$

which, by matrix multiplication, is equivalent to

$$\begin{pmatrix} x + 2y \\ 3x + 4y \end{pmatrix} = \begin{pmatrix} 1 \\ -1 \end{pmatrix}$$

Verifying (by exercise 8, section 5.2) that

$$\begin{pmatrix} 1 & 2 \\ 3 & 4 \end{pmatrix} \cdot \begin{pmatrix} x \\ y \end{pmatrix} = \begin{pmatrix} 1 \\ -1 \end{pmatrix} \quad \text{is equivalent to} \quad \begin{aligned} x + 2y &= 1 \\ 3x + 4y &= -1 \end{aligned}$$

Now we have

$$\mathbf{A} \cdot \begin{pmatrix} x \\ y \end{pmatrix} = \begin{pmatrix} 1 \\ -1 \end{pmatrix}$$

where **A** is the matrix of coefficients, $\begin{pmatrix} 1 & 2 \\ 3 & 4 \end{pmatrix}$. To solve for x and y, we must "isolate" the column vector $\begin{pmatrix} x \\ y \end{pmatrix}$. This can be achieved by multiplying both sides by $\mathbf{A}^{-1}$. Watch:

$$\mathbf{A} \cdot \begin{pmatrix} x \\ y \end{pmatrix} = \begin{pmatrix} 1 \\ -1 \end{pmatrix}$$

$$\mathbf{A}^{-1} \cdot \mathbf{A} \cdot \begin{pmatrix} x \\ y \end{pmatrix} = \mathbf{A}^{-1} \cdot \begin{pmatrix} 1 \\ -1 \end{pmatrix}$$

or

$$\mathbf{I} \cdot \begin{pmatrix} x \\ y \end{pmatrix} = \mathbf{A}^{-1} \cdot \begin{pmatrix} 1 \\ -1 \end{pmatrix}$$

and

$$\begin{pmatrix} x \\ y \end{pmatrix} = \mathbf{A}^{-1} \cdot \begin{pmatrix} 1 \\ -1 \end{pmatrix}$$

In the previous section we found $\mathbf{A}^{-1} = \begin{pmatrix} -2 & 1 \\ \frac{3}{2} & -\frac{1}{2} \end{pmatrix}$. Thus

$$\begin{pmatrix} x \\ y \end{pmatrix} = \mathbf{A}^{-1} \cdot \begin{pmatrix} 1 \\ -1 \end{pmatrix}$$

$$= \begin{pmatrix} -2 & 1 \\ \frac{3}{2} & -\frac{1}{2} \end{pmatrix} \cdot \begin{pmatrix} 1 \\ -1 \end{pmatrix} = \begin{pmatrix} -3 \\ 2 \end{pmatrix}$$

By equality of vectors, $x = -3$ and $y = 2$. A check reveals:

$$\begin{aligned} x + 2y &= 1 \\ -3 + 2(2) &= 1 \\ 1 &= 1 \checkmark \end{aligned} \qquad \text{AND} \qquad \begin{aligned} 3x + 4y &= -1 \\ 3(-3) + 4(2) &= -1 \\ -1 &= -1 \checkmark \end{aligned}$$

The same procedure applies for any system of n equations in n unknowns. An example follows.

Example 5.5.1. Solve the system:

$$\begin{aligned} 2x_1 + 3x_2 - 4x_3 &= 44 \\ x_1 + x_2 + x_3 &= 14 \\ -x_1 + 7x_2 - 10x_3 &= 75 \end{aligned}$$

Solution: The system can be represented as $\mathbf{A} \cdot \mathbf{X} = \mathbf{B}$, where

$$\mathbf{A} = \begin{pmatrix} 2 & 3 & -4 \\ 1 & 1 & 1 \\ -1 & 7 & -10 \end{pmatrix} \qquad \mathbf{X} = \begin{pmatrix} x_1 \\ x_2 \\ x_3 \end{pmatrix}$$

and

$$\mathbf{B} = \begin{pmatrix} 44 \\ 14 \\ 75 \end{pmatrix}$$

The next step is to find $\mathbf{A}^{-1}$. The reader is urged to verify that

$$\mathbf{A}^{-1} = \begin{pmatrix} \frac{17}{39} & \frac{-2}{39} & \frac{-7}{39} \\ \frac{-9}{39} & \frac{24}{39} & \frac{6}{39} \\ \frac{-8}{39} & \frac{17}{39} & \frac{1}{39} \end{pmatrix}$$

So,
$$\begin{pmatrix} x_1 \\ x_2 \\ x_3 \end{pmatrix} = \mathbf{A}^{-1} \cdot \begin{pmatrix} 44 \\ 14 \\ 75 \end{pmatrix}$$
$$= \begin{pmatrix} \frac{17}{39} & \frac{-2}{39} & \frac{-7}{39} \\ \frac{-9}{39} & \frac{24}{39} & \frac{6}{39} \\ \frac{-8}{39} & \frac{17}{39} & \frac{1}{39} \end{pmatrix} \cdot \begin{pmatrix} 44 \\ 14 \\ 75 \end{pmatrix} = \begin{pmatrix} 5 \\ 10 \\ -1 \end{pmatrix}$$

Finally,

$$x_1 = 5 \qquad x_2 = 10 \qquad x_3 = -1$$

which checks in the three original equations:

$$\begin{aligned} 2x_1 + 3x_2 - 4x_3 &= 44 \\ 2(5) + 3(10) - 4(-1) &= 44 \\ 10 + 30 + 4 &= 44 \\ 44 &= 44 \checkmark \end{aligned} \qquad \begin{aligned} x_1 + x_2 + x_3 &= 14 \\ 5 + 10 + (-1) &= 14 \\ 14 &= 14 \checkmark \end{aligned}$$

$$\begin{aligned} -x_1 + 7x_2 - 10x_3 &= 75 \\ -(5) + 7(10) - 10(-1) &= 75 \\ -5 + 70 + 10 &= 75 \\ 75 &= 75 \checkmark \end{aligned}$$

It is necessary to make two remarks here. First, we have only examined "square" systems, that is, n equations in n unknowns. In general, for systems where the number of unknowns exceeds the number of equations, there is no unique solution. Second, one question that naturally arises is, "What happens if the inverse does not exist?" For example, in

$$\begin{aligned} x + y &= 10 \\ 2x + 2y &= 20 \end{aligned}$$

the matrix of coefficients is

$$\mathbf{A} = \begin{pmatrix} 1 & 1 \\ 2 & 2 \end{pmatrix}$$

which has no inverse. A geometrical interpretation of two equations in two unknowns will illustrate this. Linear equations represent straight lines. When we have two straight lines, there are three possibilities:

1. They intersect at one point.
2. They are parallel; they do not intersect. } Here the inverse does not exist.
3. They are the same line. }

So, when the inverse does not exist, we will have no solution (parallel lines) or infinitely many solutions (coincidental lines).

Two more examples follow.

Example 5.5.2. Solve for s and t:

$$\begin{aligned} 10s - 20t &= 6 \\ s - 2t &= 11 \end{aligned}$$

Solution: We rewrite the system:

$$\begin{pmatrix} 10 & -20 \\ 1 & -2 \end{pmatrix} \cdot \begin{pmatrix} s \\ t \end{pmatrix} = \begin{pmatrix} 6 \\ 11 \end{pmatrix}$$

and try to find the inverse of the matrix of coefficients. It cannot be found, indicating there is no unique solution to the system.

Example 5.5.3. The Ptomaine Diner sells 300 hamburgers and 100 hot dogs on Monday for $185. On Tuesday they sell 100 hamburgers and 100 hot dogs for $85. Find the price of a hamburger and the price of a hot dog.

Solution: Let x_1 and x_2 denote the price of a hamburger and a hot dog, respectively. The given information translates to:

$$\begin{aligned} 300x_1 + 100x_2 &= \$185 \\ 100x_1 + 100x_2 &= \$85 \end{aligned}$$

or

$$\begin{pmatrix} 300 & 100 \\ 100 & 100 \end{pmatrix} \cdot \begin{pmatrix} x_1 \\ x_2 \end{pmatrix} = \begin{pmatrix} 185 \\ 85 \end{pmatrix}$$

Solving, we obtain

$$\begin{pmatrix} x_1 \\ x_2 \end{pmatrix} = \underbrace{\begin{pmatrix} 0.005 & -0.005 \\ -0.005 & 0.005 \end{pmatrix}}_{\mathbf{A}^{-1}} \cdot \begin{pmatrix} 185 \\ 85 \end{pmatrix} = \begin{pmatrix} .50 \\ .35 \end{pmatrix}$$

The price of a hamburger is 50¢; a hot dog costs 35¢.

The second type of application is to **cryptography,** the art of secret writing. The importance of "coding" and "decoding" is obvious in espionage work. It is also used every day by food packing plants and other industries where a code (for a date, for instance) might be stamped on a product with the intention that only "those in the know" will be able to read it.

Suppose Al wants to send the following secret message to Boris:

"MARY IS THE ONE. WATCH OUT."

We first write the message in numbers, where each letter in the alphabet has a number associated with it. For simplicity, let's use the pattern

$A = 1, B = 2, \ldots, Z = 26$. Our message now looks like this:

13 1 18 25 9 19 20 8 5 15 14 5 23 1 20 3 8 15 21 20

Now, we'll make up any convenient invertible square matrix (one that has an inverse), group our message as row vectors, and multiply. Let's use the 2×2 matrix $\mathbf{A} = \begin{pmatrix} 1 & 1 \\ 2 & 1 \end{pmatrix}$. We group the message in 1×2 vectors (why 1×2?) and multiply to get the coded message:

$$(13 \quad 1) \cdot \begin{pmatrix} 1 & 1 \\ 2 & 1 \end{pmatrix}, \qquad (18 \quad 25) \cdot \begin{pmatrix} 1 & 1 \\ 2 & 1 \end{pmatrix}, \qquad (9 \quad 19) \cdot \begin{pmatrix} 1 & 1 \\ 2 & 1 \end{pmatrix}, \qquad \text{etc.}$$

We get:

15 14 68 43 47 28 36 28 35 20 24 19 25 24 26 23 38 23 61 41

If Al wanted to go one step further, he could now convert this back to letters. Trouble arises with numbers like 68, 43, but think of the letter A as not only equal to 1 but also equal to 27 $(1 + 26)$, 53 $(1 + 52)$, -25 $(1 - 26)$, etc. Similarly, $P = 16, 42, 68, -10$, etc. Doing this, we obtain:

O N P Q U B J B I T X S Y X Z W L W I O

Boris receives the above, which looks like nonsense at first glance. However, Boris has in his possession $\mathbf{A}^{-1}$, the decoder. Boris takes

O N P Q U B J B I T X S Y X Z W L W I O

writes it as numbers, and multiplies by $\mathbf{A}^{-1} = \begin{pmatrix} -1 & 1 \\ 2 & -1 \end{pmatrix}$. The reader is urged to verify Boris' calculation:

13 1 18 25 9 19 20 8 5 15 14 5 23 1 20 3 8 15 21 20

or MARY IS THE ONE WATCH OUT

Example 5.5.4. Suppose Boris received the coded message J B U B U B G F Q R. Decode it.

Solution: We write J B U B U B G F Q R as

10 2 21 2 21 2 7 6 17 18

and use the multiplier-decoder, $\mathbf{A}^{-1}$, to obtain

-6 8 -17 19 -17 19 5 1 19 -1

But according to our "revolving" technique, -6 is equivalent to 20, -17 is equivalent to 9, etc. We have

20 8 9 19 9 19 5 1 19 25

and converting to letters we have:

THIS IS EASY

A last note suffices. One step we've added is extra and not needed for code-decode completion. It is the step where the coded numbers are changed to letters before decoding. This cannot always be done. If the coding matrix **A** is $\begin{pmatrix} \frac{1}{8} & \frac{1}{10} \\ \frac{1}{11} & \frac{1}{12} \end{pmatrix}$, the coded message will quite obviously contain fractions, making a conversion to letters impossible. The matrix $\mathbf{A} = \begin{pmatrix} 1 & 1 \\ 2 & 1 \end{pmatrix}$ will avoid this. So will the matrix $\mathbf{B} = \begin{pmatrix} 2 & 1 \\ 3 & 2 \end{pmatrix}$, but many will not.

The art of secret coding has developed into an involved and highly technical science. For further information, the reader is directed to the most interesting chapter XIV in the book *Mathematical Recreations and Essays*, by W. W. Rouse Ball (Macmillan, 1962).

EXERCISES ■ SECTION 5.5

In problems 1–10, use matrices to solve the given system, where possible.

1. $x + y = 11$
   $x - y = 3$

2. $x_1 + x_2 + x_3 = 6$
   $x_1 - x_2 = -1$
   $10x_1 + 6x_2 - x_3 = 19$

3. $x + y - z = -17$
   $x + 2y + 2z = 3$
   $7y + z = -40$

4. $10x_1 - 3x_2 = 2$
   $x_2 + x_3 = 7$
   $x_1 - x_2 + 2x_3 = -2$

5. $w + x + y + z = 11$
   $x - 2y + z = -4$
   $10w + y + 2z = 17$
   $w + 2x + 3y + 4z = 39$

6. $x + y = 7$
   $2x + 2y = 14$

7. $x + y = 7$
   $x + y = 8$

8. $a + b + c + d = 10$
   $a + 2b + c + 2d = 16$
   $3a - b - c + d = 2$
   $6a + b - c + 2d = 13$

9. $x_1 + x_2 = 9$
   $x_1 + x_3 = 4$
   $x_2 + x_3 = 5$

10. $x_1 - x_2 = 1$
    $3x_1 + 2x_2 = 23$

11. The sum of two numbers is 18. Their difference is 20. Find the numbers.
12. The E-Z Screw Company has three cost concerns: material (in units), labor (in hours), and transportation (in tons). January's total cost was \$1340, involving 100 units of material, 120 man-

hours, and 2 tons for transportation. February's cost was $3150, involving 150 units of material, 200 man-hours, and 3 tons for transportation. March's cost was $2680, involving 200 units of material, 240 man-hours, and 4 tons for transportation. Find the cost of one unit of material, one man-hour, and one ton of transportation.

13. Joe Coyne has invested his $1000 in three places: a 6%-interest savings account, a 7%-interest savings account, and a stock offering an annual dividend of 9%. The annual yield of interest and dividends is $85. The stock dividend is $59 greater than the interest sum of Joe's accounts. Find the amounts of each of Joe's investments.

In problems 14–18, use the method of the previous section to code the following message, using as your coder $\mathbf{B} = \begin{pmatrix} 2 & 1 \\ 3 & 2 \end{pmatrix}$.

14. THE EQUATION HAS BEEN SOLVED.
15. MABLE DIED.
16. THERE IS NO ANSWER FOR THIS QUESTION.
17. ERNIE IS AN ATTORNEY.
18. THIS WOULD BE EASIER TO DO WITH A HIGH-SPEED COMPUTER.

In problems 19–21, use the decoder $\mathbf{B}^{-1}$ (where $\mathbf{B}$ is given before exercise 14, above) to decode the following messages.

19. I T W I C M F K Q V
20. E Z V S
21. D D W U U R G H V E A O

In problems 22–26 we have used a 2 × 2 matrix for coding and broken our original messages into 1 × 2 vectors. Similarly, we could have used a 3 × 3 coding matrix and broken the original message into 1 × 3 vectors. Let

$$\mathbf{C} = \begin{pmatrix} 1 & 2 & 0 \\ 0 & 0 & 1 \\ 3 & 5 & 4 \end{pmatrix}$$

and code the messages in problems 14–18.

5.6 MARKOV CHAINS

A most interesting and important application of both probability and matrices to business and the social sciences is **Markov chains.** Basically, whenever we have a sequence of trials whose finitely many outcomes are independent or dependent upon at most the previous outcome, we have a Markov chain. These outcomes are called **states.** For example, consider the situation of the Caries Food Company:

A "crash" advertising campaign by the company introducing its new cereal, CHOCIT, yields the following results:

1. Of those who purchase a different kind of cereal (Brand X) one month, 20% change to CHOCIT the next month.
2. Of those who purchase CHOCIT one month, 60% will purchase it the next month.

This describes a Markov chain; the probability that someone is using CHOCIT one month depends only upon which cereal that person was using the previous month.

The question is this: "What fraction of present CHOCIT purchasers will be purchasing CHOCIT two months from now?" We delineate the situation by a tree diagram:

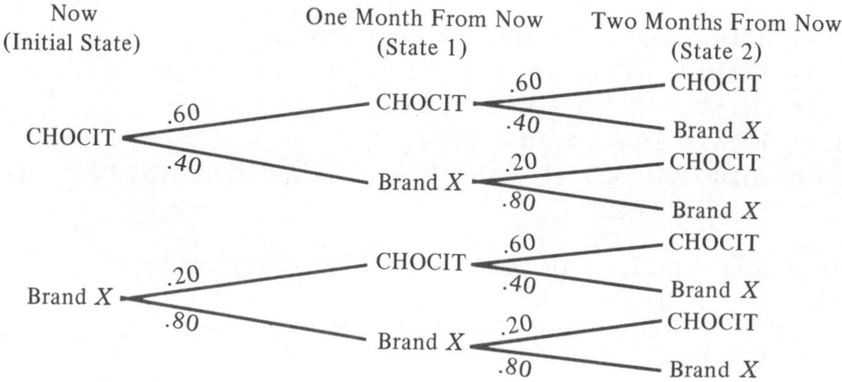

We have weighted the branches of the tree with the associated probabilities. All ".60s," for example, mean the probability of "sticking with" CHOCIT is 60%. (Verify the probabilities of each branch.)

Now, we want to know what fraction of the present CHOCIT buyers will be buying CHOCIT (or equivalently, the probability a buyer will buy CHOCIT) two months from now. The cases are indicated below.

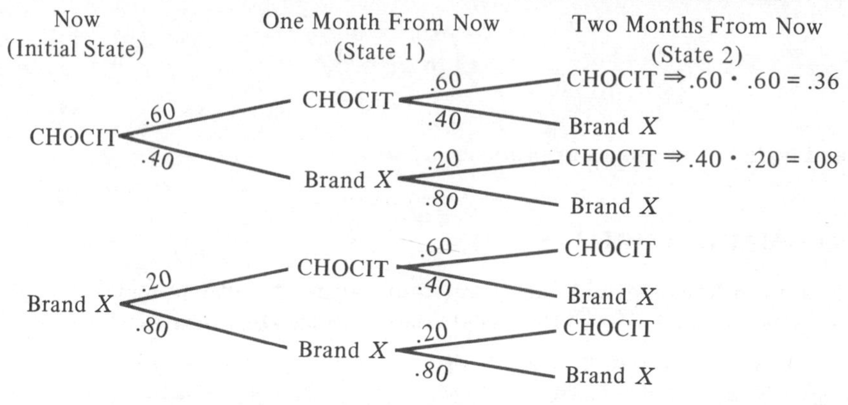

Now, 60% of 60% of the buyers will be taking the top branch of the tree diagram. So, 36% (that is, 60% · 60%) of the buyers are in that category. In two months, another 8% (that is, 40% · 20%) of the present CHOCIT buyers will be buying CHOCIT, also. So, 44% of present CHOCIT buyers will be buying CHOCIT two months from now. Or equivalently, the probability that a present CHOCIT buyer will be buying CHOCIT in two months is 44%.

Now, this has not been too difficult an operation. The difficulty arises, however, when we're asked to find the probability that a present CHOCIT purchaser will also be purchasing CHOCIT twenty months from now. A tree diagram obviously becomes inappropriate. To alleviate this, we introduce the following notation: the states of a Markov chain process are denoted by $a_1, a_2, \ldots, a_m$. A **transition probability,** p_{ij}, is the probability that the process has gone from state a_i to a_j. Finally, the **transition matrix, P**, is given by:

$$\mathbf{P} = \begin{pmatrix} p_{11} & p_{12} & \cdots & p_{1m} \\ p_{21} & & & \\ \vdots & & & \\ p_{m1} & & \cdots & p_{mm} \end{pmatrix}$$

For the cereal illustration, we have:

$$a_1 = \text{buy CHOCIT} \qquad a_2 = \text{buy Brand } X$$

p_{11} = the probability that you buy CHOCIT one month, given that you bought CHOCIT the previous month.

p_{12} = the probability that you buy Brand X one month, given that you bought CHOCIT the previous month.

p_{21} = the probability that you buy CHOCIT one month, given that you bought Brand X the previous month.

p_{22} = the probability that you buy Brand X one month, given that you bought Brand X the previous month.

$$\text{So,} \quad \mathbf{P} = \text{Now} \begin{cases} \text{CHOCIT} \\ \text{Brand } X \end{cases} \overbrace{\begin{pmatrix} \text{CHOCIT} & \text{Brand } X \\ p_{11} & p_{12} \\ p_{21} & p_{22} \end{pmatrix}}^{\text{next month}} = \begin{pmatrix} .60 & .40 \\ .20 & .80 \end{pmatrix}$$

Our answer of .44 was, notice, $p_{11} \cdot p_{11} + p_{12} \cdot p_{21}$ and if we multiply transition matrix **P** by itself, we have

$$\mathbf{P} \cdot \mathbf{P} = \mathbf{P}^2 = \begin{pmatrix} .60 & .40 \\ .20 & .80 \end{pmatrix} \cdot \begin{pmatrix} .60 & .40 \\ .20 & .80 \end{pmatrix} = \begin{pmatrix} .36 + .08 & .24 + .32 \\ .12 + .16 & .08 + .64 \end{pmatrix}$$

$$\mathbf{P}^2 = \text{Now} \begin{cases} \text{CHOCIT} \\ \text{Brand } X \end{cases} \overset{\overbrace{\text{CHOCIT} \quad \text{Brand } X}^{\text{In two months}}}{\begin{pmatrix} .44 & .56 \\ .28 & .72 \end{pmatrix}}$$

where $\mathbf{P}^2$ is the transition matrix displaying probabilities of going from states "now" to states "in two months."

Carrying this one step further, if we desired to know the probability that a present CHOCIT buyer will be buying CHOCIT in THREE months, we could compute $\mathbf{P}^3$:

$$\mathbf{P}^3 = \begin{pmatrix} .60 & .40 \\ .20 & .80 \end{pmatrix} \cdot \begin{pmatrix} .60 & .40 \\ .20 & .80 \end{pmatrix} \cdot \begin{pmatrix} .60 & .40 \\ .20 & .80 \end{pmatrix} = \begin{pmatrix} .44 & .56 \\ .28 & .72 \end{pmatrix} \cdot \begin{pmatrix} .60 & .40 \\ .20 & .80 \end{pmatrix}$$

$$\mathbf{P}^3 = \text{Now} \begin{cases} \text{CHOCIT} \\ \text{Brand } X \end{cases} \overset{\overbrace{\text{CHOCIT} \quad \text{Brand } X}^{\text{In three months}}}{\begin{pmatrix} .376 & .624 \\ .312 & .688 \end{pmatrix}}$$

THE READER SHOULD VERIFY THIS WITH A CONTINUATION OF THE TREE DIAGRAM ON PAGE 244. Furthermore, the **nth step transition matrix** is the transition matrix **P**, raised to the nth power ($\mathbf{P}^n$). Two examples follow.

Example 5.6.1. Jack either takes the bus or drives his car to work daily. He never takes the bus two days in a row; if he drives one day, the next day he is just as likely to drive as take the bus.

a. What are the states of this Markov chain?

b. Determine the transition matrix.

c. What is the probability that if Jack drives today, he will drive four days from now?

Solution:

a. The states are a_1: takes the bus; a_2: drives his car.

b.

$$\mathbf{P} = \begin{matrix} \text{Takes bus one day} \\ \text{Drives one day} \end{matrix} \overset{\begin{matrix}\text{Takes bus next day} & \text{Drives next day}\end{matrix}}{\begin{pmatrix} 0 & 1 \\ \frac{1}{2} & \frac{1}{2} \end{pmatrix}}$$

c. Here we need $\mathbf{P}^4$.

$$\mathbf{P}^4 = \begin{pmatrix} 0 & 1 \\ \frac{1}{2} & \frac{1}{2} \end{pmatrix} \cdot \begin{pmatrix} 0 & 1 \\ \frac{1}{2} & \frac{1}{2} \end{pmatrix} \cdot \begin{pmatrix} 0 & 1 \\ \frac{1}{2} & \frac{1}{2} \end{pmatrix} \cdot \begin{pmatrix} 0 & 1 \\ \frac{1}{2} & \frac{1}{2} \end{pmatrix}$$

$$\mathbf{P}^4 = \begin{matrix} a_1 \\ a_2 \end{matrix} \overset{\begin{matrix} a_1 & a_2 \end{matrix}}{\begin{pmatrix} \frac{3}{8} & \frac{5}{8} \\ \frac{5}{16} & \frac{11}{16} \end{pmatrix}}$$

So, the probability that Jack will drive four days from now, knowing that he drives today, is $\frac{11}{16}$. The other probabilities represent: $\frac{3}{8}$ is the probability Jack takes the bus in four days, given he takes the bus today; $\frac{5}{8}$ is the probability Jack drives in four days, given he takes the bus today; $\frac{5}{16}$ is the probability Jack takes the bus in four days, given he drives today.

Example 5.6.2. A small town holds an annual election for mayor. Because of the incumbent's advantage, the probability any one mayor remains in office the next year is $\frac{2}{3}$. If a Republican is presently mayor, the probability he will be beat by a Democrat is $\frac{1}{4}$, and by an Independent is $\frac{1}{12}$. If a Democrat is presently mayor, the probability he will be beat by a Republican is $\frac{1}{4}$ and by an Independent, $\frac{1}{12}$. If an Independent is presently mayor, the outcomes of being beat by a Republican or by a Democrat are equally likely. What is the probability the mayor will be a Democrat in two years if there is presently a Democrat in office?

Solution: The transition matrix is given by:

$$\mathbf{P} = \begin{matrix} \text{Presently Republican mayor} \\ \text{Presently Democratic mayor} \\ \text{Presently Independent mayor} \end{matrix} \overset{\overbrace{\begin{matrix} R & D & I \end{matrix}}^{\text{Next year}}}{\begin{pmatrix} \frac{2}{3} & \frac{1}{4} & \frac{1}{12} \\ \frac{1}{4} & \frac{2}{3} & \frac{1}{12} \\ \frac{1}{6} & \frac{1}{6} & \frac{2}{3} \end{pmatrix}}$$

We need

$$\mathbf{P}^2 = \begin{pmatrix} \frac{2}{3} & \frac{1}{4} & \frac{1}{12} \\ \frac{1}{4} & \frac{2}{3} & \frac{1}{12} \\ \frac{1}{6} & \frac{1}{6} & \frac{2}{3} \end{pmatrix} \cdot \begin{pmatrix} \frac{2}{3} & \frac{1}{4} & \frac{1}{12} \\ \frac{1}{4} & \frac{2}{3} & \frac{1}{12} \\ \frac{1}{6} & \frac{1}{6} & \frac{2}{3} \end{pmatrix}$$

$$\mathbf{P}^2 = \begin{matrix} \text{N} \\ \text{o} \\ \text{w} \end{matrix} \left\{ \begin{matrix} R \\ D \\ I \end{matrix} \right. \overset{\overbrace{\begin{matrix} R & D & I \end{matrix}}^{\text{In two years}}}{\begin{pmatrix} \frac{75}{144} & \frac{50}{144} & \frac{19}{144} \\ \frac{50}{144} & \frac{75}{144} & \frac{19}{144} \\ \frac{38}{144} & \frac{38}{144} & \frac{68}{144} \end{pmatrix}}$$

So, the probability that there will be a Democrat in office in two years, knowing there is one in office now, is $\frac{75}{144}$.

Let's return to the cereal illustration at the beginning of this section, adding the (initial) condition that 30% of the cereal purchasers are currently using CHOCIT. We obtain the tree diagram on the next page.

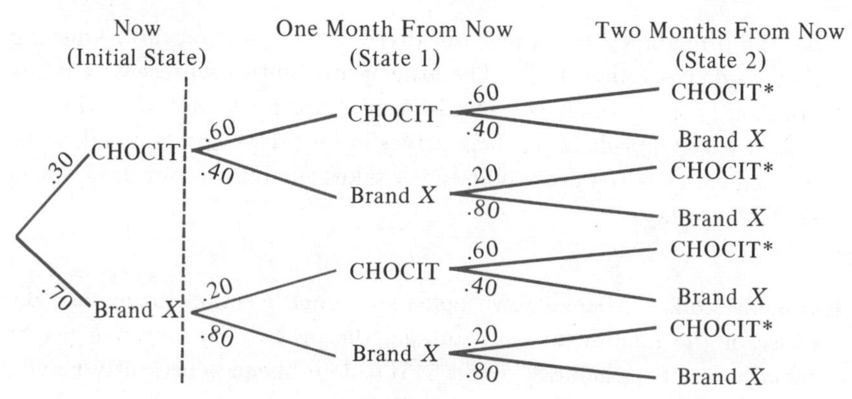

The company wants to know the percent of people that will be buying CHOCIT in two months. (Equivalently, we want to know the probability that a purchaser selected at random will be buying CHOCIT in two months.) From our tree diagram, we calculate the asterisked branches: $(.30 \cdot .60 \cdot .60) + (.30 \cdot .40 \cdot .20) + (.70 \cdot .20 \cdot .60) + (.70 \cdot .80 \cdot .20) = .328$ or 32.8%. This initial condition has altered things. We can adjust for this alteration with matrices, also. The matrix $\mathbf{P}^2$ represents that part to the right of the dotted vertical line on the tree diagram above. We can represent the initial condition (that which is to the left of the vertical dotted line) by a row vector $\mathbf{P}_0$ ($\mathbf{P}_0$ stands for "initial probability vector").

$$\mathbf{P}_0 = \begin{matrix} \text{Buys CHOCIT} & \text{Buys Brand } X \\ \text{initially} & \text{initially} \\ (.30 & .70) \end{matrix}$$

The product $\mathbf{P}_0 \cdot \mathbf{P}^2$ represents a matrix of probabilities for cereal buyers after 2 months with the initial condition consideration. So,

$$\mathbf{P}_0 \cdot \mathbf{P}^2 = \begin{matrix} \text{CHOCIT} & \text{Brand } X \\ (.30 & .70) \end{matrix} \cdot \begin{pmatrix} .44 & .56 \\ .28 & .72 \end{pmatrix}$$

$$= \begin{matrix} \text{CHOCIT} & \text{Brand } X \\ (.328 & .672) \end{matrix}$$

We summarize what has been done in this section:

1. The states of a Markov chain are denoted $a_1, a_2, \ldots, a_m$ and the probabilities p_{ij} are the probabilities of being in state a_j after being in state a_i.
2. The transition matrix $\mathbf{P}$ is

$$\mathbf{P} = \begin{pmatrix} p_{11} & \cdots & p_{1m} \\ p_{21} & & \\ \cdot & & \\ \cdot & & \\ \cdot & & \\ p_{m1} & \cdots & p_{mm} \end{pmatrix}$$

To find the probabilities of being in a certain state after n trials, we compute $\mathbf{P}^n$.

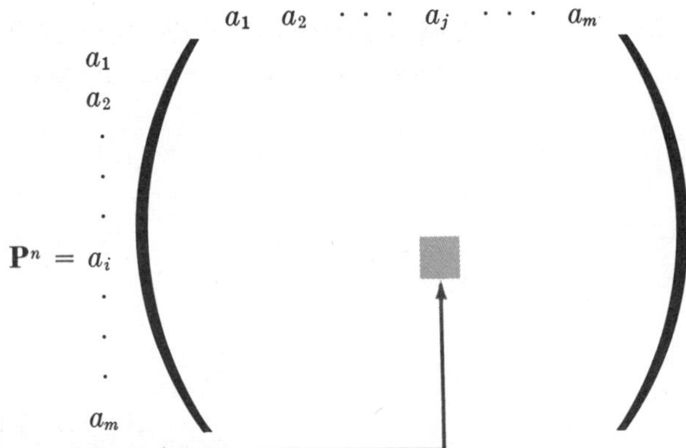

This entry is the probability that the process is in the a_jth state after n trials, given that it began in state a_i.

3. If, in addition, there is an initial condition given by $\mathbf{P}_0 = (b_1, \ldots, b_m)$, then

$$\mathbf{P}_0 \cdot \mathbf{P}^n = \begin{matrix} a_1 & \cdots & a_j & \cdots & a_m \\ (& & \blacksquare & &) \end{matrix}$$

This entry displays the probabilities of being in the a_jth state after n trials with initial condition $\mathbf{P}_0$.

Some examples follow:

Example 5.6.3. Jack, of example 5.6.1, decides before the first day of travel to roll a die and if a 6 appears he takes the bus; otherwise, he drives. What is the probability Jack is taking the bus four days from now if the form of transportation on succeeding days follows the conditions of example 5.6.1?

Solution: Here $\mathbf{P}_0 = \begin{matrix} \text{Bus} & \text{Drive} \\ (\frac{1}{6} & \frac{5}{6}) \end{matrix}$, and we have already determined

$$\mathbf{P}^4 = \begin{matrix} & \text{Bus} & \text{Drive} \\ \text{Bus} & \frac{3}{8} & \frac{5}{8} \\ \text{Drive} & \frac{5}{16} & \frac{11}{16} \end{matrix}$$

$$\mathbf{P}_0 \cdot \mathbf{P}^4 = \begin{matrix} \text{Bus} & \text{Drive} \\ (\frac{1}{6} & \frac{5}{6}) \end{matrix} \cdot \begin{pmatrix} \frac{3}{8} & \frac{5}{8} \\ \frac{5}{16} & \frac{11}{16} \end{pmatrix} = \begin{matrix} \text{Bus} & \text{Drive} \\ (\frac{31}{96} & \frac{65}{96}) \end{matrix}$$

So, the probability he will be taking the bus in four days is $\frac{31}{96}$.

Example 5.6.4. In the city of Rochester, there are three competing department stores, call them A, B, C. All three stores introduce new advertising campaigns for a year and then a study is made revealing: A lost 10% of its customers to B and 10% of its customers to C (A

kept 80%); B lost 15% of its customers to A and 20% of its customers to C (B kept 65%); C lost 5% to A and 25% to B (C kept 70%).

a. If initially A did 30% of the business, B did 50% of the business, and C did 20% of the business, give the percent breakdown after the year.
b. Suppose the results of the study pertain to future years. Which store will have increased its business the most after two years of advertising?

Solution:

a. We have

$$\mathbf{P}_0 = \overset{\begin{matrix} A & \; B & \; C \end{matrix}}{(.30 \quad .50 \quad .20)} \qquad \text{and} \qquad \mathbf{P} = \begin{matrix} & \begin{matrix} A & B & C \end{matrix} \\ \begin{matrix} A \\ B \\ C \end{matrix} & \begin{pmatrix} .80 & .10 & .10 \\ .15 & .65 & .20 \\ .05 & .25 & .70 \end{pmatrix} \end{matrix}$$

After one year, we want $\mathbf{P}_0 \cdot \mathbf{P} = \overset{\begin{matrix} A & \; B & \; C \end{matrix}}{(.325 \quad .405 \quad .270)}$. So, A has 32.5% of the business, B has 40.5%, and C has 27%.
b. Here, we need $\mathbf{P}_0 \cdot \mathbf{P}^2$:

$$\mathbf{P}_0 \cdot \mathbf{P}^2 = (.30 \quad .50 \quad .20) \cdot \begin{pmatrix} .80 & .10 & .10 \\ .15 & .65 & .20 \\ .05 & .25 & .70 \end{pmatrix} \cdot \begin{pmatrix} .80 & .10 & .10 \\ .15 & .65 & .20 \\ .05 & .25 & .70 \end{pmatrix}$$

$$\mathbf{P}_0 \cdot \mathbf{P}^2 = \overset{\begin{matrix} A & \quad B & \quad C \end{matrix}}{(.33425 \quad .36325 \quad .30250)}$$

After two years, A has increased its business 3.425%, B has lost 13.675% of its business, C has increased its business 10.25%.

A few last important properties of matrices will be examined before the chapter closes. First, a **regular transition matrix, P**, is one that can be raised to some power k, such that $\mathbf{P}^k$ contains no zero components. For example, $\begin{pmatrix} .6 & .4 \\ .2 & .8 \end{pmatrix}$ is regular because it can be raised to the first power and contains no zeros, and $\mathbf{P} = \begin{pmatrix} 0 & 1 \\ \frac{1}{2} & \frac{1}{2} \end{pmatrix}$ is regular (we saw this transition matrix in example 5.6.1) because $\mathbf{P}^4$ contains no zeros. The identity matrix, $\begin{pmatrix} 1 & 0 \\ 0 & 1 \end{pmatrix}$, is obviously not regular.

Secondly, a nonzero vector $\mathbf{v}$ is called **fixed** for $\mathbf{P}$ if the product $\mathbf{v} \cdot \mathbf{P} = \mathbf{v}$ for some regular transition matrix $\mathbf{P}$. For example, if $\mathbf{P} = \begin{pmatrix} 0 & 1 \\ \frac{1}{2} & \frac{1}{2} \end{pmatrix}$, then

$\mathbf{v} = (\frac{1}{3} \ \ \frac{2}{3})$ is a fixed vector because

$$\mathbf{v} \cdot \mathbf{P} = (\tfrac{1}{3} \ \ \tfrac{2}{3}) \cdot \begin{pmatrix} 0 & 1 \\ \frac{1}{2} & \frac{1}{2} \end{pmatrix} = (\tfrac{1}{3} \ \ \tfrac{2}{3})$$

The importance of a fixed vector is this:
If $\mathbf{P}$ is a regular transition matrix and $\mathbf{v}$ is a fixed vector for $\mathbf{P}$, then $\mathbf{P}^n$ approaches a matrix each of whose rows is $\mathbf{v}$ for large values of n.*

To illustrate this fact, return to example 5.6.1. There, $\mathbf{P} = \begin{pmatrix} 0 & 1 \\ \frac{1}{2} & \frac{1}{2} \end{pmatrix}$ and we have just seen that $\mathbf{v} = (\frac{1}{3} \ \ \frac{2}{3})$ is a fixed vector for $\mathbf{P}$. This means that $\mathbf{P}^n$, as n gets large, approaches

$$\begin{array}{c} \\ \text{Bus} \\ \text{Drive} \end{array} \begin{array}{c} \begin{array}{cc} \text{Bus} & \text{Drive} \end{array} \\ \begin{pmatrix} \frac{1}{3} & \frac{2}{3} \\ \frac{1}{3} & \frac{2}{3} \end{pmatrix} \end{array}$$

. So, after a large number of days, he'll be taking the bus about $\frac{1}{3}$ of the time and driving his car about $\frac{2}{3}$ of the time.

Example 5.6.5. Verify that $\mathbf{v} = (\frac{1}{3} \ \ \frac{1}{3} \ \ \frac{1}{3})$ is a fixed vector for the transition matrix of example 5.6.4,

$$\mathbf{P} = \begin{pmatrix} .80 & .10 & .10 \\ .15 & .65 & .20 \\ .05 & .25 & .70 \end{pmatrix}$$

Solution:

$$\mathbf{v} \cdot \mathbf{P} = (\tfrac{1}{3} \ \ \tfrac{1}{3} \ \ \tfrac{1}{3}) \cdot \begin{pmatrix} .80 & .10 & .10 \\ .15 & .65 & .20 \\ .05 & .25 & .70 \end{pmatrix} = (\tfrac{1}{3} \ \ \tfrac{1}{3} \ \ \tfrac{1}{3})$$

The significance is this: after many years, each store will end up with $33\frac{1}{3}\%$ of the business. Watch:

$$\begin{array}{rl} & \begin{array}{ccc} A & B & C \end{array} \\ \mathbf{P}_0 = & (.30 \ \ .50 \ \ .20) \end{array}$$

and

$$\mathbf{P}_0 \cdot \mathbf{P}^n = (.30 \ \ .50 \ \ .20) \cdot \begin{pmatrix} \frac{1}{3} & \frac{1}{3} & \frac{1}{3} \\ \frac{1}{3} & \frac{1}{3} & \frac{1}{3} \\ \frac{1}{3} & \frac{1}{3} & \frac{1}{3} \end{pmatrix}$$

for large values of n.

$$\begin{array}{rl} & \begin{array}{ccc} A & B & C \end{array} \\ \mathbf{P}_0 \cdot \mathbf{P}^n = & (\frac{1}{3} \ \ \frac{1}{3} \ \ \frac{1}{3}) \end{array}$$

In the next example we discover how to find fixed vectors.

* The proof of this and other results of this section are beyond the scope of this book. The reader who is interested can refer to: William Feller, *An Introduction to Probability Theory and Its Applications*, 3rd Ed., (New York: Wiley, 1968).

Example 5.6.6. **a.** If $\mathbf{P} = \begin{pmatrix} 0 & 1 \\ \frac{1}{2} & \frac{1}{2} \end{pmatrix}$, find $\mathbf{v} = (x \quad y)$ such that $\mathbf{v}$ is a fixed vector for $\mathbf{P}$.

b. If $\mathbf{P} = \begin{pmatrix} \frac{1}{4} & \frac{3}{4} \\ \frac{1}{8} & \frac{7}{8} \end{pmatrix}$, find $\mathbf{v} = (x \quad y)$ such that $\mathbf{v} \cdot \mathbf{P} = \mathbf{v}$.

Solution:

a. $(x \quad y) \cdot \begin{pmatrix} 0 & 1 \\ \frac{1}{2} & \frac{1}{2} \end{pmatrix} = (x \quad y)$

$$(\tfrac{1}{2}y \quad x + \tfrac{1}{2}y) = (x \quad y)$$

So
$$\left.\begin{aligned} \tfrac{1}{2}y &= x \\ x + \tfrac{1}{2}y &= y \end{aligned}\right\} \Rightarrow \left.\begin{aligned} -x + \tfrac{1}{2}y &= 0 \\ x - \tfrac{1}{2}y &= 0 \end{aligned}\right\}$$

The equations are equivalent; there is not a unique solution, but we have at our disposal the fact that $x + y = 1$, due to the fact that x and y are probabilities. Now,

$$\left.\begin{aligned} x + y &= 1 \\ \tfrac{1}{2}y &= x \end{aligned}\right\} \Rightarrow \begin{aligned} x &= \tfrac{1}{3} \\ y &= \tfrac{2}{3} \end{aligned}$$

Therefore, $\mathbf{v} = (\frac{1}{3} \quad \frac{2}{3})$.

Check:
$$(\tfrac{1}{3} \quad \tfrac{2}{3}) \cdot \begin{pmatrix} 0 & 1 \\ \frac{1}{2} & \frac{1}{2} \end{pmatrix} = (\tfrac{1}{3} \quad \tfrac{2}{3}) \checkmark$$

b.
$$(x \quad y) \cdot \begin{pmatrix} \frac{1}{4} & \frac{3}{4} \\ \frac{1}{8} & \frac{7}{8} \end{pmatrix} = (x \quad y)$$

$$(\tfrac{1}{4}x + \tfrac{1}{8}y \quad \tfrac{3}{4}x + \tfrac{7}{8}y) = (x \quad y)$$

$$\left.\begin{aligned} \tfrac{1}{4}x + \tfrac{1}{8}y &= x \\ \tfrac{3}{4}x + \tfrac{7}{8}y &= y \\ x + y &= 1 \end{aligned}\right\} \Rightarrow \begin{aligned} x &= \tfrac{1}{7} \\ y &= \tfrac{6}{7} \end{aligned}$$

Check:
$$(\tfrac{1}{7} \quad \tfrac{6}{7}) \cdot \begin{pmatrix} \frac{1}{4} & \frac{3}{4} \\ \frac{1}{8} & \frac{7}{8} \end{pmatrix} = (\tfrac{1}{7} \quad \tfrac{6}{7}) \checkmark$$

EXERCISES ■ SECTION 5.6.

1. Consider the matrix on page 245, $\mathbf{P} = \begin{pmatrix} .6 & .4 \\ .2 & .8 \end{pmatrix}$.

   a. Find $\mathbf{P}^5$ by direct calculation. (Hint: $\mathbf{P}^3$ and $\mathbf{P}^2$ have already been calculated in the text; $\mathbf{P}^5 = \mathbf{P}^3 \cdot \mathbf{P}^2$, of course.)
   b. Using your result of *a* above, predict what the components of $\mathbf{P}^n$ will be as n gets very large.
   c. Find a fixed vector for $\mathbf{P}$.
   d. Using your result of *c*, what is $\mathbf{P}^n$ approaching as n gets large?

2. Let $\mathbf{P} = \begin{pmatrix} .1 & .9 \\ .2 & .8 \end{pmatrix}$.

   a. Find a fixed vector for **P**.

   b. Find the value of $\mathbf{P}^n$ as n gets very large.

   c. Suppose $\mathbf{P}_0 = (.5 \quad .5)$ represents some initial condition. Find $\mathbf{P}_0 \cdot \mathbf{P}^n$.

   d. Suppose $\mathbf{P}_0 = (.99 \quad .01)$ represents some initial condition. Find $\mathbf{P}_0 \cdot \mathbf{P}^n$.

3. In example 5.6.5, use $\mathbf{P}_0 = \overset{\quad A \quad B \quad C}{(.1 \quad .9 \quad 0)}$ and find $\mathbf{P}_0 \cdot \mathbf{P}^n$. How much of the business does store C end up with even though it starts with 0% of the business?

4. Find the fixed vector for each transition matrix:

   a. $\begin{pmatrix} .2 & .8 \\ .5 & .5 \end{pmatrix}$ **b.** $\begin{pmatrix} \frac{1}{9} & \frac{8}{9} \\ \frac{6}{7} & \frac{1}{7} \end{pmatrix}$

   c. $\begin{pmatrix} \frac{1}{4} & \frac{1}{4} & \frac{1}{2} \\ \frac{1}{3} & \frac{1}{3} & \frac{1}{3} \\ 0 & \frac{1}{2} & \frac{1}{2} \end{pmatrix}$ **d.** $\begin{pmatrix} .1 & .2 & .7 \\ .3 & .4 & .3 \\ .2 & .2 & .6 \end{pmatrix}$

5. Linda estimates that the probability she passes a finite math exam is 90% if she failed the previous exam. If she passed the previous exam, she calculates her probability of passing the next exam at 60%.

   a. What are the states of this Markov chain?

   b. Establish a transition matrix **L**.

   c. Calculate $\mathbf{L}^4$, the matrix associated with the fourth exam.

   d. Suppose, in addition to the above stipulation, Linda estimates a 50% chance of passing the first exam. What is the probability she passes the fourth exam?

6. A security guard must make a "round" each evening to investigate one of three different areas of a building; call them A, B, C. He never investigates the same area twice in succession. In addition, if he investigates A he will always follow up with B. If he investigates B he is just as likely to investigate A or C on the next trip. If he investigates C he is twice as likely to investigate B as A.

   a. Determine the transition matrix **T**.

   b. Show **T** is regular.

   c. Determine a fixed vector for **T**.

   d. On Monday night he starts his job with an equal probability of investigating each area. What is the probability he will be investigating area B on Tuesday night? On Thursday night? On Friday night? In 1,000 nights?

7. Consider a coin-toss experiment for a fair coin.

    a. Determine a transition matrix **M**.
    b. Determine the fixed vector for **M**.
    c. After 100,000 tosses, what is the probability of flipping a head?

8. Consider two coins being tossed; one is fair and one is "crooked." Whenever a head occurs, the fair coin is then flipped. Whenever a tail occurs, the "crooked" coin is then flipped. For the crooked coin, $P(H) = 2/5$ and $P(T) = 3/5$.

    a. Determine a transition matrix **A**.
    b. Assuming the fair coin is flipped first, what is the probability of H on the third flip?
    c. Assuming the fair coin is flipped first, what is the probability of H on the 1,000,000 flip?

chapter 5 in review

CHAPTER SUMMARY

In this chapter we learned how to add, subtract, and multiply vectors. Operations on matrices were then performed. To add or subtract two matrices, they must be of the same dimension; for two matrices **A** and **B** to be multiplied, the number of columns of **A** must equal the number of rows of **B**. A special matrix **I**, the identity matrix, was introduced as the matrix with the property $\mathbf{A} \cdot \mathbf{I} = \mathbf{A}$ for any $m \times n$ matrix **A** (**I** is an $n \times n$ matrix). For any square matrix **B**, we found how to determine $\mathbf{B}^{-1}$, the multiplicative inverse of **B** (i.e., $\mathbf{B} \cdot \mathbf{B}^{-1} = \mathbf{I}$), using row transformations. The two applications of matrices and their inverses in section 5.5 pertained to solving systems of linear equations and to code making and breaking (cryptography). The final application of matrices and their operations came in section 5.6, where the study of Markov chains displayed a relationship between matrices and probability.

VOCABULARY

matrix
component
row vector
column vector
matrix addition
matrix multiplication
additive identity
multiplicative identity
probability matrix
transpose
inverse of a matrix
row transformations
matrix of coefficients
cryptography
Markov chain
states of a chain
transition probability
transition matrix
nth step transition matrix
regular transition matrix
fixed vector

SYMBOLS

$$\mathbf{A} = \begin{pmatrix} a_{11} & a_{12} & \cdots & a_{1n} \\ a_{21} & & & \\ \cdot & & & \cdot \\ \cdot & & & \cdot \\ \cdot & & & \cdot \\ a_{m1} & & \cdots & a_{mn} \end{pmatrix} \qquad \mathbf{I} = \begin{pmatrix} 1 & 0 & \cdots & 0 \\ 0 & 1 & \cdots & 0 \\ \cdot & & & \\ \cdot & & & \\ \cdot & & & \\ 0 & 0 & \cdots & 1 \end{pmatrix}$$

$\mathbf{A}^T$ $\quad$ $\mathbf{A}^{-1}$ $\quad$ p_{ij} $\quad$ $\mathbf{P}^n$ $\quad$ $\mathbf{P}_0$

chapter 6
Linear Programming

6.1 THE PLANE

All points in the plane can be named or located by means of a **rectangular coordinate system.** It involves the use of two number lines called **axes,** a vertical axis (usually called the y-axis) and a horizontal axis (usually called the x-axis.) These two axes intersect at a point called the **origin.** See the figure below.

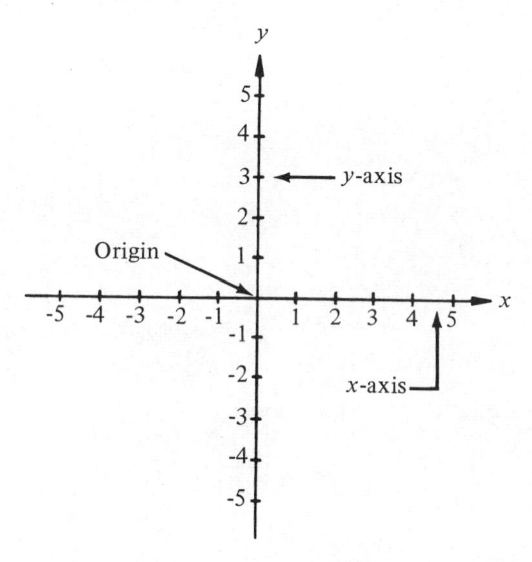

A point is located by travelling first a certain distance horizontally and then a certain distance vertically from the origin. For instance, if we travel two units to the right and three units upward, we have located the point below.

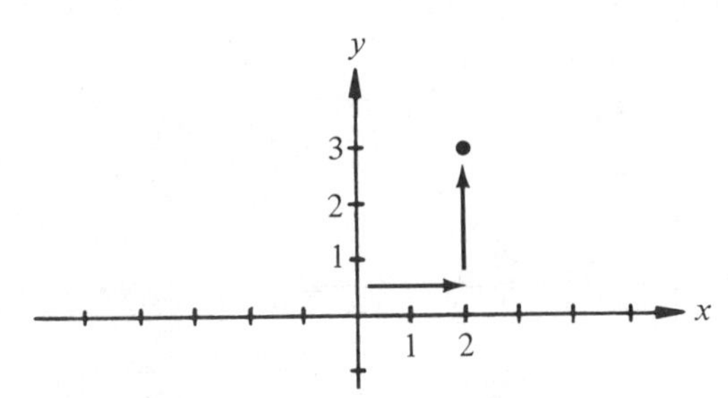

We denote the point by the ordered pair (2,3). (The term **ordered pair** is used because the point (3,2) is different from (2,3); it's the result of travelling first three units to the right and then two units upward.) Negative values indicate movement to the left (for x) or downward (for y). We depict some points below:

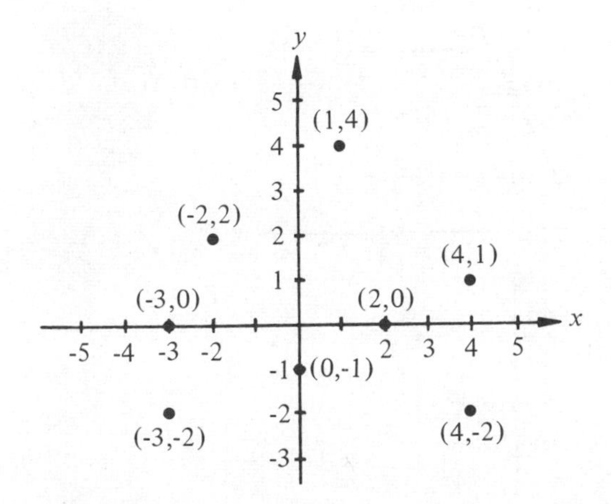

The plane with the rectangular coordinate system provides a handy means for examining "pictures" or **graphs** of certain relationships. One such relationship is an equation. Consider, for example, the relationship: "y is three more than twice x." The symbolic representation is $y = 2x + 3$ and infinitely many pairs of numbers could satisfy it. When $x = 1$, $y = 2 \cdot 1 + 3 = 5$. When $x = 0$, $y = 2 \cdot 0 + 3 = 3$. When $x = -1$, $y = 2 \cdot (-1) + 3 = 1$, etc. In table form, we have:

x	y
1	5
0	3
-1	1

We now "plot" these points on our rectangular coordinate system.

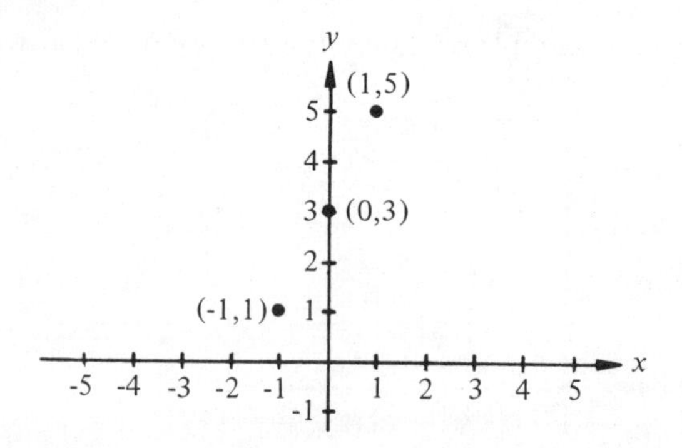

Noticing that the three points lie on a straight line, we conjecture that the graph (that is, the plotting of *ALL* points that work) of $y = 2x + 3$ is a straight line. It is. We depict that line below.

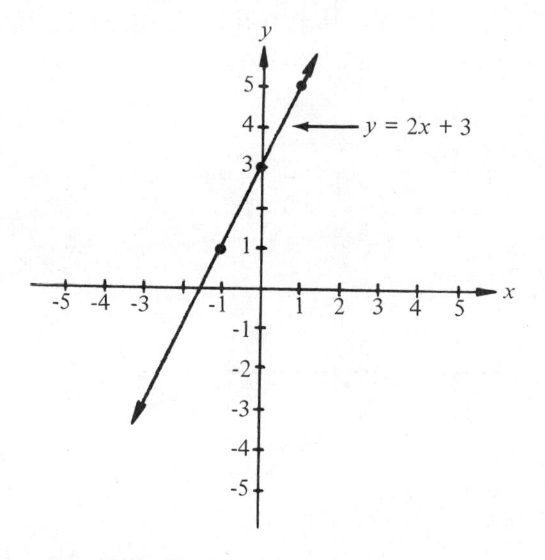

An example follows.

EXAMPLE 6.1.1. Graph: **a.** $y = x + 5$ **b.** $y = 2x$

Solution:

a. $y = x + 5$. We choose some values for x and find corresponding y values in the table below.

x	y
-3	2
0	5
1	6

We plot these points.

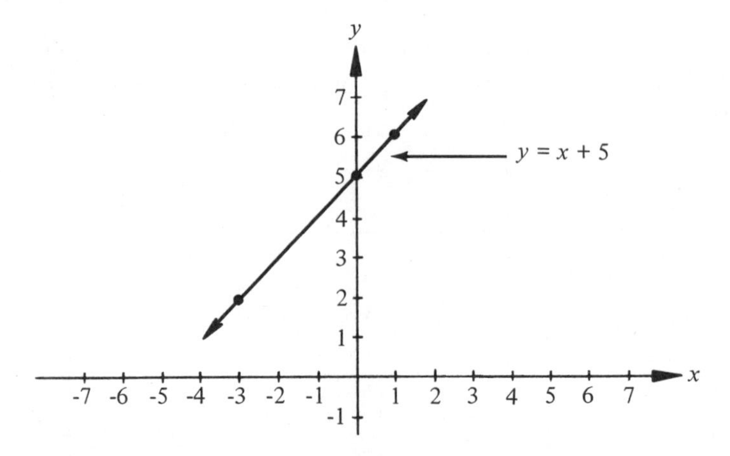

b. The same process is used for $y = 2x$:

x	y
1	2
0	0
-2	-4

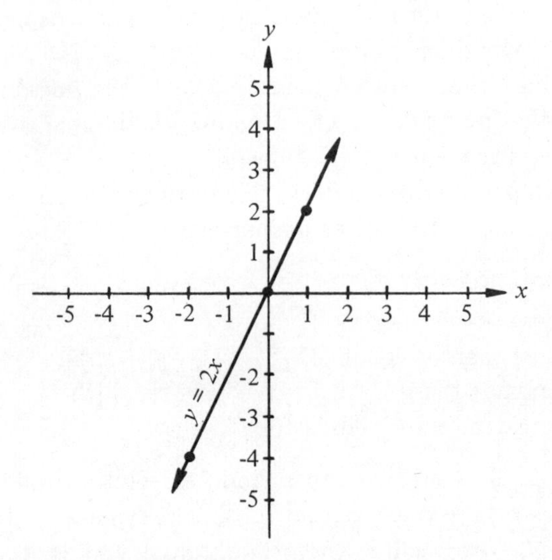

We point out here that the graph of an equation is merely a pictorial representation of a set, and the graph above of $y = 2x$ represents the set $\{(x,y) \mid y = 2x\}$. (This is **set-builder notation** and is read, "the set of all points (x,y) such that y is equal to $2x$.")

We have graphed examples of equality, the first of two kinds of relationships that will be examined here. Generally, any equation of the form $y = mx + b$ (m and b here stand for fixed numbers) will graphically be a straight line. Also, any equation of the form $y = b$ or $x = b$ (again b

stands for a number) will graphically be a horizontal straight line or a vertical straight line, respectively. For example, $\{(x,y) \mid y = 7\}$ is depicted below.

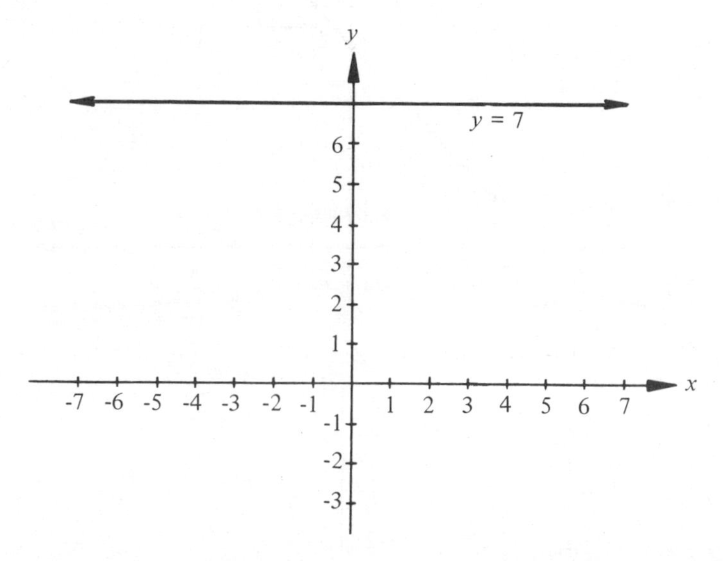

An obvious fact existing in the above illustrations is that a straight line can be thought of as "dividing" the plane into three sets: the set of all points "above" the line, the set of all points "below" the line, and the set of all points on the line. The set of all points on the line, we have already seen, represents the equality relationship.

The other relationship we shall examine in this section is the **inequality** relation. Basically, an inequality can be of four types:

Type of inequality	*Symbol*
1. is less than	$<$
2. is less than or equal to	$\leq$
3. is greater than	$>$
4. is greater than or equal to	$\geq$

When we write $y \leq 2x + 3$, we wish to include as solutions all of the points that satisfy $y = 2x + 3$ as well as those satisfying $y < 2x + 3$. Graphically, this means we include the straight line $y = 2x + 3$ plus other points. Those other points will be either all those points above the line or all those points below the line. Graphically, we have the line $y = 2x + 3$, plus all points in either the **half-plane** above or below $y = 2x + 3$. To find out which, test any point not on the line: if it satisfies the inequality, all the points in that half-plane will also; if it doesn't, choose the other half-plane. For example, choose the point (3,1), obviously in the half-plane below the line. Test to see if it satisfies the inequality in question: $y \leq 2x + 3$. It does! $(1 \leq 2 \cdot 3 + 3)$. Therefore the graph of $y \leq 2x + 3$ is:

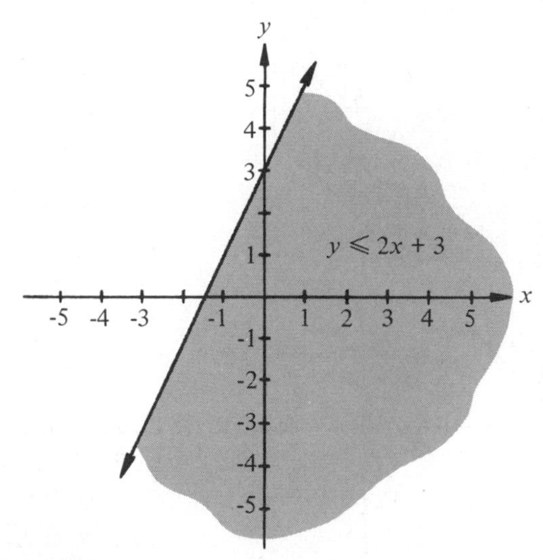

An example follows.

Example 6.1.2. Graph: **a.** $y > x + 5$ **b.** $y < -x$

Solution:

a. The line $y = x + 5$ is not included this time (why?); the line is important, though, because on the two sides of $y = x + 5$ are: (1) the set of all points where y is less than $x + 5$ and (2) the set of all points where y is greater than $x + 5$. We want the latter. To remind ourselves that the line itself is not in the solution set, we depict it with dashes.

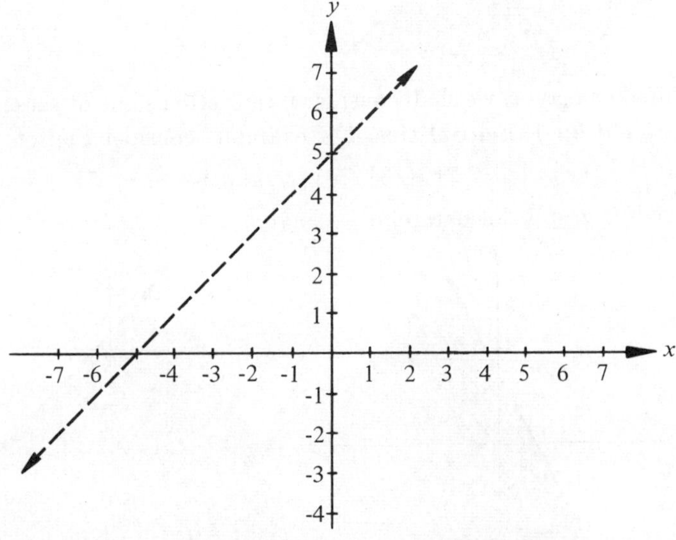

Now we test any point, say (0,0), that is below the line. It does not work, so the desired half-plane must be all the points above the dotted line. We have the graph shown on the following page.

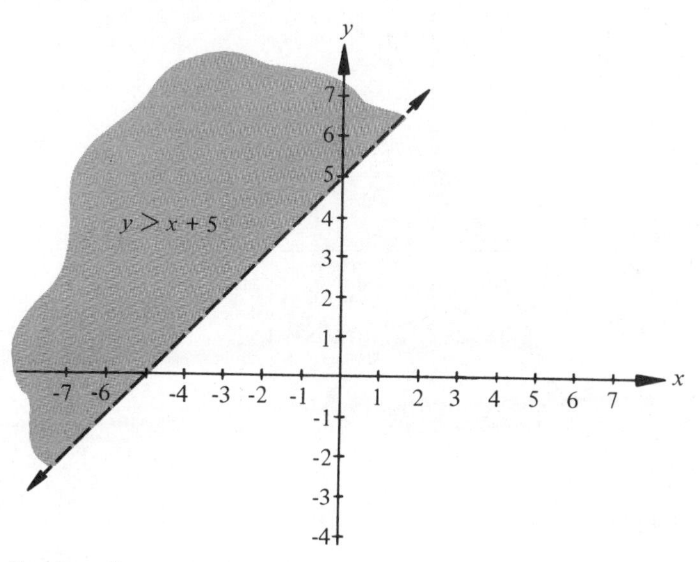

b. Following the same procedure, we have:

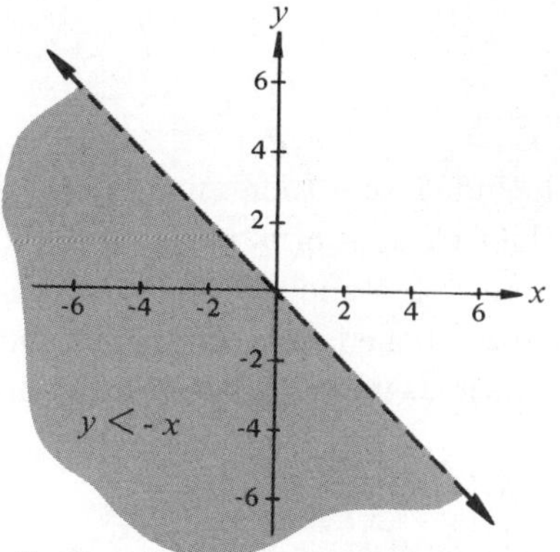

As a final endeavor, we shall consider intersecting some of these **linear** (from straight line) **inequalities.** For example, consider the set

$$\{(x,y) \mid y \leq 2x + 3\} \cap \{(x,y) \mid y \leq -x + 5\}$$

When each is graphed separately, we have:

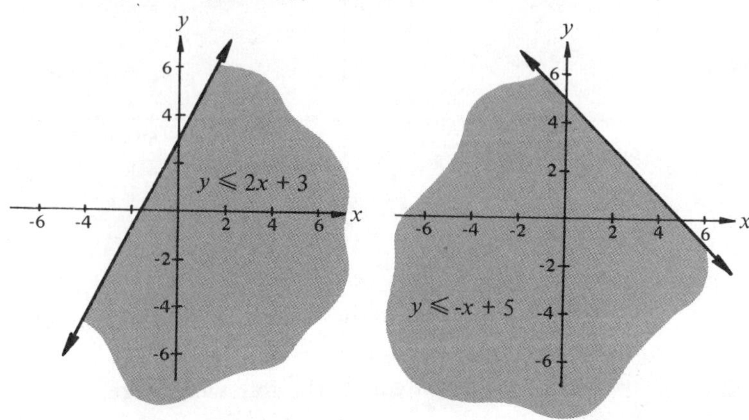

The set of points common to both graphs is the desired set and we obtain it by graphing both graphs on the same set of axes:

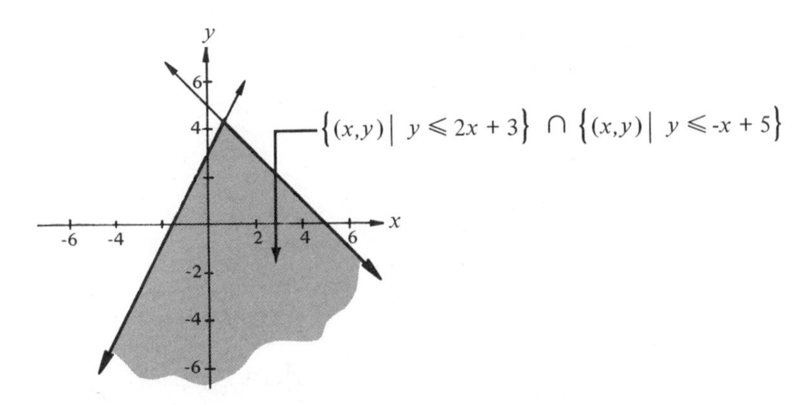

An example follows.

Example 6.1.3. Find:

$$\{(x,y) \mid y \leq 2x + 1\} \cap \{(x,y) \mid y \geq -4\} \cap \{(x,y) \mid y \leq -x\}$$

Solution: Individually, we have:

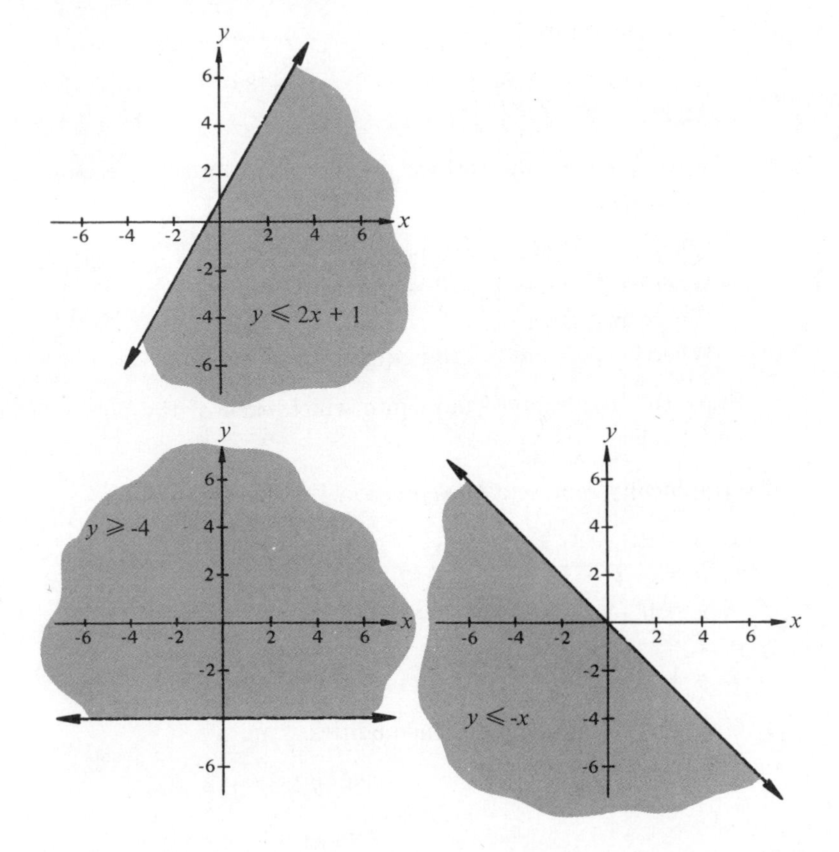

Collectively, we have:

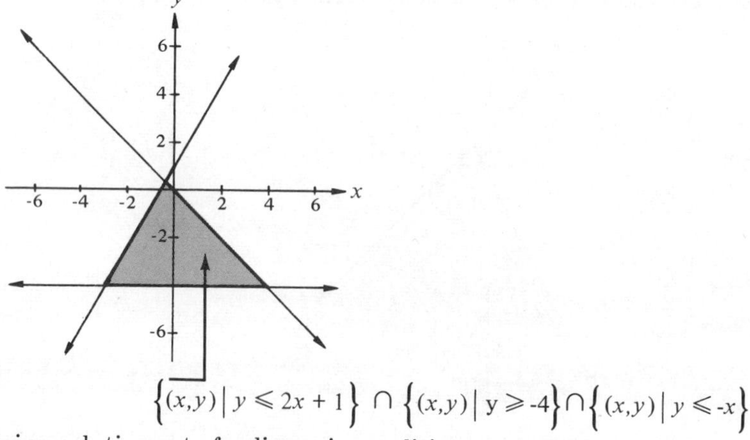

Obtaining solution sets for linear inequalities as we did in this section is one aspect of solving linear programming problems. In the next section, we take a first look at linear programming.

EXERCISES ■ SECTION 6.1.

1. Locate the points.

   a. $(5,-7)$ **b.** $(0,5)$ **c.** $(-3,0)$ **d.** $(5,15)$
   e. $(2,2)$ **f.** $(2,7)$ **g.** $(7,2)$ **h.** $(-1,-11)$

2. The two axes divide the rest of the plane into four regions called **quadrants.**

   Where $x > 0$ and $y > 0$ is quadrant I.
   Where $x < 0$ and $y > 0$ is quadrant II.
   Where $x < 0$ and $y < 0$ is quadrant III.
   Where $x > 0$ and $y < 0$ is quadrant IV.

   State the quadrant (if any) into which each of the points of exercise 1 falls.

3. Graphically represent the equation.

   a. $y = 2x + 2$ **b.** $y = x + \frac{1}{2}$
   c. $y = -x - \frac{1}{2}$ **d.** $y = 7$
   e. $y = 0$ **f.** $x = 3$
   g. $x = -2$ **h.** $x = 0$
   i. $y + x = 3$ **j.** $y = x + 1$

4. Graphically represent the inequalities.

   a. $y \leq 2x - 2$ **b.** $y > x + \frac{1}{2}$

c. $y < -x - \frac{1}{2}$ **d.** $y \geq 7$
e. $y > 0$ **f.** $x \leq 3$
g. $x \geq -2$ **h.** $x < 0$
i. $y + x > 3$ **j.** $y \leq x + 1$

5. Graphically represent the inequalities.

a. $y \leq \frac{1}{2}x + 4$ **b.** $y > \dfrac{x+1}{3}$
c. $y \geq 2x + 7$ **d.** $2x + 7 \leq y$
e. $y < x$

6. Graphically represent the region that satisfies all the given inequalities.

a. $y \leq x; y \leq -\frac{1}{2}x + 4$
b. $y \leq x; y \leq -\frac{1}{2}x + 4; y \geq 0$
c. $y \leq x; y \leq -\frac{1}{2}x + 4; y \geq 0; y \geq x - 2$

7. Graphically represent the region that satisfies all the given inequalities.

a. $x > 0; y > 0$
b. $x > 0; y > 0; x < 5$
c. $x > 0; y > 0; x < 5; y < 6$
d. $x > 0; y > 0; x < 5; y < 6; y < x + 3$
e. $x > 0; y > 0; x < 5; y < 6; y < x + 3; y > x - 3$

8. Find:

a. $\{(x,y) \mid x \leq 0\} \cap \{(x,y) \mid y \leq 0\} \cap \{(x,y) \mid y \geq -x - 4\}$
b. $\{(x,y) \mid x \leq 0\} \cap \{(x,y) \mid y \leq 0\} \cap \{(x,y) \mid y \leq -x - 4\}$
c. $\{(x,y) \mid x \leq 0\} \cap \{(x,y) \mid y \leq 0\} \cap \{(x,y) \mid y \geq -x - 4\} \cap \{(x,y) \mid y \leq -x - 3\}$

6.2 THE LINEAR PROGRAMMING PROBLEM: GRAPHICAL SOLUTION

One of the fastest growing branches of mathematics since World War II has been linear programming. Although it has its roots in solving complex military strategies, linear programming has been a most useful tool in solving problems related to business and the social sciences.

Basically, a linear programming problem involves the **objective** of either maximizing something (profit, for instance) or minimizing something (like cost), subject to certain restrictions or **constraints.** For example, suppose the profit P, for a given business venture, is given by

$$P = 20x + 30y$$

where x denotes the number of units a company can produce of item 1, selling at \$20 each, and y denotes the number of units the company can produce of item 2, selling at \$30 each. The objective is to maximize P. Suppose further that the company knows, because of consumers' demands, that they should produce at least twice as many of item 1 as item 2 and, because of the factory's physical limitations, they can produce no more than a total of 45 items. Also, there must be at least 5 units of item 1 produced. These are the constraints. We can summarize the problem below:

Problem: Maximize $P = 20x + 30y$ subject to:

$$\left.\begin{aligned} x &\geq 2y && (C1) \\ x + y &\leq 45 && (C2) \\ x &\geq 5 && (C3) \\ y &\geq 0 && (C4) \end{aligned}\right\} \text{constraints}$$

Here, $C1$ through $C4$ represent the constraints. The constraint $C4$, although not originally stated in words, is a common-sense constraint that is understood due to the makeup of the problem. (It would be impossible to produce a negative number of units of item 2, of course!)

We focus our attention on the constraints, $C1$ through $C4$. In order to find a solution (an x and a y),

$$C1 \text{ AND } C2 \text{ AND } C3 \text{ AND } C4$$

must be satisfied. We have two unknowns, x and y, and a graphical solution will be employed. Now, $C1$ is a linear inequality and is graphically represented by a set of points forming a half-plane. The same is true of $C2$, $C3$, $C4$. So, in order to satisfy $C1$ AND $C2$ AND $C3$ AND $C4$, we must find the solution set of $C1$ intersected with the solution set of $C2$ intersected with the solution set of $C3$ intersected with the solution set of $C4$. We use the following figures to illustrate this:

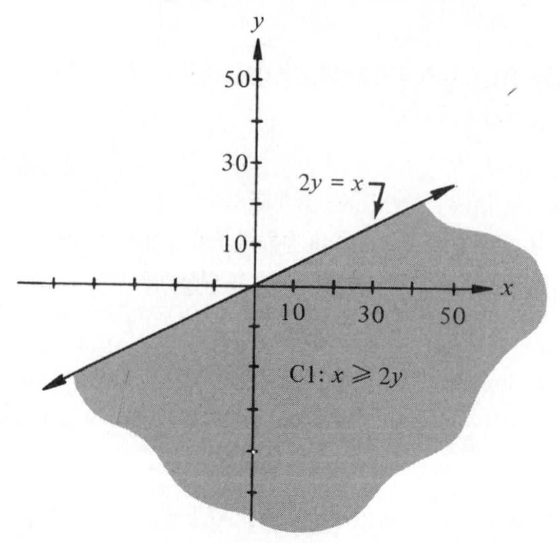

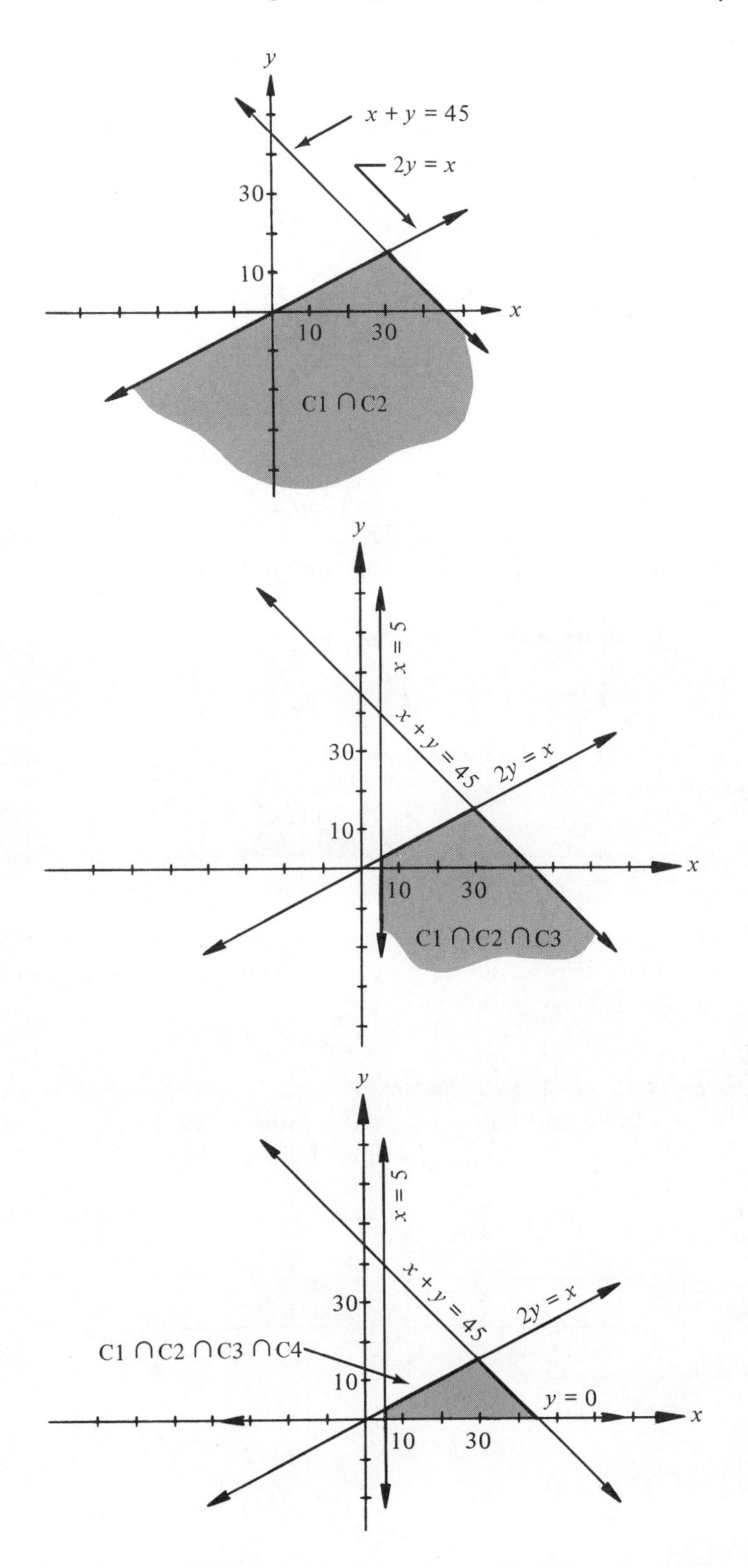

y
x + y = 45
2y = x
30
10
x
10
30
C1 ∩ C2
y
x = 5
x + y = 45
2y = x
30
10
x
10
30
C1 ∩ C2 ∩ C3
y
x = 5
x + y = 45
2y = x
30
C1 ∩ C2 ∩ C3 ∩ C4
10
y = 0
x
10
30

The region $C1 \cap C2 \cap C3 \cap C4$ forms a **convex set** of points. That is, a line segment connecting any two points on the perimeter of that figure will fall entirely within the figure.

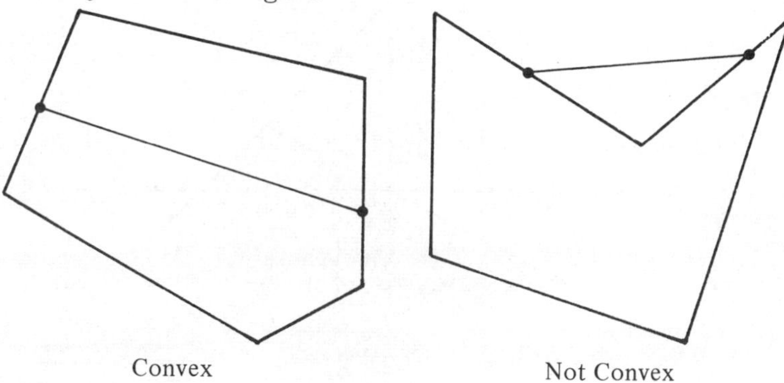

The **vertices** (or corners) of this convex set are the **feasible solutions** to the problem and one of them, the **optimal feasible solution,** *is the point at which P is maximal.*

Now, we have the figure below.

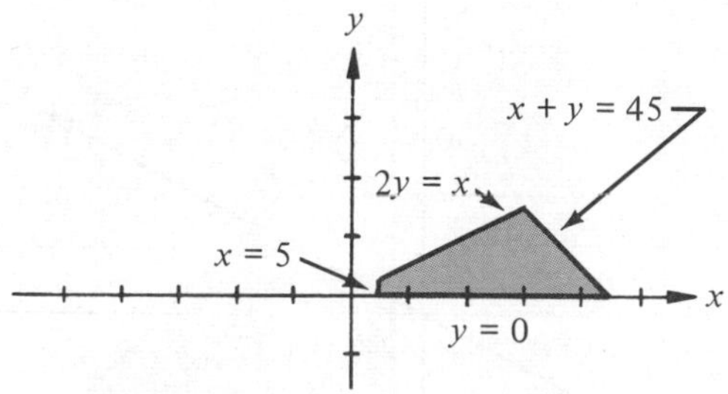

This figure represents the fact that any point (x,y) in (or on) that region will satisfy *all* constraints. We are looking for one point (x,y), in particular the one that maximizes

$$P = 20x + 30y$$

One of the vertices (feasible solutions) of that convex set will be the desired point (the optimal feasible solution). Those points (labelled A, B, C, D below) are solutions to pairs of linear equations.

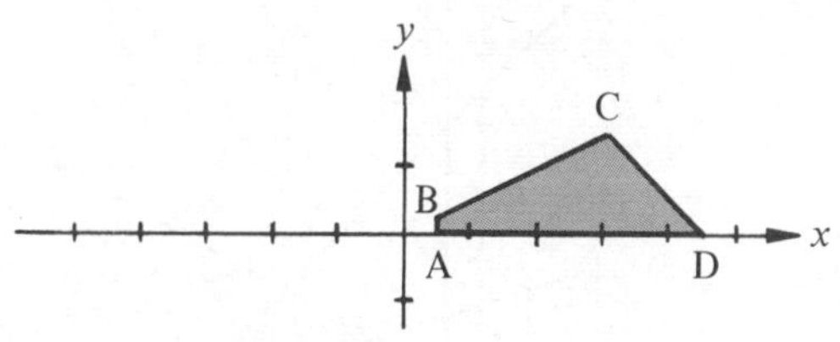

Point A is the point satisfying the system:

$$y = 0$$
$$x = 5$$

So, A is (5,0). Point B is the intersection of

$$x = 5$$

and

$$x = 2y$$

So, B is $(5,2\frac{1}{2})$. Point C is the solution of

$$x = 2y$$

and

$$x + y = 45$$

So, C is (30,15). Finally, D is the solution of

$$x + y = 45$$

and

$$y = 0$$

So, D is (45,0). We try each of the four feasible solutions in the profit equation, $P = 20x + 30y$.

Point	$20x + 30y = P$
$A = (5,0)$	$20 \cdot 5 + 30 \cdot 0 = 100$
$B = (5,2\frac{1}{2})$	$20 \cdot 5 + 30 \cdot 2\frac{1}{2} = 175$
$C = (30,15)$	$20 \cdot 30 + 30 \cdot 15 = 1050$ ←
$D = (45,0)$	$20 \cdot 45 + 30 \cdot 0 = 900$

The maximum profit (\$1050) occurs at point $C = (30,15)$. When 30 units of item 1 are produced and 15 units of item 2 are produced, the company is maximizing its profit.

Some examples follow.

Example 6.2.1. Maximize $P = 7x + 10y$ subject to:

$$x + 2y \geq 10$$
$$7x - 8y \geq -40$$
$$x + y \leq 20$$
$$4x - y \leq 40$$

Solution: We depict the convex set (the intersection of all solution sets of constraints) below:

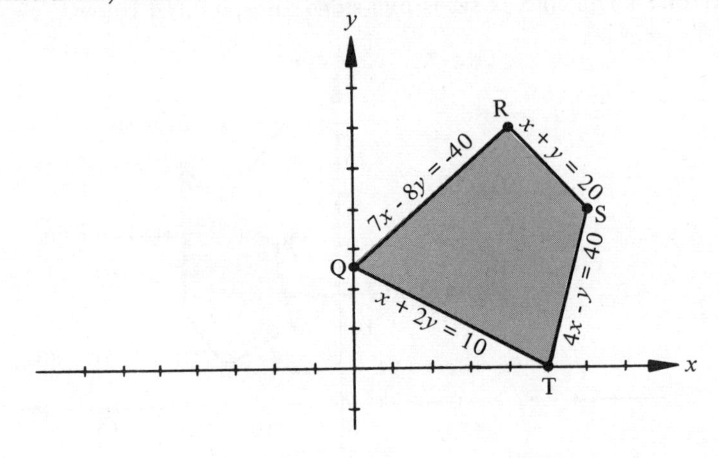

We find points Q, R, S, and T accordingly. The point Q is the solution of

$$x + 2y = 10$$

and

$$7x - 8y = -40$$

So Q is (0,5). The point R is the solution of

$$x + y = 20$$

and

$$7x - 8y = -40$$

So R is (8,12). The point S is the solution of

$$x + y = 20$$

and

$$4x - y = 40$$

So S is (12,8). The point T is the solution of

$$x + 2y = 10$$

and

$$4x - y = 40$$

So T is (10,0). We examine $P = 7x + 10y$ at each of these points.

Point	$7x + 10y = P$
$Q = (0,5)$	$7 \cdot 0 + 10 \cdot 5 = 50$
$R = (8,12)$	$7 \cdot 8 + 10 \cdot 12 = 176$
$S = (12,8)$	$7 \cdot 12 + 10 \cdot 8 = 164$
$T = (10,0)$	$7 \cdot 10 + 10 \cdot 0 = 70$

So, $P = 7x + 10y$ achieves its maximum value when $x = 8$ and $y = 12$. That value is 176.

Example 6.2.2. Minimize $T = 2x - 2y$ subject to:

$$-5x + 6y \leq 80$$
$$7x - 5y \leq 80$$
$$5x + 8y \geq 115$$
$$10x - y \leq 60$$
$$x \leq 20$$

Solution: The convex set is five-sided and pictured below.

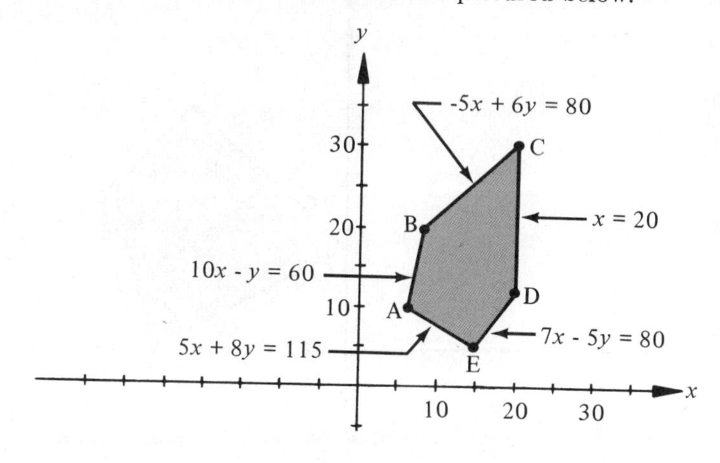

The feasible solutions are A, B, C, D, E. The point A is the solution of $5x + 8y = 115$ and $10x - y = 60$. So, $A = (7,10)$. The point B is the solution of $10x - y = 60$ and $-5x + 6y = 80$. So, $B = (8,20)$. The point C is the solution of $-5x + 6y = 80$ and $x = 20$. So, $C = (20,30)$. The point D is the solution of $x = 20$ and $7x - 5y = 80$. So, $D = (20,12)$. Finally, the point E is the solution of $7x - 5y = 80$ and $5x + 8y = 115$. So, $E = (15,5)$. To find our optimal feasible solution, we evaluate $T = 2x - 2y$ at all the feasible solutions. The smallest evaluation is our solution.

Point	$2x - 2y = T$
$A = (7,10)$	$2 \cdot 7 - 2 \cdot 10 = -6$
$B = (8,20)$	$2 \cdot 8 - 2 \cdot 20 = -24$
$C = (20,30)$	$2 \cdot 20 - 2 \cdot 30 = -20$
$D = (20,12)$	$2 \cdot 20 - 2 \cdot 12 = 16$
$E = (15,5)$	$2 \cdot 15 - 2 \cdot 5 = 20$

The equation $T = 2x - 2y$ is minimized at B. That minimal value of T is -24. So the optimal feasible solution occurs at (8,20).

Example 6.2.3. The Luxuro Company operates two factories, each producing three accessories: cigarette lighters, seat harnesses, and door handles. The following table depicts their daily output:

	Cigarette lighters	Seat harnesses	Door handles
Factory 1	100	200	400
Factory 2	100	500	200

It is known that the daily cost of operating Factory 1 is \$10,000; the daily cost of operating Factory 2 is \$20,000. To meet demands, Luxuro must produce at least 5,000 cigarette lighters, 15,000 seat harnesses, and 16,000 door handles next quarter. How many days should each factory be in production in order to minimize cost if, in addition (because of union rules), the sum of the days both factories can be open per quarter cannot exceed 100?

Solution: Let x denote the number of days Factory 1 is in operation and y the number of days Factory 2 is in operation. The problem becomes:

Minimize:	$10{,}000x + 20{,}000y$	(combined cost of operation)
Subject to:	$100x + 100y \geq 5{,}000$	(cigarette lighters)
	$200x + 500y \geq 15{,}000$	(seat harnesses)
	$400x + 200y \geq 16{,}000$	(door handles)
	$x + y \leq 100$	(union rules)
	$x \geq 0$, $y \geq 0$	common sense

We can simplify the inequalities somewhat:

$$\begin{aligned} x + y &\geq 50 \\ 2x + 5y &\geq 150 \\ 2x + y &\geq 80 \\ x + y &\leq 100 \\ x &\geq 0 \\ y &\geq 0 \end{aligned}$$

The reader is urged to verify the construction of the convex set below:

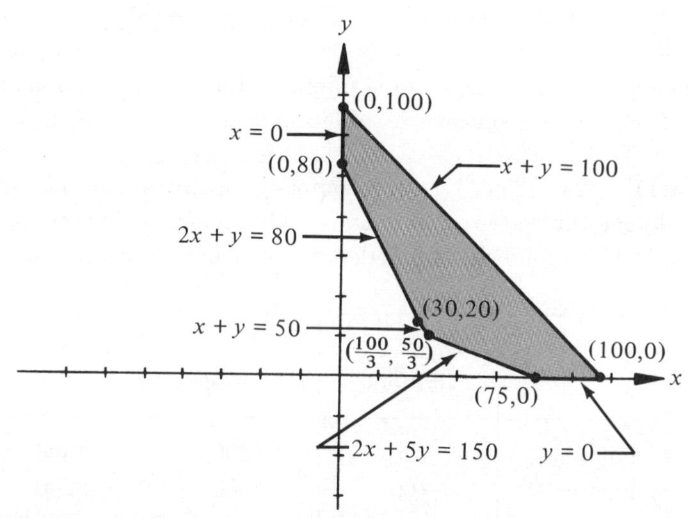

We check each of the feasible solutions below.

Point	Cost $= 10{,}000x + 20{,}000y$
(0,100)	$10{,}000 \cdot 0 + 20{,}000 \cdot 100 = \$2{,}000{,}000$
(0,80)	$10{,}000 \cdot 0 + 20{,}000 \cdot 80 = \$1{,}600{,}000$
(30,20)	$10{,}000 \cdot 30 + 20{,}000 \cdot 20 = \$700{,}000$
$(\frac{100}{3}, \frac{50}{3})$	$10{,}000 \cdot \frac{100}{3} + 20{,}000 \cdot \frac{50}{3} = \$666{,}666.67$
(75,0)	$10{,}000 \cdot 75 + 20{,}000 \cdot 0 = \$750{,}000$
(100,0)	$10{,}000 \cdot 100 + 20{,}000 \cdot 0 = \$1{,}000{,}000$

So, we see cost is minimized when Factory 1 is operating 100/3 days and Factory 2 operates 50/3 days of that quarter.

EXERCISES ■ SECTION 6.2.

1. a. Maximize $F = 10x + 10y$ subject to:

$$\begin{aligned} 3x + 2y &\leq 22 \\ y &\leq 5 \\ x &\leq 6 \\ x &\geq 0 \\ y &\geq 0 \end{aligned}$$

b. Minimize $F = 10x + 10y$ subject to the constraints of exercise 1*a*.
c. Minimize $G = -6x - 2y$ subject to the constraints of exercise 1*a*.
d. Using the constraints of exercise 1*a*, maximize $10x + y$.
e. Using the constraints of exercise 1*a*, maximize $10x - y$.
f. Using the constraints of exercise 1*a*, maximize $-x + 2y$.

2. a. Maximize $x + y$ subject to:

$$\begin{aligned} -x + 6y &\leq 30 \\ 3x + y &\leq 24 \\ x &\geq 0 \\ y &\geq 0 \end{aligned}$$

b. Minimize $x - y$ subject to the constraints of exercise 2*a*.
c. Maximize $5x + y$ subject to the constraints of exercise 2*a*.

3. a. The Embryo Toy Company manufactures two toys: wooden block sets and wooden trains. Due to limited facilities, the company cannot produce more than 40 block sets per week and cannot produce more than 60 trains per week, and the total number of toys produced cannot exceed 80. Represent the constraints of this problem as inequalities. Let x = number of block sets produced and y = number of trains produced each week.
b. Graph the constraints of exercise 3*a* to obtain the convex set and feasible solutions.
c. The Embryo Toy Company has determined that the profit for each block set is \$2.00, the profit for each train set is \$3.00. How many of each toy should the company produce to maximize profit?

4. a. Art is having a difficult time maintaining his car. His crankcase must maintain at least 2 quarts of oil, but no more than 5. His 20-quart capacity radiator must maintain at least 10 quarts of antifreeze. The sum of 5 times the number of quarts of oil and twice the number of quarts of antifreeze cannot exceed 55 because of storage problems. Represent the constraints of this problem as inequalities. Let x = number of quarts of oil and y = number of quarts of antifreeze.
b. Graph the constraints of exercise 4*a* to obtain the convex set and feasible solutions.
c. If in addition, each quart of oil costs \$.75 and each quart of antifreeze costs \$.25, minimize Art's cost.

5. The Octavius Music Company manufactures pianos and organs. Because of customers' demands, they do not make more than 30 pianos or more than 15 organs in any given month, but must produce at least 30 instruments each month. Into the production of each piano and organ goes raw materials and labor according to the table below.

	Units of raw materials	Units of labor
Each piano	1	1
Each organ	1	2

Due to costs, there is at most 35 units of raw materials and 40 units of labor available. How many units of each instrument must be produced to maximize profit if the profit for each piano is $200 and for each organ $100? What is that profit?

6. A moonlighting carpenter, Joseph, produces two types of furniture, bookcases and end tables, in his boss's shop after hours. He can produce no more than a total of nine items. In addition, he uses two tools, a lathe and a table saw, with units as described below:

	Hours on lathe	Hours on saw
Each book case	3	1
Each end table	5	3

The boss limits Joseph's hours on the lathe to 29 per week and limits the hours Joseph can spend on the saw to 15 per week. Joseph can sell each book case for a $20 profit and each end table for a $40 profit. How many items of each must he make to maximize his profit? What is that profit?

7. Use the constraints of question 6 to minimize Joseph's cost when Joseph figures each book case costs $5 in materials and each end table costs $6. What is that cost?

8. All linear programming problems discussed thus far have employed the testing of the vertices of a **bounded convex set.** It could happen that the intersection of constraints yields an unbounded convex set of points. For example, consider the following two problems where each convex set is **unbounded:**

a. Maximize $P = x + y$ subject to:
$$\begin{aligned} y - x &\leq 6 \\ y &\leq 8 \\ y &\geq 0 \\ x &\geq 0 \end{aligned}$$

b. Minimize $C = x - y$ subject to:
$$\begin{aligned} y - x &\leq 6 \\ y &\leq 8 \\ y &\geq 0 \\ x &\geq 0 \end{aligned}$$

9. **a.** Examine the points on $y = 8$ in problem 8*a*, and note that the optimal solution is also unbounded.
 b. Examine the points between (0,6) and (2,8) in problem 8*b* and see that the optimal solution need not be unique.

10. **a.** Maximize $P = x + y$ subject to:

$$\begin{aligned} y - x &\geq 6 \\ y &\leq 8 \\ y &\geq 0 \\ x &\geq 0 \end{aligned}$$

 b. Compare your results above to the results in exercise 8*a*.

6.3 THE LINEAR PROGRAMMING PROBLEM: ALGEBRAIC SOLUTION

In the previous section, all the linear programming problems we examined involved two unknowns and utilized a graphical solution. Later in this section, we shall examine a method for solving a linear programming problem involving three or more variables. The reason is because a graphical solution involving three variables is difficult and a graphical solution involving four or more variables is impossible.

First, we shall examine algebraically the graphical processes employed to solve an illustration of the previous section:

Maximize $P = 20x + 30y$ subject to:

$$\begin{aligned} x &\geq 2y && (C1) \\ x + y &\leq 45 && (C2) \\ x &\geq 5 && (C3) \\ y &\geq 0 && (C4) \end{aligned}$$

Recall that the constraints $C1$ through $C4$ individually represent half-planes and that when taken collectively, represent a convex set. The solution occurs at a vertex of that convex set and each vertex is the simultaneous solution of two of the equations:

$$\begin{aligned} x &= 2y && (E1) \\ x + y &= 45 && (E2) \\ x &= 5 && (E3) \\ y &= 0 && (E4) \end{aligned}$$

(Note, each equation above corresponds respectively to each constraint.) There are four equations; each pair of equations, when taken together, yields a point of intersection. Furthermore, there are $\binom{4}{2} = 6$ such possible pairs of equations. (For an explanation of $\binom{4}{2}$, see section 3.3.) We list them on the following page:

Possibility	Equations	Simultaneous solution point
1	$E1 \cap E2$	(30,15)
2	$E1 \cap E3$	$(5,2\frac{1}{2})$
3	$E1 \cap E4$	(0,0)
4	$E2 \cap E3$	(5,40)
5	$E2 \cap E4$	(45,0)
6	$E3 \cap E4$	(5,0)

Now, not all possibilities will be vertices of our convex sets. Some of these will not satisfy all the constraints. (The point (0,0), for example, is the point in possibility 3 above and although it is the intersection of $E1$ and $E4$, it fails in at least one constraint.) So, we check each possibility in every constraint. We have:

Possibility	Equations	Simultaneous solution point	Does this point satisfy EVERY constraint?
1	$E1 \cap E2$	(30,15)	Yes
2	$E1 \cap E3$	$(5,2\frac{1}{2})$	Yes
3	$E1 \cap E4$	(0,0)	No (violates $C3$)
4	$E2 \cap E3$	(5,40)	No (violates $C1$)
5	$E2 \cap E4$	(45,0)	Yes
6	$E3 \cap E4$	(5,0)	Yes

A "yes" in the last column indicates that the point is, in fact, a vertex of our desired convex set; the yes indicates the point is possible as solution (or is feasible). In the last step we find which feasible solution maximizes $P = 20x + 30y$. The table below completes the problem. The reader should compare it with the solution in section 6.2.

Possibility	Equations	Point	Feasible?	Value of $20x + 30y$
1	$E1 \cap E2$	(30,15)	Yes	$600 + 450 = 1050$
2	$E1 \cap E3$	$(5,2\frac{1}{2})$	Yes	$100 + 75 = 175$
3	$E1 \cap E4$	(0,0)	No	—
4	$E2 \cap E3$	(5,40)	No	—
5	$E2 \cap E4$	(45,0)	Yes	$900 + 0 = 900$
6	$E3 \cap E4$	(5,0)	Yes	$100 + 0 = 100$

Thus P reaches a maximum value of 1050 at (30,15). Hence (30,15) is the optimal feasible solution

Before going on to a problem involving three variables, we summarize and generalize the accomplishments of this section so far:

1. For the linear programming problem involving n constraints, there are n associated equations.

2. For a system with 2 variables there are $\binom{n}{2}$ possible points of intersection; for a system with r variables there will be $\binom{n}{r}$ possible points of intersection.
3. Each of those $\binom{n}{r}$ points are tested for feasibility.
4. Finally, we "plug" each feasible point into the expression that is to be maximized (or minimized).

We use these four steps in the two following examples.

Example 6.3.1. Maximize $8x + 5y$

subject to:

$$\begin{aligned} 2x + y &\leq 200 && (C1)\\ x + y &\leq 105 && (C2)\\ -10x + 9y &\leq 90 && (C3)\\ y &\geq 0 && (C4)\\ x &\leq 0 && (C5) \end{aligned}$$

Solution: There are five equations associated with each constraint.

$$\begin{aligned} 2x + y &= 200 && (E1)\\ x + y &= 105 && (E2)\\ -10x + 9y &= 90 && (E3)\\ y &= 0 && (E4)\\ x &= 0 && (E5) \end{aligned}$$

So, there are $\binom{5}{2} = 10$ possibilities. In the table below, we list these possibilities and check for feasibility.

Possibility	Equations	Point	Feasible?
1	$E1 \cap E2$	(95,10)	Yes
2	$E1 \cap E3$	$(61\frac{1}{14}, 77\frac{6}{7})$	No ($C2$ is not satisfied)
3	$E1 \cap E4$	(100,0)	Yes
4	$E1 \cap E5$	(0,200)	No ($C2$ is not satisfied)
5	$E2 \cap E3$	(45,60)	Yes
6	$E2 \cap E4$	(105,0)	No ($C1$ is not satisfied)
7	$E2 \cap E5$	(0,105)	No ($C3$ is not satisfied)
8	$E3 \cap E4$	(−9,0)	No ($C5$ is not satisfied)
9	$E3 \cap E5$	(0,10)	Yes
10	$E4 \cap E5$	(0,0)	Yes

Finally, we examine feasible solutions for $8x + 5y$ and obtain:

Possibility	Equations	Point	Feasible?	Value of $8x + 5y$ at feasible point
1	$E1 \cap E2$	$(95,10)$	Yes	$760 + 50 = 810$
2	$E1 \cap E3$	$(61\frac{1}{14},77\frac{6}{7})$	No	—
3	$E1 \cap E4$	$(100,0)$	Yes	$800 + 0 = 800$
4	$E1 \cap E5$	$(0,200)$	No	—
5	$E2 \cap E3$	$(45,60)$	Yes	$360 + 300 = 660$
6	$E2 \cap E4$	$(105,0)$	No	—
7	$E2 \cap E5$	$(0,105)$	No	—
8	$E3 \cap E4$	$(-9,0)$	No	—
9	$E3 \cap E5$	$(0,10)$	Yes	$0 + 50 = 50$
10	$E4 \cap E5$	$(0,0)$	Yes	$0 + 0 = 0$

So, the maximum of $8x + 5y$ occurs when $x = 95$ and $y = 10$. That maximum value is 810.

Example 6.3.2. Maximize $P = x + 5y + z$
subject to:

$$\begin{aligned} 3x - y + 2z &\leq 24 && (C1)\\ x + y + z &\leq 22 && (C2)\\ x + 2y + 3z &\leq 36 && (C3)\\ x &\geq 0 && (C4)\\ y &\geq 0 && (C5)\\ z &\geq 0 && (C6) \end{aligned}$$

Solution: There are six associated equations, one for each constraint. They are:

$$\begin{aligned} 3x - y + 2z &= 24 && (E1)\\ x + y + z &= 22 && (E2)\\ x + 2y + 3z &= 36 && (E3)\\ x &= 0 && (E4)\\ y &= 0 && (E5)\\ z &= 0 && (E6) \end{aligned}$$

There are three variables, so there will be $\binom{6}{3} = 20$ possibilities. We list them:

Possibility	Equations	Point	Feasible?	Value of $P = x + 5y + z$ at feasible point
1	$E1 \cap E2 \cap E3$	$(10,10,2)$	Yes	62
2	$E1 \cap E2 \cap E4$	$(0,6\frac{2}{3},15\frac{1}{3})$	No ($C3$ is not satisfied.)	—
3	$E1 \cap E2 \cap E5$	$(-20,0,42)$	No ($C3$ is not satisfied.)	—

(Poss.)	(Equations)	(Point)	(Feasible?)	(Value)
4	$E1 \cap E2 \cap E6$	$(11\frac{1}{2},10\frac{1}{2},0)$	Yes	64
5	$E1 \cap E3 \cap E4$	(0,0,12)	Yes	12
6	$E1 \cap E3 \cap E5$	(0,0,12)	Yes	12
7	$E1 \cap E3 \cap E6$	(12,12,0)	No ($C2$ is not satisfied.)	—
8	$E1 \cap E4 \cap E5$	(0,0,12)	Yes	12
9	$E1 \cap E4 \cap E6$	$(0,-24,0)$	No ($C5$ is not satisfied.)	—
10	$E1 \cap E5 \cap E6$	(8,0,0)	Yes	8
11	$E2 \cap E3 \cap E4$	$(0,30,-8)$	No ($C6$ is not satisfied.)	—
12	$E2 \cap E3 \cap E5$	(15,0,7)	No ($C1$ is not satisfied.)	—
13	$E2 \cap E3 \cap E6$	(8,14,0)	Yes	78
14	$E2 \cap E4 \cap E5$	(0,0,22)	No ($C1$ is not satisfied.)	—
15	$E2 \cap E4 \cap E6$	(0,22,0)	No ($C3$ is not satisfied.)	—
16	$E2 \cap E5 \cap E6$	(22,0,0)	No ($C1$ is not satisfied.)	—
17	$E3 \cap E4 \cap E5$	(0,0,12)	Yes	12
18	$E3 \cap E4 \cap E6$	(0,18,0)	Yes	90
19	$E3 \cap E5 \cap E6$	(36,0,0)	No ($C1$ is not satisfied.)	—
20	$E4 \cap E5 \cap E6$	(0,0,0)	Yes	0

So, the optimal solution occurs when $x = 0$, $y = 18$, $z = 0$. The maximum value of P at that point is 90.

The obvious length of the preceeding example may render the reader a bit cautious at this point—and rightly so! This kind of tediousness is justified only as an attempt to further familiarize the reader with the linear programming problem and a method of solution. Actually, a much more prudent means of solution is to employ a computer, and the technique used by many computer programmers is the **simplex method,** described in the next two sections.

EXERCISES ■ SECTION 6.3.

For exercises 1–7, redo exercises 1–7 of section 6.2 using the methods described in this section.

8. Minimize $x + y$ subject to:

$$\begin{aligned} x &\geq 2y \\ x + y &\leq 45 \\ x &\geq 5 \\ y &\geq 0 \end{aligned}$$

9. Maximize $P = 7x + 10y$ subject to:

$$\begin{aligned} x + 2y &\geq 10 \\ 7x - 8y &\geq -40 \\ x + y &\leq 20 \\ 4x - y &\leq 40 \end{aligned}$$

(Compare with example 6.2.1.)

10. Minimize $2x - y$ subject to:

$$\begin{aligned} 2x + y &\leq 200 \\ x + y &\leq 105 \\ -10x + 9y &\leq 90 \\ y &\geq 0 \\ x &\geq 0 \end{aligned}$$

The feasible solutions to this problem were obtained in example 6.3.1.

11. Minimize $x - y - z$ subject to:

$$\begin{aligned} 3x - y + 2z &\leq 24 \\ x + y + z &\leq 22 \\ x + 2y + 3z &\leq 36 \\ x &\geq 0 \\ y &\geq 0 \\ z &\geq 0 \end{aligned}$$

6.4 THE SIMPLEX METHOD

In this section we will examine another method for solving linear programming problems. In particular, we will solve *maximum* problems. Minimum problems will be discussed in the next section.

The maximum problem must first be in the form: Maximize $P = \mathbf{C} \cdot \mathbf{X}$ subject to:

$$\begin{aligned} \mathbf{AX} &\leq \mathbf{B}^* \\ \mathbf{X} &\geq \mathbf{O} \end{aligned} \tag{1}$$

where $\mathbf{C}$ is a row vector of n coefficients, $\mathbf{X}$ is a column vector of n variables, $\mathbf{A}$ is an $m \times n$ matrix of coefficients, and $\mathbf{B}$ is a column vector of m constants. For example, the problem: Maximize $P = 20x + 30y$ subject to:

* Note: The inequalities must appear in exactly this form. If an inequality appears as $x \geq 2y$, it can be rewritten to fit the form above: $-x + 2y \leq 0$. Also, we assume nonnegativity of all constants in $\mathbf{B}$ and all variables in $\mathbf{X}$.

$$\begin{aligned} -x + 2y &\le 0 \\ x + y &\le 45 \\ x &\ge 0 \\ y &\ge 0 \end{aligned} \tag{2}$$

This system is in the form of (1). Here, $\mathbf{X} = \begin{pmatrix} x \\ y \end{pmatrix}$, $\mathbf{C} = (20 \quad 30)$,

$$\mathbf{A} = \begin{pmatrix} -1 & 2 \\ 1 & 1 \end{pmatrix} \quad \text{and} \quad \mathbf{B} = \begin{pmatrix} 0 \\ 45 \end{pmatrix}$$

We now transform (1) to a system where the constraints $\mathbf{A} \cdot \mathbf{X} \le \mathbf{B}$ are converted from inequalities into equations. We do this by introducing **slack variables.** That is, we insert "dummy" variables to "take up the slack" of the inequalities. For example, the inequality

$$-x + 2y \le 0$$

is equivalent to

$$-x + 2y + s_1 = 0$$

where $s_1 \ge 0$ is some slack variable. If we insert two such variables in (2), our system looks like: Maximize $P = 20x + 30y$ subject to:

$$\begin{aligned} -x + 2y + s_1 &= 0 \\ x + y + s_2 &= 45 \\ x &\ge 0 \\ y &\ge 0 \end{aligned} \tag{2'}$$

Next, we represent the system (2′) in the **initial tableau,** below:

x	y	s_1	s_2	**B**
−1	2	1	0	0
1	1	0	1	45
−20	−30	0	0	0

indicators

The elements in the bottom row (except the last number) are called **indicators**; they're the negatives of the coefficients of the variables in P.*

* Actually, the last row in the tableau can be interpreted as $P - 20x - 30y = 0$.

Example 6.4.1. Construct the initial tableau for the following system: Maximize $P = 8x + 10y + z$ subject to:

$$\begin{aligned} x + y + z &\leq 5 & (C1) \\ y &\leq 3 & (C2) \\ y &\leq x & (C3) \\ x &\geq 0 & (C4) \\ y &\geq 0 & (C5) \\ z &\geq 0 & (C6) \end{aligned}$$

Solution: First notice that constraint $C3$ is not in the form of (1). The inequality $y \leq x$ is, however, equivalent to $-x + y \leq 0$. Next, we incorporate slack variables into the constraints:

$$\begin{aligned} x + y + z + s_1 &= 5 \\ y + s_2 &= 3 \\ -x + y + s_3 &= 0 \end{aligned}$$

The tableau is pictured below:

x	y	z	s_1	s_2	s_3	**B**
1	1	1	1	0	0	5
0	1	0	0	1	0	3
−1	1	0	0	0	1	0
−8	−10	−1	0	0	0	0

The construction of the initial tableau is the first step in the procedure of the simplex method. The second step is to locate the **pivot entry** in the tableau. The pivot entry is found in the following manner:

1. Locate the most negative indicator. The column containing it is called the **pivotal column.**
2. Divide each POSITIVE entry of the pivotal column into the corresponding element in the last column. The row associated with the smallest resulting quotient is called the **pivotal row.**
3. The pivot entry (or simply *pivot*) is the element in the pivotal column and pivotal row.

For an example, consider the tableau below:

x	y	s_1	s_2	**B**
−1	2	1	0	0
1	1	0	1	45
−20	−30	0	0	0
	↑ pivotal column			

Since -30 is the most negative indicator, we have located the pivotal column. Next, we examine the results of dividing each positive entry of the pivotal column into the corresponding element in the last column:

x	y	s_1	s_2	**B**	
-1	2	1	0	0	$0 \div 2 = 0$
1	1	0	1	45	$45 \div 1 = 45$
-20	-30	0	0	0	
	↑ pivotal column				

The results are 0 and 45. Since 0 is the smallest, the pivotal row is the first row. We circle the pivot 2, below:

x	y	s_1	s_2	**B**	
-1	2	1	0	0	← pivotal row
1	1	0	1	45	
-20	-30	0	0	0	
	↑ pivotal column				

The third step in the simplex method is to perform row transformations (like those described in section 5.4) on the tableau, forming a new tableau. We perform them by "converting" the pivot to 1 (if p is the pivot, multiply the pivotal row by $1/p$) and "converting" all other elements in the pivotal column to zero by using row transformations. THE SECOND AND THIRD STEPS ARE REPEATED UNTIL ALL INDICATORS ARE NONNEGATIVE.

We summarize the simplex method as described above in the four steps A through D, below:

STEP A. Establish the initial tableau.

STEP B. Determine the pivot.

STEP C. Using row transformations, convert the pivot to 1 and then eliminate (convert to 0) all other elements in the pivotal column.

STEP D. Repeat steps B and C until all indicators are nonnegative. When this occurs, the resulting tableau is called the **terminal tableau** and, with proper interpretation, can give us our desired maximum.

We put it all together in the following example.

Example 6.4.2. Maximize $P = 5x_1 + 3x_2$ subject to:

$$\begin{aligned} 4x_1 + 2x_2 &\leq 60 \\ x_1 + x_2 &\leq 20 \\ x_1 &\geq 0 \\ x_2 &\geq 0 \end{aligned}$$

Solution: STEP *A*. There are two slack variables, s_1 and s_2.

x_1	x_2	s_1	s_2	**B**
4	2	1	0	60
1	1	0	1	20
−5	−3	0	0	0

STEP *B*. The first column is the pivotal column (because $-5 < -3$) and the first row is the pivotal row. Hence, 4 is the pivot.
STEP *C*. Using 4 as our pivot, we perform row transformations to convert the initial tableau

x_1	x_2	s_1	s_2	**B**
4	2	1	0	60
1	1	0	1	20
−5	−3	0	0	0

into

x_1	x_2	s_1	s_2	**B**
1	$\frac{1}{2}$	$\frac{1}{4}$	0	15
0	$\frac{1}{2}$	$-\frac{1}{4}$	1	5
0	$-\frac{1}{2}$	$\frac{5}{4}$	0	75

STEP *D*. Now with $\frac{1}{2}$ as our pivot (why?), we transform the tableau above into the terminal tableau below:

x_1	x_2	s_1	s_2	**B**
1	0	$\frac{1}{2}$	−1	10
0	1	$-\frac{1}{2}$	2	10
0	0	1	1	80

To solve the problem, all that is left is to interpret the terminal tableau. The last row represents the following equation:

$$P + s_1 + s_2 = 80 \quad \text{or} \quad P = 80 - s_1 - s_2$$

Since we desire P to be maximum, it is obvious we want $s_1 = 0$ and $s_2 = 0$. Accordingly, the maximum value of P is 80. Also, the first row of the terminal tableau is represented by:

$$x_1 + \tfrac{1}{2}s_1 - s_2 = 10$$

With $s_1 = s_2 = 0$, we get $x_1 = 10$. Similarly, the second row yields $x_2 = 10$. That is, when $x_1 = 10$ and $x_2 = 10$, the maximum value of P is 80.

Example 6.4.3. Maximize $P = 8x + 5y$ subject to:

$$\begin{aligned} 2x + y &\leq 200 \\ x + y &\leq 105 \\ -10x + 9y &\leq 90 \\ x &\geq 0 \\ y &\geq 0 \end{aligned}$$

Solution: STEP A. The initial tableau is:

x	y	s_1	s_2	s_3	**B**
(2)	1	1	0	0	200
1	1	0	1	0	105
−10	9	0	0	1	90
−8	−5	0	0	0	0

STEP B. The pivot is 2, circled above.

STEP C.

x	y	s_1	s_2	s_3	**B**
1	$\frac{1}{2}$	$\frac{1}{2}$	0	0	100
0	($\frac{1}{2}$)	$-\frac{1}{2}$	1	0	5
0	14	5	0	1	1090
0	−1	4	0	0	800

STEP D. Repeat the process with the new pivot, $\frac{1}{2}$.

x	y	s_1	s_2	s_3	**B**
1	0	1	−1	0	95
0	1	−1	2	0	10
0	0	19	−28	1	950
0	0	3	2	0	810

According to this terminal tableau, 810 is the maximum value of P; it occurs when $x = 95$ and $y = 10$. (The same rationale as the previous example can be applied: $P + 3s_1 + 2s_2 = 810$ implies $P = 810 - 3s_1 - 2s_2$. So, $s_1 = s_2 = 0$. The top row declares $x + s_1 - s_2 = 95$, but $s_1 = s_2 = 0$, so $x = 95$. Similarly, $y = 10$.) The reader should compare this solution with the solution of example 6.3.1.

Before ending this section with a final example, we remark that the techniques described in this section apply to many "maximum" linear programming problems but not to several special cases that may prevail. (For example, we assume the elements in the last column of the initial tableau are all nonnegative.) For the reader curious about such special cases and other conditions that we have imposed, the following reference may be very useful: *The Theory of Linear Economic Models* by D. Gale (New York: McGraw-Hill, 1960).

Example 6.4.4. Maximize $P = x + 5y + z$ subject to:

$$\begin{aligned} 3x - y + 2z &\leq 24 \\ x + y + z &\leq 22 \\ x + 2y + 3z &\leq 36 \\ x &\geq 0 \\ y &\geq 0 \\ z &\geq 0 \end{aligned}$$

Solution: We set up the initial tableau and circle the pivot.

x	y	z	s_1	s_2	s_3	**B**
3	−1	2	1	0	0	24
1	1	1	0	1	0	22
1	(2)	3	0	0	1	36
−1	−5	−1	0	0	0	0

Pivoting on the 2 yields:

x	y	z	s_1	s_2	s_3	**B**
$\frac{7}{2}$	0	$\frac{7}{2}$	1	0	$\frac{1}{2}$	42
$\frac{1}{2}$	0	$-\frac{1}{2}$	0	1	$-\frac{1}{2}$	4
$\frac{1}{2}$	1	$\frac{3}{2}$	0	0	$\frac{1}{2}$	18
$\frac{3}{2}$	0	$\frac{13}{2}$	0	0	$\frac{5}{2}$	90

We stop here because all indicators are nonnegative. By the bottom row, $P + \frac{3}{2}x + \frac{13}{2}z + \frac{5}{2}s_3 = 90$ or $P = 90 - \frac{3}{2}x - \frac{13}{2}z - \frac{5}{2}s_3$ and it is obvious (since $x, z, s_3 \geq 0$) that P is maximized when $x = 0$, $z = 0$, $s_3 = 0$. The value of y under these conditions is 18, according to row three. This solution checks with the one found in the previous section (example 6.3.2) and is much less tedious!

EXERCISES ■ SECTION 6.4.

1. Establish the initial tableaux for the following linear programming problems.

   a. Maximize $P = 10x + 5y$ subject to:
$$\begin{aligned} 5x + 2y &\leq 6 \\ 7x + 3y &\leq 20 \\ x &\geq 0 \\ y &\geq 0 \end{aligned}$$

   b. Maximize $P = 10x + 5y$ subject to:
$$\begin{aligned} x + 3y &\leq 30 \\ 4x + 7y &\leq 100 \\ x &\geq 0 \\ y &\geq 0 \end{aligned}$$

   c. Maximize $P = 10x + 5y$ subject to:
$$\begin{aligned} x + 3y &\leq 30 \\ 4x + 7y &\leq 100 \\ x - 18 &\leq 0 \\ x &\geq 0 \\ y &\geq 0 \end{aligned}$$

   d. Maximize $P = x_1 + 2x_2 + 10x_3$ subject to:
$$\begin{aligned} 3x_1 + 2x_2 &\leq 14 \\ x_1 + 10x_3 &\leq 102 \\ x_1 + x_2 + 5x_3 &\leq 56 \\ x_3 &\leq 10 \\ x_1 &\geq 0 \\ x_2 &\geq 0 \\ x_3 &\geq 0 \end{aligned}$$

2. Locate the pivot for each tableaux:

   a.

x	y	s_1	s_2	**B**
1	3	1	0	30
4	7	0	1	100
−10	−5	0	0	0

   b.

x_1	x_2	x_3	s_1	s_2	s_3	s_4	**B**
1	6	3	1	0	0	0	175
1	4	4	0	1	0	0	150
1	0	5	0	0	1	0	85
0	0	1	0	0	0	1	15
−10	−20	−1	0	0	0	0	0

c.

x_1	x_2	x_3	x_4	s_1	s_2	s_3	s_4	s_5	**B**
1	−3	2	5	1	0	0	0	0	21
5	0	0	2	0	1	0	0	0	13
0	2	0	−1	0	0	1	0	0	0
1	2	3	4	0	0	0	1	0	30
0	0	0	1	0	0	0	0	1	4
−5	−5	−6	−8	0	0	0	0	0	0

In exercises 3–9, solve the linear programming problems using the simplex method.

3. Maximize $P = 10x + 5y$ subject to:

$$\begin{aligned} x + 3y &\leq 30 \\ 4x + 7y &\leq 100 \\ x &\geq 0 \\ y &\geq 0 \end{aligned}$$

4. Maximize $P = 10x + 5y$ subject to:

$$\begin{aligned} x + 3y &\leq 30 \\ 4x + 7y &\leq 100 \\ x &\leq 18 \\ x &\geq 0 \\ y &\geq 0 \end{aligned}$$

5. Maximize $P = x_1 + 2x_2 + 10x_3$ subject to:

$$\begin{aligned} 3x_1 + 2x_2 &\leq 14 \\ x_1 + 10x_3 &\leq 102 \\ x_1 + x_2 + 5x_3 &\leq 56 \\ x_2 &\leq 4 \\ x_3 &\leq 10 \\ x_1 &\geq 0 \\ x_2 &\geq 0 \\ x_3 &\geq 0 \end{aligned}$$

6. Maximize $P = x_1 + 2x_2 + 10x_3$ subject to:

$$\begin{aligned} 3x_1 + 2x_2 &\leq 14 \\ x_1 + 10x_3 &\leq 102 \\ x_1 + x_2 + 5x_3 &\leq 56 \\ x_3 &\leq 10 \\ x_1 &\geq 0 \\ x_2 &\geq 0 \\ x_3 &\geq 0 \end{aligned}$$

7. The Embryo Toy Company manufactures two toys: wooden block sets and wooden trains. Due to limited facilities, the company cannot produce more than 40 block sets per week and cannot produce more than 60 trains per week, and the total number of toys produced cannot exceed 80. Further, the profit for each block set is

$2.00 and the profit for each train set is $3.00. How many of each toy should the company produce to maximize profit? What maximum profit is achieved?

8. A moonlighting carpenter, Joseph, produces two types of furniture, bookcases and end tables, in his boss's shop after hours. He can produce no more than nine items. In addition, he uses two tools, a lathe and a table saw, with units as described below:

	Hours on lathe	Hours on saw
Each bookcase	3	1
Each end table	5	3

The boss limits Joseph's hours on the lathe to 29 per week and limits the hours Joseph can spend on the saw to 15 per week. Joseph can sell each bookcase for a $20 profit and each end table for a $40 profit. How many items of each must he make to maximize his profit? What is that profit?

9. An apprentice carpenter working in a co-op estimates his time spent on manufacturing three items according to the table below. Because others must use the tools, his time is limited to at most 90 hours per month on the lathe, at most 15 hours per month on the saw, and at most 55 hours per month on the sander. In addition, because of consumers' demands, he knows he must make at least three times as many candlesticks as bookcases. Maximize the carpenter's profit.

Project	No. of each	Hours on lathe	Hours on saw	Hours on sander	Profit on each
candlestick	x	5	1	2	$1
end table	y	2	1	3	$5
bookcase	z	1	0	1	$2

6.5 DUALITY AND THE MINIMUM PROBLEM

In sections 6.2 and 6.3 we learned how to solve linear programming problems where it was desired either to maximize something or to minimize something. In those two sections we employed geometric and algebraic approaches. Section 6.4 introduced the simplex method as a shorter way of solving a "maximum" problem. In the present section, we will show how the simplex method can be used to solve a "minimum" linear programming problem. To apply the simplex method, we "convert" the minimum problem to its maximum **dual.**

In general, we transform a minimum linear programming problem into its dual according to the following scheme:

Minimum problem	*Its dual*
Minimize $Q = b_1y_1 + b_2y_2 + \cdots + b_ny_n$	Maximize $P = c_1x_1 + c_2x_2 + \cdots + c_mx_m$
subject to:	subject to:
$a_{11}y_1 + a_{12}y_2 + \cdots + a_{1n}y_n \geq c_1$	$a_{11}x_1 + a_{21}x_2 + \cdots + a_{m1}x_m \leq b_1$
$a_{21}y_1 + a_{22}y_2 + \cdots + a_{2n}y_n \geq c_2$	$a_{12}x_1 + a_{22}x_2 + \cdots + a_{m2}x_m \leq b_2$
$\vdots$	$\vdots$
$a_{m1}y_1 + a_{m2}y_2 + \cdots + a_{mn}y_n \geq c_m$	$a_{1n}x_1 + a_{2n}x_2 + \cdots + a_{mn}x_m \leq b_n$
$y_1, y_2, \ldots, y_n \geq 0$	$x_1, x_2, x_3, \ldots, x_m \geq 0$

In matrix notation we can write:

Minimum problem

Minimize $Q = \mathbf{B} \cdot \mathbf{Y}$
subject to $\mathbf{A} \cdot \mathbf{Y} \geq \mathbf{C}$
$\mathbf{Y} \geq \mathbf{O}$

where $\mathbf{B} = (b_1 \quad b_2 \quad \cdots \quad b_n)$

$$\mathbf{Y} = \begin{pmatrix} y_1 \\ y_2 \\ \cdot \\ \cdot \\ \cdot \\ y_n \end{pmatrix} \qquad \mathbf{C} = \begin{pmatrix} c_1 \\ c_2 \\ \cdot \\ \cdot \\ \cdot \\ c_m \end{pmatrix}$$

$$\mathbf{A} = \begin{pmatrix} a_{11} & a_{12} & \cdots & a_{1n} \\ a_{21} & a_{22} & \cdots & a_{2n} \\ \cdot & & & \\ \cdot & & & \\ \cdot & & & \\ a_{m1} & a_{m2} & \cdots & a_{mn} \end{pmatrix}$$

Its dual

Maximize $P = \mathbf{C}^T \cdot \mathbf{X}$
subject to $\mathbf{A}^T \cdot \mathbf{X} \leq \mathbf{B}^T$ (1)
$\mathbf{X} \geq \mathbf{O}$

where $\mathbf{C}^T$ is the transpose of $\mathbf{C}$

$$\mathbf{X} = \begin{pmatrix} x_1 \\ x_2 \\ \cdot \\ \cdot \\ \cdot \\ x_m \end{pmatrix}$$

$\mathbf{A}^T$ is the transpose of $\mathbf{A}$, and $\mathbf{B}^T$ is the transpose of $\mathbf{B}$. (Recall that the transpose of a matrix is formed by interchanging rows and columns.)*

The above scheme partially illustrates the **duality theorem,** which further states that P's maximum is equal to Q's minimum.

The following example illustrates the scheme less abstractly.

Example 6.5.1. Form the dual of the problem: Minimize $Q = 60y_1 + 20y_2$ subject to:

$$4y_1 + y_2 \geq 5$$
$$2y_1 + y_2 \geq 3$$
$$y_1 \geq 0$$
$$y_2 \geq 0$$

* See exercises 25, 26 of section 5.3.

Solution: Here, $\mathbf{B} = (60 \quad 20)$, $\mathbf{A} = \begin{pmatrix} 4 & 1 \\ 2 & 1 \end{pmatrix}$, $\mathbf{Y} = \begin{pmatrix} y_1 \\ y_2 \end{pmatrix}$, and $\mathbf{C} = \begin{pmatrix} 5 \\ 3 \end{pmatrix}$.

The dual can be obtained by introducing $\mathbf{X} = \begin{pmatrix} x_1 \\ x_2 \end{pmatrix}$ and using matrix multiplication:

$$\mathbf{A}^T = \begin{pmatrix} 4 & 2 \\ 1 & 1 \end{pmatrix} \qquad \mathbf{B}^T = \begin{pmatrix} 60 \\ 20 \end{pmatrix} \qquad \mathbf{C}^T = (5 \quad 3)$$

$$\mathbf{C}^T \cdot \mathbf{X} = 5x_1 + 3x_2$$

$$\mathbf{A}^T \cdot \mathbf{X} = \begin{pmatrix} 4x_1 + 2x_2 \\ x_1 + x_2 \end{pmatrix}$$

So, according to (1), we have the dual: Maximize $P = 5x_1 + 3x_2$ subject to:

$$\begin{aligned} 4x_1 + 2x_2 &\leq 60 \\ x_1 + x_2 &\leq 20 \\ x_1 &\geq 0 \\ x_2 &\geq 0 \end{aligned}$$

To SOLVE the minimum linear programming problem, we solve the corresponding dual. That is, Q's minimum is P's maximum. The following example illustrates this fact further.

Example 6.5.2. Minimize $Q = 200y_1 + 105y_2 + 90y_3$ subject to:

$$\begin{aligned} 2y_1 + y_2 - 10y_3 &\geq 8 \\ y_1 + y_2 + 9y_3 &\geq 5 \\ y_1 &\geq 0 \\ y_2 &\geq 0 \\ y_3 &\geq 0 \end{aligned}$$

Solution: The reader should verify that the dual is: Maximize $P = 8x_1 + 5x_2$ subject to:

$$\begin{aligned} 2x_1 + x_2 &\leq 200 \\ x_1 + x_2 &\leq 105 \\ -10x_1 + 9x_2 &\leq 90 \\ x_1 &\geq 0 \\ x_2 &\geq 0 \end{aligned}$$

Next, we proceed as in the previous section by constructing the initial tableau.

x_1	x_2	s_1	s_2	s_3	$\mathbf{B}^T$
2	1	1	0	0	200
1	1	0	1	0	105
−10	9	0	0	1	90
−8	−5	0	0	0	0

We let the reader verify that the terminal tableau is:*

x_1	x_2	s_1	s_2	s_3	$\mathbf{B}^T$
1	0	1	−1	0	95
0	1	−1	2	0	10
0	0	19	−28	1	950
0	0	3	2	0	810

The maximum value of $P = 8x_1 + 5x_2$ is 810 when $x_1 = 95$, $x_2 = 10$. Furthermore, (by the duality theorem), the minimum value of Q is 810. The last three indicators will be the values of y_1, y_2, y_3, where

$$Q = 200y_1 + 105y_2 + 90y_3$$

Thus when $y_1 = 3$, $y_2 = 2$, and $y_3 = 0$, $Q = 810$, its smallest possible value according to the constraints. We end this chapter by solving the problem below.

Example 6.5.3. Minimize $Q = 4y_1 + 3y_2$ subject to:

$$\begin{aligned} y_1 + 10y_2 &\geq 10 \\ y_1 - y_2 &\geq 5 \\ y_2 &\geq 1 \\ y_1 &\geq 0 \\ y_2 &\geq 0 \end{aligned}$$

Solution: We solve the dual: Maximize $P = 10x_1 + 5x_2 + x_3$ subject to:

$$\begin{aligned} x_1 + x_2 &\leq 4 \\ 10x_1 - x_2 + x_3 &\leq 3 \\ x_1 &\geq 0 \\ x_2 &\geq 0 \\ x_3 &\geq 0 \end{aligned}$$

The initial tableau is:

x_1	x_2	x_3	s_1	s_2	$\mathbf{B}^T$
1	1	0	1	0	4
10	−1	1	0	1	3
−10	−5	−1	0	0	0

* The verification is simple if you look at example 6.4.3.

Pivoting on 10 yields:

x_1	x_2	x_3	s_1	s_2	$\mathbf{B}^T$
0	$\frac{11}{10}$	$-\frac{1}{10}$	1	$-\frac{1}{10}$	$\frac{37}{10}$
1	$-\frac{1}{10}$	$\frac{1}{10}$	0	$\frac{1}{10}$	$\frac{3}{10}$
0	-6	0	0	1	3

Pivoting on $\frac{11}{10}$ yields:

x_1	x_2	x_3	s_1	s_2	$\mathbf{B}^T$
0	1	$-\frac{1}{11}$	$\frac{10}{11}$	$-\frac{1}{11}$	$\frac{37}{11}$
1	0	$\frac{1}{11}$	$\frac{1}{11}$	$\frac{1}{11}$	$\frac{7}{11}$
0	0	$-\frac{6}{11}$	$\frac{60}{11}$	$\frac{5}{11}$	$\frac{255}{11}$

Pivoting on $\frac{1}{11}$ yields the terminal tableau:

x_1	x_2	x_3	s_1	s_2	$\mathbf{B}^T$
1	1	0	1	0	4
11	0	1	1	1	7
6	0	0	6	1	27

We interpret the terminal tableau and find that $Q = 27$ and occurs when $y_1 = 6$, $y_2 = 1$, the solution to our problem.

EXERCISES ■ SECTION 6.5.

1. Convert each linear programming problem to its dual.

a. Minimize $Q = y_1 + 3y_2$ subject to:

$$\begin{aligned} y_1 + y_2 &\geq 7 \\ 3y_1 + y_2 &\geq 17 \\ y_2 &\geq 2 \\ y_1 &\geq 0 \\ y_2 &\geq 0 \end{aligned}$$

b. Minimize $Q = y_1 + 10y_2 + 20y_3$ subject to:

$$\begin{aligned} y_1 + y_2 + y_3 &\geq 21 \\ 6y_1 - y_3 &\geq 28 \\ y_1 + 9y_2 &\geq 69 \\ y_1 &\geq 6 \\ y_1 &\geq 0 \\ y_2 &\geq 0 \\ y_3 &\geq 0 \end{aligned}$$

2. a. Solve the problem of 1*a*, above.
   b. Solve the problem of 1*b*, above.

3. Use the simplex method to solve:

   Minimize $Q = 2x + 4y$ subject to:

$$\begin{aligned} 2x + y &\geq 5 \\ 4x - y &\leq -2 \\ 6x &\geq 3 \\ x &\geq 0 \\ y &\geq 0 \end{aligned}$$

   (Hint: the second constraint has to be rewritten.)

4. Use the simplex method to solve:

   Minimize $Q = 5x + 7y + 10z$ subject to:

$$\begin{aligned} x + y + z &\geq 15 \\ -x + 2y + 4z &\geq 47 \\ y + 2z &\geq 24 \\ x &\geq 0 \\ y &\geq 0 \\ z &\geq 0 \end{aligned}$$

chapter 6 in review

CHAPTER SUMMARY

In this chapter we took a first look at the rectangular coordinate system and graphed linear inequalities and systems of linear inequalities as a prelude to graphically solving linear programming problems. A linear programming problem involves finding an optimal feasible solution (either a maximum or minimum) of a linear expression subject to certain restrictions (called constraints). To graphically solve such a problem, we formed a convex set of points that satisfied all constraints, found the vertices of that convex set, and chose the vertex that maximized (or minimized) the particular expression. The algebraic solution was similar: we found all feasible solutions (the vertices) and chose the optimal feasible solution. For large problems (i.e., those with either a large number of constraints or a large number of variables), both the algebraic and graphical methods became either extremely tedious or physically impossible. The simplex method of section 6.4 solved the maximum problems more handily; it gave a procedure for solving the problem. Minimum problems to be solved by the simplex method must first be converted to their duals, as observed in section 6.5.

VOCABULARY

rectangular coordinate system
origin
set-builder notation
half-plane
quadrants
objective
constraints
convex set
feasible solution
optimal feasible solution
bounded convex set
simplex method
slack variables
initial tableau
indicators
pivot
pivotal column
pivotal row
terminal tableau
dual

SYMBOLS

$\{(x,y)|y = x + 2\}$

$<$

$\leq$

$>$

$\geq$

x_1	x_2	x_3	s_1	s_2	**B**
a_{11}	a_{12}	a_{13}	1	0	b_1
a_{21}	a_{22}	a_{23}	0	1	b_2
$-c_1$	$-c_2$	$-c_3$	0	0	0

chapter 7
The Theory of Games

7.1 INTRODUCTION

The theory of games had its formal beginning in the 1920s with the work of the mathematician John von Neumann (1903–1957). After this introduction into the world of mathematics, game theory did not create much excitement until World War II. At that time, when the United States was hard put to solve many war problems, specialists of all types worked on solutions. One result was the use of game theory to analyze human and social events. More recently, game theory has found applications in business, political science, psychology, sociology—the list keeps growing. Hence, a **game** really refers to possible strategies by the **players.** A player might be a person, a business, a whole country, or even Mother Nature.

7.2 TWO-PERSON ZERO-SUM GAMES

Consider a game of guessing colors. Player I picks either red or green and Player II tries to guess which choice was made. The players agree to the following:

Player I picks red and Player II guesses red, whereby I pays II a sum of $2.

Player I picks red and Player II guesses green, whereby II pays I a sum of $5.

Player I picks green and Player II guesses green, whereby I pays II a sum of $3.

Player I picks green and Player II guesses red, whereby II pays I a sum of $4.

This game has two players and we refer to such a game as a **two-person game.** In addition, all the money lost by one player is gained by the other. A game with this property is referred to as a **zero-sum game.** We can summarize the four possible game outcomes by a 2×2 matrix:

$$\begin{array}{cc} & \text{Player II} \\ & \begin{array}{cc} R & G \end{array} \\ \text{Player I} \begin{array}{c} R \\ G \end{array} & \begin{pmatrix} -2 & 5 \\ 4 & -3 \end{pmatrix} \end{array}$$

This element signifies that Player I picked green and Player II guessed green and thus I paid II $3.

Notice:

1. "Player I" (called the **row player**) is written to the left of the matrix.
2. "Player II" (called the **column player**) is written above the matrix.
3. The matrix elements are always the gains by the row player (a minus sign indicates a loss).

Any element in the matrix above is called a **payoff** and the matrix itself is called the **payoff matrix** or the **game matrix.**

Example 7.2.1. Joni and Jim agree to play a game whereby Joni flips a coin after which Jim flips a coin. When Jim matches Joni, he wins $1. When Jim does not match Joni, he loses $2. Set up a payoff matrix where Joni is the row player. (Remember to list all elements as gains by the row player.)

Solution: We begin by writing the payoff matrix:

$$\begin{array}{cc} & \text{Jim} \\ & \begin{array}{cc} H & T \end{array} \\ \text{Joni} \begin{array}{c} H \\ T \end{array} & \begin{pmatrix} & \\ & \end{pmatrix} \end{array}$$

When Jim matches Joni, he wins $1. In order to display this as a gain by Joni (who is the row player), we must write:

$$\begin{array}{cc} & \text{Jim} \\ & \begin{array}{cc} H & T \end{array} \\ \text{Joni} \begin{array}{c} H \\ T \end{array} & \begin{pmatrix} -1 & \\ & -1 \end{pmatrix} \end{array}$$

Furthermore, when Jim fails to match Joni, he loses $2. So we have:

$$\begin{array}{cc} & \text{Jim} \\ & \begin{array}{cc} H & T \end{array} \\ \text{Joni} \begin{array}{c} H \\ T \end{array} & \begin{pmatrix} -1 & 2 \\ 2 & -1 \end{pmatrix} \end{array}$$

As we examine more games, we will assume they are played many times. The reason is that we are interested in what can happen over a long period of time. After all, each player has two **strategies** available (for example, choose red or choose green). The **row player** chooses a strategy by selecting a row of the game matrix and then the **column player** chooses a strategy by selecting a column of the game matrix. The intersection of the row and the column chosen by the players determines the payoff. (Remember, the payoffs are listed as amounts received by the row player; a minus sign means a loss by the row player.) Sometimes the payoff will be to the row player and other times it will be to the column player. So when many games are played, a few questions arise:

1. What strategy is best for each player?
2. Which player does the game favor (if either)?
3. If a certain player is favored, how much will be gained?

The answering of these questions constitutes our main goal of this chapter. We proceed with an example.

Example 7.2.2. Consider the two-person zero-sum game defined by the following matrix:

$$\text{Row player} \quad \overset{\text{Column player}}{\begin{pmatrix} 2 & 4 \\ -3 & -1 \end{pmatrix}}$$

The game is played by the row player selecting a row and the column player selecting a column. The intersection of these choices gives the payoff. (Remember, payoffs are listed as gains by the row player. A minus sign signifies a loss by the row player.)

a. Determine each player's best strategy.
b. Who does the game favor?

Solution:

a. It would be wise for the row player always to choose row one, since he cannot possibly lose with that choice. Further, the column player should then select column one, because that will cause his loss to be as small as possible.
b. When the best strategies are employed by each player, the payoff is $2 (to the row player). Hence this game surely favors the row player.

The solution in part *a* indicates that when a player is about to select a strategy, he must survey the choices available to him and then make a choice with the hope of maximizing his gain or minimizing his loss. When each player selects the best strategy in the game of example 7.2.2, we have seen that the row player wins $2. This amount is called the **value of the game.** Generally, the term "value" refers to the payoff that results when each player chooses his best strategy. The value of any game can be:

1. positive (game favors the row player),
2. negative (game favors the column player),
3. zero (game favors neither player—it is a fair game).

The game in example 7.2.2 is illustrative of a special type, where there is a matrix element (the 2 in this case) that is both minimum in its row and maximum in its column. This element is called a **saddle point.** A game with a saddle point is said to be a game of **pure strategies** (or a **strictly determined game**). It can be shown that when a game has a saddle point, the row containing the saddle point is the best strategy for the row player and the column containing the saddle point is the best strategy for the column player. The saddle point itself is the value of the game. We now present some more examples.

Example 7.2.3. Determine which (if any) of the games represented by the matrices below are games of pure strategy. For those that are, state each player's best strategy, state the value of the game, and tell who the game favors.

a. Column player

Row player $\begin{pmatrix} 2 & -1 \\ 4 & 6 \end{pmatrix}$

b. Column player

Row player $\begin{pmatrix} 4 & 1 \\ -1 & -1 \end{pmatrix}$

c. Column player

Row player $\begin{pmatrix} -2 & 5 \\ 4 & -3 \end{pmatrix}$

d. Column player

Row player $\begin{pmatrix} -1 & -2 \\ -3 & 3 \end{pmatrix}$

e. Column player

Row player $\begin{pmatrix} 2 & 0 & 1 \\ 4 & -1 & -2 \end{pmatrix}$

Solution:

a. The minimum value in row one is -1. The minimum value in row two is 4. We now check to see whether -1 is maximum in its column—it is not. We now check to see whether 4 is maximum in its column—it is.

$$\begin{pmatrix} 2 & -1 \\ 4 & 6 \end{pmatrix}$$

So this is a game of pure strategies. The row player should choose row two, and the column player should choose column one. The value of this game is 4 (because 4 is the saddle point). Hence, each time the game is played, the row player will win \$4 (presuming the players employ their best strategies). This game favors the row player.

b. The element in the first row, second column position, namely 1, is a saddle point.

$$\begin{pmatrix} 4 & \textcircled{1} \\ -1 & -1 \end{pmatrix}$$

So, this is a game of pure strategies. The row player should choose row one and the column player should choose column two. The value of the game is 1 and hence the row player is favored.

c. The minimum value in row one is -2. The minimum value in row two is -3.

$$\begin{pmatrix} -2 & 5 \\ 4 & -3 \end{pmatrix}$$

However, -2 is not the maximum value in its column, and -3 is not the maximum value in its column. Thus this game matrix has no saddle point. The game is not one of pure strategies. This type of situation will be studied later in sections 7.3 and 7.4.

d. The minimum value in row one is -1. The minimum value in row two is -3. We see that -1 is the maximum value in its column, whereas -3 is not the maximum value in its column. So, -1 is a saddle point for this game matrix:

$$\begin{pmatrix} \textcircled{-1} & -2 \\ -3 & 3 \end{pmatrix}$$

This is a game of pure strategies, where the row player should choose row one and the column player should choose column one. The value of the game is -1 and hence the game is favorable to the column player.

e. The matrix under consideration is

$$\begin{array}{cc} & \text{Column player} \\ \text{Row player} & \begin{pmatrix} 2 & 0 & 1 \\ 4 & -1 & -2 \end{pmatrix} \end{array}$$

The game is such that the row player has two strategies available and the column player has three strategies available. We call this a 2×3 game. For the purposes of this example, we deal with it in the same manner as those above. The minimum value in row one is 0. The minimum value in row two is -2. Here 0 is the maximum value in its column, while -2 is not the maximum value in its column. Hence, 0 is a saddle point for this game matrix:

$$\begin{pmatrix} 2 & \textcircled{0} & 1 \\ 4 & -1 & -2 \end{pmatrix}$$

This is a game of pure strategies, where the row player should choose row one and the column player should choose column two. The value of the game is 0 and hence neither player is favored—the game is fair.

EXAMPLE 7.2.4. Consider a simplified football game between the Lords and the Lancers. The Lords' offensive team has two different offensive plays: pass or run. The Lancers' defensive team knows this, so they employ just two basic defensive plays: a passing defense and a running defense. When the Lords' offensive team passes against the Lancers' defensive team, which is employing a running defense, an average gain of 9 yards results. When the Lords' offensive team runs against the Lancers' defensive team, which is employing a passing defense, an average gain of 6 yards results. When the Lancers' defense matches the Lords' offensive play, there is no gain or loss.

a. Set up a game matrix for this example with the Lords as row players.
b. Decide whether this game is a game of pure strategies.

Solution:

a. The answer is:

		Lancers' defense	
		Pass	Run
Lords' offense	Pass	0	9
	Run	6	0

b. This matrix has no saddle point, so it is not a game of pure strategies. As already mentioned, games of this type will be discussed in sections 7.3 and 7.4.

The reader should note that the matrix elements in this example are not monetary gains. We wish to point this out because the concept of a gain or a loss can be measured in many ways, depending upon the setting of the problem. This will become more apparent as we proceed.

The present section has introduced the reader to some of the terminology of game theory. We also saw how to determine a saddle point and the associated best strategies. As pointed out, the occurrence of a saddle point indicates that a game is one of pure strategies. It happens that if a saddle point exists, it is unique. However, a matrix game may have that saddle point occurring in more than one location. Consider the following payoff matrix:

$$\begin{pmatrix} \circled{2} & 3 & 4 \\ \circled{2} & 2 & 5 \\ -1 & 4 & -3 \end{pmatrix}$$

The saddle point is 2, occurring in the circled locations.

EXERCISES ■ SECTIONS 7.1 AND 7.2.

1. Players A and B agree to play a game whereby each flips a coin. Player A wins a quarter when the coins match. Player B wins a

quarter when the coins do not match. Set up the payoff matrix for this game. Let A be the row player.

2. Consider the coin-tossing game between Arlo and Bob in the beginning of section 4.8 (Expectation) on page 190. Set up the payoff matrix for such a game and let Arlo be the row player. (List the payoffs in cents.)
3. David and Joan agree to play a game whereby each throws out either one or two fingers. David agrees to pay Joan the sum (in dollars) of the total number of fingers shown. Set up the game matrix and let David be the row player.
4. Two armies (Red and Green) are engaged in war games. The Red Army has two ammunition supplies (small and large). The large supply is three times as important as the small supply. (It has three times as much explosive power.) The Green Army is planning a raid on the Red Army's ammunition supply and the Red Army can repel an attack at only one location. Let "1" denote the importance (or value) of the small supply. Then "3" will represent the importance of the large supply. Set up the game matrix and let the Red Army be the row player.
5. You are a manufacturer of leather goods and need to decide whether or not to invest in new sewing and punching equipment. If the forthcoming year is a good one, a decision to invest will be worth \$8,000. A decision not to invest will cost you \$2,000 (in loss of customers). If the forthcoming year is a recession year and you invest in the new equipment, you will lose \$5,000. If you do not invest and there is a recession year, you will break even. Set up a payoff matrix for this game, where you are the row player and Mother Nature (the conditions that control the future economic states) is the column player.
6. Two competing businesses, Infinity Corporation and Finite Products, are contemplating building new outlet stores in Berwick, Pa. The basic decision to be made is whether to build in the city or in the suburbs. The game matrix appears below:

		Finite	
		City	Suburbs
Infinity	City	1	−2
	Suburbs	3	−4

The matrix elements represent amounts (in thousands of dollars) either gained or lost by one business from or to the other as a result of the decisions as to building location.

   a. If Infinity builds in the city and Finite builds in the suburbs, who gains more business from the other and how much?

b. If both companies build in the city, describe the outcome in terms of the information of this exercise.

c. Is this game strictly determined?

d. If so, what is the value of the game?

7. Determine which of the following two-person zero-sum games are games of pure strategies (i.e., strictly determined). For those that are, determine the best strategies, the value of the game, and who the game favors.

a. Row player, Column player $\begin{pmatrix} 1 & 0 \\ 0 & -1 \end{pmatrix}$ **b.** Row player, Column player $\begin{pmatrix} 5 & 2 \\ -2 & -2 \end{pmatrix}$

c. Row player, Column player $\begin{pmatrix} 2 & 5 \\ 4 & -1 \end{pmatrix}$ **d.** Row player, Column player $\begin{pmatrix} -2 & 0 & 1 \\ -3 & -2 & 4 \end{pmatrix}$

e. Row player, Column player $\begin{pmatrix} 0 & -5 & 6 \\ 2 & -4 & 4 \\ 1 & -3 & 2 \end{pmatrix}$

8. Which of the following two-person zero-sum games are games of pure strategies? For those that are, determine the best strategies, the value of the game, and who the game favors.

a. $\begin{pmatrix} 10 & -3 \\ 2 & -1 \end{pmatrix}$ **b.** $\begin{pmatrix} 0 & 1 \\ 2 & 3 \end{pmatrix}$

c. $\begin{pmatrix} -7 & 0 \\ 3 & 2 \\ 4 & -1 \end{pmatrix}$ **d.** $\begin{pmatrix} 4 & 3 & 2 & 1 & -2 \\ -1 & 2 & 3 & -1 & -1 \end{pmatrix}$

9. Consider the following game matrix:

$$\begin{pmatrix} -1 & -2 \\ 2 & -1 \\ 3 & -1 \\ 1 & -3 \end{pmatrix}$$

a. What is the value of the game?

b. Exhibit the best strategies for the row player and the column player. The reader should note that the value of the game is unique but the best strategy is not.

7.3 TWO-PERSON ZERO-SUM GAMES WITH MIXED STRATEGIES

The previous section dealt with two-person zero-sum games, most of which were games with pure strategies. However, consider the 2×2 game described by the matrix

$$\begin{matrix} & R & G \\ \begin{matrix} R \\ G \end{matrix} & \multicolumn{2}{c}{\begin{pmatrix} -2 & 5 \\ 4 & -3 \end{pmatrix}} \end{matrix}$$

which was looked at in the very beginning of section 7.2. This matrix has no saddle point and hence is not a game of pure strategies. In order to deal with this game, let us use some common sense. The row player would be wise to choose row one, because there exists the chance of winning \$5 or losing \$2 as opposed to the possible choice of row two, where only \$4 can be won or \$3 can be lost. However, the row player should not always pick row one, because the column player would get smart and start always to pick column one, thus causing the row player continually to lose. So the row player must pick row one sometimes and row two other times. By this we mean the row player must use a strategy that is *mixed*. Thus, the column player will also need a mixed strategy. We refer to a game like this as one of **mixed strategies.** The manner of mixing that the row player chooses should be random so that the column player will not be able to detect any observable pattern. (This is discussed more in section 7.4.)

In order to analyze the above game when mixed strategies are employed, let us assume that the row player decides to choose row one 40% of the time and row two 60% of the time. Also, let us assume that the column player chooses column one 30% of the time and column two 70% of the time. The probability that the row player chooses row one AND the column player chooses column one is (.4)(.3) = .12. So, the expected payoff for this situation is (.12)(−\$2) = −\$.24. (The reader must recall the concept of expectation from section 4.8.) Performing similar computations leads to the table below.

Choice of strategies				
Row player	Column player	Payoff	Probability	Expected payoff
row one	column one	−\$2	(.4)(.3) = .12	−\$.24
row one	column two	\$5	(.4)(.7) = .28	\$1.40
row two	column one	\$4	(.6)(.3) = .18	\$.72
row two	column two	−\$3	(.6)(.7) = .42	−\$1.26
			Sum: 1.00	Sum: \$ .62

The positive result shows the game favors the row player.

Hence, the expected payoff of this game when the players choose the strategies mentioned above is \$.62. Recalling that expectation is an average (section 4.8), we can interpret this result to mean that if the players played many games (with the chosen strategies), ON THE AVERAGE the row player would win \$.62 per game.

Example 7.3.1. Using the game matrix in the above illustration, find the expected payoff when the row player chooses row one 70% of the time and row two 30% of the time. Further, assume the column player chooses column one 50% of the time and column two 50% of the time.

Solution: The reader should check the computations below.

Choice of strategies				
Row player	Column player	Payoff	Probability	Expected payoff
row one	column one	−\$2	.35	−\$.70
row one	column two	\$5	.35	\$1.75
row two	column one	\$4	.15	\$.60
row two	column two	−\$3	.15	−\$.45
			Sum: 1.00	Sum: \$1.20

The positive result shows the game favors the row player.

Thus with the strategies employed, we see that the expected payoff is \$1.20. So, in the long run the row player is favored and will win \$1.20 per game ON THE AVERAGE.

Next we will outline a way of computing this expected payoff (E) using multiplication of matrices and thus quickening computation time Consider the matrix

$$\mathbf{A} = \begin{pmatrix} a_{11} & a_{12} \\ a_{21} & a_{22} \end{pmatrix}$$

which represents a 2×2 game. Let:

p_1 be the percent of time (probability) that the row player chooses row one,

p_2 be the percent of time (probability) that the row player chooses row two,

q_1 be the percent of time (probability) that the column player chooses column one,

q_2 be the percent of time (probability) that the column player chooses column two.

Form the row vector $\mathbf{P} = (p_1 \quad p_2)$ for the row player's strategy. Form the column vector $\mathbf{Q} = \begin{pmatrix} q_1 \\ q_2 \end{pmatrix}$ for the column player's strategy. The reader can verify by checking the table in example 7.3.1 that the expected payoff E is given by:

$$E = p_1q_1a_{11} + p_1q_2a_{12} + p_2q_1a_{21} + p_2q_2a_{22}$$

Using the matrices **P**, **A**, and **Q** we can write:

$$E = \mathbf{PAQ}^*$$

(The reader is asked to verify this in exercise 3 of this section.)

Example 7.3.2. Compute the expected payoff for the game in example 7.3.1 by using matrix multiplication, $E = \mathbf{PAQ}$.

Solution: $\mathbf{P} = (.7 \quad .3)$,

$$\mathbf{A} = \begin{pmatrix} -2 & 5 \\ 4 & -3 \end{pmatrix} \quad \text{and} \quad \mathbf{Q} = \begin{pmatrix} .5 \\ .5 \end{pmatrix}$$

So,

$$E = (.7 \quad .3)\begin{pmatrix} -2 & 5 \\ 4 & -3 \end{pmatrix}\begin{pmatrix} .5 \\ .5 \end{pmatrix} = (-1.4 + 1.2 \quad 3.5 - .9)\begin{pmatrix} .5 \\ .5 \end{pmatrix}$$

$$= (-.2 \quad 2.6)\begin{pmatrix} .5 \\ .5 \end{pmatrix} = -.1 + 1.30 = +1.20$$

Hence, the expected payoff is \$1.20 (to the row player), which, of course, checks with the solution in example 7.3.1.

Example 7.3.3. Consider the game matrix

$$\mathbf{A} = \begin{pmatrix} 2 & 0 & -5 \\ -4 & 1 & 2 \end{pmatrix}$$

Assume:

the row player chooses row one 40% of the time,
the row player chooses row two 60% of the time,
the column player chooses column one 20% of the time,
the column player chooses column two 30% of the time,
the column player chooses column three 50% of the time.

Find the expected payoff (E).

Solution: First form the row vector of the row player's strategy, $\mathbf{P} = (.4 \quad .6)$. Next form the column vector of the column player's strategy,

$$\mathbf{Q} = \begin{pmatrix} .2 \\ .3 \\ .5 \end{pmatrix}$$

* $E = \mathbf{PAQ}$ is an abbreviated way of writing $E = \mathbf{P} \cdot \mathbf{A} \cdot \mathbf{Q}$.

Then $E = \mathbf{PAQ} = (.4 \quad .6)\begin{pmatrix} 2 & 0 & -5 \\ -4 & 1 & 2 \end{pmatrix}\begin{pmatrix} .2 \\ .3 \\ .5 \end{pmatrix}$

$$= (-1.6 \quad .6 \quad -.8)\begin{pmatrix} .2 \\ .3 \\ .5 \end{pmatrix} = -3.2 + .18 - .40$$

Therefore $E = -.54$. Hence this game, with the strategies as given, favors the column player. ON THE AVERAGE, when many games are played, the column player will receive a payoff of .54.

Example 7.3.4. Consider the game matrix

$$\mathbf{A} = \begin{pmatrix} 2 & 1 \\ -1 & 2 \end{pmatrix}$$

Find the expected payoff (E) for the strategies below and determine who the game favors.

a. $\mathbf{P} = (.5 \quad .5),\ \mathbf{Q} = \begin{pmatrix} .5 \\ .5 \end{pmatrix}$ **b.** $\mathbf{P} = (.5 \quad .5),\ \mathbf{Q} = \begin{pmatrix} .2 \\ .8 \end{pmatrix}$

c. $\mathbf{P} = (.1 \quad .9),\ \mathbf{Q} = \begin{pmatrix} .9 \\ .1 \end{pmatrix}$

Solution:

a. $E = \mathbf{PAQ} = (.5 \quad .5)\begin{pmatrix} 2 & 1 \\ -1 & 2 \end{pmatrix}\begin{pmatrix} .5 \\ .5 \end{pmatrix} = (.5 \quad 1.5)\begin{pmatrix} .5 \\ .5 \end{pmatrix} = .25 + .75$

Therefore $E = 1$. The game favors the row player.

b. $E = \mathbf{PAQ} = (.5 \quad .5)\begin{pmatrix} 2 & 1 \\ -1 & 2 \end{pmatrix}\begin{pmatrix} .2 \\ .8 \end{pmatrix} = (.5 \quad 1.5)\begin{pmatrix} .2 \\ .8 \end{pmatrix} = .1 + 1.2$

Thus $E = 1.3$ and the game favors the row player.

c. $E = \mathbf{PAQ} = (.1 \quad .9)\begin{pmatrix} 2 & 1 \\ -1 & 2 \end{pmatrix}\begin{pmatrix} .9 \\ .1 \end{pmatrix} = (-.7 \quad 1.9)\begin{pmatrix} .9 \\ .1 \end{pmatrix}$

$= -.63 + .19$

Therefore $E = -.44$. The game favors the column player.

The results of this example, as well as some others of this section, make it apparent that the same game matrix can yield different expected payoffs, depending upon the strategies employed by each of the players. In the next section we will inquire into the question of what percent of the time each row (or column) should be chosen in order to maximize the gain (or minimize the loss) of a player.

We will now close with an example that illustrates that the methods developed in this section to find the expected payoff with games having mixed strategies can also be applied to games of pure strategies.

Example 7.3.5. Consider the game of pure strategies defined by the matrix

$$\begin{pmatrix} -4 & 3 \\ -2 & -1 \end{pmatrix}$$

Find:

a. the saddle point,
b. the strategies that should be used by each player,
c. the payoff using the methods of section 7.2,
d. the expected payoff using the methods of this section.

Solution:

a. The element -2 is the saddle point (it is the only element that is both minimum in its row and maximum in its column).
b. The row player should always choose row two. The column player should always choose column one.
c. The payoff is the saddle point, namely, -2.
d. Since the row player should always choose row two, $\mathbf{P} = (0 \quad 1)$. Since the column player should always choose column one, $\mathbf{Q} = \begin{pmatrix} 1 \\ 0 \end{pmatrix}$.

So $$E = (0 \quad 1)\begin{pmatrix} -4 & 3 \\ -2 & -1 \end{pmatrix}\begin{pmatrix} 1 \\ 0 \end{pmatrix} = (-2 \quad -1)\begin{pmatrix} 1 \\ 0 \end{pmatrix}$$

Thus $E = -2$, which, of course, checks with the solution in part *c*.

EXERCISES ■ SECTION 7.3.

1. Consider the payoff matrix

$$\begin{pmatrix} 1 & -2 \\ -3 & 4 \end{pmatrix}$$

The elements are dollars. Assume the row player will choose row one 20% of the time and row two 80% of the time. Further, assume that the column player will choose column one 40% of the time and column two 60% of the time.

a. Form the row vector **P** of the row player's strategy.
b. Form the column vector **Q** of the column player's strategy.
c. Find the expected payoff, E, using matrix multiplication.
d. Who does this game favor and by what amount, on the average?

2. Using the game matrix of exercise 1, answer the four questions of of that exercise and assume:

the row player chooses row one 10% of the time and row two 90% of the time, the column player chooses column one 90% of the time and column two 10% of the time.

3. Given $\mathbf{P} = (p_1 \quad p_2)$,

$$\mathbf{A} = \begin{pmatrix} a_{11} & a_{12} \\ a_{21} & a_{22} \end{pmatrix} \quad \text{and} \quad \mathbf{Q} = \begin{pmatrix} q_1 \\ q_2 \end{pmatrix}$$

Show that: $\mathbf{PAQ} = p_1q_1a_{11} + p_1q_2a_{12} + p_2q_1a_{21} + p_2q_2a_{22}$.

4. Suppose that two people have agreed to play a game defined by the game matrix

$$\begin{pmatrix} -1 & 3 \\ 1 & -2 \end{pmatrix}$$

Also, presume a spy has told the row player that the column player intends always to choose column two. $\therefore \mathbf{Q} = \begin{pmatrix} 0 \\ 1 \end{pmatrix}$ Determine which of the strategies, $(.5 \quad .5)$ or $(.4 \quad .6)$, would be the better choice for the row player.

5. Consider the game of pure strategies defined by

$$\begin{pmatrix} 0 & -1 \\ 3 & -2 \end{pmatrix}$$

Determine the payoff using the method of example 7.3.5*d*.

6. Consider the coin-tossing game between Arlo and Bob in the beginning of section 4.8 (Expectation) on page 190. The payoff matrix for that game is:

$$\begin{array}{cc} & \text{Bob} \\ & \begin{array}{cc} H & T \end{array} \\ \text{Arlo} \begin{array}{c} H \\ T \end{array} & \begin{pmatrix} -100 & 30 \\ 30 & 30 \end{pmatrix} \end{array}$$

The elements are listed in cents. This is a game of mixed strategies, but neither player can actually control the percent of the time each row (or column) is chosen, because the game is one of tossing coins. So, when many games are played,

$$\mathbf{P} = (\tfrac{1}{2} \quad \tfrac{1}{2}) \quad \text{and} \quad \mathbf{Q} = \begin{pmatrix} \tfrac{1}{2} \\ \tfrac{1}{2} \end{pmatrix}$$

Knowing this, compute the expected payoff, E. Check your solution against the result on page 191.

7. For the game matrix $\begin{pmatrix} 2 & 1 & 1 \\ -1 & 2 & 3 \end{pmatrix}$, compute the expected payoff for the strategies below. In each case, tell who the game favors.

a. $\mathbf{P} = (0 \quad 1)$, $\mathbf{Q} = \begin{pmatrix} 0 \\ 0 \\ 1 \end{pmatrix}$ **b.** $\mathbf{P} = (\tfrac{1}{2} \quad \tfrac{1}{2})$, $\mathbf{Q} = \begin{pmatrix} \tfrac{1}{3} \\ \tfrac{1}{3} \\ \tfrac{1}{3} \end{pmatrix}$

c. $\mathbf{P} = (0 \quad 1)$, $\mathbf{Q} = \begin{pmatrix} \frac{2}{3} \\ \frac{1}{3} \\ 0 \end{pmatrix}$ d. $\mathbf{P} = (\frac{3}{4} \quad \frac{1}{4})$, $\mathbf{Q} = \begin{pmatrix} \frac{1}{4} \\ \frac{3}{4} \\ 0 \end{pmatrix}$

e. $\mathbf{P} = (.1 \quad .9)$, $\mathbf{Q} = \begin{pmatrix} .8 \\ .1 \\ .1 \end{pmatrix}$

8. Suppose a game matrix is given by: $\begin{pmatrix} -2 & 2 \\ 0 & -4 \end{pmatrix}$. Assume that the column player will use the strategy given by $\mathbf{Q} = \begin{pmatrix} \frac{1}{2} \\ \frac{1}{2} \end{pmatrix}$. Determine what strategy the row player should employ so that this game will be fair. (Hint: let $\mathbf{P} = (p \quad 1 - p)$; then compute E, set it equal to zero, and solve for p.)

7.4 DETERMINATION OF OPTIMAL STRATEGIES

In the previous section we made the observation (see example 7.3.4) that a given game matrix will yield different expected payoffs for different strategy choices. In this section it is our desire to find **optimal strategies** for two-person zero-sum **2 × 2 games.** The optimal strategies are the strategies that should be chosen in order to maximize gain or minimize loss. That is, for the game matrix $\mathbf{A} = \begin{pmatrix} a_{11} & a_{12} \\ a_{21} & a_{22} \end{pmatrix}$, with the row player's strategy given by $\mathbf{P} = (p_1 \quad p_2)$ and the column player's strategy given by $\begin{pmatrix} q_1 \\ q_2 \end{pmatrix}$, we wish to know the values of p_1, p_2, q_1, and q_2 that will cause the expected payoff (E) to be maximum.

We will state below (without proof) the values that give the optimal strategies for each player.

1. For the row player, $p_1 = \dfrac{a_{22} - a_{21}}{a_{11} + a_{22} - a_{21} - a_{12}}$

$$p_2 = \frac{a_{11} - a_{12}}{a_{11} + a_{22} - a_{21} - a_{12}}$$

2. For the column player, $q_1 = \dfrac{a_{22} - a_{12}}{a_{11} + a_{22} - a_{21} - a_{12}}$

$$q_2 = \frac{a_{11} - a_{21}}{a_{11} + a_{22} - a_{21} - a_{12}}$$

It is also true that when these values of p_1, p_2, q_1, and q_2 are used,

$$E = \mathbf{PAQ} = \frac{a_{11} \cdot a_{22} - a_{21} \cdot a_{12}}{a_{11} + a_{22} - a_{21} - a_{12}}$$

This quantity is called the **value** of the game.

We next will show the use of these formulas as well as some ways to remember them.

Example 7.4.1. For the game matrix $\begin{pmatrix} 1 & -1 \\ -3 & 5 \end{pmatrix}$, find:

a. each player's optimal strategy,
b. the value of the game,
c. who the game favors.

Solution:

a.
$$p_1 = \frac{a_{22} - a_{21}}{a_{11} + a_{22} - a_{21} - a_{12}} = \frac{5 - (-3)}{5 + 1 - (-3) - (-1)} = \frac{8}{10}$$
$$p_2 = \frac{a_{11} - a_{12}}{a_{11} + a_{22} - a_{21} - a_{12}} = \frac{1 - (-1)}{10} = \frac{2}{10}$$
$$q_1 = \frac{a_{22} - a_{12}}{a_{11} + a_{22} - a_{21} - a_{12}} = \frac{5 - (-1)}{10} = \frac{6}{10}$$
$$q_2 = \frac{a_{11} - a_{21}}{a_{11} + a_{22} - a_{21} - a_{12}} = \frac{1 - (-3)}{10} = \frac{4}{10}$$

So, $\mathbf{P} = (.8 \quad .2)$, $\mathbf{Q} = \begin{pmatrix} .6 \\ .4 \end{pmatrix}$. Thus the row player should play row one 80% of the time and row two 20% of the time. The column player should play column one 60% of the time and column two 40% of the time.

b. $E = \dfrac{5 - 3}{10} = \dfrac{2}{10}$

c. This game favors the row player and, on a per-game basis, he will gain an average of 2/10.

In the previous solution, there are some things to be pointed out.

1. In all five formulas for p_1, p_2, q_1, q_2, and E, the denominators are the same, $a_{11} + a_{22} - a_{21} - a_{12}$.
2. The numerators of the formulas for p_1, p_2, q_1, q_2 can be found by writing the game matrix $\begin{pmatrix} a_{11} & a_{12} \\ a_{21} & a_{22} \end{pmatrix}$ and then, from the element a_{22}, draw two arrows as follows:

$$\begin{pmatrix} a_{11} & a_{12} \\ a_{21} & a_{22} \end{pmatrix} \quad \text{and} \quad \begin{pmatrix} a_{11} & a_{12} \\ a_{21} & a_{22} \end{pmatrix}$$

   The differences along the arrows are $a_{22} - a_{21}$, $a_{11} - a_{12}$, $a_{22} - a_{12}$, $a_{11} - a_{21}$. These differences respectively form the numerators of p_1, p_2, q_1, and q_2.
3. The numerator of the formula for E can be found by taking the

differences along the diagonals of the game matrix in the order indicated below:

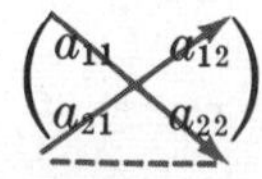

Following the arrows and taking the difference of the two products gives $a_{11} \cdot a_{22} - a_{21} \cdot a_{12}$, which is the numerator of the formula for E.

4. It is not necessary to know the formula for E. The value of a game can be computed from $E = \mathbf{PAQ}$, once $\mathbf{P}$ and $\mathbf{Q}$ have been determined. For example, return the parts a and b of the solution to example 7.4.1.

$$E = (.8 \quad .2)\begin{pmatrix} 1 & -1 \\ -3 & 5 \end{pmatrix}\begin{pmatrix} .6 \\ .4 \end{pmatrix} = (.2 \quad .2)\begin{pmatrix} .6 \\ .4 \end{pmatrix}$$

$$= .12 + .08 = .2 = \frac{2}{10}$$

Example 7.4.2. For the game matrix $\begin{pmatrix} 2 & 1 \\ -4 & 3 \end{pmatrix}$, find

a. each player's optimal strategy,
b. the value of the game,
c. who the game favors.

Solution: We write: $\begin{pmatrix} 2 & 1 \\ -4 & 3 \end{pmatrix}$ and $\begin{pmatrix} 2 & 1 \\ -4 & 3 \end{pmatrix}$.

a. $p_1 = \dfrac{3 - (-4)}{2 + 3 - (-4) - 1} = \dfrac{7}{8}$

$p_2 = \dfrac{2 - 1}{8} = \dfrac{1}{8}$

$q_1 = \dfrac{3 - 1}{8} = \dfrac{2}{8}$

$q_2 = \dfrac{2 - (-4)}{8} = \dfrac{6}{8}$

So $\mathbf{P} = (\frac{7}{8} \quad \frac{1}{8})$ and $\mathbf{Q} = \begin{pmatrix} \frac{2}{8} \\ \frac{6}{8} \end{pmatrix}$. Thus the row player should play row one 7/8 of the time and row two 1/8 of the time. The column player should play column one 2/8 of the time and column two 6/8 of the time.

b. Write $\begin{pmatrix} 2 & 1 \\ -4 & 3 \end{pmatrix}$ Thus $E = \dfrac{6 - (-4)}{8} = \dfrac{10}{8}$. Alternatively, we can write:

$$E = \mathbf{PAQ} = (\tfrac{7}{8} \quad \tfrac{1}{8})\begin{pmatrix} 2 & 1 \\ -4 & 3 \end{pmatrix}\begin{pmatrix} \frac{2}{8} \\ \frac{6}{8} \end{pmatrix} = (\tfrac{10}{8} \quad \tfrac{10}{8})\begin{pmatrix} \frac{2}{8} \\ \frac{6}{8} \end{pmatrix}$$

$$= \frac{20}{64} + \frac{60}{64} = \frac{80}{64} = \frac{10}{8}$$

c. This game favors the row player since the result for E is positive.

In this example we found the optimal strategies of each player to be given by $\mathbf{P} = (\frac{7}{8} \quad \frac{1}{8})$ and $\mathbf{Q} = \begin{pmatrix} \frac{2}{8} \\ \frac{6}{8} \end{pmatrix}$. So, for instance, the row player should play row one 7/8 of the time and row two 1/8 of the time. However, in doing so, he should not have any set pattern, because the column player might discover it and take advantage of it. So the row player should make random selections of rows totaling $(\frac{7}{8} \quad \frac{1}{8})$. He could do so by taking eight identical slips of paper, putting a "1" on seven of them and a "2" on the eighth. Then all the slips could be placed in a hat, and each time a row is to be selected, the row player could select one slip from the eight available. In the long run, row one would be played 7/8 of the time and row two would be played 1/8 of the time. The column player should also perform random selections of the columns. He could use eight slips of paper with a "1" on two of them and a "2" on the other six in order to achieve his optimal strategy, $\begin{pmatrix} \frac{2}{8} \\ \frac{6}{8} \end{pmatrix}$.

In the next example, we deal with a game of pure strategies that allows us to conclude an important fact.

Example 7.4.3. For exercise 6 of section 7.2., find the optimal strategies and the value of the game.

Solution: The payoff matrix is

		Finite	
		City	Suburbs
Infinity	City	1	−2 ← a saddle point
	Suburbs	3	−4

This game is one of pure strategies, since it has a saddle point, so the row player should always pick row one and the column player should always pick column two:

$$\mathbf{P} = (1 \quad 0) \qquad \mathbf{Q} = \begin{pmatrix} 0 \\ 1 \end{pmatrix}$$

The value of the game is the saddle point, −2. The reader is cautioned against trying to use the formulas given at the beginning of this section for games of pure strategies. For instance, if we try to calculate p_1, we get:

$$p_1 = \frac{a_{22} - a_{21}}{a_{11} + a_{22} - a_{21} - a_{12}} = \frac{-4 - (3)}{1 + (-4) - (3) - (-2)}$$
$$= \frac{-7}{1 - 4 - 3 + 2} = \frac{7}{4}$$

which is clearly not possible.

The remainder of this section will consist of two examples that illustrate some techniques developed thus far in the chapter.

Example 7.4.4. There are two candidates for a school board election, Rebecca and Carlos. There are two important issues: teacher's salary and money for the sports program. A game matrix appears below.

		Carlos	
		Salaries	Sports
Rebecca	Salaries	3	−2
	Sports	−2	4

The positive entries indicate a strength for Rebecca and the negative entries a weakness for Rebecca (equivalently, a strength for Carlos). Find:

a. the optimal strategies for each player,

b. the value of the game.

Solution: Since this is a game of mixed strategies, we may proceed with the methods of this section.

a. We write: $\begin{pmatrix} 3 & -2 \\ -2 & 4 \end{pmatrix}$ and $\begin{pmatrix} 3 & -2 \\ -2 & 4 \end{pmatrix}$.

So,

$$p_1 = \frac{4 - (-2)}{3 + 4 - (-2) - (-2)} = \frac{6}{11}$$

$$p_2 = \frac{3 - (-2)}{11} = \frac{5}{11}$$

$$q_1 = \frac{4 - (-2)}{11} = \frac{6}{11}$$

$$q_2 = \frac{3 - (-2)}{11} = \frac{5}{11}$$

Rebecca's optimal strategy is given by $\mathbf{P} = (\frac{6}{11} \;\; \frac{5}{11})$. Carlos's optimal strategy is given by $\mathbf{Q} = \begin{pmatrix} \frac{6}{11} \\ \frac{5}{11} \end{pmatrix}$.

b. $E = \mathbf{PAQ} = (\frac{6}{11} \;\; \frac{5}{11}) \begin{pmatrix} 3 & -2 \\ -2 & 4 \end{pmatrix} \begin{pmatrix} \frac{6}{11} \\ \frac{5}{11} \end{pmatrix}$

$$= (\tfrac{8}{11} \;\; \tfrac{8}{11}) \begin{pmatrix} \frac{6}{11} \\ \frac{5}{11} \end{pmatrix} = \tfrac{48}{121} + \tfrac{40}{121} = \tfrac{88}{121} = \tfrac{8}{11}$$

So $E = 8/11$, implying that Rebecca is favored in this game; she is the stronger candidate on the basis of the two issues considered.

Example 7.4.5. Consider exercise 5 of section 7.2. (The reader is urged to reread this.) The game matrix is

		Mother Nature	
		Good year	Recession year
You	Invest	8,000	−5,000
	Do not invest	−2,000	0

Suppose that government experts agree that the probability of having a good year is $\frac{3}{4}$. What course of action should you follow?

Solution: The government experts' prediction for a good year gives us:

$$\mathbf{Q} = \begin{pmatrix} \frac{3}{4} \\ \frac{1}{4} \end{pmatrix}$$

Of course, you now want to choose a strategy that will be optimal. Let $\mathbf{P} = (p \quad 1 - p)$, and compute $E = \mathbf{PAQ}$.

$$E = (p \quad 1 - p)\underbrace{\begin{pmatrix} 8 & -5 \\ -2 & 0 \end{pmatrix}}_{\text{these are thousands of dollars}}\begin{pmatrix} \frac{3}{4} \\ \frac{1}{4} \end{pmatrix} = (\underbrace{8p - 2(1 - p)} \quad \underbrace{-5p})\begin{pmatrix} \frac{3}{4} \\ \frac{1}{4} \end{pmatrix}$$

Notice that these are just the two elements of this 1×2 vector.

$$= (8p - 2 + 2p \quad -5p)\begin{pmatrix} \frac{3}{4} \\ \frac{1}{4} \end{pmatrix} = (10p - 2 \quad -5p)\begin{pmatrix} \frac{3}{4} \\ \frac{1}{4} \end{pmatrix}$$
$$= \tfrac{3}{4}(10p - 2) - \tfrac{1}{4}(5p) = \tfrac{30}{4}p - \tfrac{6}{4} - \tfrac{5}{4}p = \tfrac{25}{4}p - \tfrac{6}{4}$$

Now, knowing that $E = \frac{25}{4}p - \frac{6}{4}$, you next want to make E as big as possible, since you are the row player. In this case, E will be a maximum when p assumes its maximum value of 1. Hence, $\mathbf{P} = (1 \quad 0)$ and you should ways choose row one. The word "always" for this problem is actually somewhat meaningless because this game will only be played once (you buy the new equipment or you don't). Nonetheless, since your optional strategy is $(1 \quad 0)$, you should choose to buy the new equipment.

Before closing, we summarize the facts for 2 × 2 games (two-person zero-sum) that have been used thus far to find optimal strategies. Let the game be represented by $\mathbf{A} = \begin{pmatrix} a_{11} & a_{12} \\ a_{21} & a_{22} \end{pmatrix}$.

1. Look for a saddle point (minimum in its row and maximum in its column).

Pure-strategy game:

2. When a saddle point exists,
   a. the row player should always play the row containing the saddle point;
   b. the column player should always play the column containing the saddle point;
   c. the value of the game is given by the saddle point.

Mixed-strategy game

3. When a saddle point does not exist,
   a. the row player should use the strategy given by $\mathbf{P} = (p_1 \quad p_2)$, where the values of p_1 and p_2 are given on page 310;
   b. the column player should use the strategy given by $\mathbf{Q} = \begin{pmatrix} q_1 \\ q_2 \end{pmatrix}$, where the values of q_1 and q_2 are given on page 310;
   c. the value of the game is given by $E = \mathbf{PAQ}$.

EXERCISES ■ SECTION 7.4.

1. For each of the game matrices below, find:
   (1) the optimal strategies of each player,
   (2) the value of the game,
   (3) who the game favors.

   a. $\begin{pmatrix} -1 & 3 \\ 2 & -1 \end{pmatrix}$ b. $\begin{pmatrix} 1 & 0 \\ -1 & 2 \end{pmatrix}$

   c. $\begin{pmatrix} 3 & -2 \\ 1 & -4 \end{pmatrix}$ d. $\begin{pmatrix} 2 & -3 \\ -3 & 4 \end{pmatrix}$

   e. $\begin{pmatrix} 2 & -2 \\ -5 & 5 \end{pmatrix}$

2. For exercise 1*a* above, describe how you would insure the randomness of each player's strategy.
3. Consider the school board election in example 7.4.4. Suppose that as a result of heavy news-media campaigning, the strengths of the candidates, Rebecca and Carlos, have shifted somewhat and are now given by:

		Carlos	
		Salaries	Sports
Rebecca	Salaries	2	−1
	Sports	−2	5

   a. Find the optimal strategies for each player.
   b. Find the value of the game.
   c. From Carlos's viewpoint, has the situation improved?

4. Suppose you inherit \$10,000 and you desire to invest it. After careful research, you decide either to put it in a savings account or to buy stocks. Which of these will turn out to be a better deal depends upon the economic situation next year (there are two possibilities—good or poor). If next year is good, savings account and stock gains will be 6% and 8%, respectively. If next year is bad, a savings

account investment will yield a 5% gain, whereas a stock purchase will end up in a loss of 2%.

a. Set up a game matrix where you play the rows against Mother Nature.
b. If each economic situation is equally likely, what should your strategy be? (Hint: let $\mathbf{P} = (p \quad 1 - p)$.)
c. What is your expected payoff in dollars?

The reader should note that we are somewhat limited in this problem by having only two states of nature (two possible economic situations). Realistically, we might have more; say, excellent, good, fair, poor. In the next problem we allow such possibilities.

5. Suppose that in exercise 4, we introduce four possible future economic states—excellent, good, fair, poor—which experts estimate to have the following respective chances of occurrence: $\frac{2}{10}$, $\frac{3}{10}$, $\frac{4}{10}$, $\frac{1}{10}$. The game matrix appears below.

		Mother Nature			
		Excellent	Good	Fair	Poor
You	Savings account	6	6	5	4
	Stocks	8	6	−3	−6

a. Determine your best strategy. (Hint: let $\mathbf{P} = (p \quad 1 - p)$.)
b. What is your expected payoff in dollars?

7.5 MORE ON OPTIMAL STRATEGIES IN GAMES LARGER THAN 2 × 2

The previous section showed us how to find optimal strategies for two-person zero-sum 2 × 2 games. The present section will deal with finding optimal strategies for two-person zero-sum games that are not 2 × 2. We will examine only mixed-strategy games here because the solution of any pure-strategy game has already been presented in section 7.2.

Consider the 3 × 4 game with this payoff matrix:

$$\mathbf{A} = \begin{pmatrix} 10 & 14 & -7 & -9 \\ 1 & 3 & 4 & 3 \\ 1 & -1 & 1 & -4 \end{pmatrix}$$

Since this matrix has no saddle point, let us begin to examine it by considering the row player's viewpoint. Here, the row player has three choices. However, he should never choose row three in preference to row two in matrix **A** because each element of row two is greater than or equal to the corresponding element in row three ($1 = 1$, $3 > -1$, $4 > 1$, $3 > -4$).

In a situation like this, where each element of one row r is GREATER THAN or EQUAL TO each corresponding element of another row r', we say that row r **dominates** row r' (r' is referred to as a **recessive row**). The point here is that since row three is recessive, there is no sense keeping it because the row player will never choose it (we presume that players are rational). Hence, we can represent the game by a **reduced matrix:**

$$\begin{pmatrix} 10 & 14 & -7 & -9 \\ 1 & 3 & 4 & 3 \end{pmatrix}$$

Next, let us look at the game from the column player's viewpoint. He has a choice of four columns. Since the matrix elements represent gains to the row player, the column player (being rational) would never choose column three in preference to column four, because doing so would tend to increase the row player's gain (since $-9 < -7$ and $3 < 4$). In a situation like this, where each element of one column c is LESS THAN or EQUAL TO each corresponding element of another column c', we say that column c dominates column c' (c' is referred to as a **recessive column**). For the 2×4 matrix above, column four dominates column three. Also, column one dominates column two ($10 < 14$ and $1 < 3$). Eliminating the two recessive columns (two and three) leaves a 2×2 reduced matrix:

$$\mathbf{A}_r = \begin{pmatrix} 10 & -9 \\ 1 & 3 \end{pmatrix}$$

(The symbol $\mathbf{A}_r$ is used when ALL recessive rows and columns have been eliminated.) Thus, what started out as a 3×4 game has now been reduced to a 2×2 game, which can be analyzed by methods of the previous sections. For the illustration above, where

$$\mathbf{A}_r = \begin{pmatrix} 10 & -9 \\ 1 & 3 \end{pmatrix}$$

we now find optimal strategies and the value of the game. (Note that $\mathbf{A}_r$ does not have a saddle point, since $\mathbf{A}$ did not have one.)

$$p_1 = \frac{3 - 1}{10 + 3 - 1 - (-9)} = \frac{2}{21}$$

$$p_2 = \frac{10 - (-9)}{21} = \frac{19}{21}$$

$$q_1 = \frac{3 - (-9)}{21} = \frac{12}{21}$$

$$q_2 = \frac{10 - 1}{21} = \frac{9}{21}$$

These values of p_1, p_2, q_1, q_2 are the strategy results that pertain to $\mathbf{A}_r$.

However, by referring back to the original matrix **A** (we have darkened $\mathbf{A}_r$'s elements for emphasis),

$$\begin{pmatrix} \mathbf{10} & 14 & -7 & \mathbf{-9} \\ \mathbf{1} & 3 & 4 & \mathbf{3} \\ 1 & -1 & 1 & -4 \end{pmatrix}$$

we can interpret the values of p_1, p_2, q_1, q_2 to mean:

the row player should
- play row one 2/21 of the time,
- play row two 19/21 of the time,
- never play row three (it is recessive).

the column player should
- play column one 12/21 of the time,
- never play columns two or three (they are recessive),
- play column four 9/21 of the time.

Having obtained the optimal strategies, we next find the value of the game. With the original game now described by $\mathbf{A}_r$, the row player's strategy is $\mathbf{P}_r = (\frac{2}{21} \;\; \frac{19}{21})$, and the column player's strategy is $\mathbf{Q}_r = \begin{pmatrix} \frac{12}{21} \\ \frac{9}{21} \end{pmatrix}$. The value of the game is $E = \mathbf{P}_r\mathbf{A}_r\mathbf{Q}_r$.* So we find:

$$E = (\tfrac{2}{21} \;\; \tfrac{19}{21}) \begin{pmatrix} 10 & -9 \\ 1 & 3 \end{pmatrix} \begin{pmatrix} \frac{12}{21} \\ \frac{9}{21} \end{pmatrix}$$

$$= (\tfrac{39}{21} \;\; \tfrac{39}{21}) \begin{pmatrix} \frac{12}{21} \\ \frac{9}{21} \end{pmatrix} = \tfrac{819}{441} = \tfrac{13}{7}$$

Thus when many games are played, the row player will gain 13/7 units (of some type) per game (on the average).

Example 7.5.1. Consider the mixed strategies game defined by the matrix

$$\mathbf{A} = \begin{pmatrix} 3 & -5 & -4 \\ -2 & 1 & 2 \\ -1 & 1 & 2 \end{pmatrix}$$

a. Eliminate any recessive rows.
b. Eliminate any recessive columns.
c. If the reduced matrix is 2 × 2, find the optimal strategies and the value of the game.

* $\mathbf{P}_r$ and $\mathbf{Q}_r$ denote the players' strategies relative to $\mathbf{A}_r$. Note, we could write $E = \mathbf{PAQ}$, where $\mathbf{P} = (\frac{2}{21} \;\; \frac{19}{21} \;\; 0)$ and $\mathbf{Q} = \begin{pmatrix} \frac{12}{21} \\ 0 \\ 0 \\ \frac{9}{21} \end{pmatrix}$.

Solution:

a. Row three dominates row two $(-1 > -2,\ 1 = 1,\ 2 = 2)$. The reduced matrix is:

$$\begin{pmatrix} 3 & -5 & -4 \\ -1 & 1 & 2 \end{pmatrix}$$

b. Looking at the 2×3 reduced matrix in (*a*) we see that column two dominates column three $(-5 < -4,\ 1 < 2)$. Hence, by eliminating the recessive column (three), we get a 2×2 reduced matrix:

$$\mathbf{A}_r = \begin{pmatrix} 3 & -5 \\ -1 & 1 \end{pmatrix}$$

c. The 2×2 matrix $\mathbf{A}_r = \begin{pmatrix} 3 & -5 \\ -1 & 1 \end{pmatrix}$ determines the following values:

$$p_1 = \frac{1 - (-1)}{3 + 1 - (-1) - (-5)} = \frac{2}{10}$$

$$p_2 = \frac{3 - (-5)}{10} = \frac{8}{10}$$

$$q_1 = \frac{1 - (-5)}{10} = \frac{6}{10}$$

$$q_2 = \frac{3 - (-1)}{10} = \frac{4}{10}$$

Hence $\mathbf{P}_r = (\frac{2}{10} \quad \frac{8}{10})$, $\mathbf{Q}_r = \begin{pmatrix} \frac{6}{10} \\ \frac{4}{10} \end{pmatrix}$. In terms of the original game (represented by $\mathbf{A}$),

the row player should:
- play row one 20% of the time,
- never play row two (since it is recessive),
- play row three 80% of the time.

the column player should:
- play column one 60% of the time,
- play column two 40% of the time,
- never play column three (it is recessive).

The value of the game will be:

$$\begin{aligned} E = \mathbf{P}_r\mathbf{A}_r\mathbf{Q}_r &= (\tfrac{2}{10} \quad \tfrac{8}{10}) \begin{pmatrix} 3 & -5 \\ -1 & 1 \end{pmatrix} \begin{pmatrix} \frac{6}{10} \\ \frac{4}{10} \end{pmatrix} \\ &= (-\tfrac{2}{10} \quad -\tfrac{2}{10}) \begin{pmatrix} \frac{6}{10} \\ \frac{4}{10} \end{pmatrix} \\ &= -\tfrac{20}{100} = -.2, \end{aligned}$$

so when many games are played, the column player will gain .2 (units) per game (on the average).

The previous example had a 3×3 game that reduced to a 2×2 game after elimination of recessive rows and columns. However, this type of reduction is not always possible, as the next example shows.

Example 7.5.2. For the mixed-strategy matrix game

$$\mathbf{A} = \begin{pmatrix} 2 & 1 & 8 & 4 \\ 7 & 2 & -1 & 0 \\ 1 & -2 & 0 & 4 \end{pmatrix}$$

eliminate all recessive rows and columns.

Solution: In

$$\begin{pmatrix} 2 & 1 & 8 & 4 \\ 7 & 2 & -1 & 0 \\ 1 & -2 & 0 & 4 \end{pmatrix}$$

row one dominates row three ($2 > 1$, $1 > -2$, $8 > 0$, $4 = 4$), so row three is recessive. We now write:

$$\begin{pmatrix} 2 & 1 & 8 & 4 \\ 7 & 2 & -1 & 0 \end{pmatrix}$$

Now, column two dominates column one ($1 < 2$, $2 < 7$), so column one is recessive. There are no other recessive columns. So the reduced matrix is:

$$\mathbf{A}_r = \begin{pmatrix} 1 & 8 & 4 \\ 2 & -1 & 0 \end{pmatrix}$$

This, of course, cannot be further reduced. Consequently, we have no method for finding optimal strategies in this situation. (In section 7.6 we will solve this problem.)

Let us summarize the results of this section. When dealing with mixed-strategy matrix games (larger than 2×2), we have found that recessive rows and recessive columns should be eliminated. When the reduced matrix $\mathbf{A}_r$ is 2×2, the optimal strategies and game value can be found (see the summary at the end of section 7.4). When the reduced matrix is not 2×2 we, as yet have no technique for finding optimal strategies.

EXERCISES ■ SECTION 7.5.

1. In each of the mixed-strategy matrix games below, eliminate recessive rows and columns in order to obtain a reduced matrix that is 2×2 (if possible).

a. $\begin{pmatrix} 2 & 0 & 1 \\ -1 & 3 & 4 \end{pmatrix}$ **b.** $\begin{pmatrix} 2 & 0 & -1 & 5 \\ 6 & -1 & 4 & 5 \end{pmatrix}$

c. $\begin{pmatrix} 2 & -3 \\ 1 & 4 \\ -1 & -2 \end{pmatrix}$ d. $\begin{pmatrix} 2 & 0 & 0 & 2 \\ -1 & 3 & 4 & 3 \\ 6 & -3 & -4 & 2 \end{pmatrix}$

e. $\begin{pmatrix} 2 & 0 & 0 & 2 \\ -1 & 3 & 4 & 3 \\ 6 & 3 & -4 & 2 \end{pmatrix}$ f. $\begin{pmatrix} 2 & 1 & -3 & -4 \\ 1 & 0 & -4 & -5 \\ -4 & -3 & 6 & 5 \\ -5 & -6 & 5 & 4 \end{pmatrix}$

g. $\begin{pmatrix} 2 & -3 & 4 & -5 \\ -3 & 4 & -5 & 6 \end{pmatrix}$

2. For the matrix game given by $\mathbf{A} = \begin{pmatrix} 2 & 0 & -1 & 5 \\ 6 & 3 & 4 & 5 \end{pmatrix}$,
   a. find optimal strategies for each player,
   b. find the value of the game and who is favored.

3. For the matrix game given by $\mathbf{A} = \begin{pmatrix} 2 & 1 & -4 \\ -1 & -3 & 5 \end{pmatrix}$,
   a. find optimal strategies for each player,
   b. find the value of the game and who is favored.

4. For the matrix game given by $\mathbf{A} = \begin{pmatrix} 1 & 4 & -3 \\ -2 & -3 & -4 \\ -3 & 6 & 5 \end{pmatrix}$,
   a. find optimal strategies for each player,
   b. find the value of the game and who is favored.
   c. Suppose the column player elects to play column one 50% of the time and column three 50% of the time. What strategy will be best for the row player?

7.6 OPTIMAL STRATEGIES FOR $2 \times m$ AND $m \times 2$ GAMES

When a matrix game cannot be reduced to a 2×2 matrix by eliminating recessive rows and columns, we as yet have no method of finding optimal strategies. In this section, we will develop a graphical method for finding optimal strategies for either **$2 \times m$ or $m \times 2$ games** $(m > 2)$ of mixed strategies.

Consider the 2×3 game whose matrix is

$$\mathbf{A} = \begin{pmatrix} 1 & 8 & 4 \\ 2 & -1 & 0 \end{pmatrix}$$

which is not reducible. Let us look at the game from the row player's viewpoint. He has two choices and is looking for his optimal strategy.

Represent the row player's strategy by $\mathbf{P} = (p \quad 1-p)$. Next we consider the expected gain, E, by the row player for each of the column player's choices:

If the column player picks column one,

$$E = \mathbf{PAQ} = (p \quad 1-p)\begin{pmatrix} 1 & 8 & 4 \\ 2 & -1 & 0 \end{pmatrix}\begin{pmatrix} 1 \\ 0 \\ 0 \end{pmatrix} = p + 2(1-p) = 2 - p$$

If the column player picks column two,

$$E = \mathbf{PAQ} = (p \quad 1-p)\begin{pmatrix} 1 & 8 & 4 \\ 2 & -1 & 0 \end{pmatrix}\begin{pmatrix} 0 \\ 1 \\ 0 \end{pmatrix} = 8p - 1(1-p) = 9p - 1$$

If the column player picks column three,

$$E = \mathbf{PAQ} = (p \quad 1-p)\begin{pmatrix} 1 & 8 & 4 \\ 2 & -1 & 0 \end{pmatrix}\begin{pmatrix} 0 \\ 0 \\ 1 \end{pmatrix} = 4p$$

We next create a rectangular coordinate system with E (the row player's gains) on the vertical axis and p on the horizontal axis. In this coordinate system we graph each of the three equations:

$$\left.\begin{array}{l} \textbf{1.}\ E = 2 - p \\ \textbf{2.}\ E = 9p - 1 \\ \textbf{3.}\ E = 4p \end{array}\right\}\ \text{Note that each of these is a straight line.}$$

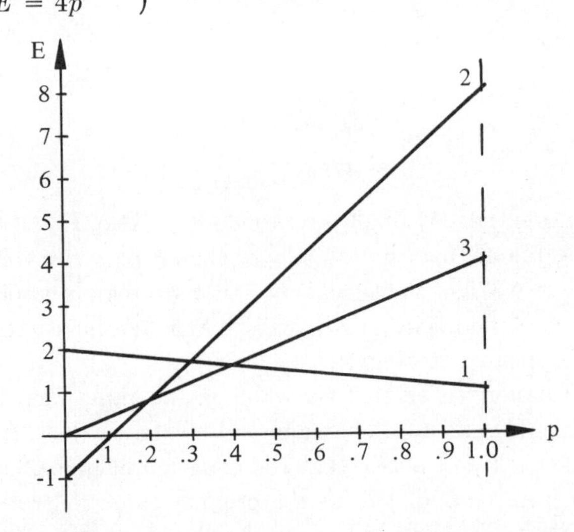

We graph each of the equations only for values of p between 0 and 1 (why?). As we view these three lines, bear in mind that they result from considering the column player to be playing either column one, column two, or column three. Actually, the column player will be employing a mixed strategy in hopes of keeping E as small as possible. With that in mind, the column player will be somewhere "on" the darkened portion (called the **strategy line**) of the graph below.

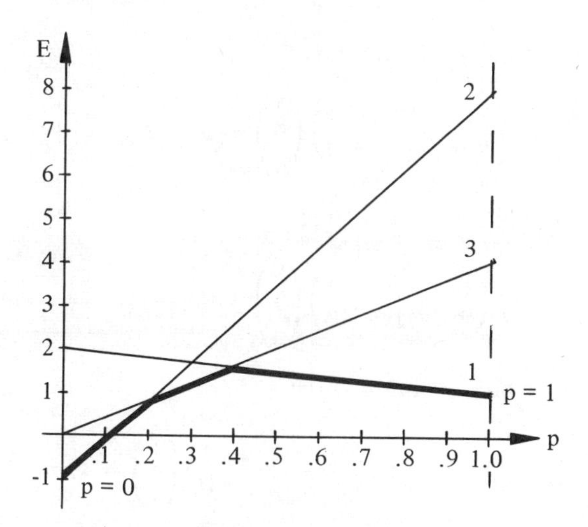

The strategy line is formed by taking the lowest path from $p = 0$ to $p = 1$. The row player wants to make E as large as possible, so he will now pick the highest point on the strategy line. That point lies at the intersection of lines (1) and (3), whose equations we now solve simultaneously.

$$E = 2 - p \qquad (1)$$
$$E = 4p \qquad (3)$$

So

$$4p = 2 - p$$
$$5p = 2$$
$$p = 2/5$$

Substituting this into either (1) or (3) gives $E = 8/5$. Also, $\mathbf{P} = (\frac{2}{5} \quad \frac{3}{5})$.

Thus far we have found that the row player should play row one 2/5 of the time and row two 3/5 of the time. When that strategy is employed, the value of the game (the maximum E) will be 8/5. We must still find the column player's optimal strategy.

Notice in the discussion above that the value of the game was the intersection of the lines (1) and (3). The value did not lie on line (2). The reader should check back and notice that the equation of line (2) ($E = 9p - 1$) originates from column two of the original game matrix. This means that the column player should never choose column two for purposes of achieving his optimal strategy. So, in the original 2×3 matrix game we do not need column two:

$$\begin{pmatrix} 1 & \not{8} & 4 \\ 2 & -\not{1} & 0 \end{pmatrix}$$

The remaining four elements form the matrix $\begin{pmatrix} 1 & 4 \\ 2 & 0 \end{pmatrix}$. We can use the formulas for q_1 and q_2 in section 7.4 to find the column player's strategy.

$$q_1 = \frac{0 - 4}{1 + 0 - 2 - 4} = \frac{-4}{-5} = \frac{4}{5}$$

$$q_2 = \frac{1 - 2}{-5} = \frac{-1}{-5} = \frac{1}{5}$$

So,

$$\mathbf{Q} = \begin{pmatrix} \frac{4}{5} \\ 0 \\ \frac{1}{5} \end{pmatrix}$$

thus implying that the column player should play column one 4/5 of the time, column two should never be played, and column three should be played 1/5 of the time.

Example 7.6.1. Find the optimal strategies and the value of the game for the matrix game:

$$\begin{pmatrix} 3 & 4 & -1 & -3 \\ 1 & -2 & 2 & 3 \end{pmatrix}$$

Solution: Note there is no saddle point nor are there any recessive columns. Next, we compute possible gains made by the row player.

If the column player picks column one, the row player's gain, E, is:

$$E = (p \quad 1 - p) \begin{pmatrix} 3 & 4 & -1 & -3 \\ 1 & -2 & 2 & 3 \end{pmatrix} \begin{pmatrix} 1 \\ 0 \\ 0 \\ 0 \end{pmatrix} = 2p + 1 \qquad (1)$$

If the column player picks column two, the row player's gain, E, is:

$$E = (p \quad 1 - p) \begin{pmatrix} 3 & 4 & -1 & -3 \\ 1 & -2 & 2 & 3 \end{pmatrix} \begin{pmatrix} 0 \\ 1 \\ 0 \\ 0 \end{pmatrix} = 6p - 2 \qquad (2)$$

If the column player picks column three, the row player's gain, E, is:

$$E = (p \quad 1 - p) \begin{pmatrix} 3 & 4 & -1 & -3 \\ 1 & -2 & 2 & 3 \end{pmatrix} \begin{pmatrix} 0 \\ 0 \\ 1 \\ 0 \end{pmatrix} = -3p + 2 \qquad (3)$$

If the column player picks column four, the row player's gain, E, is:

$$E = (p \quad 1-p)\begin{pmatrix} 3 & 4 & -1 & -3 \\ 1 & -2 & 2 & 3 \end{pmatrix}\begin{pmatrix} 0 \\ 0 \\ 0 \\ 1 \end{pmatrix} = -6p + 3 \qquad (4)$$

Next, each of those four equations is graphed and the strategy line is drawn.

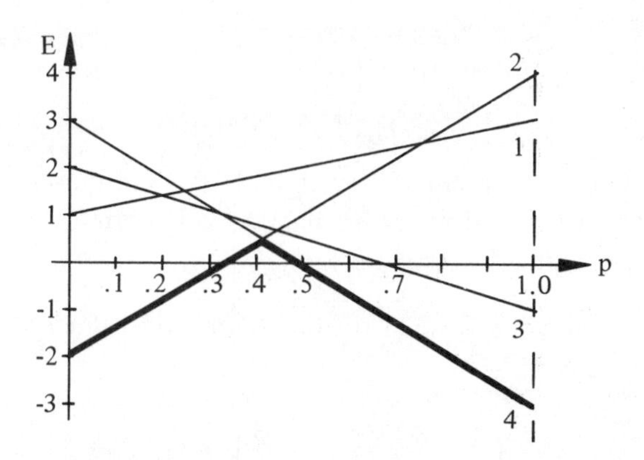

The highest point on the strategy line will give us p and E. Lines (2) and (4) determine the highest point.

$$E = 6p - 2 \qquad (2)$$
$$E = -6p + 3 \qquad (4)$$

So,

$$6p - 2 = -6p + 3$$
$$12p = 5$$
$$p = \tfrac{5}{12}$$

and $E = 6/12 = 1/2$, the value of the game. Also, $\mathbf{P} = (\tfrac{5}{12} \quad \tfrac{7}{12})$. Thus the row player should play row one 5/12 of the time and row two 7/12 of the time. Next, we compute the column player's optimal strategy by using the fact that the value of the game came from equations (2) and (4). Equations (1) and (3) were not used to get the game value. Hence, the column player should not choose columns one or three of the original game matrix. We delete those below:

$$\begin{pmatrix} \not{3} & 4 & -\not{1} & -3 \\ \not{1} & -2 & \not{2} & 3 \end{pmatrix}$$

The remaining elements form a 2×2 matrix:

$$\begin{pmatrix} 4 & -3 \\ -2 & 3 \end{pmatrix}$$

$$q_1 = \frac{3 - (-3)}{4 + 3 - (-2) - (-3)} = \frac{6}{12} = \frac{1}{2}$$

$$q_2 = \frac{4 - (-2)}{12} = \frac{6}{12} = \frac{1}{2}$$

Thus,
$$\mathbf{Q} = \begin{pmatrix} 0 \\ \frac{1}{2} \\ 0 \\ \frac{1}{2} \end{pmatrix}$$

The column player should never play columns one or three and should play each of columns two and four 50% of the time.

The two games just solved were each $2 \times m$. Next we will examine a game that is $m \times 2$.

Example 7.6.2. Find the optimal strategies and the value of the 4×2 game whose matrix is:

$$\begin{pmatrix} 3 & 2 \\ 4 & -1 \\ -1 & 4 \\ -2 & 5 \end{pmatrix}$$

Solution: Notice there is no saddle point. Further, there are no recessive rows. The row player has four choices and the column player has two choices. We represent the column player's strategy by $\begin{pmatrix} q \\ 1 - q \end{pmatrix}$. Next we will compute the row player's gain, E, for each of his possible row choices.

If the row player picks row one his gain, E, is:

$$E = (1 \quad 0 \quad 0 \quad 0)\begin{pmatrix} 3 & 2 \\ 4 & -1 \\ -1 & 4 \\ -2 & 5 \end{pmatrix}\begin{pmatrix} q \\ 1 - q \end{pmatrix} = q + 2 \qquad (1)$$

If the row player picks row two his gain, E, is:

$$E = (0 \quad 1 \quad 0 \quad 0)\begin{pmatrix} 3 & 2 \\ 4 & -1 \\ -1 & 4 \\ -2 & 5 \end{pmatrix}\begin{pmatrix} q \\ 1 - q \end{pmatrix} = 5q - 1 \qquad (2)$$

If the row player picks row three his gain, E, is:

$$E = (0 \quad 0 \quad 1 \quad 0)\begin{pmatrix} 3 & 2 \\ 4 & -1 \\ -1 & 4 \\ -2 & 5 \end{pmatrix}\begin{pmatrix} q \\ 1 - q \end{pmatrix} = -5q + 4 \qquad (3)$$

If the row player picks row four his gain, E, is:

$$E = (0 \quad 0 \quad 0 \quad 1)\begin{pmatrix} 3 & 2 \\ 4 & -1 \\ -1 & 4 \\ -2 & 5 \end{pmatrix}\begin{pmatrix} q \\ 1-q \end{pmatrix} = -7q + 5 \qquad (4)$$

Next, we graph each of the four equations in a coordinate system with E on the vertical scale and q on the horizontal scale. We graph each equation only for values of q between 0 and 1.

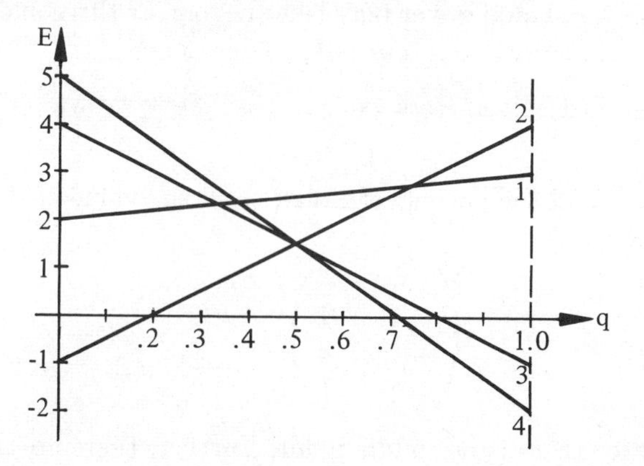

Observe these four lines and recall that they resulted from considering the row player to be playing either row one, row two, row three, or row four. Actually, the row player would employ a mixed strategy to make E as large as possible. Hence, he will prefer to be somewhere "on" the strategy line (darkened portion) of the graph below:

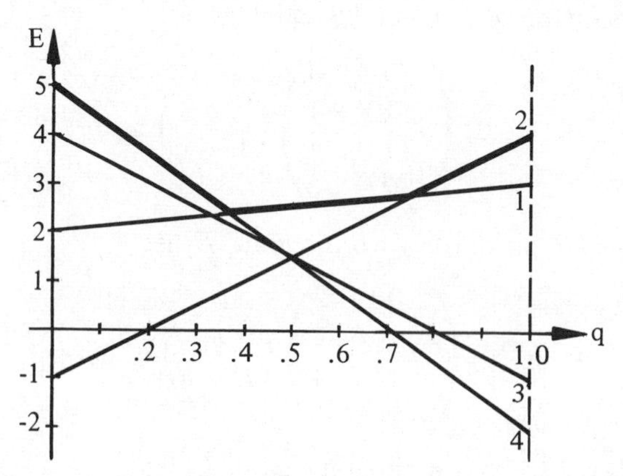

The strategy line is formed by taking the highest path from $q = 0$ to $q = 1$. The column player will be interested in making E as small as possible, so he will pick the lowest point on the strategy line. That point is the intersection of lines (1) and (4). The equations of these lines are now solved simultaneously.

$$E = q + 2 \qquad (1)$$
$$E = -7q + 5 \qquad (4)$$

So,

$$q + 2 = -7q + 5$$
$$8q = 3$$
$$q = \tfrac{3}{8}$$

Thus $\mathbf{Q} = \begin{pmatrix} \frac{3}{8} \\ \frac{5}{8} \end{pmatrix}$. Also, $E = 2\frac{3}{8}$. Thus the column player should play column one 3/8 of the time and column two 5/8 of the time. The value of the game is $2\frac{3}{8}$. Also, since lines (2) and (3) were not used to find the value of the game, the row player should never play rows two or three. Deleting those rows in the original game matrix

$$\begin{pmatrix} 3 & 2 \\ \cancel{4} & \cancel{-1} \\ \cancel{-1} & \cancel{4} \\ -2 & 5 \end{pmatrix}$$

leaves us with $\begin{pmatrix} 3 & 2 \\ -2 & 5 \end{pmatrix}$. The formulas for p_1 and p_2 in section 7.4 can now be used to find the row player's strategy.

$$p_1 = \frac{5 - (-2)}{3 + 5 - (-2) - 2} = \frac{7}{8}$$

$$p_2 = \frac{3 - 2}{8} = \frac{1}{8}$$

So, $\mathbf{P} = (\frac{7}{8} \quad 0 \quad 0 \quad \frac{1}{8})$, thus indicating that the row player should play row one 7/8 of the time, row four 1/8 of the time, and rows two and three should never be played.

At this point it would be wise for the reader to reread examples 7.6.1 and 7.6.2 to review and compare the methods of solution. Notice that in the graphical solution of a $2 \times m$ game (where E is graphed against p), the strategy line is the LOWEST PATH from $p = 0$ to $p = 1$. The value of a $2 \times m$ game is at the HIGHEST POINT on the strategy line. In the graphical solution of an $m \times 2$ game (where E is graphed against q), the strategy line is the HIGHEST PATH from $q = 0$ to $q = 1$. The value of an $m \times 2$ game is at the LOWEST POINT on the strategy line.

It is possible that in mixed-strategy $2 \times m$ games or $m \times 2$ games, there will be no unique solution for the optimal strategies, although the value of the game will be unique. The next example will illustrate such a situation.

Example 7.6.3. Find optimal strategies and the value of the game for the matrix game:

$$\begin{pmatrix} 2 & -4 \\ -1 & -1 \\ -8 & 2 \end{pmatrix}$$

Solution: This game has no saddle point and there are no recessive rows. Let the column player's strategy be $\mathbf{Q} = \begin{pmatrix} q \\ 1-q \end{pmatrix}$. If the row player chooses row one, his gain, E, will be:

$$E = (1 \quad 0 \quad 0)\begin{pmatrix} 2 & -4 \\ -1 & -1 \\ -8 & 2 \end{pmatrix}\begin{pmatrix} q \\ 1-q \end{pmatrix} = 6q - 4 \tag{1}$$

Similarly, when the row player plays rows two or three, his gains are given respectively by:

$$E = -1 \tag{2}$$

$$E = -10q + 2 \tag{3}$$

We graph the three equations and mark the strategy line below.

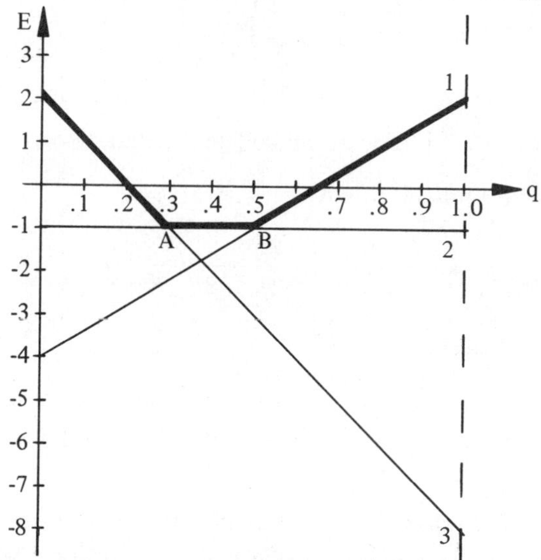

There is no single lowest point on the strategy line. The flat portion (between points A and B) is of concern to us.

At point A $\begin{cases} q = .3 \\ E = -1 \end{cases}$ (This is found by solving the equations of lines (2) and (3) simultaneously.)

At point B $\begin{cases} q = .5 \\ E = -1 \end{cases}$ (This is found by solving the equations of lines (1) and (2) simultaneously.)

Anywhere along the line segment AB, the value of the game, E, is unique and equal to -1. However, q varies from .3 (at point A) to .5 (at point B). This means that the column player can elect to play column one anywhere between 30% and 50% of the time as an optimal strategy. (The rest of the time column two should be played.) It still remains to find the row player's optimal strategy. We leave it to the reader in exercise 5 to show that the row player's optimal strategy is always to choose row two.

We remark here that **m × n matrix games** (m, $n > 2$) can also be solved, if enough recessive rows or recessive columns can be found to reduce the game to either $2 \times m$ or $m \times 2$. Exercises 6 and 8 illustrate this point.

We close this chapter by mentioning that there are other aspects of game theory the reader might find interesting. For instance, it is possible to solve $m \times n$ games that do not have saddle points and do not reduce to either $2 \times m$ or $m \times 2$ games. This can be done using the simplex method of chapter 6. Another variation of game theory is to consider **non-zero-sum games** (i.e., games where one player's losses are not equal to the other player's winnings). We can also deal with games with more than two players. For example, three-person games where two players form a coalition against the third have been investigated and employed in economics. The reader who is interested in additional information on game theory is referred to: *Introduction to the Theory of Games*, by J. C. C. McKinsey (McGraw-Hill Book Company, 1952).

EXERCISES ■ SECTION 7.6.

1. Find optimal strategies and the value of the mixed-strategy game whose payoff matrix is:

$$\begin{pmatrix} 2 & 0 & 1 \\ -1 & 4 & 3 \end{pmatrix}$$

2. Find optimal strategies and the value of the game represented by:

$$\begin{pmatrix} 5 & -4 & 3 & -2 \\ -2 & 3 & -1 & 0 \end{pmatrix}$$

3. For the game matrix

$$\begin{pmatrix} 2 & -1 \\ 0 & 4 \\ 1 & 3 \end{pmatrix}$$

find the optimal strategies and the value of the game.

4. For the game matrix

$$\begin{pmatrix} 5 & -3 \\ -4 & 3 \\ 3 & -1 \\ -2 & 1 \end{pmatrix}$$

find the optimal strategies and the value of the game.

5. Consider example 7.6.3. Verify that the row player's optimal strategy is always to select row two. (Hint: There are three cases to consider: (1) what happens at point A, (2) what happens at point B, (3) what happens between points A and B.)

6. Consider the 3×4 matrix game:

$$\begin{pmatrix} 2 & 1 & 8 & 4 \\ 1 & 0 & 6 & -2 \\ 3 & 2 & -1 & 0 \end{pmatrix}$$

a. Eliminate recessive rows and columns.

b. Find optimal strategies and the value of the game.

7. Consider the 3×4 matrix game:

$$\begin{pmatrix} 2 & -3 & 3 & 1 \\ -1 & -2 & 0 & 4 \\ 5 & -4 & 4 & -6 \end{pmatrix}$$

What are the optimal strategies and game value?

8. Consider the 3×3 matrix game:

$$\begin{pmatrix} -4 & 3 & 6 \\ 1 & -3 & 1 \\ -1 & 0 & 2 \end{pmatrix}$$

What are the optimal strategies and game value?

9. A cosmetic company (A) decides to put advertisements on three television networks. Their major competitor (B) will advertise on radio, in newspapers, and in magazines. The economics of this situation are given below (the matrix entries represent conjectured sales gained by company A from company B in hundreds of thousands of dollars).

		Company B		
		Radio	Newspaper	Magazines
Company A	Network 1	5	2	3
	Network 2	−2	3	0
	Network 3	−5	0	1

Find the optimal strategy for each company and the game value.

10. A music store (A) stocks guitars and zithers. A new competitor (B) on the same street hopes to gain business by offering a greater variety (guitars, zithers, and dulcimers). The economics of this situation is displayed below (matrix entries are in thousands of dollars).

		Store B		
		Guitars	Zithers	Dulcimers
Store A	Guitars	−2	−3	−1
	Zithers	−1	1	−2

Find the optimal strategies and the value of the game.

11. Refer to exercise 9 and suppose that because of advertising costs, company A must advertise 50% of the time on Network 1, 40% of the time on Network 2, and 10% of the time on Network 3. Also, suppose that company B (because of previous agreements) must do 50% of its advertising in magazines.
    a. Determine how B can best choose to advertise on radio and in newspapers.
    b. Determine the game value when B makes the best choice possible. Hint: $\mathbf{Q} = \begin{pmatrix} q \\ .5 - q \\ .5 \end{pmatrix}$.

chapter 7 in review

CHAPTER SUMMARY

In this chapter we studied games in the mathematical sense: a situation where the players employ strategies against one another. All games were two-person (exactly two players) zero-sum (what one player wins, the other loses) games. The payoffs were represented by a game matrix, and in strictly determined games (games of pure strategies), one component in that matrix (called the saddle point) was the value of the game (the row player's expectation). In games of mixed strategies we used the formula $E = \mathbf{PAQ}$. In section 7.4 the ways of determining optimal strategies for each player in 2×2 games were found and in section 7.5 we utilized row and column reductions of the game matrix to reduce the matrix to a 2×2 game, if possible. Finally, we solved $2 \times m$ and $m \times 2$ games by means of a graphical solution employing an analysis of the strategy line.

VOCABULARY

game	value of a game	recessive
two-person game	saddle point	dominant
zero-sum game	pure strategy	reduced matrix
payoff	strictly determined game	$2 \times m$ game
game matrix	mixed strategy	$m \times 2$ game
strategies	optimal strategy	strategy line
row player	2×2 game	$m \times n$ game
column player		

SYMBOLS

$\mathbf{P} = (p_1 \quad p_2)$

$\mathbf{Q} = \begin{pmatrix} q_1 \\ q_2 \end{pmatrix}$

$E = \mathbf{PAQ}$

$\mathbf{A}_r$

r'

c'

$\mathbf{P}_r$

$\mathbf{Q}_r$

chapter 8
Statistics

8.1 INTRODUCTION

The branch of mathematics known as **statistics** is a discipline that is usually organized into two major areas: **descriptive statistics** and **inference statistics.** Descriptive statistics is concerned with the collection, organization, and interpretation of numerical information (called **data**). Inference statistics makes use of data for testing certain hypotheses and making estimates.

We live in a world where there are some three billion people. All sorts of problems naturally arise in the areas of business and economics, education, food production and distribution, housing, politics, waste disposal, etc., that require experiments, studies, surveys, or polls to obtain information. When this information is numerical, these investigations necessarily involve the tools and methods of statistics, without which we can never maintain or improve certain communications among people of earth. It is the purpose of this chapter to give the reader a taste of descriptive statistics.

8.2 THE LANGUAGE OF STATISTICS

The main purpose of statistics is to gain information about a group of people or things (called the **population**), often without examining the whole group. We do this by selecting a **sample,** which is just a smaller number of items from the population. How do we select a sample? How large should a sample be? These are questions that will not be answered right now. The reasons for sampling might be manyfold.

Suppose, in canvassing public opinion, you are assigned to determine the percent of American voters that favors legalized free abortions upon request. Here, quite obviously, is a situation where it would be physically impossible to collect all of the voters' opinions. Sampling is the solution here. Gallup and Harris polls are commonly used in sampling the opinions of people.

The sampling process can be destructive, as in the testing of flashcubes coming off an assembly line. To test a given day's production of flashcubes would destroy all of them, so here we pick a sample.

Food inspection by government inspectors necessarily involves a sampling process. Let's face it, hiring, training, and paying inspectors to examine every cut of beef, pork, or poultry would be economically prohibitive.

From the three previous illustrations pointing out the necessity of sampling, we can see that when a sample is drawn from a population, we hope that it has characteristics similar to the entire population—otherwise we may be led astray. For example, suppose a government inspector examines a lot of 300 chickens by drawing a sample of 30 of them, and he finds all 30 in his sample to be perfectly healthy birds. He would then quite likely infer the lot of 300 to be acceptable. However, what an unfortunate situation it would be if all 270 of the unexamined birds were sickly.

Most sampling processes involve **random sampling,** meaning that each item in the population has an equal chance (or probability) of being chosen. Then, in the above illustration, it would be very unlikely for the inspector to get all 30 "good birds" in his sample. Whole books have been written on random sampling. It is something we will not get into extensively, but let us look at a couple of examples.

Example 8.2.1. An automobile dealer wishes to determine what makes of cars are driven by the 400 Monroe Community College faculty members. Define the population and tell how a random sample of size 25 might be drawn.

Solution: Since the car dealer is interested only in Monroe Community College faculty members, the population consists of just those 400 individuals. In order to draw a sample of 25 and be assured that each member of our population has an equal chance of ending up in that sample (thus making it random), the car dealer could take 400 identical slips of paper and put each faculty member's name on one slip. Then after thorough mixing of the slips, a blindfolded selection of 25 of them could be made.

This example might sound rather silly and quite time-consuming, but the importance of obtaining a random sample cannot be overemphasized. In fact, the information from more than one random sample from the same population can be combined to get a more detailed and accurate representation of the population.

In example 8.2.1 above, we used a lottery-type selection process. Such a process can be simulated by the use of a *random digit table*. (See Table IV in Appendix III.) Viewing this table reveals an array of 0s, 1s, 2s, . . . , 9s. (A table of random digits can be constructed by using random electronic processes; computers are a great aid.) To select a random sample of size 25 from a population of 400, we first give each of the 400 people a three-digit code name: 001, 002, 003, . . . , 399, 400. Next we do two things:

1. randomly select a starting point in the table (point with your eyes closed, for instance);
2. read out 25 three-digit numbers (reading down or across or diagonally or in some predetermined manner).

If you came to any three-digit number that is not in your original list (001, 002, . . . , 400), disregard it. Also disregard (after the first time) any three-digit numbers that appear more than once.

Example 8.2.2. A clean-air advocate has accused the City Transit Company (CTC) of polluting the air with their 85 buses that have gasoline-burning engines. She gets CTC's permission to test for emission pollutants and with the available money, decides to draw a sample of size 15. Define the population and describe how she might obtain a random sample of size 15.

Solution: Here the population consists of all 85 buses in the city. The buses could be assigned the numbers 01, 02, . . . , 85. Then two-digit selections could be made from Table IV until we had our sample of 15.

Once a sample is selected, some aspect of it, called the **response variable,** is examined, yielding either qualitative or quantitative information. If you select 20 people and note their hair color, you will have qualitative information, called **attribute data.** On the other hand, if you measure their heights you will generate quantitative information, called **numerical data.** Actually, numerical data can be either of two types. A measuring experiment, like above, yields **continuous variable data** (or just *continuous data*). Suppose, however, that you count the number of teeth in each person's mouth. Here is a different thing, a counting experiment: it yields **discrete variable data** (or just *discrete data*). The essential distinction is that when measuring you can come up with any whole number or decimal part, whereas when counting you come up with only whole numbers. (A person can have five teeth or 19 teeth, but not 14.2 teeth.)

Example 8.2.3. Determine whether the data that would be obtained from each of the response variables below is attribute, continuous variable, or discrete variable data:

a. the number of typographical errors on pages typed by a secretary;
b. the time it takes for each of 100 light bulbs to burn out;
c. the width of each of the pages in this chapter;
d. the taste of rotten apples in a barrel;
e. the number of rotten apples in a barrel;
f. the weight of each of the rotten apples in a barrel.

Solution:

a. discrete	**b.** continuous	**c.** continuous
d. attribute	**e.** discrete	**f.** continuous

EXERCISES ■ SECTION 8.2.

1. Suppose the New York Civil Liberties Union desires to know what percent of New York State Voters favors reinstatement of the death penalty for the crime of murder. Define the population of interest.
2. A biochemist wishes to determine the effect of vitamin C, either in preventing colds or reducing their severity, on children ages 14 and younger. Define the population.
3. A population numbers 1200 and we are to draw a sample of size 100 from it. Describe how we could proceed.
4. A manufacturer wants to test one out of every six carburetors that are coming off an assembly line. Describe how this could be done randomly.
5. A sociologist desires to learn more about drug usage by high school students. An anonymous questionnaire is composed and copies are left in the library for anyone to fill out. How do you feel about this procedure?
6. In order to determine how the residents of a community feel about a proposed railroad underpass, a newsman questions people on a downtown street corner. Does anything seem wrong about such an approach?
7. Suppose you want to predict the outcome of a presidential election and your sample is taken from telephone directories and lists of automobile owners. How do you feel about this method?
   Note: In 1936 a sample was drawn as just described and the poll results incorrectly predicted that A. H. Landon would become president of the U.S.

8. Suppose you want to predict the outcome of a presidential election and your sample is taken by putting questionnaires in boxes of popcorn sold in the theaters in the U.S. How do you feel about this method?
Note: Since 1948 this method has lead to a correct prediction of the outcome of each presidential election!!

9. Determine whether the data that would be obtained from each of the response variables below is attribute, continuous variable, or discrete variable data:

a. scores on the first exam for this course;
b. the number of "cuts" taken by students in a mathematics class;
c. the number of hours spent studying statistics each week;
d. the difficulty of your first exam;
e. inches of rainfall for each of the months of 1974 in Penfield, New York;
f. the number of days it rained 0.1 inches (or more) for each of the months of 1973 in Penfield, New York;
g. number of defective thermometers coming off an assembly line each day;
h. percent of the total yearly energy consumption in the U.S. by each of the fifty states.

8.3 MEASURES OF CENTRAL TENDENCY

We are now ready to begin our study of some of the methods of descriptive statistics. Depending upon the nature of an investigation or experiment, a sample might be *small* (30 items or less) or it might be *large* (more than 30 items). This section and the next one will deal with small samples for ease of introduction and computation.

Measures of central tendency are certain numerical values that tend to locate or describe the middle region of a set of data. Many people use the word "average" (although we advise against it here) to describe this effect.

Suppose a dentist's five morning patients have had cavities in 9, 2, 6, 7, and 1 teeth, respectively. Then the **mean** number of cavities is defined to be

$$\frac{9 + 2 + 6 + 7 + 1}{5} = 5$$

To find the *mean* of a set of data, we add all the pieces of data (x) together and then divide by the number of pieces of data (i.e., the sample size, n):

$$\text{the mean} = \bar{x} = \frac{\Sigma x}{n}$$

Here we are using the summation symbol, Σ, which is described in Appendix II. We have omitted the index of the summation and will continue to do so, since all summations will be over the entire set of data.

Note that the mean of a set of data takes into account each individual item; if one sample value changes, so will the mean. (How?) In fact, the dependency of the mean on all the data might explain why many educators like to use the mean as a numerical description of a student's performance in a course. For any set of data, the mean will be larger than the smallest sample value and smaller than the largest sample value. In other words a mean will be somewhat centrally located (hence the name: measure of central tendency).

However, the central location can be somewhat deceiving. Suppose the owner of a small factory wishes to advertise for new employees at his plant, where the annual salaries (x) are:

	x
	7,500
	8,000
	7,000
	10,000
	100,000 (the owner's salary)
$\Sigma x =$	132,500

Here, the owner can advertise the mean salary to be $\bar{x} = \dfrac{\$132,500}{5} =$ \$26,500—true, but quite obviously a statistical deception.

For situations like this, we need another measure of central tendency to describe the data.

The **median** ($\tilde{x}$) is the sample value in the middle position when the data has been **ranked** (arranged in order, lowest to highest or vice versa). So, for the salaries above we can write:

x (original or raw data)	x (ranked data)
7,500	7,000
8,000	7,500
7,000	8,000 = median = $\tilde{x}$
10,000	10,000
100,000	100,000

So in this case the median salary is \$8,000—clearly a much better description of what a prospective worker is likely to earn.

Consider the case where a set of data has an even number of items.

x (raw data)		x (ranked data)
2		2
4		3
5	We first rank the data.	3
3		4
3		5
9		9

Then we go to the two middle measurements and take their halfway point. Hence in this case the median would be $\tilde{x} = \dfrac{3 + 4}{2} = 3.5.$

Another measure of central tendency that can be useful is the **mode,** which is the most frequently occurring piece of data. In the illustration immediately above, since the 3 occurs twice and all other sample values occur only once, the mode is 3. There can be situations when no mode exists, like for the data below.

x	
2	Here there is no mode because each number occurs equally often.
4	
1	
6	
7	

There are also situations when two or more modes exist for a given set of data.

x (raw)	x (ranked)	
5	4	Here there are two modes, 5 and 9.
9	5	
5	5	
8	5	
9	6	
10	8	
5	9	
9	9	
6	9	
4	10	

Sometimes more than one mode occurs when the sample contains a combination of what is really different sets of data. Suppose the above sample is really a list of shoe sizes sold one day at a shoe store. The occurrence of the two modes could be because the data is a mixture of boy's shoe sales (small feet) and men's shoe sales (big feet).

An advantage of the mode is that it can be determined in the case of either attribute or numerical data. Consider the survey results below taken from a questionnaire sampling reading habits of people who receive a health magazine called *Health Now*. The question asked was, "Do you read the food advertisements in *Health Now?*" The results were as follows:

always:	55	The data here are the attributes of always, sometimes, seldom, and never. The most frequent response is "sometimes."
sometimes:	383	
seldom:	202	
never:	20	

We say that the modal response is "sometimes."

EXAMPLE 8.3.1. Determine the mean, median, and mode for the data 2, 2, 1, 4, 3, 7, 5.

Solution:

x (raw)		x (ranked)	
2		1	
2	mean $= \bar{x} = \dfrac{\Sigma x}{n}$	2	mode $= 2$
1		2	
4	$\bar{x} = \frac{24}{7}$	3	median $= \tilde{x} = 3$
3		4	
7	$\bar{x} = 3\frac{3}{7}$	5	
5		7	
$\Sigma x = 24$			

In this example the mean is $3\frac{3}{7}$. Often we write our answer in decimal form, so here $\bar{x} = 3.428571 \cdots$, which points out the need for a round-off rule. In this chapter we will round off to one more decimal place than present in the original data. So, 3.428571 gets rounded off to $\bar{x} = 3.4$. When it becomes necessary to round off a 5, we agree to round to the even value (3.25 becomes 3.2 whereas 3.35 gets written as 3.4).

EXAMPLE 8.3.2. A social worker at lunch overheard the comment, "Once on welfare, always on welfare." Enraged, she returned to work immediately and wrote to the United States Department of Health, Education, and Welfare for some numerical facts. In her answer she received the following data:

Length of Time on Welfare (% of Families)

Less than six months	17.4%
Six months to one year	17.8%
One year to two years	20.8%
Two years to three years	12.2%
Three years to five years	13.7%
Five years to ten years	11.6%
Ten years or more	6.1%

Source: A pamphlet published by the U.S. Department of Health, Education, and Welfare.

a. What percent of families have been on welfare one year or less?
b. What percent of families have been on welfare two years or less?
c. What percent of families have been on welfare five years or more?
d. What is the modal length of time of families on welfare?
e. What is the median length of time of families on welfare?
f. Do you agree with the statement, "Once on welfare, always on welfare"?

Solution:

a. From reading the tabulated data, the percent of families on welfare one year or less is 17.4% + 17.8% = 35.2%.
b. 17.4% + 17.8% + 20.8% = 56%
c. 11.6% + 6.1% = 17.7%
d. The largest percent is 20.8%. Thus the time period most frequently occurring (20.8% of the time) is one to two years; this is the modal length of time.
e. Recall that the median is the measurement in the middle position of a set of data. In other words, one-half (or 50%) of the data is less than it. By examining the answers to parts *a* and *b*, we see that the median is somewhere between one and two years. (There is no way to be any more definite here.)
f. Certainly not: only 6.1% of families are on welfare for ten years or more—and some of these may even get off sometime.

As we have seen in this section, statistics can be used to describe data in effective ways. We can also misuse statistics, as we pointed out back on page 340.

Now we look at another deceptive tactic used in statistics. Consider the three sets of data on the following page where x, y, and z represent the number of damaged spots on each of five tires (from three different manufacturers) that were subjected to rough road tests.

Manufacturer A	Manufacturer B	Manufacturer C
x	y	z
1	1	1
1	2	2
3	3	2
4	4	2
6	4	9
Here $\bar{x} = 3$, $\tilde{x} = 3$, mode $= 1$	Here $\bar{y} = 2.8$, $\tilde{y} = 3$, mode $= 4$	Here $\bar{z} = 3.2$, $\tilde{z} = 2$, mode $= 2$

To a prospective buyer of tires, manufacturer A says: "On the average, my tires show the fewest damaged spots." (Note, A's tires have the smallest mode.) Manufacturer B says: "On the average, my tires show the fewest damaged spots." (Note, B's tires have the smallest mean.) Manufacturer C claims: "On the average, my tires show the fewest damaged spots." (C's tires have the smallest median.) In a way, all three manufacturers are speaking the truth, but only by misusing the word "average," which has little meaning these days. Pick up a newspaper on almost any day and somewhere in it you'll see the word "average"—put up your guard! To be precise about things, the manufacturers should not even use the word "average"; rather, they should refer to the particular measure of central tendency they are using.

EXERCISES ■ SECTION 8.3.

1. Determine the mean, median, and mode of the data 2, 3, 1, 2, 4.
2. Determine the mean, median, and mode of the data 3, 2, 4, 1, 7, 6, 5, 8.
3. Determine the mean, median, and mode of the data 2, 5, 6, 2, 6, 9, 4, 4.
4. Give an example of a set of data where the mean, median, and mode are all equal.
5. Give an example of a set of data where the mean and median are equal, but the mode is larger.
6. Give an example of a set of data where the mean is larger than the median and, in turn, the median is larger than the mode.
7. Consider the set of data 1, 2, 2, 4, 6, 8, 10. Suppose the 10 were

to be changed to an 11. What effect (increases, decreases, remains the same) would this have on:

a. the mean,
b. the median,
c. the mode?

8. A large modular-home construction company has 60 employees. Thirty of the employees make $4.00 per hour, 20 of the employees make $5.00 per hour, and 10 of the employees make $6.00 per hour.

a. What is the mean hourly wage?
b. What is the median hourly wage?

Problems 9 and 10 involve the percent of increase and percent of decrease of a quantity, two concepts useful for comparing sets of data. Suppose the number 20 changes to the number 30. The **percent of increase** is

$$\frac{\text{the change}}{\text{the original}} = \frac{10}{20} = 50\%$$

Suppose the number 30 changes to the number 20. The **percent of decrease** is

$$\frac{\text{the change}}{\text{the original}} = \frac{10}{30} = 33\tfrac{1}{3}\%$$

9. For the data below, in each case calculate the percent of increase. (Round off to the nearest tenth.)

	Jan. 1969	Jan. 1973	% of increase
Pork chops (per lb.)	$1.04	$1.40	
Lettuce (per head)	.31	.39	
Frying chicken (per lb.)	.39	.44	
Haddock (per lb.)	.69	1.16	
Potatoes (10 lb.)	.75	1.03	
Apples (per lb.)	.23	.25	
Coffee (per lb.)	.76	.96	
Butter (per lb.)	.84	.87	

10. A corporation, in an annual report to its stockholders, shows the Consolidated Statement of Earnings (amounts shown are in thousands) that appears on the following page. For each of the eleven items, calculate the **percent of change** (use a plus sign for in crease and a minus sign for decrease) from June 11, 1973, to June 17, 1974.

Sales	For period ending: *June 11, 1973*	*June 17, 1974*	*% of change*
Sales to:			
Customers in U.S.	$950,000	$1,100,000	
Customers outside the U.S.	540,000	600,000	
TOTAL SALES.	1,490,000	1,700,000	
Costs			
Costs of goods sold.	700,000	800,000	
Sales, advertising, and administration expenses.	320,000	290,000	
TOTAL COSTS.	1,020,000	1,090,000	
Earnings			
Earnings from operations. .	400,000	300,000	
Interest income.	15,000	20,000	
Total earnings before taxes	415,000	320,000	
Income taxes.	200,000	170,000	
Net earnings.	215,000	150,000	

8.4 MEASURES OF VARIATION

A teacher of psychology gave a "pop quiz" to each of her two small group seminar classes in developmental psychology. The results obtained were:

x (ranked scores from class A)	y (ranked scores from class B)
2	4
4	5
6	6
8	7
10	8

The means, $\bar{x}$ and $\bar{y}$, are both equal to 6 here, yet the data looks different because it is not spread out for class A the way it is for class B. The medians $\tilde{x}$ and $\tilde{y}$ are also both the same (each is 6), and there are no modes. So here we have two sets of data and none of the measures of central tendency we talked about in section 8.3 helps us to distinguish one set from the other. **Measures of variation** are certain numerical values that will help us to describe how much a set of data is "spread out."

The measure of variation that is easiest to calculate is the **range,** which is the largest sample value minus the smallest sample value. For a formula we will write:

$$R = H - L$$

Applying this to the "pop quiz" scores, we see:

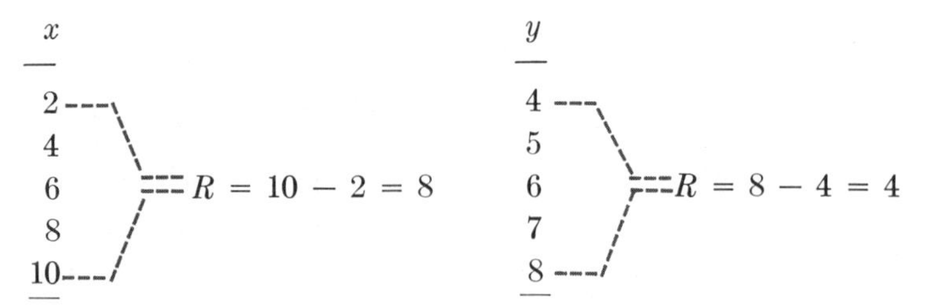

Hence, the greater the spread or variability of the data, the larger will be its range. Inversely, when the spread of data is small its range will also be small. When the range is calculated, just the highest and lowest sample values are needed. This, of course, makes it very easy to compute. At the same time, however, when we are calculating the range, all but the largest and smallest pieces of data are being neglected. Sometimes this makes the range a bit deceptive.

Suppose that over the last two weeks, daily readings have been taken of the coliform (a type of bacteria) levels at a local beach and that the readings (in units per million parts of water) ranged from 2,000 to 50,000. ($\therefore R = 50{,}000 - 2{,}000 = 48{,}000$.) If an acceptable level of the coliform count for swimming is 2,200, the above range from 2,000 to 50,000 sounds really bad. But what if the data collected really was:

	x (coliform count)		
day 1	2,000	day 8	2,100
day 2	2,200	day 9	2,100
day 3	2,050	day 10	2,015
day 4	2,300	day 11	2,010
day 5	50,000	day 12	2,000
day 6	2,170	day 13	2,000
day 7	2,000	day 14	2,000

Now a much clearer picture is presented. Sure, the range is 48,000 (from 2,000 to 50,000), but of the 14 days tested, it was safe to swim on all but two of those days (the fourth and fifth).

Two measures of variation that take into account each piece of data are the **variance** (v) and the **standard deviation** (s). The variance is defined to be

$$v = \frac{\Sigma(x - \bar{x})^2}{n - 1}$$

(We'll illustrate the use of this in a moment.) The standard deviation is given by

$$s = \sqrt{v}$$

Thus to get the standard deviation we just take the positive square root of the variance. Table V in Appendix III is a table of square roots that we'll discuss later. Let us now deal with the variance.

The formula is $v = \frac{\Sigma(x - \bar{x})^2}{n - 1}$. First, notice the parentheses. As is always the case, we "work inside the parentheses first." Notice the exponent. It applies only to the parentheses, not to the summation. So, to use this formula, we first compute $x - \bar{x}$ (piece of data minus the mean of data) for each piece of data. Next, each of those differences (or *deviations*) is squared. After that, these squares are all summed and lastly we divide by $n - 1$ (one less than the sample size). We will now sequentially illustrate the process for the pop quiz data of class A referred to in the beginning of this section.

x		$x - \bar{x}$	$(x - \bar{x})^2$
2	Remember, $\bar{x} = 6$	−4	16
4		−2	4
6		0	0
8		2	4
10		4	16
			40

Sample size is $n = 5$, so, $n - 1 = 4$. Therefore $v = \frac{40}{4} = 10$.

Example 8.4.1. Compute the variance of the pop quiz data for class B referred to at the beginning of this section.

Solution:

y		$y - \bar{y}$	$(y - \bar{y})^2$
4	$\bar{y} = 6$	−2	4
5		−1	1
6		0	0
7		1	1
8		2	4
			10

So $v = \frac{10}{4} = 2.5$.

Summarizing our results:

x		y	
2	$v = 10$	4	$v = 2.5$
4		5	
6		6	
8		7	
10		8	

Notice that the data having more spread also has the larger variance. Roughly speaking, the variance is a measure of the variability of data. There are many applications of the variance, but for where we're going, we will have more need for the standard deviation. So first we consider square roots.

The square root table (Table V on p. 425) looks like this:

n	n^2	$\sqrt{n}$	$\sqrt{10n}$
1.00	1.00	1.00	3.16228
1.01	.	.	.
1.02	.	.	.
.	.	.	.
.	.	.	.
.	.	.	.
10.00			

Note: in this table n represents the number whose square root you desire.

When a number is between 1 and 10, locate it in the "n column" and then read across to the "$\sqrt{n}$ column" and read its square root. (For directions dealing with numbers not between 1 and 10, the reader is strongly urged to read the explanation at the beginning of Table V.)

EXAMPLE 8.4.2. For the pop quiz data of classes A and B, determine the standard deviations.

Solution:

x		y	
2		4	
4		5	
6	$v = 10$	6	$v = 2.5$
8		7	
10		8	

$s = \sqrt{10} = 3.16228$
This we round off, so
$s = 3.2$.

$s = \sqrt{2.5} = 1.58114$
Rounding off here gives
$s = 1.6$.

Right now, the standard deviation is just a "thing" to be calculated. It's only meaning at this time is that it, too, like the range and variance, is a measure of the spread or variation of a set of data. There is another formula that can be used to calculate the variance and it is easier to use, because it involves fewer arithmetic operations. If x is a piece of data and n is the sample size,

$$v = \frac{n(\Sigma x^2) - (\Sigma x)^2}{n(n - 1)}$$

Example 8.4.3. Find the variance and standard deviation of the data 2, 4, 6, 8, 10.

Solution:

x	x^2
2	4
4	16
6	36
8	64
10	100
$\Sigma x = 30$	$\Sigma x^2 = 220$

$$v = \frac{n(\Sigma x^2) - (\Sigma x)^2}{n(n-1)} = \frac{5(220) - (30)^2}{5(4)}$$

$$= \frac{1100 - 900}{20} = \frac{200}{20}$$

$$v = 10$$

Also $s = \sqrt{v} = \sqrt{10} = 3.2$

These check with answers obtained earlier in this section.

Now that we have discussed how to calculate some measures of central tendency and measures of variation, it is time to learn more about them. We already know that measures of central tendency describe the central region of a set of data. Furthermore, the measures of variation describe the amount of spread for a set of data. The range does this quite lucidly; it is just the difference between the largest and smallest numbers. However, the standard deviation, no doubt, seems quite vague at this stage. It turns out that the standard deviation for a set of data can be any number, zero or more.

Consider this set of data:

x
2
2
2
2
2

This data does not exhibit any spread at all. What is its standard deviation?

x	x^2
2	4
2	4
2	4
2	4
2	4
10	20

$$v = \frac{n(\Sigma x^2) - (\Sigma x)^2}{n(n-1)}$$

$$= \frac{5(20) - (10)^2}{5(4)} = \frac{100 - 100}{20}$$

$$= \frac{0}{20} = 0$$

Hence, $s = \sqrt{0} = 0$ also. So when data has no "spread," the standard deviation is zero. As soon as the data does exhibit some amount of variability, there will be a nonzero value of the standard deviation. (This is seen in any of the illustrations and examples of this section.) Just how large the standard deviation is depends on the amount of spread that the data exhibits.

The mean and the standard deviation of a set of data can be used together to gain information about the population from which the data was taken. First consider the example below.

Example 8.4.4. The following measurements represent the amount of milk dispensed (in ounces) by an automatic milk dispenser: 8.2, 8.1, 8.0, 7.8, 8.3. Calculate the mean and standard deviation.

Solution:

x	x^2
8.2	67.24
8.1	65.61
8.0	64.00
7.8	60.84
8.3	68.89
40.4	326.58

It is possible to use Table V to obtain these squares: locate the number whose square you want in the "n column" and read its square in the "n^2 column."

$$\bar{x} = \frac{\Sigma x}{n} = \frac{40.4}{5} = 8.08$$

(Remember, we carry one more decimal than present in the original data.)

$$v = \frac{n(\Sigma x^2) - (\Sigma x)^2}{n(n-1)} = \frac{5(326.58) - (40.4)^2}{5(4)}$$

$$= \frac{1632.90 - 1632.16}{20} = \frac{.74}{20} = .0370$$

$$s = \sqrt{.0370} \approx .19$$

Any dispensing machine always has a certain amount of variation. People who work with machines try to keep them adjusted and in good working order. To do that, they must regularly check their functioning. The previous example shows us the numerical results of such a check. A decision must now be made: we must decide on the basis of this one sample whether:

1. the dispensing process is under control, or
2. the dispensing process is out of control.

Basically, the answer here depends on how much fluctuation (in ounces dispensed) the milk company is willing to tolerate.

A possibility

Suppose the company has decided to say that the process is under control when all the data lies within one standard deviation of either side of the mean.

$$\bar{x} + s = 8.08 + .19 = 8.27$$
$$\bar{x} - s = 8.08 - .19 = 7.89$$

Now rank the original data.

7.8
. . . $\bar{x} - s = 7.89$
8.0
. . . $\bar{x} = 8.08$
8.1
8.2
. . . $\bar{x} + s = 8.27$
8.3

This is an interval that is one standard deviation either side of the mean (often called *one standard deviation about the mean*).

Notice that not all the data lies within one standard deviation about the mean. Hence, the process is out of control. The machine needs adjusting.

Another possibility

Suppose the company has decided to say that the process is under control when at least 80% of the data lies within two standard deviations of either side of the mean.

$$\bar{x} + 2s = 8.08 + .38 = 8.46$$
$$\bar{x} - 2s = 8.08 - .38 = 7.70$$

Again we rank the data.

. . . $\bar{x} - 2s = 7.70$
7.8
8.0
. . . $\bar{x} = 8.08$
8.1
8.2
8.3
. . . $\bar{x} + 2s = 8.46$

This is two standard deviations either side of the mean (often called *two standard deviations about the mean*).

Now we see that all the data lies within two standard deviations about the mean. (Certainly then, we have at least 80% of it there.) Thus, the process is under control.

These two discussions point out that BEFORE ANY DECISION CAN BE MADE, IT IS FIRST NECESSARY TO LAY DOWN SOME BASIS UPON WHICH TO MAKE THAT DECISION. Keep that in mind; it is very important in any decision-making process.

Example 8.4.5. A machine that makes certain bolts for parts of a space station's life support system is to be tested because the bolts' diameters are of critical importance. Further, since the element of human life is at stake, the manufacturing plant has decided to declare the process under control when a sample of size ten is such that:

1. the sample mean differs by no more than 0.1 mm (either way) from 6 mm (which is the desired size), and also
2. all data lies within one-half standard deviation of the mean.

A sample of size ten is collected and gives 6.00, 5.98, 5.99, 6.00, 6.02, 6.01, 6.00, 6.01, 6.00, 6.01. Should the process be declared under control?

Solution:

x (mm)	x^2
6.00	36.0000
5.98	35.7604
5.99	35.8801
6.00	36.0000
6.02	36.2404
6.01	36.1201
6.00	36.0000
6.01	36.1201
6.00	36.0000
6.01	36.1201
60.02	360.2412

$$\bar{x} = \frac{\Sigma x}{n} = \frac{60.02}{10} = 6.002 \text{ mm}$$

$$v = \frac{n(\Sigma x^2) - (\Sigma x)^2}{n(n-1)} = \frac{10(360.2412) - (60.02)^2}{10(9)}$$

$$= \frac{3{,}602.412 - 3{,}602.4004}{90} = \frac{.0116}{90} = .000129$$

So,
$$s = \sqrt{.000129} \approx .011$$

First note that condition 1 above is satisfied, since $\bar{x} = 6.002$ mm. (We are well within the 0.1-mm tolerance.) Secondly,

$$\bar{x} + \tfrac{1}{2}s = 6.002 + .006 = 6.008 \text{ mm}$$
$$\bar{x} - \tfrac{1}{2}s = 6.002 - .006 = 5.996 \text{ mm}$$

Now compare each piece of data with the interval of one-half standard deviation about the mean, which extends from 5.996 mm to 6.008 mm. There are six pieces of data not in this interval. We must therefore conclude that the bolt manufacturing process is NOT under control.

Testing of objects as in the example above is quite usual in factory settings. Often, repeated testing on a weekly, daily, or even hourly basis is common in order to keep as close a check on production as is necessary.

EXERCISES ■ SECTION 8.4.

1. Calculate the range, variance, and standard deviation for the data 3, 5, 4, 1, 2, 3. Use $v = \dfrac{\Sigma(x - \bar{x})^2}{n - 1}$ to compute the variance.

2. Calculate the range, variance, and standard deviation for the data 4, 6, 1, 3, 2. Use $v = \dfrac{\Sigma(x - \bar{x})^2}{n - 1}$ to compute the variance.

The reader should notice the difficulty encountered in exercise 2 when calculating the variance, due to the fact that the mean contains a decimal. This same difficulty did not occur in exercise 1, where the mean was a whole number.

3. Calculate the variance of the data in exercise 2 using

$$v = \frac{n(\Sigma x^2) - (\Sigma x)^2}{n(n - 1)}$$

Here the reader should notice the increased ease of computation due to using $v = \dfrac{n(\Sigma x^2) - (\Sigma x)^2}{n(n - 1)}$. Also notice that when you use this formula, the mean is not needed. Most people prefer to use this formula.

4. Table V can be used to square numbers not between 1 and 10 by neglecting the decimal point there. For example, to find $(12.3)^2$, locate 123 in the "n column." Then read 15129 out of the "n^2 column." Now figure this way: you know that $12^2 = 144$, so $(12.3)^2 = 151.29$. Use this procedure to get the variance and standard deviation of the data 12.1, 13.0, 12.5, 12.4, 12.4, 12.6.
5. Can the standard deviation of any set of data ever be negative? Why?
6. A foreman at a bakery periodically checks the weights of loaves of bread. He does so by taking a sample of five loaves every three hours and computing the mean of that sample. The dough dispensing machine will be judged under control when the mean differs from 16 ounces by no more than 0.5 ounces (either way). During one particular three-hour period, the sample weights were 16.0, 16.1, 15.2, 16.8, and 16.2 ounces. What is the decision to be made here?
7. A machine makes mechanical timing devices that are supposed to measure a period of 5 minutes. The process will be deemed under control when:

    1. The mean of a sample of size four differs from the desired time by 0.2 minutes or less.
    2. At least 75% of the data lies within one standard deviation about the mean.

    The sample yields 4.92, 5.12, 5.02, 5.06 minutes. Is this process under control?
8. An experimental psychologist has been studying the effect of

human attention on learning times of monkeys for certain tasks. Five monkeys in group I were given attention every hour for three hours preceding the task to be learned. Five monkeys in group II received no attention for three hours preceding the task to be learned. The psychologist decides (for a number of reasons) that she will choose to believe that the attention reduces the learning time if the mean learning time of monkeys in group II minus the mean learning time of monkeys in group I is a minute or more. The data below gives the learning times for each group.

Group I (min.)	*Group II (min.)*
21	20
25	24
24	28
20	20
19	18

What decision will the psychologist make on the basis of this one run of the experiment?

9. For the data 2, 3, 5, 4, 1, compute the intervals:

a. one standard deviation about the mean ($\bar{x} \pm s$);
b. two standard deviations about the mean ($\bar{x} \pm 2s$);
c. three standard deviations about the mean ($\bar{x} \pm 3s$).

10. A percent of a quantity refers to a part of it, relative to some whole quantity. For instance, in a classroom consisting of 20 people (15 women, 5 men), the fraction of women is 15/20:

$$20\overline{\smash{)}15.00}^{\;.75}$$

Thus, there is 75% women. Use this idea in conjunction with the data of exercise 9 in order to determine the percent of data present in:

a. $\bar{x} \pm s$
b. $\bar{x} \pm 2s$
c. $\bar{x} \pm 3s$

11. Earlier in this section we dealt with the sample 4, 5, 6, 7, 8 and found its standard deviation to be 1.58. In exercise 9 above we had 1, 2, 3, 4, 5 and there the standard deviation was also 1.58. On that basis, either guess or calculate the standard deviation of the data 37, 38, 39, 40, 41.

As a result of doing problem 11, the reader should realize that s truly is a measure of variation. That is, s really depends exclusively on the spread of the numbers relative to each other.

12. Find the standard deviation of the data 502, 499, 503, 500. (Hint: read and do exercise 11 first.)
13. A teacher gave an hour examination to twenty-five students. The results are shown below. (These are percents.)

98	85	74	51	83
71	83	70	72	92
70	70	75	61	71
92	72	76	60	75
77	79	79	80	76

Here $\bar{x} = 75.7$, $s = 10.2$. The teacher decides to define the "C range" as scores falling in $\bar{x} \pm s$ (inclusive). The "*B* range" is any score larger than $\bar{x} + s$ up through $\bar{x} + 2s$. The "A range" is any score larger than $\bar{x} + 2s$. The "D range" is any score less than $\bar{x} - s$ down through $\bar{x} - 3s$. The "F range" is any score less than $\bar{x} - 3s$.

a. Determine the letter grade received by each of the twenty-five students.
b. What percent of students received a C?
c. What percent of students passed (D or better)?
d. What percent of students did better than a C?

14. For 1973 at the Dime and Dollar Savings Bank, balances were noted at the end of the year and it was found that $\bar{x} = \$230.56$ and $s = \$50.10$. For 1974 it was found that $\bar{x} = \$210.78$ and $s = \$70.10$.

a. Find the percent of decrease in the mean from 1973 to 1974.
b. Find the percent of increase in the standard deviation from 1973 to 1974.

The reader should think about what it means to have $\bar{x}$ decrease while s increases from 1973 to 1974 in exercise 14.

15. Derive the formula

$$v = \frac{n(\Sigma x^2) - (\Sigma x)^2}{n(n-1)}$$

from the definition of variance. (Hint: Expand $(x - \bar{x})^2$. Also use the fact that $\Sigma \bar{x}^2 = n\bar{x}^2$.)

8.5 PICTORIAL WAYS OF DISPLAYING DATA

In this section we will present some graphical ways of displaying data. When a sample is large (more than 30 items) it can be cumbersome to work with, so we commonly *classify* (or *group*) the data. Simply speaking, this process involves choosing some classes and then determining into which class each piece of data falls. We'll illustrate this procedure below.

In the beginning of section 4.3, we considered the experiment of tossing a die 600 times. Actually, we were selecting a sample there (a large one) of size $n = 600$ from the population of all possible tosses of the die. A portion of the table that summarizes that information is represented below.

Sample points	*Frequency* (f)
1	97
2	94
3	102
4	99
5	105
6	103

Notice that $\Sigma f = 600 = n$.

A table like this one is called a **frequency table.** It shows how the number of occurrences (frequency) is distributed among the various possibilities of 1 through 6. We can draw a picture of this **frequency distribution** as shown below.

Frequency Distribution for 600 Tosses of a Die

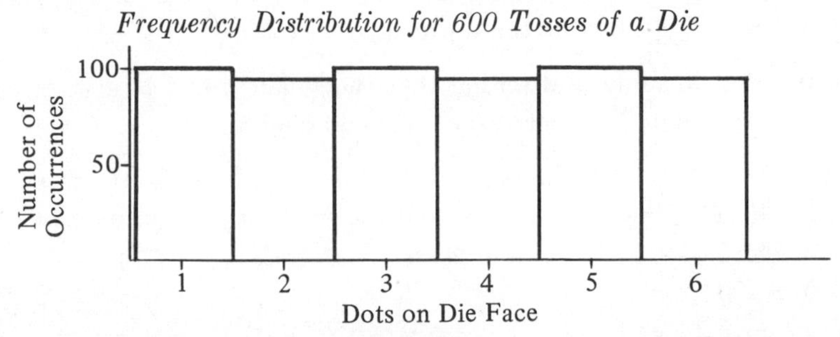

A picture like this, referred to as a **histogram,** has a label on each scale as well as a title. The vertical scale depicts the frequency. The horizontal scale has six equal-width classes (or categories), each one being defined by its midpoint (1 or 2 or 3 or 4 or 5 or 6). We could equally well define each class by its **class boundaries** (the beginning and end of the class). Then our frequency table would be:

Class boundaries	*Frequency*
.5–1.5	97
1.5–2.5	94
2.5–3.5	102
3.5–4.5	99
4.5–5.5	105
5.5–6.5	103

This might seem unnecessary, but when you first collect a large sample it is not classified until you choose some class boundaries and then determine how much data is in each class. For instance, suppose an efficiency expert is investigating the number of "breaks" taken (in a two-week period) by each of 50 workers in a plant, each of whom is allowed as many four-minute breaks as desired. (This does not count lunch periods.) The data collected appears below.

63	63	64	92	93
99	128	67	148	140
100	128	103	88	74
47	97	91	103	78
101	61	70	108	85
115	127	91	126	93
100	108	104	123	94
86	111	106	117	95
114	136	104	116	111
113	140	104	114	108

Now it is time to select class boundaries. There are no hard and fast rules here, but may statisticians seem to:

1. choose somewhere between six and twelve classes and,
2. choose all classes the same width.

Bearing that in mind and noting that the smallest and largest pieces of data are 47 and 148, respectively, we could choose:

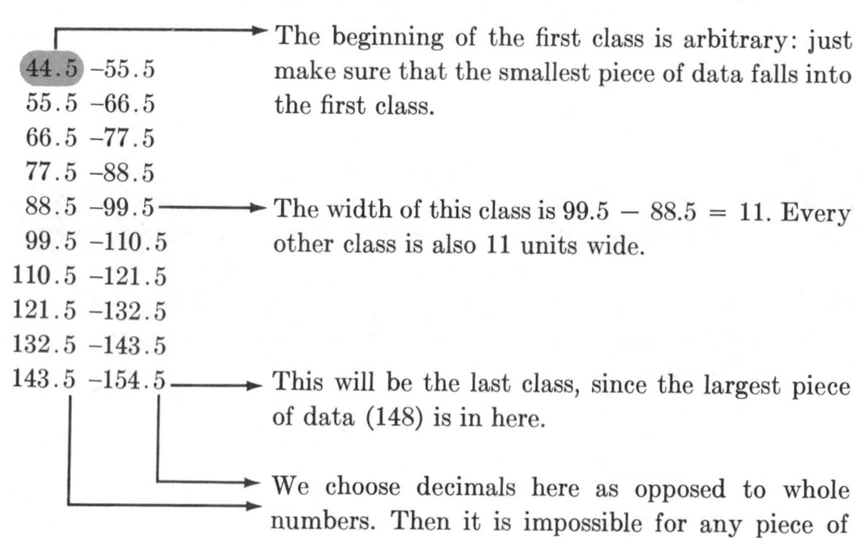

Next, it is time to determine how much data is in each class. Referring to the original list of raw data, the 63 lies in the second class (55.5–66.5), the 99 lies in the fifth class (88.5–99.5), the 100 lies in the sixth class (99.5–110.5), etc. We keep track of these facts with tally marks (/):

Class boundaries	*Tally*
44.5–55.5	/
55.5–66.5	////
66.5–77.5	///
77.5–88.5	////
88.5–99.5	/////////
99.5–110.5	////////////
110.5–121.5	////////
121.5–132.5	/////
132.5–143.5	///
143.5–154.5	/

Now, by counting the tally marks for a class, we get each frequency, yielding the table below.

Class boundaries	*Frequency*
44.5–55.5	1
55.5–66.5	4
66.5–77.5	3
77.5–88.5	4
88.5–99.5	9
99.5–110.5	12
110.5–121.5	8
121.5–132.5	5
132.5–143.5	3
143.5–154.5	1
	50

The histogram appears below.

Histogram Showing Distribution of Four-Minute Breaks Taken by 50 Workers at a Factory

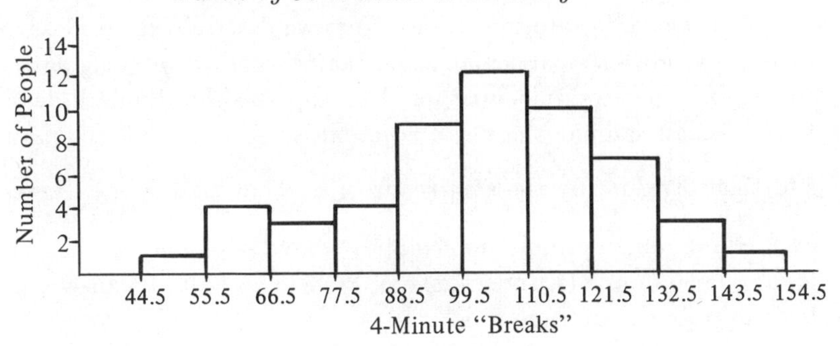

Now that the data has been pictured, there are many purposes the histogram can serve. Perhaps management considers 133 or more breaks to be excessive. A glance at the histogram shows a total frequency of four measurements (three from the class 132.5–143.5 and one from the class 143.5–154.5) to be 133 or more. Thus, 4/50 or 8% of the workers (in our sample) are taking an excessive number of breaks. Whether this 8% figure is acceptable to management or not would involve many economic factors.

Another use of a histogram (or frequency table) is computing the mean and the standard deviation of the data. We do this by assuming that all of the data within any particular class lies at the center of that class, called the **class mark.** To determine the class mark you simply find the midpoint of the class under consideration. As an example, for the class 44.5 to 55.5, the class mark is 50 $\left(= \frac{44.5 + 55.5}{2}\right)$. Computing the other class marks gives:

We use x to denote class mark with large samples: →

x (class mark)	f (frequency)
50	1
61	4
72	3
83	4
94	9
105	12
116	8
127	5
138	3
149	1

Notice that the class marks are 11 units apart from each other, just as the classes were 11 units wide.

This also is called a frequency table.

Recall that this discussion started with a sample ($n = 50$) of numbers of four-minute breaks taken by people working at a factory. We then classified that data into ten classes. Next we determined the class marks, because within any particular class, that's where we assume data to be. Remember now, we're looking for the mean and the standard deviation. With a small sample, the mean is found by summing all the data (Σx) and then dividing by the sample size (n). That is, $\bar{x} = \frac{\Sigma x}{n}$. Here, however, our frequency table immediately above says that we have as our data: one 50, four 61s, three 72s, etc. We can add all fifty pieces of data, but multiplying will save time.

This is x — This is f — This is xf

$$
\begin{aligned}
50 \times 1 &= 50\\
61 \times 4 &= 244\\
72 \times 3 &= 216\\
83 \times 4 &= 332\\
94 \times 9 &= 846\\
105 \times 12 &= 1260\\
116 \times 8 &= 928\\
127 \times 5 &= 635\\
138 \times 3 &= 414\\
149 \times 1 &= 149\\
\hline
&\ 5074
\end{aligned}
$$

5074 ← This is Σxf and represents the sum of all our data.

So, now the mean is given by $\bar{x} = \frac{\Sigma xf}{n}$. Therefore

$$\bar{x} = \frac{5074}{50} = 101.48 \approx 101.5$$

Notice the similarity of formulas:

1. For small samples, $\bar{x} = \frac{\Sigma x}{n}$ → Here x is the value of the various pieces of data.

2. For large samples, $\bar{x} = \frac{\Sigma xf}{n}$ → Here x is the class mark, where we assume the data lies.

A similar comparison can be made between the formulas for variance of small and large samples.

For small samples, $v = \frac{n(\Sigma x^2) - (\Sigma x)^2}{n(n-1)}$.

For large samples, $v = \frac{n(\Sigma x^2 f) - (\Sigma xf)^2}{n(n-1)}$.

Below we present the computations for variance, which the reader should follow through carefully.

x	f	x^2	xf	x^2f
50	1	2500	50	2500
61	4	3721	244	14884
72	3	5184	216	15552
83	4	6889	332	27556
94	9	8836	846	79524
105	12	11025	1260	132300
116	8	13456	928	107648
127	5	16129	635	80645
138	3	19044	414	57132
149	1	22201	149	22201
	50		5074	539,942
	$\Sigma f = n$		Σxf	$\Sigma x^2 f$

15552 → This, for instance, is 5184 × 3.
x^2 f

$$v = \frac{n(\Sigma x^2 f) - (\Sigma xf)^2}{n(n-1)} = \frac{50(539{,}942) - (5074)^2}{50(49)}$$

$$= \frac{26{,}997{,}100 - 25{,}745{,}476}{2450} = \frac{1{,}251{,}624}{2450}$$

$$= 510.8669$$

So, $s = 22.6$ (by using Table V).

The mean and the standard deviation have been calculated here by assuming the data lies at the class marks. This assumption can cause the values of $\bar{x}$ and s to differ slightly from the results you would get by calculating them from the original set of raw data. However, the time saved is usually considered to be more important. Some examples follow:

Example 8.5.1. A biologist wishes to determine the "rate of uptake" into living tissues of some radioactive material. An experiment is performed and data is collected and classified as shown below. (CPM is counts per minute by a geiger counter.)

Class boundaries (CPM)	*Frequency*
1716.5–1737.5	3
1737.5–1758.5	4
1758.5–1779.5	10
1779.5–1800.5	8
1800.5–1821.5	5
1821.5–1842.5	2

Find the mean CPM.

Solution: We need the class marks. The first is

$$\frac{1716.5 + 1737.5}{2} = \frac{3454}{2} = 1727$$

Also, noting that each class is 21 units wide tells us that the class marks are 21 units apart.

x	f	xf
1727	3	5181
1748	4	6992
1769	10	17690
1790	8	14320
1811	5	9055
1832	2	3664
	$32 = n$	$56{,}902 = \Sigma xf$

So here, $\bar{x} = 56{,}902/32 = 1778.2$.

Example 8.5.2. A manufacturer of car batteries has tested one type of battery to see how long it takes to discharge when the electrical load is two normal headlights. A frequency table appears below. The variable x represents the time in hours it takes the battery to discharge.

x (hours)	f
3.0	7
3.5	10
4.0	12
4.5	6
5.0	4
5.5	0
6.0	1

a. Draw a histogram.
b. Calculate $\bar{x}$ and s.

Solution:

a. *Histogram Showing Discharge Time in Hours of 40 Batteries*

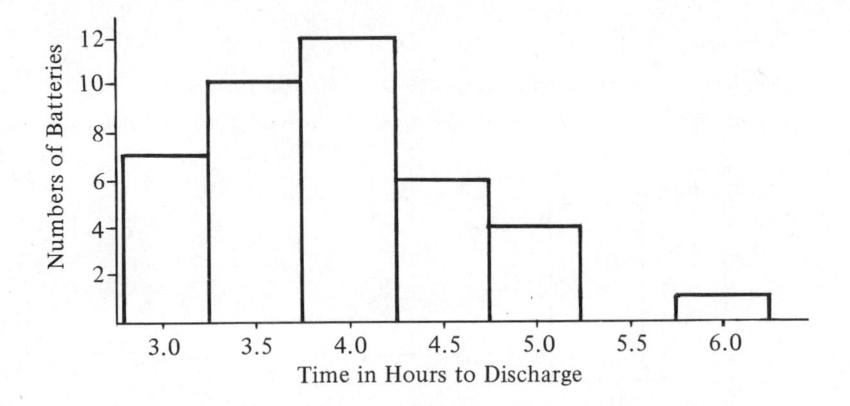

Notice we have used the class marks to label the horizontal scale.

b.

x	f	xf	x^2	x^2f
3.0	7	21	9.00	63.0
3.5	10	35	12.25	122.5
4.0	12	48	16.00	192.0
4.5	6	27	20.25	121.5
5.0	4	20	25.00	100.0
5.5	0	0	30.25	0
6.0	1	6	36.00	36.0
	$40 = n$	$157 = \Sigma xf$		$635.0 = \Sigma x^2f$

$$\bar{x} = \frac{\Sigma xf}{n} = \frac{157}{40} = 3.925 \approx 3.92$$

$$v = \frac{n(\Sigma x^2f) - (\Sigma xf)^2}{n(n-1)} = \frac{40(635) - (157)^2}{(40)(39)}$$

$$= \frac{25{,}400 - 24{,}649}{1560} = \frac{751}{1560} = 0.481410$$

Thus, $s = \sqrt{.481410} = 0.22.$

For any of the large samples discussed so far, we have computed the mean or the standard deviation. However, we are not always interested in those two statistics. Consider the data below, which gives the percent of time that a particular type of weapon was used in the act of murder in the year 1972. A histogram of the data follows:

Type of weapon	*Frequency* (% of total)
hand gun	54%
rifle	5%
shotgun	7%
cutting or stabbing	19%
other weapon (club, poison, etc.)	7%
personal weapon (hands, fists, feet, etc.)	8%

Source: Uniform Crime Reports by the FBI for 1972.

Histogram Showing Types of Weapons Used in the Act of Murder in 1972

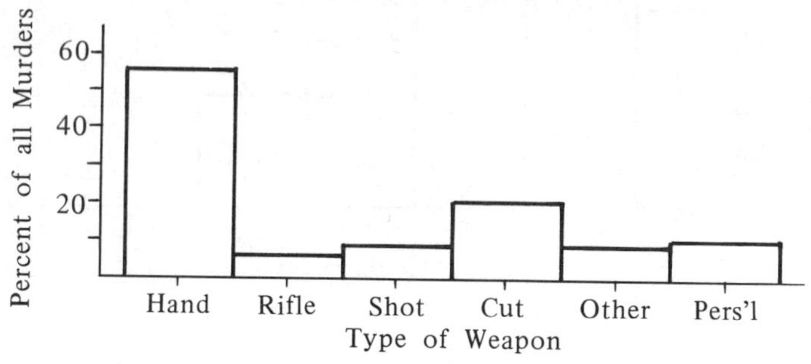

It would make no sense whatever in this setting to ask for either $\bar{x}$ or s, since the type of weapon is not numerical information. We do, however, often speak of a **modal class,** which is the class with the greatest frequency. So "handgun" is the modal class in the above situation. It is this class that might be of most interest to someone in the fields of criminal justice or law legislation.

This section has shown us how to organize data (by classifying it) and also how to compute the mean and the standard deviation from a frequency table. When we have large samples it is common to compute the percent of data lying in each of the intervals $\bar{x} \pm s$, $\bar{x} \pm 2s$, $\bar{x} \pm 3s$. (The reason will be understood shortly.) We'll now do that for the data pertaining to four-minute breaks taken by a sample of 50 factory workers; we discussed this at the beginning of this section. Repeated below is the frequency table and histogram for that illustration.

Histogram Showing Distribution of Four-Minute Breaks Taken by 50 Workers at a Factory

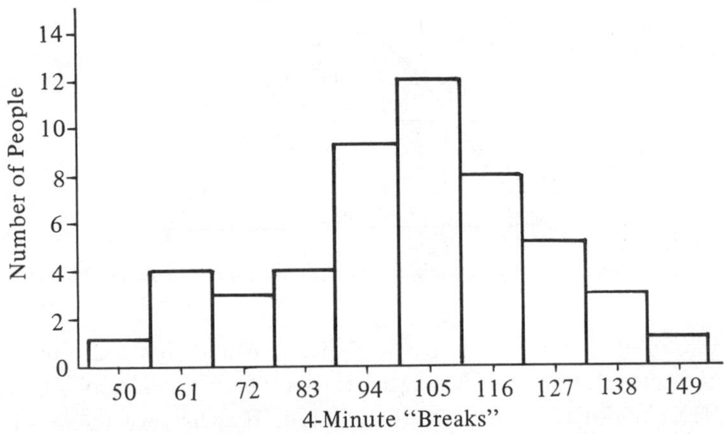

We previously calculated $\bar{x} = 101.5$, $s = 22.6$. Therefore:

$$\left.\begin{aligned} \bar{x} + s &= 124.1 \\ \bar{x} - s &= 78.9 \end{aligned}\right\} \bar{x} \pm s$$

$$\left.\begin{aligned} \bar{x} + 2s &= 146.7 \\ \bar{x} - 2s &= 56.3 \end{aligned}\right\} \bar{x} \pm 2s$$

$$\left.\begin{aligned} \bar{x} + 3s &= 169.3 \\ \bar{x} - 3s &= 33.7 \end{aligned}\right\} \bar{x} \pm 3s$$

Now, recall that the data is assumed to lie at the class marks. Thus if a class mark is within any of these intervals, all the data in that class will be included in that interval. In the interval 78.9 to 124.1 there is

$$\frac{4 + 9 + 12 + 8}{50} = \frac{33}{50} = 66\% \text{ of all the data}$$

In the interval 56.3 to 146.7 there is $\dfrac{4 + 3 + 4 + 9 + 12 + 8 + 5 + 3}{50}$

$= \dfrac{48}{50} = 96\%$ of all the data.

In the interval 33.7 to 169.3 there is 100% of the data.

Statisticians have dealt with large samples like this latter illustration since the end of the 1800s, when Sir Francis Galton (first cousin to Charles Darwin) collected a large amount of data concerning the results of planting different sizes of peas. Since then, many distributions have been investigated. One such distribution is the **normal distribution,** which has the properties listed below.

1. It is symmetric to the left and right of the mean.
2. There is 68% of the data contained within $\bar{x} \pm s$.
3. There is 95% of the data contained within $\bar{x} \pm 2s$.
4. There is 99.7% of the data contained within $\bar{x} \pm 3s$.

It looks like this:

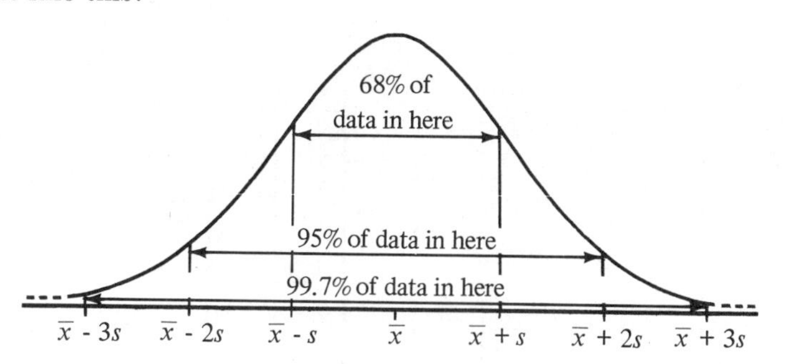

In real life situations there are few distributions that turn out to be perfectly normal; however, it is often the case that the data of a sample is APPROXIMATELY normally distributed. By this, we mean that we "come close" to meeting the four properties mentioned. Unfortunately, there is no agreement among people who are involved with statistics as to what "come close" means. (It often depends on how critical a study is being made.) So we will say that if:

1. the distribution "looks" symmetrical on each side of the mean, and
2. each of the calculated percents in $\bar{x} \pm s$, $\bar{x} \pm 2s$, $\bar{x} \pm 3s$ is within three percent points of the desired 68%, 95%, and 99.7% figures,

then the distribution is approximately normal.

Now let us check on our distribution of four-minute breaks. There we had

66% of the data in $\bar{x} \pm s$
96% of the data in $\bar{x} \pm 2s$
100% of the data in $\bar{x} \pm 3s$

Thus, we can comment that this distribution is APPROXIMATELY normal.

EXAMPLE 8.5.3. For the frequency table below, use the facts that $\bar{x} = 3.3$ and $s = 1.7$ and test to see whether the distribution of data is normal.

Class boundaries	f
.5–1.5	6
1.5–2.5	8
2.5–3.5	10
3.5–4.5	7
4.5–5.5	6
5.5–6.5	1
6.5–7.5	1
7.5–8.5	1

Solution: Note that $\Sigma f = 40 = n$, and

$\bar{x} \pm s$ is 1.6 to 5.0,
$\bar{x} \pm 2s$ is -0.1 to 6.7,
$\bar{x} \pm 3s$ is -1.8 to 8.4.

→Here we have $\dfrac{8 + 10 + 7 + 6}{40} = \dfrac{31}{40} = 78\%$ of the data.

→Here we have $\dfrac{6 + 8 + 10 + 7 + 6 + 1}{40} = \dfrac{38}{40} = 95\%$ of the data.

→Here we have $\dfrac{40}{40} = 100\%$ of the data.

We have 78% of the data within $\bar{x} \pm s$. Hence this distribution is not approximately normal.

EXAMPLE 8.5.4. Using the properties of a normal distribution, find the approximate percent in each of the regions below.

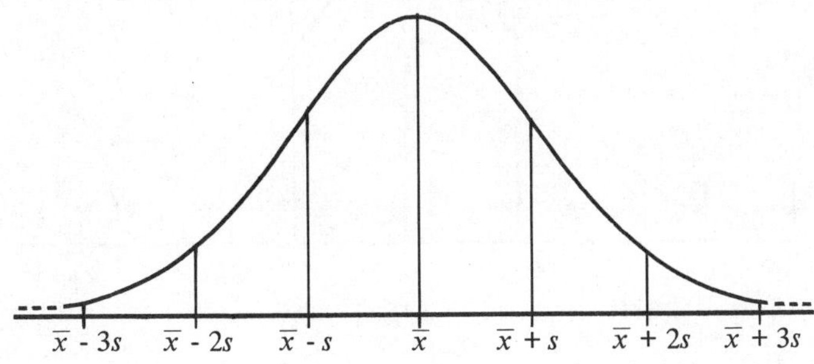

Solution: Between $\bar{x} \pm s$ there is 68% of data. Within $\bar{x} \pm 2s$, there is 95%. The difference is 27%, which must be split: half into $\bar{x} + s$ to $\bar{x} + 2s$ and half into $\bar{x} - s$ to $\bar{x} - 2s$. Reasoning in a similar way, we find:

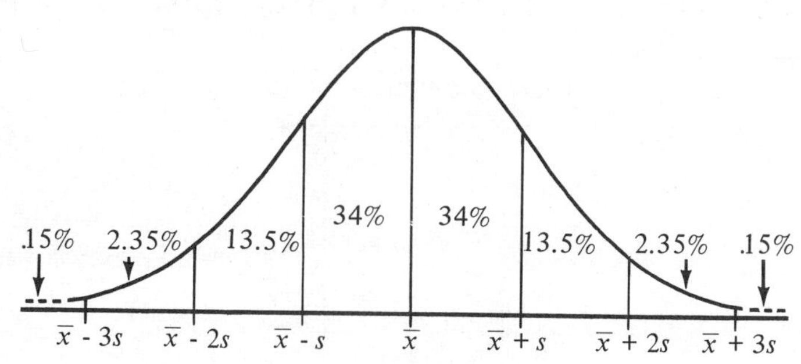

EXAMPLE **8.5.5.** Suppose we have found that the time it takes monkeys to learn a certain task is approximately normally distributed, with a mean of 145 seconds and a standard deviation of 15 seconds.

a. Display this information pictorially.
b. What percent of monkeys take somewhere between 130 seconds and 160 seconds to learn the task?
c. What percent of monkeys take more than 130 seconds to learn the task?
d. A monkey is considered to be a "genius" if he can learn the task in 100 seconds or less. What percent of monkeys are geniuses?
e. Out of a group of 1000 monkeys, how many can we expect to be geniuses?
f. What learning time is such that any monkey taking at least this time will put it in the top 2.5% of monkeys, timewise? (These are the monkeys that take the longest amount of time to learn.)

Solution:

a.

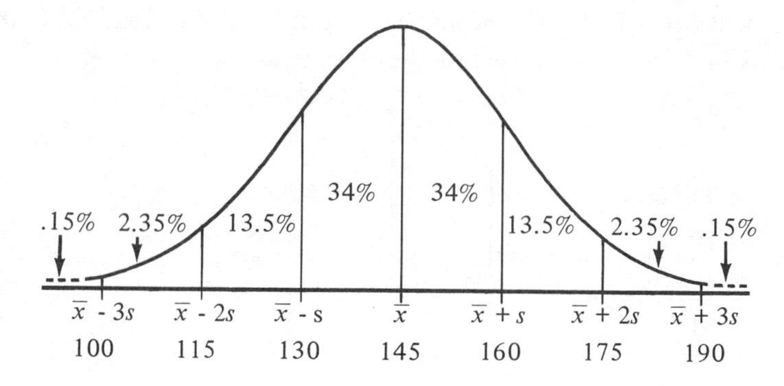

b. Reading from the picture above gives 68% of the monkeys between 130 seconds and 160 seconds.
c. This is the sum of the percents to the right of the 130-second position: we get 84%.
d. This is left of the 100-second position: we get .15%.
e. This is $.15\% \times 1000 = .0015 \times 1000 = 1.5$. Hence 1 or 2 monkeys could be expected to be geniuses out of a group of 1000.
f. We want the learning time at position * below.

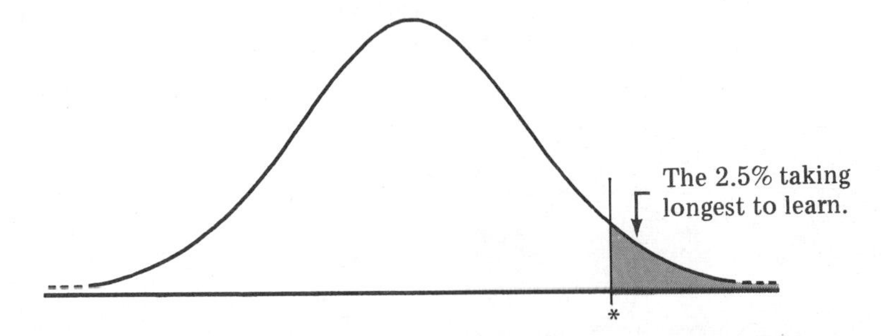

A comparison of this picture with the one in part *a* above gives a result of 175 seconds.

This has been a long and involved section. We will briefly recount some of the highlights. When dealing with large samples it is beneficial to classify or group the data. The class boundaries chosen are arbitrary. Nevertheless, it is standard to make sure that:

1. the smallest piece of data falls into the first class;
2. all classes are of the same width;
3. there are between six and twelve classes.

Once a frequency table is obtained, a histogram is often drawn to "picture" the data. We also demonstrated how to calculate $\bar{x}$, v, and s. The mean and the standard deviation can be used in conjunction with either the frequency table or the histogram to test a distribution to see whether it is normal (or approximately so). The kinds of questions that become answerable when we have a normal distribution (or an approximate one) were illustrated in example 8.5.5. In section 8.7 we will further investigate the normal distribution.

EXERCISES ■ SECTION 8.5.

1. Consider the sample of four-minute breaks first discussed at the beginning of this section. Choose classes of 46.5–56.5, 56.5–66.5, 66.5–76.5, etc. Classify the sample of size 50 into these classes and

draw a histogram. Compare your result here with the result on page 359 and note the similar shapes.

2. Take the classified data in exercise 1 above and calculate the mean. Compare it with the mean already calculated on page 361. The reader should notice that this change of classes hardly affects the value of the calculated mean.

3. The purpose of this exercise is to introduce the reader to another kind of pictorial method of displaying data: a **frequency polygon.** Consider the frequency table of example 8.5.2. Over each class mark, place a dot at the frequency value of that class. For example, the first class looks like this:

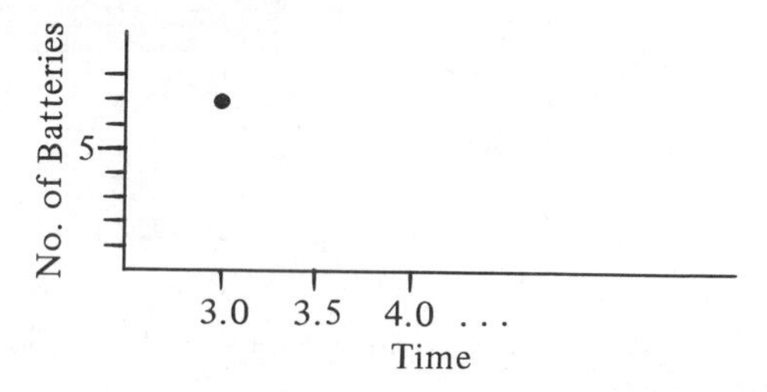

Connect the dots, label both scales, attach a title to the picture, and you have a frequency polygon. Complete the frequency polygon for example 8.5.2.

4. Draw a frequency polygon for the data in example 8.5.3.

5. Draw a frequency polygon for the classified data below.

Class boundaries	*Frequency*
6.5–13.5	1
13.5–20.5	0
20.5–27.5	5
27.5–34.5	8
34.5–41.5	9
41.5–48.5	6

A distribution like this one, where a large number of measurements lies on the right, is said to be **skewed** to the left. Of course, distributions can be skewed to the right, also.

6. This section did not make mention of the median. However, it can be found fairly easily. Recall from section 8.3 that the median is the middle measurement or position when the data is ranked. Any set of data can be ranked (given enough time), so the median can be found in the middle position. Consider the data of exercise 5

on the previous page. Since we assume the data lies at the class marks we have:

10	31	31	38	45
24	31	31	38	45
24	31	38	38	45
24	31	38	38	45
24	31	38	38	45
24	31	38	45	

Now, what is the median?

7. The data below represents the number of cigarettes still smoked each day by 40 people who are trying to quit smoking.

x	f
1	5
2	8
3	10
4	5
5	4
6	3
7	2
8	1
9	1
10	1

a. Draw a histogram.
b. Draw a frequency polygon.
c. What is the modal class?
d. What is the median?
e. Calculate the mean.

8. Calculate $\bar{x}$ and v and s using the frequency table in exercise 5 on the previous page.

9. **a.** Using $\bar{x}$ and s from exercise 8, calculate each of the intervals $\bar{x} \pm s$, $\bar{x} \pm 2s$, $\bar{x} \pm 3s$.
b. Determine the percent of data lying in each of the intervals from part *a* by referring to exercise 5.
c. Is the data approximately normally distributed? (In order to answer this question, you must compare your results in part *b* with the properties of a normal distribution.)

10. The various parts of this question refer to the data of exercise 7.

a. Calculate $\bar{x}$ and s.
b. Determine $\bar{x} \pm s$, $\bar{x} \pm 2s$, $\bar{x} \pm 3s$.
c. Find the percent of data lying in each of the intervals of part *b*.
d. Is the data approximately normally distributed?

11. At a fireman's training academy, the time that it takes new recruits to learn to drive a 40-foot hook-and-ladder truck has been investigated for a sample of size 50 and the results obtained are shown below.

Class boundaries (time in hours)	*Frequency*
.25–1.75	1
1.75–3.25	3
3.25–4.75	6
4.75–6.25	8
6.25–7.75	13
7.75–9.25	9
9.25–10.75	6
10.75–12.25	3
12.25–13.75	1

a. Draw a histogram. **b.** Calculate $\bar{x}$ and s.
c. Calculate each of the intervals $\bar{x} \pm s$, $\bar{x} \pm 2s$, $\bar{x} \pm 3s$.
d. Find the percent of data lying in each of the intervals of part *c*.
e. Is the data approximately normally distributed?

12. Use the frequency table of exercise 11 for this exercise.

a. Draw a histogram.
Answer questions *b* and *c* by reading the answer directly off the histogram.
b. A new recruit is considered to be a "fast learner" if he can learn to drive the 40-foot truck in 3 hours and 15 minutes or less. What percent of recruits are "fast learners"?
c. What percent of recruits can learn to drive the truck in 10 hours and 45 minutes or less?

13. The diagram on the next page is a histogram that classifies by type of activity the number of law enforcement officers killed for the years 1963–1972. (Source: Uniform Crime Reports for 1972 issued annually by the FBI.) Notice that the rectangles are drawn horizontally; this makes the labelling of the vertical scales a bit easier. For each of the ten categories, compute the percent of change from the period 1963–1967 to the period 1968–1972.

Law Enforcement Officers Killed
(by type of activity)
1963–1972

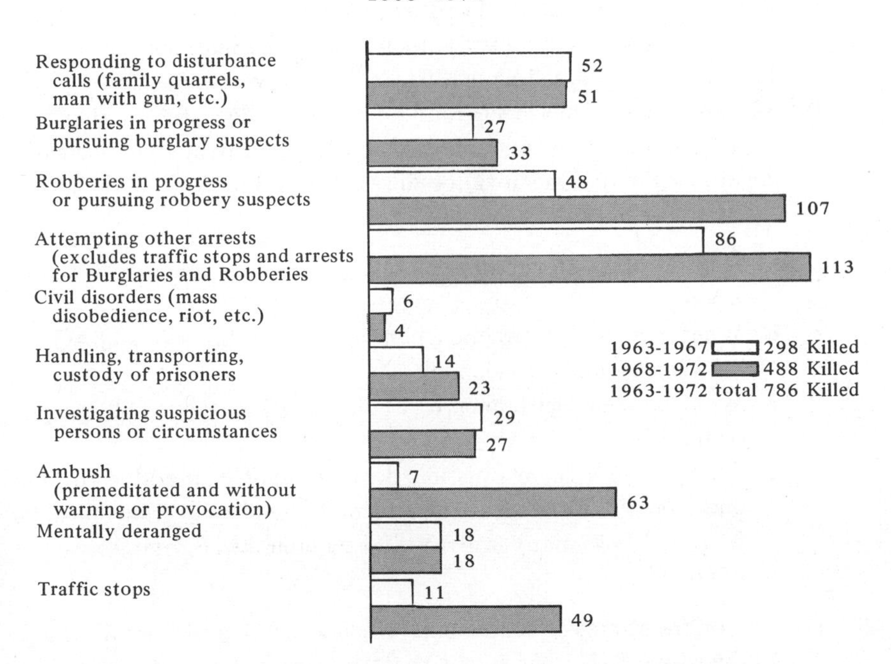

14. The table below classifies by race the total arrests made in the U.S. (Source: 1972 Uniform Crime Reports by FBI.)

Race	*Number of arrests*
White	4,664,220
Negro	1,847,566
Indian	130,375
Chinese	4,234
Japanese	1,821
All others	58,734
	6,706,950

a. Compute the percent of arrests for each race in this survey. $\left(\text{For example: percent of arrests for Chinese is } \frac{4234}{6{,}706{,}950} \approx .0006 \text{ or } .06\%.\right)$

b. Draw a histogram with the frequency scale (vertical) depicting the percents and the horizontal scale, the races.

c. What is the modal class?

A histogram like the one in exercise 14*b*, where the *relative frequency* (a percent) is displayed instead of the frequency, is called a **relative frequency histogram.**

15. For the data of exercise 11, draw a relative frequency histogram. Notice that the shape here is identical to the histogram drawn in 11*a*; merely the labelling is different on the vertical scale.
16. Suppose that it has been found from extended studies that the breath-holding time (x) of navy frogmen is approximately normally distributed with a mean ($\bar{x}$) of 80 seconds and a standard deviation (s) of 10 seconds.

    a. Display this information pictorially (as was done in example 8.5.5*a*).
    b. What percent of frogmen can hold their breaths somewhere between 70 and 90 seconds?
    c. What percent of frogmen can hold their breaths for 90 seconds or more?
    d. Survival on a secret mission depends upon being able to hold one's breath for at least 100 seconds. Out of a group of 1000 frogmen, how many will we find capable of surviving a secret mission?

17. One of the exercises that a soccer coach has his players do is the 100-yard dash. He has found that the time it takes players to run the 100-yard dash is normally distributed (approximately) with mean of 10.2 seconds and a standard deviation of .8 seconds.

    a. Display this information pictorially.
    b. The coach gets extremely upset at players who are "too slow." In his own mind, the coach defines "too slow" as 11 seconds or more. With what percent of the players will the coach be extremely upset?
    c. As far as the coach is concerned, a player must be able to run the 100-yard dash in 10.2 seconds or less to qualify for the halfback position. From this point of view, what percent of players will be able to qualify for the halfback position?

18. A filling gauge is designed to fill 100-gallon tanks with liquid propane gas. From previous testing of such gauges, during the manufacturing phase it has been found that the number of gallons dispensed is normally distributed with a mean of 99 gallons and a standard deviation of 0.3 gallons. "Overfill" is a very critical item in this situation, so it has been decided ahead of time to declare the filling process out of control when the gauge allows the dispensing of 99.9 gallons or more. What percent of the gauges can we expect to be out of control at any one moment?

8.6 MEASURES OF LOCATION

The previous section, where we classified data, allowed us to handle rather large samples with reasonable ease. Once all the data is grouped into classes, the whole sample is effectively reduced to a frequency table. This process does cause some loss of detail. On the other hand, when we do not put the data of large samples into groups, we can get hopelessly lost trying to juggle all the pieces of raw data. Moreover, the importance of one piece of data can easily be overshadowed by all of the others. **Measures of location** are certain numerical values that "locate" one piece of data among the entire sample. There are two measures of location discussed in this section: the **percentiles** and the **z-values.**

Percentiles are numbers that divide a sample into hundredths. For instance, the thirty-second percentile (denoted P_{32}) is that position BELOW which 32% of the data lies. The sample below,

29	15	11	41	43
29	16	12	48	43
30	21	13	50	45
42	19	34	51	48
55	15	38	49	60
55	14	38	56	60
49	28	85	60	60
40	24	80	60	58
40	21	61	61	59
11	30	40	71	82

when ranked, looks like this

11	21	40	49	60
11	24	40	49	60
12	28	40	50	60
13	29	41	51	60
14	29	42	55	61
15	30	43	55	61
15	30	43	56	71
16	34	45	58	80
19	38	48	59	82
21	38	48	60	85

Now we are looking for the position below which 32% of the data lies. First we need to know how much of the data constitutes 32%. Since the sample has 50 items, 32% of 50 is $(.32)(50) = 16.0$. The 16 means that we must count 16 pieces of data from the low end. Doing that (in the ranked list, of course) gives $P_{32} = 30$.

Example 8.6.1. For the data in the above illustration find:

a. P_8 **b.** P_{51} **c.** P_{33}

Solution:

a. First we ask how much data constitutes 8%. That is 8% of 50 or (.08)(50) = 4 pieces of data. So, $P_8 = 13$.
b. We need 51% of 50 = (.51)(50) = 25.5 pieces of data. So here we count to halfway between the 25th and 26th pieces of data: $P_{51} = 42.5$.
c. 33% of 50 = 16.5. So, $P_{33} = 30$. Notice that P_{32} and P_{33} are the same. Sometimes that happens.

When a set of data has been classified and the raw data is no longer available, we must "translate" the frequency table into data form. (Remember, data is assumed to be at the class marks.) Consider the table from example 8.5.3.

Class boundaries	*f*
.5–1.5	6
1.5–2.5	8
2.5–3.5	10
3.5–4.5	7
4.5–5.5	6
5.5–6.5	1
6.5–7.5	1
7.5–8.5	1

The data contained herein is:

1	1	2	3	3	4	4	5
1	2	2	3	3	4	5	5
1	2	2	3	3	4	5	6
1	2	2	3	3	4	5	7
1	2	3	3	4	4	5	8

Now find P_{30}. We need 30% of the data in this sample of size 40, and 30% of 40 = (.30)(40) = 12. So P_{30} is the twelfth piece of data, or 2. Not only is $P_{30} = 2$, but so are $P_{17}, P_{18}, \ldots$, up through P_{36}. This is an unfortunate occurrence that takes place when you find percentiles from a frequency table. Just be aware of it!

One use of percentiles is in reporting a raw score and its associated percentile from some set of data (usually large). For instance, suppose a class of students takes an examination and one particular student receives a 78. What does that mean? Nothing, of course, until you know some other things, for example:

1. What were the lowest and highest possible scores?
2. What percentile corresponds to a score of 78?

This last item is especially important if the instructor of the class grades students on the basis of their relative standing in the class. We explore this idea further in the following example.

Example 8.6.2. The data below is a set of exam scores from 62 students in a sociology class. (It has already been ranked.)

21	54	62	76	86	95	110
22	55	63	76	87	96	111
22	56	64	76	88	98	
29	57	65	76	89	99	
38	58	68	77	90	100	
41	59	73	77	90	102	
45	59	73	79	91	104	
46	60	74	83	92	104	
46	60	75	84	93	105	
54	61	76	86	94	106	

The instructor decides to give letter grades according to the scheme below:

"A" for a raw score from P_{90} through P_{100}
"B" for a raw score from P_{79} to P_{90}
"C" for a raw score from P_{48} to P_{79}
"D" for a raw score from P_{21} to P_{48}
"F" for a raw score from P_1 to P_{21}

Find the "cutoffs" in terms of raw scores.

Solution: We need to compute P_{21}, P_{48}, P_{79}, P_{90}.

21% of 62 = (.21)(62) = 13.02 (13th piece of data), so $P_{21} = 56$.
48% of 62 = (.48)(62) = 29.76 (30th piece of data), so $P_{48} = 76$.
79% of 62 = (.79)(62) = 48.98 (49th piece of data), so $P_{79} = 93$.
90% of 62 = (.90)(62) = 55.80 (56th piece of data), so $P_{90} = 102$.

We can conclude:

"A" will be for a raw score from 102 through 111,
"B" will be for a raw score from 93 to 102,
"C" will be for a raw score from 76 to 93,
"D" will be for a raw score from 56 to 76,
"F" will be for a raw score from 21 to 56.

We will now talk about the other measure of location, namely, the z-score. We define this by:

$$z = \frac{\text{value of a piece of data} - \text{mean of the data}}{\text{standard deviation of the data}}$$

In terms of symbols used earlier,

$$z = \frac{x - \bar{x}}{s}$$

(When s does not divide $x - \bar{x}$ exactly, ALWAYS round it to two decimal places.) Notice that in order to determine a z-value, we must first calculate the mean and standard deviation of the sample with which we are dealing. Let us consider the sample of four-minute breaks discussed in the previous section ($\bar{x} = 101.5$ and $s = 22.6$). What is the z-value for a worker who takes 121 four-minute breaks? Here $z = \dfrac{121 - 101.5}{22.6} = .86$ (rounded to two decimal places).

Example 8.6.3. For a given sample, it has been calculated that $\bar{x} = 12$ and $s = 2$. Find the z-value that corresponds to each of the following x-values:

a. $x = 15$ **b.** $x = 11$ **c.** $x = 12$
d. x is one standard deviation above the mean.
e. x is three standard deviations below the mean.

Solution:

a. $z = \dfrac{x - \bar{x}}{s} = \dfrac{15 - 12}{2} = \dfrac{3}{2} = 1.50$

b. $z = \dfrac{11 - 12}{2} = -\dfrac{1}{2} = -.50$ (Notice that z can be negative.)

c. $z = \dfrac{12 - 12}{2} = 0$ (Notice that at the mean, $z = 0$.)

d. One standard deviation above the mean is $\bar{x} + s = 12 + 2 = 14$, so, $z = \dfrac{14 - 12}{2} = 1$.

e. Three standard deviations below the mean is $\bar{x} - 3s = 12 - 6 = 6$, so, $z = \dfrac{6 - 12}{2} = \dfrac{-6}{2} = -3$.

Now by examining each of the parts of this example, it should become clear that z is just the NUMBER OF STANDARD DEVIATIONS THAT A PARTICULAR SAMPLE ITEM LIES EITHER ABOVE (z is positive) OR BELOW (z is negative) THE MEAN.

It is also possible to start a calculation with a z-value and obtain an x-value. Recall,

$$z = \frac{x - \bar{x}}{s}$$

and a bit of algebraic manipulation shows:

$$zs = x - \bar{x}$$
$$x = \bar{x} + zs$$

Example 8.6.4. Using the information from example 8.6.3, find:

a. x, when $z = 3$.
b. x, when $z = -.50$
c. The value of x that is two standard deviations above the mean.

Solution:

a. $x = \bar{x} + zs = 12 + (3)(2) = 18$. Note: 18 is three standard deviations above the mean.
b. $x = \bar{x} + zs = 12 + (-.50)(2) = 12 - 1 = 11$. Note: 11 is one-half of a standard deviation below the mean.
c. Two standard deviations above the mean implies $z = 2$, so $x = \bar{x} + zs = 12 + (2)(2) = 16$.

We have seen in section 8.5 that sometimes the data in a large sample is approximately normally distributed. In such cases, it is possible to draw a picture showing the various percents of data as they are distributed into each of the intervals $\bar{x} \pm s$, $\bar{x} \pm 2s$, $\bar{x} \pm 3s$. However, now we can add more information to that picture by showing the various z-values.

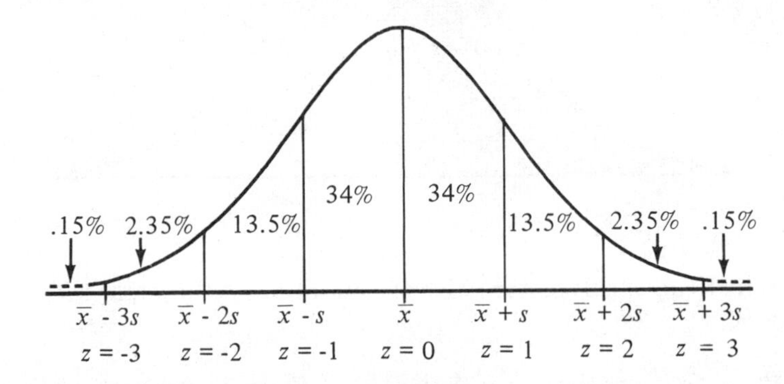

So, we can say that for a normal distribution:
68% of the measurements lie between $z = \pm 1$.
95% of the measurements lie between $z = \pm 2$.
99.7% of the measurements lie between $z = \pm 3$.
Percentiles can also be related to a normal distribution, as we now point out in a final example.

Example 8.6.5. We reconsider the information of example 8.5.5, where learning times (x) of monkeys for a certain task were approximately normally distributed with a mean ($\bar{x}$) of 145 seconds and a standard deviation (s) of 15 seconds. Find:

a. P_{50}
b. P_{84}
c. P_{80}

Solution: We have:

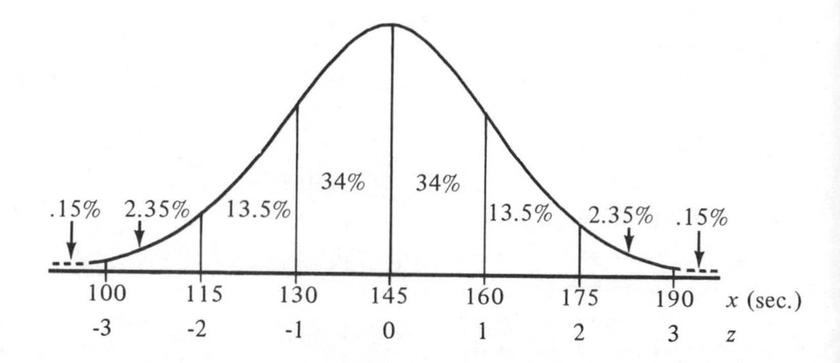

a. P_{50} is the position such that 50% of the data is less than (lies to the left of) it. That must look like this:

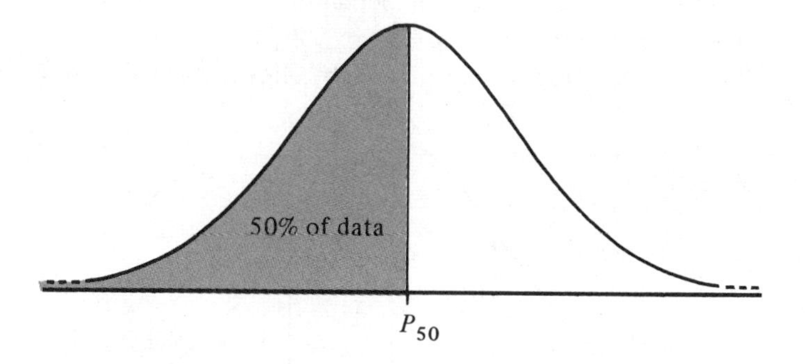

A comparison of this picture with the one above it shows that $P_{50} = 145$ seconds.

b. P_{84} is that position such that 84% of the data lies to the left of it. So, a picture shows:

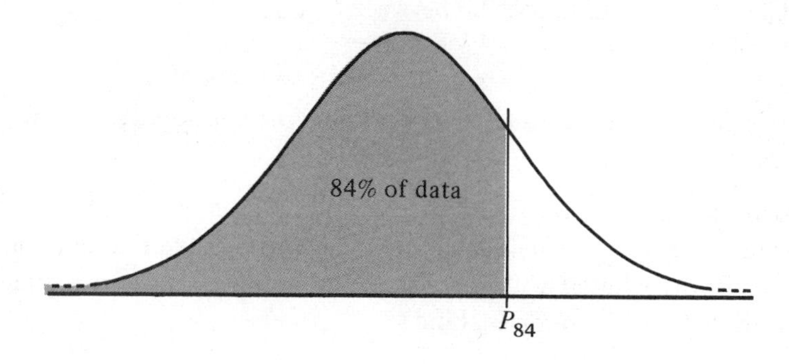

Comparing this picture with the one above shows that P_{84} is right at $z = 1$ (since we have a normal distribution). So, $P_{84} = 160$ seconds.

c. To find P_{80} we have a picture looking like this:

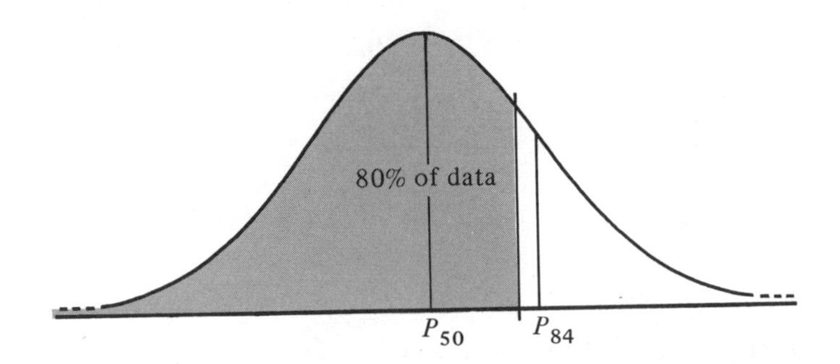

P_{80} would be about here; there is no way to determine its exact value without additional study. All we can say here is that P_{80} is somewhere between 145 seconds and 160 seconds. In the next section we will return to this problem.

EXERCISES ■ SECTION 8.6.

1. Use the sample below to obtain:

a. P_{50} **b.** P_{94}
c. P_{13} **d.** P_2

13	15	20	21	22
2	14	17	7	25
3	19	19	20	25
4	18	18	21	24
4	16	18	29	26
6	7	12	30	26
8	9	7	3	27
10	10	30	29	28

2. Some people working with statistics use things called **quartiles** (there are three of them), denoted Q_1, Q_2, and Q_3. The meaning is:

$$Q_1 = P_{25}$$
$$Q_2 = P_{50}$$
$$Q_3 = P_{75}$$

For the sample in exercise 1 above, find the value of each of the three quartiles.

3. Some people working with statistics use things called **deciles,** denoted D_1, D_2, D_3, . . . , D_9. Their meaning is:

$$D_1 = P_{10}$$
$$D_2 = P_{20}$$
$$D_3 = P_{30}$$
$$\vdots$$
$$D_9 = P_{90}$$

For the sample in exercise 1, find D_1, D_4, and D_9.

4. Use the information and solution in example 8.5.2 to answer the questions below.

   a. Find the z-value for an x-value of 3.5.
   b. Find the z-value for a discharge time of 5 hours and 30 minutes.
   c. Find the z-value for a discharge time of 3 hours and 50 minutes.
   d. Find the x-value for a z-value of $+2.00$.
   e. Find the x-value for a z-value of -1.00.
   f. Find the x-value for a z-value of -1.28.
   g. Find the x-value for a discharge time that is 3.5 standard deviations above the mean.

5. Use the information in example 8.6.5 to find:

   a. P_{16}
   b. P_{33}

6. An instructor of Swahili grades the students by noting the number of standard deviations that a score is either above or below the mean. She gives the same test to two different classes and finds:

   Class I has $\bar{x} = 80$ and $s = 5$;
   Class II has $\bar{x} = 83$ and $s = 3$.

   In which class would a score of 85 have a higher standing (if either)?

8.7 THE NORMAL DISTRIBUTION

In section 8.5 we first discussed the normal distribution. In that section, and the one following it, we learned that a normal distribution looks like this:

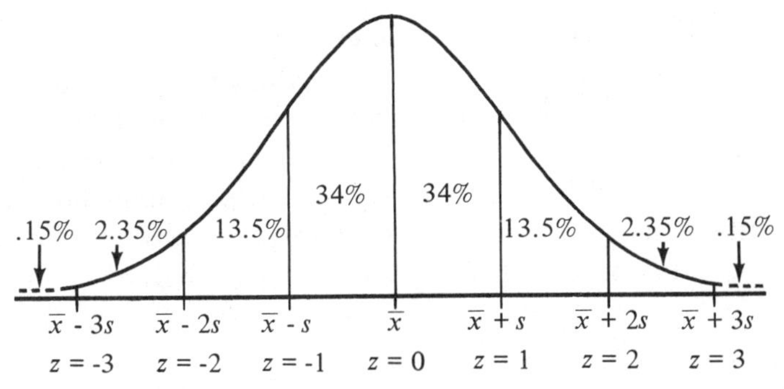

The percents shown, of course, refer to percents of data within each of the intervals determined by the vertical lines at the integral (whole number) values of z. So far, however, we have not given any method for finding the percent of data between two nonintegral z-values. This type of problem is solved by using Table VI in Appendix III.

For example, suppose we set out to determine the percent of data between $z = 0$ and $z = 2.34$. Read down the left-hand column in Table VI to the row beginning with 2.3. Find the intersection of this row with the column headed .04.

	Second decimal place in z						
z	.00	.01	.02	.03	.04	.05	. . .
.0							
.1							
.2							
.							
.							
.							
2.3					.4904		
.							
.							
.							

The table indicates that the amount of data from $z = 0$ to $z = 2.34$ is .4904. There are various interpretations:

1. The percent of data from $z = 0$ to $z = 2.34$ is 49.04%.
2. The amount of AREA under the normal distribution curve from $z = 0$ to $z = 2.34$ is .4904 (or 49.04%). The area under the whole normal distribution is 1 (or 100%).
3. The PROBABILITY that a z-value picked at random lies between 0 and 2.34 is .4904. In symbols we write, $P(0 < z < 2.34) = .4904$.

Notice the three possible ways of interpreting a value read from this table. It can be thought of as a percent of data, an area, or a probability.

Here is another illustration. We want to find the area under the normal distribution curve from $z = 0$ to $z = -1.28$. Table VI does not contain negative z-values, but because of the symmetry of a normal distribution on each side of the mean, we can find the area from $z = 0$ to $z = +1.28$; the answer will be the same. We get .3997. In symbols,

$$P(-1.28 < z < 0) = .3997$$

Example 8.7.1. Using Table VI, find:

a. $P(-1.28 < z < 2.34)$ **b.** $P(-1.00 < z < +1.00)$
c. $P(z < 1.50)$ **d.** $P(z < -1.50)$ **e.** $P(z > 1.28)$

Solution:

a. $P(-1.28 < z < 2.34)$ is needed. A picture is shaded below with the required area.

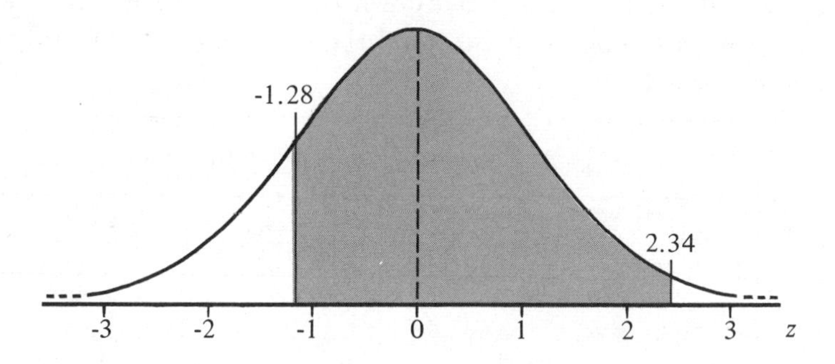

We do this problem in two parts. The area from $z = -1.28$ to $z = 0$ will be added to the area from $z = 0$ to $z = 2.34$. Thus:

$$P(-1.28 < z < 2.34) = .3997 + .4904 = .8901$$

b. $P(-1.00 < z < +1.00)$ is displayed below.

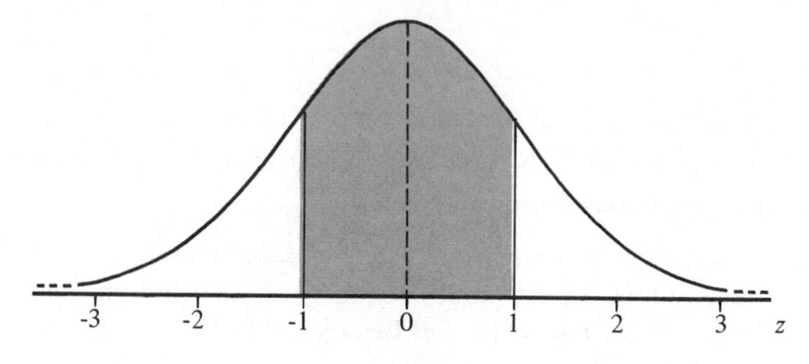

The area from $z = 0$ to $z = +1$ is .3413. So, by symmetry,

$$P(-1.00 < z < +1.00) = .3413 + .3413 = .6826.$$

Hence, within one standard deviation of the mean there is 68.26% of the data (for a normal distribution). In section 8.5, where the properties of a normal distribution were listed, we rounded this to 68% for simplicity's sake.

c. $P(z < 1.50)$ is displayed below.

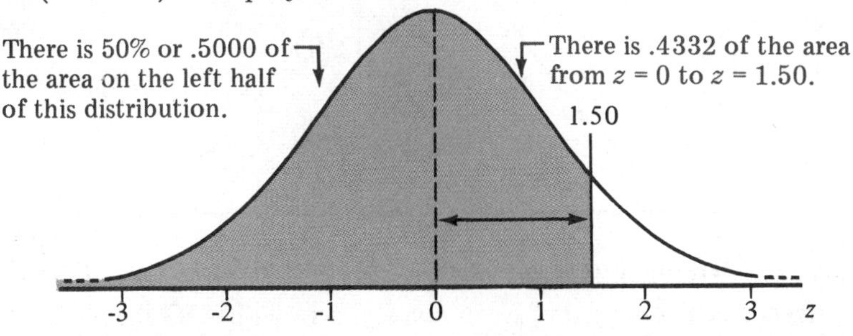

Adding the two areas gives us $P(z < 1.50) = .5000 + .4332 = .9332$

d. $P(z < -1.50)$ is depicted below.

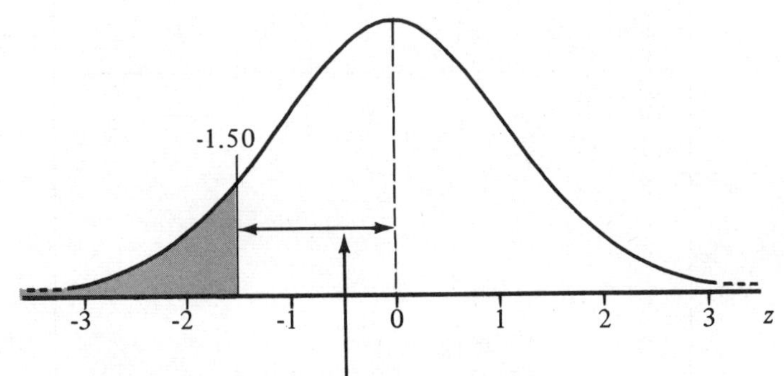

There is .4332 of the area here. We subtract this from .5000 to get

$$P(z < -1.50) = .0668$$

e. $P(z > 1.28)$ is shown below.

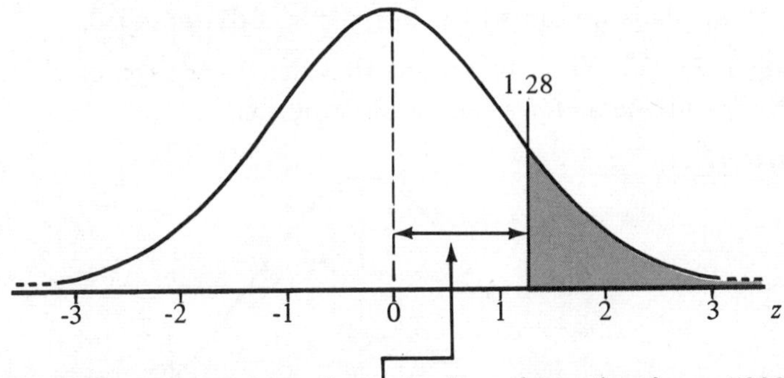

There is .3997 amount of area here. Again, subtracting from .5000 gives

$$P(z > 1.28) = .1003$$

We have been using Table VI to find an area, knowing z at the start of the problem. It is also possible to determine a z-value, knowing the area. For instance, suppose we have an area as shown below.

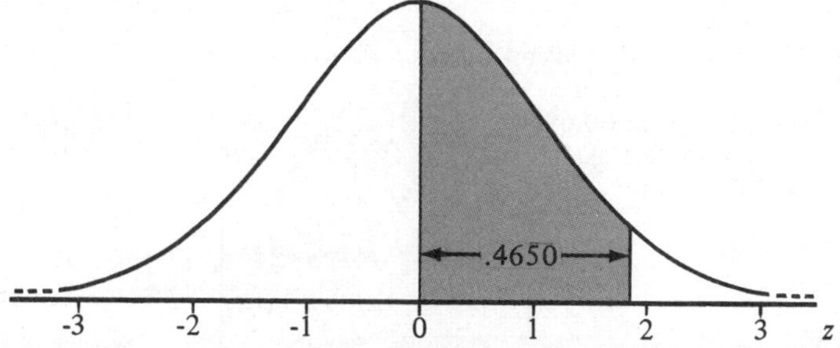

Clearly the left boundary of this area is at $z = 0$, but what is the value of z at the right boundary? Scan the areas in Table VI for .4650. You'll see this:

	Second decimal place in z					
z	.00	.01	.02	.03	.04	. . .
.0						
.1						
.2						
.						
.						
.						
1.8		.4649	.4656			
.						
.						
.						

Pick the area closest to .4650, which is .4649. The associated z-value is 1.81. In symbols we can write: $P(0 < z < 1.81) = .4650$.

Example 8.7.2. Find the z-value that defines the top or highest 5% of the measurements for a normal distribution.

Solution:

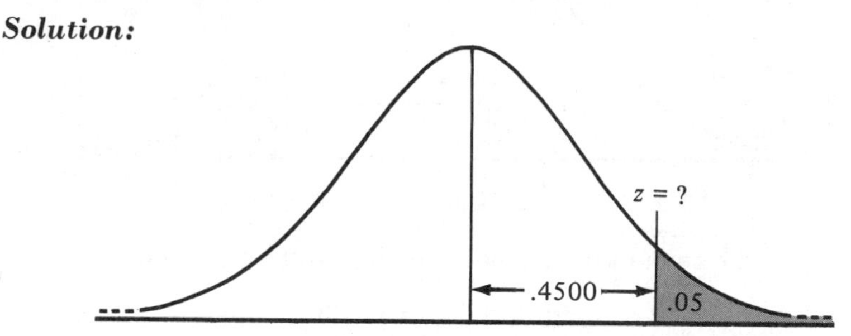

We look up .4500 in Table VI. There are two areas, each equally close to .4500. (They are .4495 and .4505.) The associated z-values are 1.64 and 1.65. So, as our solution, we write:

$$z = 1.645$$

EXAMPLE 8.7.3. Find the two z-values that contain the middle 90% of the measurements for a normal distribution.

Solution:

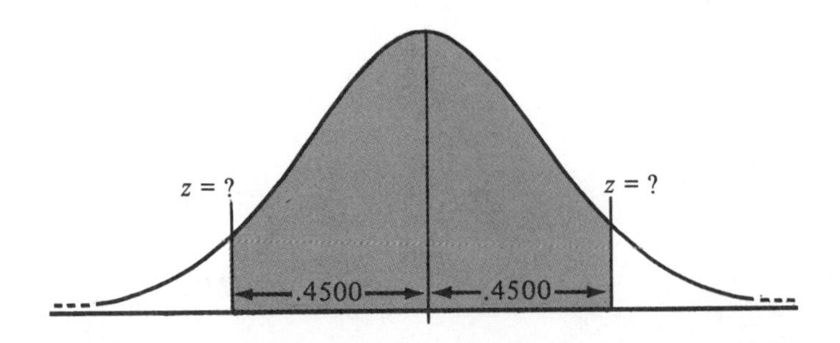

On each side of the center we will have 45% of the measurements. So the z-values on the left and right are -1.645 and $+1.645$, respectively.

Now that we have seen how to use Table VI, we can present some applications of the normal distribution. For the rest of this section, POPULATIONS of various data will be dealt with. Further, we will be dealing with populations whose shape is normal (or at least approximately normal) and whose mean and standard deviation are known. Since populations (as opposed to samples) will be considered, new symbols for the mean and standard deviation are used:

$$\mu = \text{population mean}$$
$$\sigma = \text{population standard deviation}$$

From section 8.6, $z = \dfrac{\text{value of a piece of data} - \text{mean of the data}}{\text{standard deviation of the data}}$. So, in this section, $z = \dfrac{x - \mu}{\sigma}$ will be used.

Consider the population of male students at a large college. Suppose study has shown their heights (x) to be approximately normally distributed with mean of 70 inches and standard deviation of 3 inches. What percent of male students at this college have heights between 65 inches and 77 inches? The picture on the following page can be drawn.

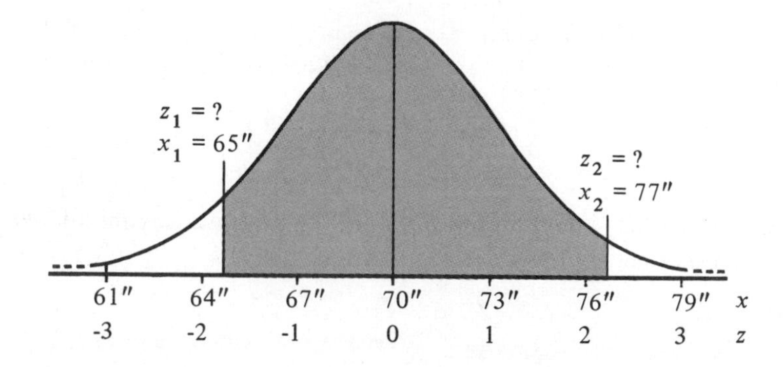

We must find the values of z_1 and z_2; then we can use Table VI to obtain our answer.

$$z_1 = \frac{x_1 - \mu}{\sigma} = \frac{65'' - 70''}{3''} = -1.67$$

$$z_2 = \frac{x_2 - \mu}{\sigma} = \frac{77'' - 70''}{3''} = +2.33$$

The percent of data between 65″ and 77″ is:

$$P(-1.67 < z < 2.33) = .4525 + .4901 = .9426 = 94.26\%$$

So, 94.26% of the male students at the college have heights between 65 inches and 77 inches.

Two examples follow.

Example 8.7.4. A machine operation produces gears whose diameter (x) is approximately normally distributed with mean 8.52 cm and standard deviation .02 cm.

a. Suppose required specifications dictate that these gears have a diameter of 8.52 cm plus or minus .05 cm. What percent of the gears being produced will be acceptable?
b. Under the specifications of part *a*, what percent of the gears will be unacceptable?
c. If each day there are 1000 gears produced, how many gears can we expect to be unacceptable on a daily basis?
d. Suppose the loss of money on an unacceptable gear (includes material cost, employees salaries, overhead expenses, etc.) is $3.52. What is the weekly loss on all the gears?

Solution:

a. From the given information we can draw the distribution curve:

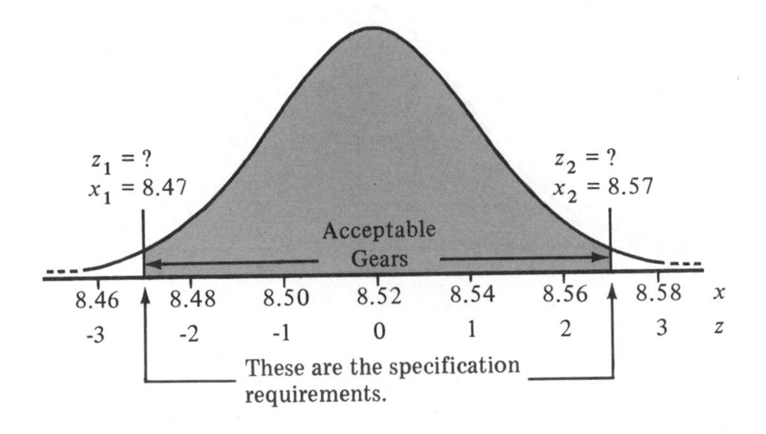

$$z_1 = \frac{8.47 - 8.52}{.02} = \frac{-.05}{.02} = -2.50$$

$$z_2 = \frac{8.57 - 8.52}{.02} = \frac{.05}{.02} = +2.50$$

So, the percent of acceptable gears is $P(-2.50 < z < +2.50) = .4938 + .4938 = .9876 = 98.76\%$.

b. $100\% - 98.76\% = 1.24\%$

c. $(1000)(1.24\%) = (1000)(.0124) = 12.4$. Thus about 12 or 13 gears would be unacceptable per day.

d. The number of unacceptable gears per week is $(5)(12.4) = 62.0$, presuming that the machine is run five days each week. Thus the money lost per week on the gears is $(62)(\$3.52) = \$218.24$.

EXAMPLE 8.7.5. A psychologist, studying learning times and memory, tested many people to see how long it took each to memorize a sequence of 12 digits. She found the time (x) it took to be approximately normally distributed with a mean of 50 seconds and a standard deviation of 20 seconds.

a. What is the probability that a person picked at random from this population takes 22 or fewer seconds to memorize the 12-digit sequence?

b. What percent of people from this population will take 22 or fewer seconds?

c. The psychologist wants to identify the 10% of the population that took the longest. What minimum time defines this group?

Solution:

a. We need $P(x \leq 22)$. When $x = 22$, $z = \dfrac{22 - 50}{20} = -1.40$. So, $P(z \leq -1.40)$ is shaded as shown on the following page.

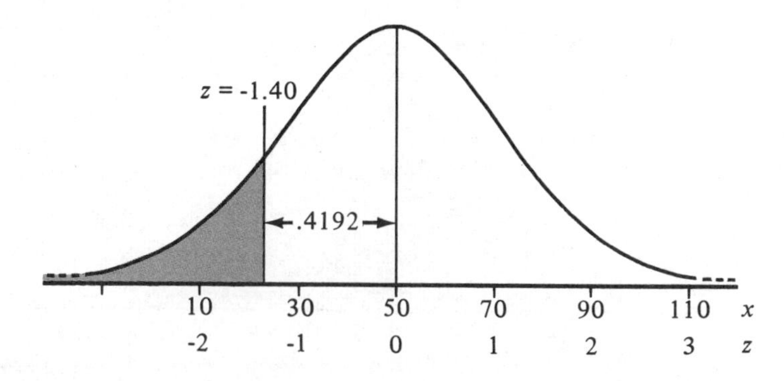

Thus $P(z \leq -1.40) = .5000 - .4192 = .0808$.

b. 8.08%

c. The 10% taking the longest time fall in the shaded region shown at the right-hand end.

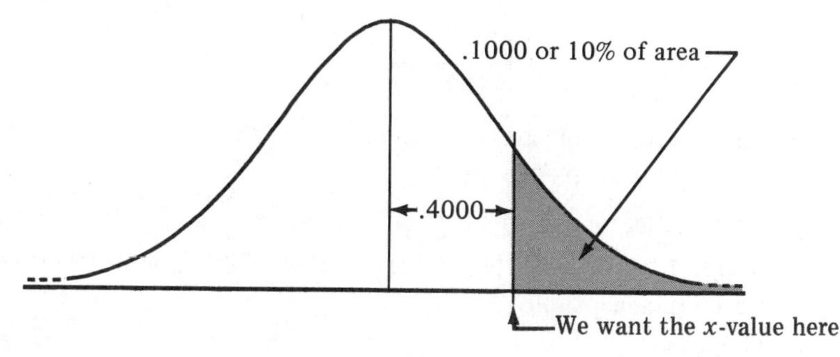

Table VI gives us a z-value of $+1.28$ at the boundary of the area. Further, since

$$z = \frac{x - \mu}{\sigma}$$

we have

$$x = \mu + z\sigma = 50 + (1.28)(20)$$
$$= 50 + 25.6 = 75.6 \text{ seconds}$$

EXERCISES ■ SECTION 8.7.

1. For a normal distribution, find the area in each of the intervals:

   a. between $z = -2.12$ and $z = 3.25$;
   b. between $z = 1.75$ and $z = 2.23$;
   c. between $z = -2.56$ and $z = -1.18$;
   d. to the right of $z = 2.02$;
   e. to the right of $z = -2.02$;
   f. to the left of $z = 1.28$;
   g. to the left of $z = -2.98$.

2. For a normal distribution, find:

   a. $P(0 < z < 2.28)$
   b. $P(-2.11 < z < 1.28)$
   c. $P(-2.11 < z < -1.78)$
   d. $P(-2 < z < +2)$
   e. $P(z > 2.28)$
   f. $P(z < 1.96)$

3. The height (x) of U.S. males is approximately normally distributed with a mean of 69 inches and a standard deviation of 3 inches.

   a. What percent of U.S. males are taller than seven feet?
   b. Knowing that the U.S. male population is about 100,000,000, find the number of U.S. males that are taller than seven feet.

4. The number of ounces (x) of a cereal, CHOCIT, dispensed into 16-ounce boxes is normally distributed with mean 16.2 ounces and standard deviation 0.2 ounce.

   a. What percent of boxes will actually have a minimum of 16 ounces of CHOCIT?
   b. What percent of boxes will be underfilled?

5. Solve part *c* of example 8.6.5.
6. The score on a placement test in mathematics administered to college freshman is normally distributed with a mean of 100 and a standard deviation of 10.

   a. What is the probability that a college freshman picked at random scored higher than 125?
   b. What is the probability that a college freshman picked at random scored between 86 and 111?
   c. What score is at the 80th percentile (P_{80})?
   d. What percent of college freshmen taking this placement test obtained a score higher than 2.5 standard deviations above the mean?

7. An urban renewal agency conducted a study to determine the number of new multiple-family dwellings being constructed per 1000 families in large urban areas. They found the number of new units (x) to be approximately normally distributed with mean 25 and standard deviation 8.

   a. What percent of large urban areas are constructing at least 20 multiple-family dwellings?
   b. Consider the 5% of large urban areas that are doing the least amount of construction. What is the most number of multiple-family dwellings being constructed for that group?

8. The Olfactory Waste Disposal Plant burns trash and garbage in a high-temperature incinerator. They are usually running behind schedule and they have found that the number of days that elapse (x) between delivery date and incineration date is normally distributed with a mean of 10 days and a standard deviation equal to 3 days.

   a. What percent of trash and garbage deliveries "sit around" for a week or more before incineration?
   b. What percent of trash and garbage deliveries get burned within one day of the delivery date?

chapter 8 in review

CHAPTER SUMMARY

In this chapter we discussed how randomly to select a sample from a population and found that data collected from such a sample could be attribute data or numerical data (continuous or discrete). We examined three measures of central tendency: the mean, median, and mode. The three measures of variation (or "spread") we studied were range, variance, and standard deviation. Two formulas for variance were employed:

$$v = \frac{\Sigma(x - \bar{x})^2}{n - 1} \quad \text{and} \quad v = \frac{n(\Sigma x^2) - (\Sigma x)^2}{n(n - 1)}$$

The standard deviation was applied to decision-making, where criteria were set up and a process could be determined out of control or under control. In section 8.5 we saw pictorial ways of displaying data, including histograms and frequency polygons. We studied one particular distribution, the normal distribution, as well as some of its properties. Two measures of location, percentiles and z-values, were examined in section 8.6. In section 8.7 we returned to a study of the normal distribution, and there we learned how to find various areas under the normal curve. This method was then applied to some real-life situations.

VOCABULARY

statistics
descriptive statistics
inference statistics
data
population
sample
random sampling
response variable
attribute data
numerical data
continuous data
discrete data
measures of central tendency
mean
median
ranked data
mode
per cent of change
measures of variation
range
variance
standard deviation
frequency table
frequency distribution
histogram
class boundaries
class mark
modal class
normal distribution
skewed distribution
measures of location
percentile
z-values

SYMBOLS

$\bar{x}$
Σ
$\tilde{x}$
v
s
$\sqrt{}$
z
P_{32}
Q_3
D_9
μ
σ

chapter 9
Models

9.1 INTRODUCTION

Mathematics has been defined as "the science of numbers and their operations, interrelations, combinations, generalizations, and abstractions and of space configurations and their structure, measurement, transformations and generalizations."* Mathematics can be very theoretical and abstract or it can be very applied and practical. In this chapter we will indicate how mathematics is used in practical ways to help create and describe **models.** By a model we simply mean a representation of something.

Consider a child's plastic model of a cruise ship. We immediately realize that such a model boat is quite different from a real one. The model is plastic and held together by glue. Such construction would hardly be appropriate for a real cruise ship that would undergo ocean travel. Further, the model typically would not have the inner intricacies (plumbing, heating, electrical system, dining room, berths, etc.) that a real ship would have. However, with all its shortcomings, people still recognize the model as a representation of a cruise ship. The model lacks many of the characteristics of the real boat, but it does do what it is supposed to—namely, look like a real boat. Think about many of the other models you have either seen or used: a model train, a model of a molecule in a chemistry laboratory, a model of our solar system, a globe, an architect's model of an urban renewal project, a monkey used as a model of a human being in a psychology laboratory. All of these models are similar in that they represent (in a simple fashion) some object of interest. A model often

* *Webster's Seventh New Collegiate Dictionary* (Springfield, Mass.: G. and C. Merriam Company, 1971).

does not have all of the characteristics of the object it represents. However, it usually is still useful in certain ways.

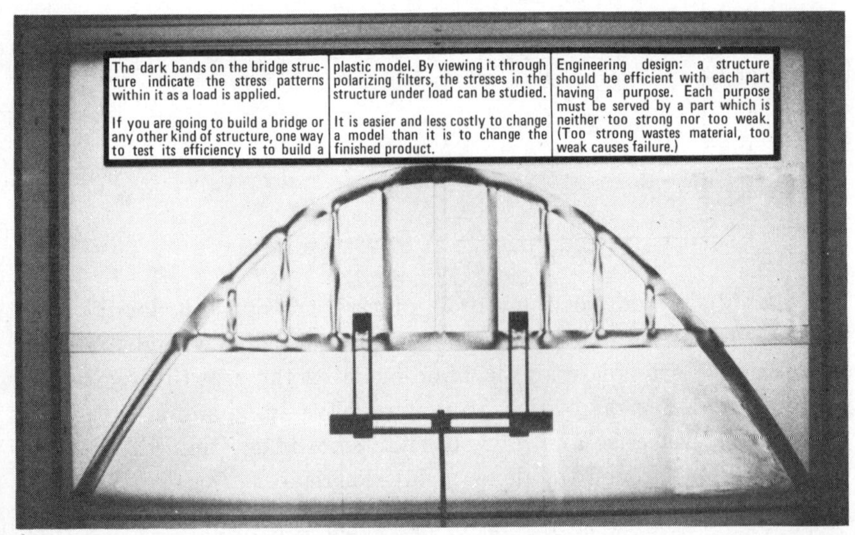

Courtesy of Ontario Science Center, Toronto

Models in mathematics are representations of objects or situations. We will see that such models are also simpler than the "real thing," but nonetheless useful.

9.2 EXAMPLES OF MODELING WITH MATHEMATICS

Growth Process Models. Many people are accustomed to thinking of growth as a *linear process.* A quantity grows linearly when it increases by a constant amount in a constant time period. Consider a child who grows two inches each year.

Age	*Height*
5	40″
6	42″
7	44″
8	46″
9	48″
10	50″
11	52″

This is an example of a **linear growth process** because there is a constant increase in height (two inches) for each constant time period (one year). If we let

x = age in years $\qquad$ y = height in inches

then the information in the table above has the following graph:

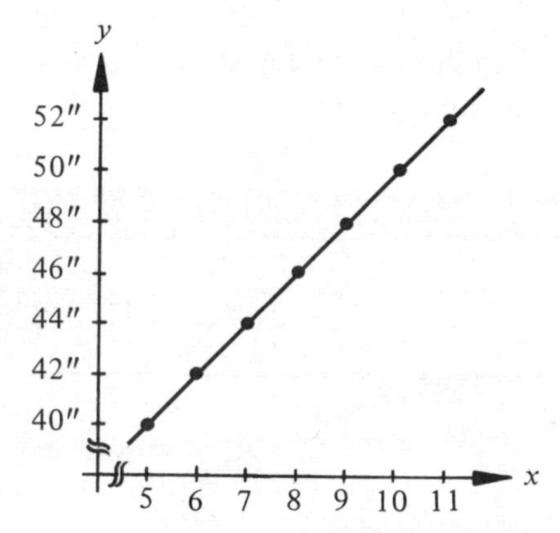

This graph is a model of the growth process. Of course, it doesn't really look like a growing child, but it does represent the age-height relationship. We can also create an algebraic model of the growth process. The graph above is a straight line, so it must have an equation of the form $y = mx + b$, where m and b are constants to be determined. We know the line passes through the point (5,40). Therefore,

$$40 = m(5) + b \qquad \text{or} \qquad 5m + b = 40 \tag{1}$$

We also know the line passes through the point (8,46). Therefore,

$$46 = m(8) + b \qquad \text{or} \qquad 8m + b = 46 \tag{2}$$

Equations (1) and (2) can now be solved simultaneously for m and b. We obtain $m = 2$ and $b = 30$, so the equation of the line is

$$y = 2x + 30$$

This last equation will give the height (y) for each age (x). Thus it is an algebraic model for the growth process of a child who grows two inches each year.

Not all growth processes are linear, as our next illustration shows.

The protozoan Glaucoma, which reproduces by binary fission, divides as frequently as every three hours. Thus in the course of a day it could become a "six greats grand-parent" and the progenitor of 510 descendents.*

* *Guinness Book of World Records*, Norris and Ross McWhirter (New York: Bantam Books, Inc., 1972), p. 84.

This reproduction process gives the table below.

Time elapsed (hours)	*Number of Glaucoma*
0	2
3	4
6	8
9	16
12	32
15	64
18	128
21	256
24	512

In graphical form, with x equal to the time elapsed and y equal to the number of Glaucoma, we obtain the graph below:

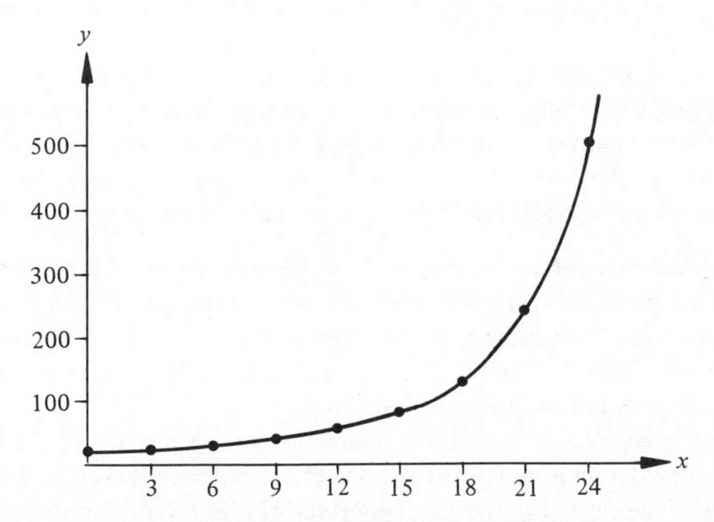

This process is referred to as **exponential growth.** A quantity grows exponentially when it increases by a constant percentage of the whole during a constant time period. In the illustration above, the constant time period is three hours and the constant percentage is a doubling of the previous period's number of Glaucoma. The graph above is a model of the exponential growth of the number of Glaucoma. An algebraic model would be: $y = 2^{x/3+1}$.

In recent years, social scientists have become extremely interested in the exponential growth model because it represents the growth of the number of people in any population. If we consider the world population and its present growth rate, we find that the population on our planet doubles about every 33 years.* It appears graphically on the following page.

* This is based on 1970 figures, when the world population was 3.6 billion.

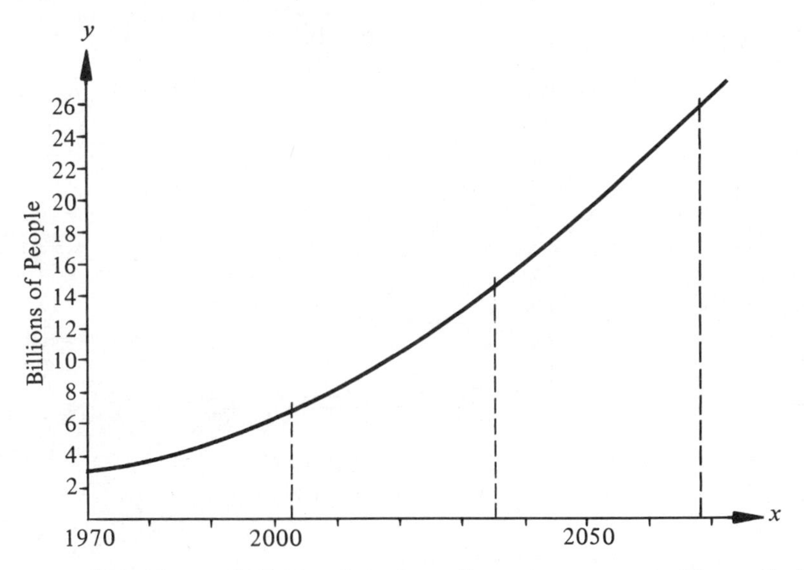

Since our planet has a finite land surface, there are many problems that accompany our population growth—for example, economic growth, food production, energy consumption, carbon dioxide concentration in the atmosphere, nuclear wastes, . . . the list grows daily. In order to better understand and solve the problems, other models have been created.*

Probability Models. In chapter 4 we defined and examined the concept of a binomial experiment. Such an experiment is usable as a model for many real-life situations. (See examples 4.9.3, 4.9.4, 4.9.5, and 4.9.6.) We will now consider another typical application of the binomial probability model in **lot acceptance sampling.**

Suppose an electrical-products manufacturing firm makes light bulbs and knows that one out of every ten of its bulbs is defective (on the average). The electrical firm has a contract with a distributor of electrical products that will buy a particular lot of twelve bulbs, provided there are no more than two defective ones. What is the probability that the distributor will buy this particular shipment?

The distributor will tolerate two or fewer defective bulbs. Here, "success" means a defective bulb and $n = 12$, $p = \frac{1}{10}$, $q = \frac{9}{10}$.

$$\begin{aligned} P(\text{buy}) &= P(0) + P(1) + P(2) \\ &= \binom{12}{0}(.1)^0(.9)^{12} + \binom{12}{1}(.1)^1(.9)^{11} + \binom{12}{2}(.1)^2(.9)^{10} \\ &= .282 + .377 + .230 \text{ (from Table III)} \\ &= .889 \end{aligned}$$

* *World Dynamics*, by Jay W. Forrester (Cambridge: Wright-Allen Press, 1971), describes a model of the world that is being used to learn more about five areas of concern: accelerating industrialization, rapid population growth, widespread malnutrition, depletion of nonrenewable resources, and a deteriorating environment.

Our binomial probability model solves this problem, but in actual situations of lot acceptance, the order is often for a much larger number of bulbs than twelve. Suppose that the distributor is interested in buying a very large quantity of bulbs, provided a sample of size 100 has twelve or fewer defectives. Now the computation becomes quite tedious:

$$P(\text{buy}) = P(0) + P(1) + P(2) + \cdots + P(12)$$
$$= \binom{100}{0}(.1)^0(.9)^{100} + \binom{100}{1}(.1)^1(.9)^{99} + \cdots + \binom{100}{12}(.1)^{12}(.9)^{88}$$

At this stage we have the choice of performing the computation above or using a different model. We choose the latter.

One can approximate a binomial probability model with a **normal probability model** (i.e., a normal distribution). It is necessary to know only that the discrete variable x in a binomial experiment is approximately normally distributed with mean and standard deviation given by:

$$\mu = np$$
$$\sigma = \sqrt{npq}^*$$

Continuing, we now solve the above problem using the normal probability model.

$$\mu = np = 100(\tfrac{1}{10}) = 10$$
$$\sigma = \sqrt{npq} = \sqrt{(100)(\tfrac{1}{10})(\tfrac{9}{10})} = 3$$

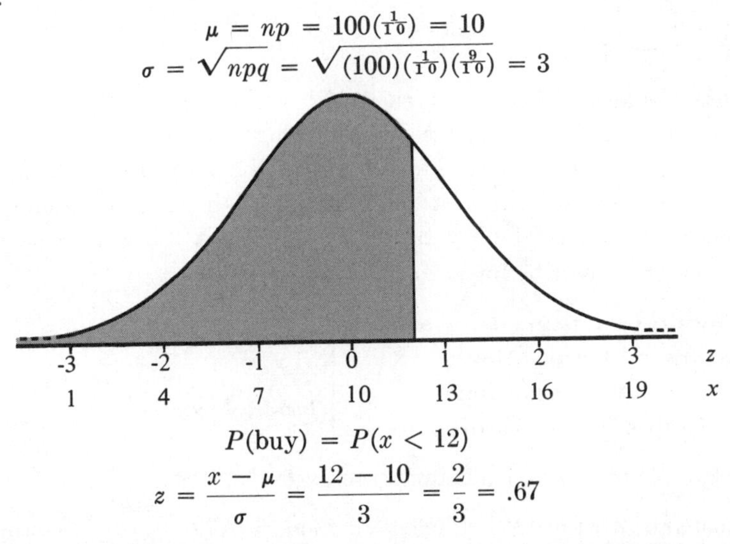

$$P(\text{buy}) = P(x < 12)$$
$$z = \frac{x - \mu}{\sigma} = \frac{12 - 10}{3} = \frac{2}{3} = .67$$

So, $P(\text{buy}) = P(z < .67) = .5000 + .2486 = .7486$. Thus, 74.86% of the time a lot of 100 bulbs will be accepted. Equivalently, 25.14% of the time they will be rejected (assuming the quality of production remains the same).

* The approximation referred to is reasonably decent provided $\mu \pm 2\sigma$ lies within the bounds 0 to n.

The reader should note the ease of computation involved in this example using the normal probability model. This substitute model is useful when n is large and we would otherwise have to sum many binomial probabilities.

A Personality Model. One of the most innovative approaches to modern psychotherapy is known as **transactional analysis**, which was originated by Dr. Eric Berne (1910–1970). The main interest in transactional analysis is the study of **ego states.** Each person has three such states, called *Parent*(P), *Adult*(A), and *Child*(C). These states constitute a model of a personality. The Parent refers to the patterns of behavior derived from parental figures. In this state a person behaves just as one (or both) of his parents did when he was little. The Adult refers to the state in which a person considers and appraises his environment in an objective fashion based on past experience. The Child ego state in a person refers to behavior patterns that were typical of the person when he or she was a young child (birth to five years). A diagram of a personality looks like this:

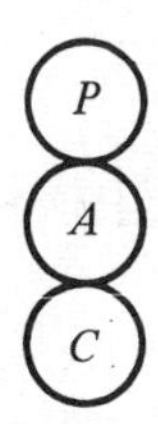

Transactional analysis is concerned with verbal transactions (a stimulus by one person and a response by the other) that take place when two people confront each other. The purpose of the analysis is to learn which part of each person (Parent, Adult, Child) is originating each stimulus and response. There will be a total of six ego states, three in each person involved. We describe four types of transactions:

1. **complementary transactions** } *one-level transactions*
2. **crossed transactions** }
3. **angular transactions** } *two-level transactions*
4. **duplex transactions** }

An example of a complementary transaction is:

Husband to wife: "What smells so good, dear?" (stimulus)

Wife to husband: "Homemade whole wheat bread—it's almost ready to come out of the oven!" (response)

The stimulus here is from the husband's Adult to the wife's Adult. The response is from the wife's Adult to the husband's Adult. We have the diagram on the facing page.

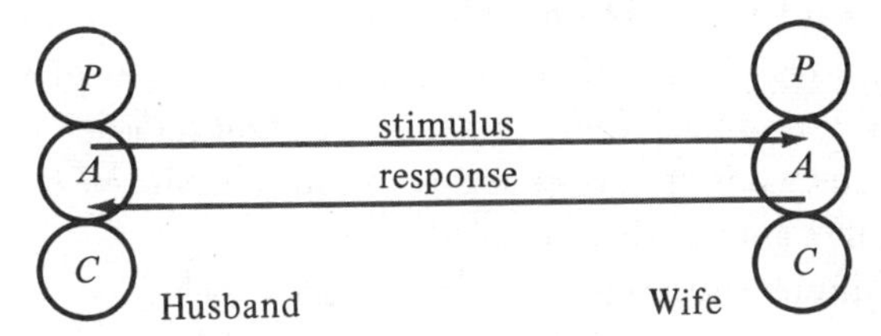

As long as transactions are complementary (as indicated by the response arrow retracing in opposite direction the same path that the stimulus arrow took) communication can proceed indefinitely. A direct application of the $r \cdot s$ rule (see sections 2.7, 3.2) determines that there are nine possible complementary transactions.

An example of a crossed transaction is:

Husband to wife: "Do you know where my belt is?" (stimulus)

Wife to husband: "Why do you always have to blame me for everything?" (response)

The stimulus here (a request for information) is from the husband's Adult to the wife's Adult. However, the response is from the wife's Child to the husband's Parent. A diagram is shown below.

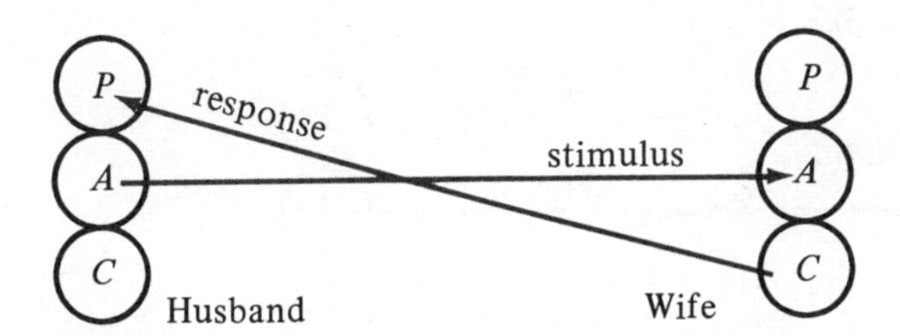

When transactions become crossed like this one, communication is inhibited or breaks down. There are 72 types of crossed transactions possible.

An example of an angular transaction is partially diagrammed below.

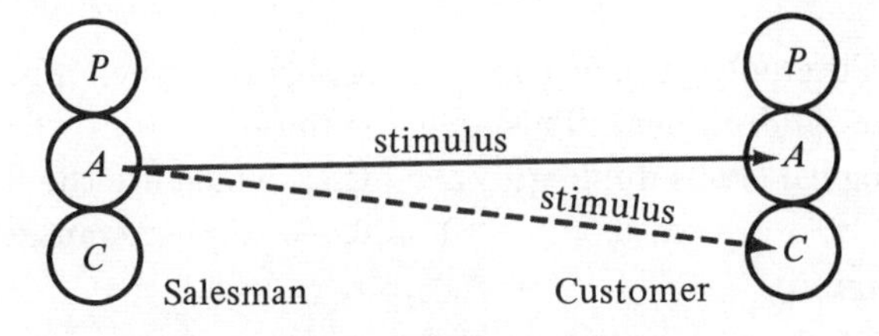

Imagine a salesman who makes a rational-sounding sales pitch (presumably Adult-to-Adult) that is really designed to appeal to another ego state (say the Child). In the diagram this stimulus is represented by a solid line (AA) and a dotted line (AC). The dotted line and the solid line both always come from the same ego state. For a given ego state, there are three possibilities for the solid line, and for each solid line, there are two possible dotted lines. Thus there are $3 \cdot 3 \cdot 2 = 18$ types of angular transactions where the response is to the dotted line and another 18 types of angular transactions where the response is to the solid line. Thus there are 36 types of angular transactions possible.

An example of a duplex transaction is diagrammed below.

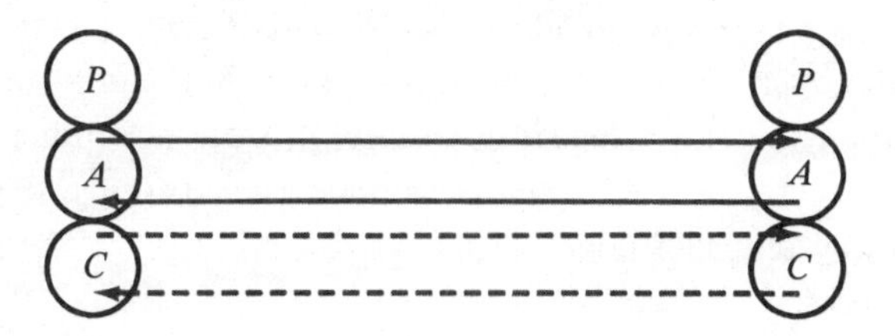

In this type of transaction there are two levels. There is an underlying psychological level (dotted lines) and a social level (solid lines). A specific illustration follows:

Husband to wife: "Where did you hide the aspirin?" (stimulus)

"Where is the aspirin?" This question merely seeks information. It is *AA*.

The word hide carries with it an element of criticism. It is *PC*.

The stimulus looks like this:

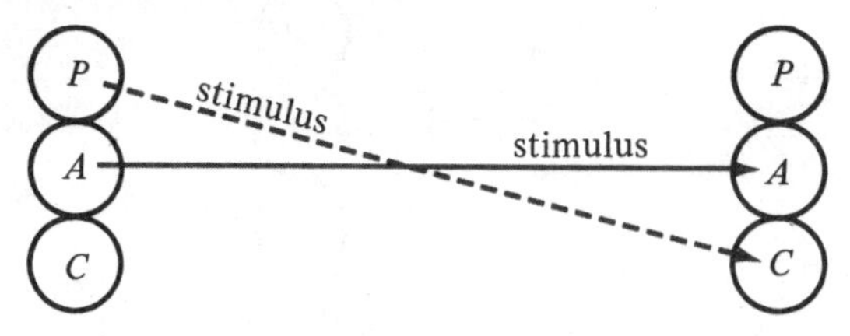

The wife's response can be parallel to the solid line: "I hid them next to the vitamin C tablets." } This is complementary.

Or, the wife's response can be parallel to the dotted line: "What are you—blind?" } This is crossed.

There are 6,561 different types of transactions possible (including complementary, crossed, and duplex). This includes those types in which the social and psychological levels duplicate each other, which are the 81 one-level transactions. So we have $6{,}561 - 81 = 6{,}480$ duplex transactions. In total we have found:

9	complementary transactions	one-level
72	crossed transactions	one-level
36	angular transactions	two-level
6,480	duplex transactions	two-level
6,597	possible types of transactions	

The purpose of this discussion is to point out that there exists a rigorous personality model with a known number of transactions possible. This model (as defined by Parent, Adult, and Child) allows for all possible forms of social behavior. It is also quite effective because, so far, no example has been uncovered from among millions of possible interchanges that can not be dealt with by this model.*

Voting Models. There are many types of elections for people who hold office. Here we will look at various voting models in use and point out some differences among them.

Winner by plurality. Consider a common model for voting for various candidates, where there is to be a single winner. The model is described by the statement: *each elector votes for the candidate he prefers and the candidate with the greatest number of votes wins.* We analyze this statement logically. For the sake of closer examination, let us imagine four candidates a, b, c, d and thirteen electors. Presume each elector has ranked the candidates in order of preference as shown:

$$\begin{matrix} a & a & a & b & b & b & b & c & c & c & d & d & d \\ d & d & d & a & a & a & a & a & a & a & a & a & a \\ c & c & c & c & c & c & c & d & d & d & c & c & c \\ b & b & b & d & d & d & d & b & b & b & b & b & b \end{matrix}$$

Here, candidate a is considered best by three electors and second by all the rest. Yet, by the above method, candidate b would win even though nine of the thirteen electors consider him to be the worst candidate.

Winner by simple majority. This model is a popular one for the election of a candidate. The model is described by the statement: *each elector votes for the one candidate preferred and the candidate with a simple majority (at least 50% of the votes) wins.* Consider four candidates a, b, c, d and thirteen electors who have ranked the candidates in order of preference as shown:

$$\begin{matrix} a & a & a & a & a & a & b & b & b & b & b & b & b \\ c & d & c & d & c & d & a & a & a & a & a & a & a \\ d & c & d & c & d & c & c & d & c & d & c & d & c \\ b & b & b & b & b & b & d & c & d & c & d & c & d \end{matrix}$$

Here, candidate a is seen to be the best by almost half of the voters and never less than second best. However, candidate b will win by this method, even though nearly half of the voters consider him the worst.

* Much of the material for this model can be found in two works: *I'm OK—You're OK*, by Thomas A. Harris (New York: Avon Books, 1969). *What Do You Say After You Say Hello*, by Eric Berne (New York: Bantam Books, 1973).

Winner by elimination. With this model, the weakest candidates are eliminated first—the philosophy being that the strongest candidate will be left as winner. The model is described by the statement: *each elector names the one candidate he prefers and the one with the fewest votes is dropped from consideration. The process is repeated until a single winner is left.* Consider the possible ranking of four candidates by thirteen electors shown below:

$$\begin{array}{ccccccccccccc}
a & a & b & b & b & b & c & c & c & c & d & d & d \\
b & c & a & a & a & a & a & a & a & a & a & a & a \\
d & d & d & d & c & d & b & b & b & d & c & c & b \\
c & b & c & c & d & c & d & d & d & b & b & b & c
\end{array}$$

Here, candidate a is preferred by two electors and is never placed lower than second by all the others. However, with the voting model described above, candidate a is immediately eliminated. The preferences left are:

$$\begin{array}{ccccccccccccc}
b & c & b & b & b & b & c & c & c & c & d & d & d \\
d & d & d & d & c & d & b & b & b & d & c & c & b \\
c & b & c & c & d & c & d & d & d & b & b & b & c
\end{array}$$

Next, candidate d is eliminated with only three votes. We have:

$$\begin{array}{ccccccccccccc}
b & c & b & b & b & b & c & c & c & c & c & c & b \\
c & b & c & c & c & c & b & b & b & b & b & b & c
\end{array}$$

Candidate b is now eliminated, leaving c as the winner (even though, initially, c was placed last by five electors and next to last by three electors).

Each of the three models for a voting procedure described above seems reasonable, but when subjected to a hypothetical test (as models frequently are), each exhibits weaknesses. The final voting model we will examine gives various points to candidates as they appear in a ranked list.

Winner by ranked order. The model is described by the following three statements. *Presume there are n candidates and each elector ranks them in his order of preference. The first choice is then assigned n − 1 points, the second, n − 2 points, etc. After all electors have submitted their ranked lists, points are totaled and the candidate with the most number of points wins.* We now test this model on each of the previous three voting examples to determine the ranked-order winner.

													Points for each candidate			
													a	*b*	*c*	*d*
a	*a*	*a*	*b*	*b*	*b*	*b*	*c*	*c*	*c*	*d*	*d*	*d*	9	12	9	9
d	*d*	*d*	*a*	*a*	*a*	*a*	*a*	*a*	*a*	*a*	*a*	*a*	20	0	0	6
c	*c*	*c*	*c*	*c*	*c*	*c*	*d*	*d*	*d*	*c*	*c*	*c*	0	0	10	3
b	*b*	*b*	*d*	*d*	*d*	*d*	*b*	*b*	*b*	*b*	*b*	*b*	0	0	0	0
													29	12	19	18

Here, candidate a is the winner instead of b (who wins with the plurality model).

													a	b	c	d
a	a	a	a	a	a	b	b	b	b	b	b	b	18	21	0	0
c	d	c	d	c	d	a	a	a	a	a	a	a	14	0	6	6
d	c	d	c	d	c	c	d	c	d	c	d	c	0	0	7	6
b	b	b	b	b	b	d	c	d	c	d	c	d	0	0	0	0
													32	21	13	12

Here, candidate a is the winner instead of b (who wins with the simple majority model).

													a	b	c	d
a	a	b	b	b	b	c	c	c	c	d	d	d	6	12	12	9
b	c	a	a	a	a	a	a	a	a	a	a	a	22	2	2	0
d	d	d	d	c	d	b	b	b	d	c	c	b	0	4	3	6
c	b	c	c	d	c	d	d	d	b	b	b	c	0	0	0	0
													28	18	17	15

Here, candidate a is the winner instead of c (who wins with the elimination model).

Thus we have seen that different models can lead to different winners. Whether one model is better than another can be answered only by carefully considering what factors we deem important for creating a winner in an election.

EXERCISES ■ SECTION 9.2.

1. Consider the numerical information below.

x	y
2	3
4	6
6	9
8	12
10	15
12	18
14	21

   a. Create a graphical model for this information.
   b. Is this a linear growth process?
   c. Create an algebraic model for this information.

2. Consider the data on the following page, which was collected by measuring corresponding centigrade (x) and Fahrenheit (y) temperatures.

x	y
-25	-13
0	32
20	68
100	212

a. Create a graphical model for this data.
b. Create an algebraic model for this data.

3. You are offered a job whereby the salary is 2¢ the first day, 4¢ the second day, 8¢ the third day, 16¢ the fourth day, etc.

a. Create a graphical model for this situation.
b. Create an algebraic model for this situation. Let x = number of days, y = salary for the xth day.
c. How much will you make on the tenth day?
d. How much will you have made after ten days of work?
e. How much will you make on the 29th day?
f. How much will you have made after 29 days of work? (Hint: see Appendix I, exercise 4.)

4. Show that when the protozoan Glaucoma divides every three hours, it becomes the progenitor of 510 descendents during a 24-hour period.

5. Consider the data below.

x	y
1	3
3	9
5	27
7	81
9	243

a. Create a graphical model of the data.
b. Is this an exponential or linear growth process?
c. Create an algebraic model of the data.

6. A neighborhood's population consists of 30% black people and 70% white people. A sociologist is in the process of selecting a random sample of people from this neighborhood. The first ten people selected are all white. What is the probability of this happening by chance?

7. A binomial experiment has $n = 500$, $p = \frac{1}{2}$. Find $P(x < 20)$ by using the normal probability model.

8. A salesman has found that the probability of a sale is .2 (on the average). If the salesman approaches 50 customers, what is the probability of making at least 15 sales? (Use the normal probability approximation for this binomial experiment.)

9. Verify that the number of complementary transactions for Eric Berne's personality model is nine.

10. Verify that the number of crossed transactions for Eric Berne's personality model is 72.

11. Verify that the number of transactions (including complementary, crossed, and duplex) for Berne's personality model is 6,561.

12. Nine electors will elect one winner from candidates a, b, c. They rank the candidates as shown:

$$\begin{array}{ccccccccc} a & a & a & a & a & b & b & b & b \\ b & b & b & b & b & c & c & c & c \\ c & c & c & c & c & a & a & a & a \end{array}$$

a. Determine the winner by plurality.
b. Determine the winner by simple majority.
c. Determine the winner by elimination.
d. Determine the winner by ranked order.

13. Eleven electors will elect one winner from candidates a, b, c, d, e. They rank the candidates as shown:

$$\begin{array}{ccccccccccc} b & b & b & b & b & c & c & d & e & e & e \\ a & a & c & c & c & b & a & a & d & c & c \\ d & d & a & a & a & d & b & b & b & b & b \\ c & c & d & d & d & a & d & c & a & a & d \\ e & e & e & e & e & e & e & e & c & d & a \end{array}$$

a. Determine the winner by plurality.
b. Determine the winner by simple majority.
c. Determine the winner by elimination.
d. Determine the winner by ranked order.

chapter 9 in review

CHAPTER SUMMARY

This chapter presented descriptions of some models. A model is simply a representation of something. Models allow us to view complex things in simpler ways. In this chapter we dealt with four very different models. We considered the growth process models and noted that both algebraic and graphical models were possible. We illustrated both linear and exponential growth processes. The binomial and normal probability models were dealt with next and we saw how the normal probability model can be used as an approximation to the binomial probability model. Dr. Eric Berne's personality model illustrated a rigorous, well-defined model, a milestone for investigation of human behavior. Lastly, the voting models were a verbal description of the situation they represented.

VOCABULARY

model
linear growth process
exponential growth process
lot acceptance sampling
ego states
transactions (four types)
winner by plurality
winner by simple majority
winner by elimination
winner by ranked order

SYMBOLS

$\mu = np$

$\sigma = \sqrt{npq}$

appendix I
Mathematical Induction

A principle used commonly as a method of proof is **the principle of mathematical induction.** It is used to prove a statement involving n, where n is an integer allowed to vary. For example, the statement

$$4 + 8 + 12 + \cdots + 4n = 2n(n + 1) \qquad (n \geq 1)$$

is true for all $n \geq 1$ and we wish to prove it. To do so we use the principle of mathematical induction:

STEP 1: Show the statement is true for $n = 1$;

STEP 2: Show that if the statement is true for $n = k$, then it is true for $n = k + 1$.

(Notice that the principle of mathematical induction enables us to examine all $n \geq 1$. That is, the statement is true for $n = 1$ by step 1, and by step 2 it will be true for $n = 2$; again by step 2 it is true for $n = 3$, and so on.)

We examine: $4 + 8 + 12 + \cdots + 4n = 2n(n + 1)$

STEP 1: For $n = 1$ we have

$$4 = 2 \cdot 1(1 + 1) \qquad \text{or} \qquad 4 = 4$$

The statement is true for $n = 1$.

STEP 2: Now we assume the statement is true for $n = k$ and try to show it is true for $n = k + 1$.

Assume: $\qquad 4 + 8 + 12 + \cdots + 4k = 2k(k + 1) \qquad (*)$

We want to show:

$$4 + 8 + 12 + \cdots + 4k + 4(k+1) = 2(k+1)(k+2)$$

If we add $4(k+1)$ to both sides of (*) we have:

$$\begin{aligned} 4 + 8 + 12 + \cdots + 4k + 4(k+1) &= 2k(k+1) + 4(k+1) \\ &= 2k^2 + 2k + 4k + 4 \\ &= 2k^2 + 6k + 4 \\ &= 2(k^2 + 3k + 2) \\ &= 2(k+1)(k+2) \end{aligned}$$

which was to be shown.

Example I.1. Show by the principle of mathematical induction that

$$1 + 3 + 5 + \cdots + (2n - 1) = n^2$$

Solution:

STEP 1: When $n = 1$ we have $(2 \cdot 1) - 1 = 1^2$, which is true.
STEP 2: Assume $1 + 3 + 5 + \cdots + (2k - 1) = k^2$ and show

$$1 + 3 + 5 + \cdots + (2k-1) + [2(k+1) - 1] = (k+1)^2$$

We have

$$1 + 3 + 5 + \cdots + (2k - 1) = k^2$$

Now add $[2(k+1) - 1] = 2k + 1$ to the equation:

$$\begin{aligned} 1 + 3 + 5 + \cdots + (2k-1) + (2k+1) &= k^2 + (2k+1) \\ &= (k+1)^2 \end{aligned}$$

which was to be shown.

Example I.2. **a.** Show by the principle of mathematical induction that

$$1 + 2 + 3 + \cdots + n = \frac{n(n+1)}{2}$$

b. Find the sum of the integers from 1 to 100.

Solution:

a. STEP 1: For $n = 1$ we have $1 = \frac{1(2)}{2}$, which is true.

STEP 2: Assume $1 + 2 + 3 + \cdots + k = \frac{k(k+1)}{2}$ and show

$$1 + 2 + 3 + \cdots + k + (k+1) = \frac{(k+1)(k+2)}{2}$$

We have $1 + 2 + 3 + \cdots + k = \frac{k(k+1)}{2}$. Adding $k + 1$ to both sides:

$$\begin{aligned} 1 + 2 + 3 + \cdots + k + (k+1) &= \frac{k(k+1)}{2} + (k+1) \\ &= \frac{k(k+1)}{2} + \frac{2k+2}{2} \\ &= \frac{k^2 + k + 2k + 2}{2} \\ &= \frac{k^2 + 3k + 2}{2} \\ &= \frac{(k+1)(k+2)}{2} \end{aligned}$$

which was to be shown.

b. From part *a*, $1 + 2 + 3 + \cdots + 100 = \frac{100(101)}{2} = 5050$.

We shall employ the principle of mathematical induction to prove the *binomial theorem:*

For any positive integer n:

$$\begin{aligned} (x+y)^n = \binom{n}{0} x^n + \binom{n}{1} x^{n-1}y + \cdots + \binom{n}{r-1} x^{n-r+1}y^{r-1} \\ + \binom{n}{r} x^{n-r}y^r + \cdots + \binom{n}{n-1} xy^{n-1} + \binom{n}{n} y^n \end{aligned}$$

Proof: STEP 1: For $n = 1$ (and, in fact, for $n = 2, 3, 4, 5$), the verification consists of algebraic manipulations, which have been performed in section 3.5.

STEP 2: Assume it true for $n = k$:

$$\begin{aligned} (x+y)^k = \binom{k}{0} x^k + \binom{k}{1} x^{k-1}y + \cdots + \binom{k}{r-1} x^{k-r+1}y^{r-1} \\ + \binom{k}{r} x^{k-r}y^r + \cdots + \binom{k}{k-1} xy^{k-1} + \binom{k}{k} y^k \end{aligned}$$

We must show it is true for $n = k + 1$, so we multiply both sides by $(x + y)$. We obtain:

$$\begin{aligned} (x+y)^{k+1} = x\Bigg[\binom{k}{0} x^k + \binom{k}{1} x^{k-1}y + \cdots + \binom{k}{r-1} x^{k-r+1}y^{r-1} \\ + \binom{k}{r} x^{k-r}y^r + \cdots + \binom{k}{k-1} xy^{k-1} + \binom{k}{k} y^k\Bigg] \\ + y\Bigg[\binom{k}{0} x^k + \binom{k}{1} x^{k-1}y + \cdots + \binom{k}{r-1} x^{k-r+1}y^{r-1} \\ + \binom{k}{r} x^{k-r}y^r + \cdots + \binom{k}{k-1} xy^{k-1} + \binom{k}{k} y^k\Bigg] \end{aligned}$$

Multiplying on the right-hand side and collecting terms, we have:

$$\begin{aligned}(x + y)^{k+1} = \binom{k}{0} x^{k+1} &+ \left[\binom{k}{0} + \binom{k}{1}\right] x^k y + \cdots \\ &+ \left[\binom{k}{r-1} + \binom{k}{r}\right] x^{k-r+1} y^r + \cdots \\ &+ \left[\binom{k}{k-1} + \binom{k}{k}\right] x y^k + \binom{k}{k} y^{k+1}\end{aligned}$$

Now, since $\binom{k}{r-1} + \binom{k}{r} = \binom{k+1}{r}$ [see exercise 12, section 3.5] and $\binom{k}{0} = \binom{k+1}{0} = \binom{k}{k} = \binom{k+1}{k+1}$, we have:

$$\begin{aligned}(x + y)^{k+1} = \binom{k+1}{0} x^{k+1} &+ \binom{k+1}{1} x^k y + \cdots \\ &+ \binom{k+1}{r} x^{k+1-r} y^r + \cdots \\ &+ \binom{k+1}{k} x y^k + \binom{k+1}{k+1} y^{k+1}\end{aligned}$$

which was to be shown.

EXERCISES ■ APPENDIX I.

Prove each of the following by the principle of mathematical induction.

1. $2 + 4 + 6 + \cdots + 2n = n(n + 1)$
2. $20 + 29 + 38 + \cdots + [20 + 9(n - 1)] = \dfrac{n[40 + 9(n - 1)]}{2}$
3. $1^2 + 2^2 + 3^2 + \cdots + n^2 = \dfrac{n(n + 1)(2n + 1)}{6}$
4. $2 + 4 + 8 + 16 + \cdots + 2^n = 2^{n+1} - 2$
5. $6 + 10 + 14 + \cdots + (4n + 2) = n(2n + 4)$
6. $1^3 + 2^3 + 3^3 + \cdots + n^3 = \left[\dfrac{n(n + 1)}{2}\right]^2$
7. *Both* step 1 *and* step 2 of the principle of mathematical induction must be verified in the proof.
   - **a.** Disprove: $2 + 4 + 6 + \cdots + 2n = (4n - 2)$ by showing one of the two steps fails.
   - **b.** Disprove: $2 + 4 + 6 + \cdots + 2n = n(n + 1) + 1$ by showing one of the two steps fails.

appendix II
Summation Notation

There are many occasions in various areas of mathematics when we want to add a set of numbers. If $x_1 = 2$, $x_2 = 8$, $x_3 = 1$, and $x_4 = 3$, then $x_1 + x_2 + x_3 + x_4 = 2 + 8 + 1 + 3 = 14$. It is convenient to use a notation that symbolizes any summation process. The Greek capital letter sigma, Σ, which is read "summation of," will be used to accomplish this. In the above illustration, $x_1 + x_2 + x_3 + x_4$ can be written $\sum_{i=1}^{4} x_i$ and read "summation of the x_i's from $i = 1$ to $i = 4$." So, $\sum_{i=1}^{4} x_i = 14$. (The "i" is called the index of summation; sometimes j or k is used.) If there are five values of x, we write $\sum_{i=1}^{5} x_i = x_1 + x_2 + x_3 + x_4 + x_5$.

Example II.1. Write out each of the following summations.

a. $\sum_{i=1}^{3} x_i$ **b.** $\sum_{i=1}^{4} x_i y_i$ **c.** $\sum_{i=1}^{3} x_i^2$ **d.** $\sum_{i=1}^{n} x_i f_i$

Solution:

a. $\sum_{i=1}^{3} x_i = x_1 + x_2 + x_3$ **b.** $\sum_{i=1}^{4} x_i y_i = x_1 y_1 + x_2 y_2 + x_3 y_3 + x_4 y_4$

c. $\sum_{i=1}^{3} x_i^2 = x_1^2 + x_2^2 + x_3^2$

d. $\sum_{i=1}^{n} x_i f_i = x_1 f_1 + x_2 f_2 + x_3 f_3 + \cdots + x_n f_n$

Example II.2. Symbolize each of the summations below using summation notation.

a. $x_1 + x_2 + x_3 + x_4 + x_5 + x_6 + x_7$
b. $x_1f_1 + x_2f_2 + x_3f_3$
c. $x_1^2 + x_2^2 + x_3^3 + x_4^2 + x_5^2$
d. $(x_1 + y_1) + (x_2 + y_2) + (x_3 + y_3) + (x_4 + y_4)$

Solution:

a. $\sum_{i=1}^{7} x_i$ **b.** $\sum_{i=1}^{3} x_if_i$ **c.** $\sum_{j=1}^{5} x_j^2$ **d.** $\sum_{i=1}^{4} (x_i + y_i)$

Example II.3. Let $x_1 = 2$, $x_2 = 4$, $x_3 = 3$. Compute:

a. $\sum_{i=1}^{3} x_i$ **b.** $\sum_{i=1}^{3} x_i^2$ **c.** $\left(\sum_{i=1}^{3} x_i\right)^2$

Solution:

a. $\sum_{i=1}^{3} x_i = 2 + 4 + 3 = 9$

b. $\sum_{i=1}^{3} x_i^2$ means "sum the squares of the x_i's." Therefore

$$\sum_{i=1}^{3} x_i^2 = 2^2 + 4^2 + 3^2 = 4 + 16 + 9 = 29$$

c. We want $\left(\sum_{i=1}^{3} x_i\right)^2$. As is always the case with parentheses, we must work inside the parentheses first, then square after. So,

$$\left(\sum_{i=1}^{3} x_i\right)^2 = (2 + 4 + 3)^2 = 9^2 = 81$$

The reader should compare the answers to parts *b* and *c* above and note that

$$\sum_{i=1}^{3} x_i^2 \neq \left(\sum_{i=1}^{3} x_i\right)^2$$

EXERCISES ■ APPENDIX II.

1. Write out each of the following summations.

a. $\sum_{i=1}^{5} x_i$ **b.** $\sum_{i=1}^{3} x_iy_i$ **c.** $\sum_{i=1}^{4} x_i^2f_i$ **d.** $\sum_{i=1}^{3} (x_i + y_i)$

2. Symbolize each of the summations below in summation notation.

a. $x_1 + x_2 + x_3 + x_4 + x_5 + x_6 + x_7 + x_8$
b. $x_1f_1 + x_2f_2 + x_3f_3 + x_4f_4 + x_5f_5$
c. $x_1^2 + x_2^2 + x_3^2$

3. Let $x_1 = 1$, $x_2 = 3$, $x_3 = 1$, $x_4 = 2$. Find:

a. $\sum_{i=1}^{4} x_i$ **b.** $\sum_{i=1}^{4} x_i^2$ **c.** $\left(\sum_{i=1}^{4} x_i\right)^2$

4. Let $x_1 = 2$, $x_2 = -3$, $x_3 = 1$. Find:

a. $\sum_{i=1}^{3} x_i$ **b.** $\sum_{i=1}^{3} x_i^2$ **c.** $\left(\sum_{i=1}^{3} x_i\right)^2$

5. Let $x_1 = 1$, $x_2 = 2$, $x_3 = 4$, $f_1 = 1$, $f_2 = 3$, $f_3 = 2$. Find:

a. $\sum_{i=1}^{3} x_if_i$ **b.** $\sum_{i=1}^{3} x_i^2f_i$ **c.** $\left(\sum_{i=1}^{3} x_if_i\right)^2$

appendix III
Tables

Table I: Factorials

n	$n!$
0	1
1	1
2	2
3	6
4	24
5	120
6	720
7	5,040
8	40,320
9	362,880
10	3,628,800
11	39,916,800
12	479,001,600
13	6,227,020,800
14	87,178,291,200
15	1,307,674,368,000
16	20,922,789,888,000
17	355,687,428,096,000
18	6,402,373,705,728,000
19	121,645,100,408,832,000
20	2,432,902,008,176,640,000

Table II: Binomial Coefficients

n	$\binom{n}{0}$	$\binom{n}{1}$	$\binom{n}{2}$	$\binom{n}{3}$	$\binom{n}{4}$	$\binom{n}{5}$	$\binom{n}{6}$	$\binom{n}{7}$	$\binom{n}{8}$	$\binom{n}{9}$	$\binom{n}{10}$
0	1										
1	1	1									
2	1	2	1								
3	1	3	3	1							
4	1	4	6	4	1						
5	1	5	10	10	5	1					
6	1	6	15	20	15	6	1				
7	1	7	21	35	35	21	7	1			
8	1	8	28	56	70	56	28	8	1		
9	1	9	36	84	126	126	84	36	9	1	
10	1	10	45	120	210	252	210	120	45	10	1
11	1	11	55	165	330	462	462	330	165	55	11
12	1	12	66	220	495	792	924	792	495	220	66
13	1	13	78	286	715	1287	1716	1716	1287	715	286
14	1	14	91	364	1001	2002	3003	3432	3003	2002	1001
15	1	15	105	455	1365	3003	5005	6435	6435	5005	3003
16	1	16	120	560	1820	4368	8008	11440	12870	11440	8008
17	1	17	136	680	2380	6188	12376	19448	24310	24310	19448
18	1	18	153	816	3060	8568	18564	31824	43758	48620	43758
19	1	19	171	969	3876	11628	27132	50388	75582	92378	92378
20	1	20	190	1140	4845	15504	38760	77520	125970	167960	184756

If necessary, use the identity $\binom{n}{k} = \binom{n}{n-k}$.

Reprinted from John E. Freund, *Statistics, A First Course*, Prentice-Hall, Inc., Englewood Cliffs, N.J., 1970, p. 313.

Table III: Binomial Probabilities

$$\left[\binom{n}{x} \cdot p^x q^{n-x}\right]$$

n	x	.01	.05	.10	.20	.30	.40	.50	.60	.70	.80	.90	.95	.99	x
2	0	980	902	810	640	490	360	250	160	090	040	010	002	0+	0
	1	020	095	180	320	420	480	500	480	420	320	180	095	020	1
	2	0+	002	010	040	090	160	250	360	490	640	810	902	980	2
3	0	970	857	729	512	343	216	125	064	027	008	001	0+	0+	0
	1	029	135	243	384	441	432	375	288	189	096	027	007	0+	1
	2	0+	007	027	096	189	288	375	432	441	384	243	135	029	2
	3	0+	0+	001	008	027	064	125	216	343	512	729	857	970	3
4	0	961	815	656	410	240	130	062	026	008	002	0+	0+	0+	0
	1	039	171	292	410	412	346	250	154	076	026	004	0+	0+	1
	2	001	014	049	154	265	346	375	346	265	154	049	014	001	2
	3	0+	0+	004	026	076	154	250	346	412	410	292	171	039	3
	4	0+	0+	0+	002	008	026	062	130	240	410	656	815	961	4
5	0	951	774	590	328	168	078	031	010	002	0+	0+	0+	0+	0
	1	048	204	328	410	360	259	156	077	028	006	0+	0+	0+	1
	2	001	021	073	205	309	346	312	230	132	051	008	001	0+	2
	3	0+	001	008	051	132	230	312	346	309	205	073	021	001	3
	4	0+	0+	0+	006	028	077	156	259	360	410	328	204	048	4
	5	0+	0+	0+	0+	002	010	031	078	168	328	590	774	951	5
6	0	941	735	531	262	118	047	016	004	001	0+	0+	0+	0+	0
	1	057	232	354	393	303	187	094	037	010	002	0+	0+	0+	1
	2	001	031	098	246	324	311	234	138	060	015	001	0+	0+	2
	3	0+	002	015	082	185	276	312	276	185	082	015	002	0+	3
	4	0+	0+	001	015	060	138	234	311	324	246	098	031	001	4
	5	0+	0+	0+	002	010	037	094	187	303	393	354	232	057	5
	6	0+	0+	0+	0+	001	004	016	047	118	262	531	735	941	6
7	0	932	698	478	210	082	028	008	002	0+	0+	0+	0+	0+	0
	1	066	257	372	367	247	131	055	017	004	0+	0+	0+	0+	1
	2	002	041	124	275	318	261	164	077	025	004	0+	0+	0+	2
	3	0+	004	023	115	227	290	273	194	097	029	003	0+	0+	3
	4	0+	0+	003	029	097	194	273	290	227	115	023	004	0+	4
	5	0+	0+	0+	004	025	077	164	261	318	275	124	041	002	5
	6	0+	0+	0+	0+	004	017	055	131	247	367	372	257	066	6
	7	0+	0+	0+	0+	0+	002	008	028	082	210	478	698	932	7
8	0	923	663	430	168	058	017	004	001	0+	0+	0+	0+	0+	0
	1	075	279	383	336	198	090	031	008	001	0+	0+	0+	0+	1
	2	003	051	149	294	296	209	109	041	010	001	0+	0+	0+	2
	3	0+	005	033	147	254	279	219	124	047	009	0+	0+	0+	3
	4	0+	0+	005	046	136	232	273	232	136	046	005	0+	0+	4
	5	0+	0+	0+	009	047	124	219	279	254	147	033	005	0+	5
	6	0+	0+	0+	001	010	041	109	209	296	294	149	051	003	6
	7	0+	0+	0+	0+	001	008	031	090	198	336	383	279	075	7
	8	0+	0+	0+	0+	0+	001	004	017	058	168	430	663	923	8

Table III (*Continued*)

n	x	.01	.05	.10	.20	.30	.40	.50	.60	.70	.80	.90	.95	.99	x
								p							
9	0	914	630	387	134	040	010	002	0+	0+	0+	0+	0+	0+	0
	1	083	299	387	302	156	060	018	004	0+	0+	0+	0+	0+	1
	2	003	063	172	302	267	161	070	021	004	0+	0+	0+	0+	2
	3	0+	008	045	176	267	251	164	074	021	003	0+	0+	0+	3
	4	0+	001	007	066	172	251	246	167	074	017	001	0+	0+	4
	5	0+	0+	001	017	074	167	246	251	172	066	007	001	0+	5
	6	0+	0+	0+	003	021	074	164	251	267	176	045	008	0+	6
	7	0+	0+	0+	0+	004	021	070	161	267	302	172	063	003	7
	8	0+	0+	0+	0+	0+	004	018	060	156	302	387	299	083	8
	9	0+	0+	0+	0+	0+	0+	002	010	040	134	387	630	914	9
10	0	904	599	349	107	028	006	001	0+	0+	0+	0+	0+	0+	0
	1	091	315	387	268	121	040	010	002	0+	0+	0+	0+	0+	1
	2	004	075	194	302	233	121	044	011	001	0+	0+	0+	0+	2
	3	0+	010	057	201	267	215	117	042	009	001	0+	0+	0+	3
	4	0+	001	011	088	200	251	205	111	037	006	0+	0+	0+	4
	5	0+	0+	001	026	103	201	246	201	103	026	001	0+	0+	5
	6	0+	0+	0+	006	037	111	205	251	200	088	011	001	0+	6
	7	0+	0+	0+	001	009	042	117	215	267	201	057	010	0+	7
	8	0+	0+	0+	0+	001	011	044	121	233	302	194	075	004	8
	9	0+	0+	0+	0+	0+	002	010	040	121	268	387	315	091	9
	10	0+	0+	0+	0+	0+	0+	001	006	028	107	349	599	904	10
11	0	895	569	314	086	020	004	0+	0+	0+	0+	0+	0+	0+	0
	1	099	329	384	236	093	027	005	001	0+	0+	0+	0+	0+	1
	2	005	087	213	295	200	089	027	005	001	0+	0+	0+	0+	2
	3	0+	014	071	221	257	177	081	023	004	0+	0+	0+	0+	3
	4	0+	001	016	111	220	236	161	070	017	002	0+	0+	0+	4
	5	0+	0+	002	039	132	221	226	147	057	010	0+	0+	0+	5
	6	0+	0+	0+	010	057	147	226	221	132	039	002	0+	0+	6
	7	0+	0+	0+	002	017	070	161	236	220	111	016	001	0+	7
	8	0+	0+	0+	0+	004	023	081	177	257	221	071	014	0+	8
	9	0+	0+	0+	0+	001	005	027	089	200	295	213	087	005	9
	10	0+	0+	0+	0+	0+	001	005	027	093	236	384	329	099	10
	11	0+	0+	0+	0+	0+	0+	0+	004	020	086	314	569	895	11
12	0	886	540	282	069	014	002	0+	0+	0+	0+	0+	0+	0+	0
	1	107	341	377	206	071	017	003	0+	0+	0+	0+	0+	0+	1
	2	006	099	230	283	168	064	016	002	0+	0+	0+	0+	0+	2
	3	0+	017	085	236	240	142	054	012	001	0+	0+	0+	0+	3
	4	0+	002	021	133	231	213	121	042	008	001	0+	0+	0+	4
	5	0+	0+	004	053	158	227	193	101	029	003	0+	0+	0+	5
	6	0+	0+	0+	016	079	177	226	177	079	016	0+	0+	0+	6
	7	0+	0+	0+	003	029	101	193	227	158	053	004	0+	0+	7
	8	0+	0+	0+	001	008	042	121	213	231	133	021	002	0+	8
	9	0+	0+	0+	0+	001	012	054	142	240	236	085	017	0+	9

Table III (*Continued*)

n	x	.01	.05	.10	.20	.30	.40	.50	.60	.70	.80	.90	.95	.99	x
								p							
12	10	0+	0+	0+	0+	0+	002	016	064	168	283	230	099	006	10
	11	0+	0+	0+	0+	0+	0+	003	017	071	206	377	341	107	11
	12	0+	0+	0+	0+	0+	0+	0+	002	014	069	282	540	886	12
13	0	878	513	254	055	010	001	0+	0+	0+	0+	0+	0+	0+	0
	1	115	351	367	179	054	011	002	0+	0+	0+	0+	0+	0+	1
	2	007	111	245	268	139	045	010	001	0+	0+	0+	0+	0+	2
	3	0+	021	100	246	218	111	035	006	001	0+	0+	0+	0+	3
	4	0+	003	028	154	234	184	087	024	003	0+	0+	0+	0+	4
	5	0+	0+	006	069	180	221	157	066	014	001	0+	0+	0+	5
	6	0+	0+	001	023	103	197	209	131	044	006	0+	0+	0+	6
	7	0+	0+	0+	006	044	131	209	197	103	023	001	0+	0+	7
	8	0+	0+	0+	001	014	066	157	221	180	069	006	0+	0+	8
	9	0+	0+	0+	0+	003	024	087	184	234	154	028	003	0+	9
	10	0+	0+	0+	0+	001	006	035	111	218	246	100	021	0+	10
	11	0+	0+	0+	0+	0+	001	010	045	139	268	245	111	007	11
	12	0+	0+	0+	0+	0+	0+	002	011	054	179	367	351	115	12
	13	0+	0+	0+	0+	0+	0+	0+	001	010	055	254	513	878	13
14	0	869	488	229	044	007	001	0+	0+	0+	0+	0+	0+	0+	0
	1	123	359	356	154	041	007	001	0+	0+	0+	0+	0+	0+	1
	2	008	123	257	250	113	032	006	001	0+	0+	0+	0+	0+	2
	3	0+	026	114	250	194	085	022	003	0+	0+	0+	0+	0+	3
	4	0+	004	035	172	229	155	061	014	001	0+	0+	0+	0+	4
	5	0+	0+	008	086	196	207	122	041	007	0+	0+	0+	0+	5
	6	0+	0+	001	032	126	207	183	092	023	002	0+	0+	0+	6
	7	0+	0+	0+	009	062	157	209	157	062	009	0+	0+	0+	7
	8	0+	0+	0+	002	023	092	183	207	126	032	001	0+	0+	8
	9	0+	0+	0+	0+	007	041	122	207	196	086	008	0+	0+	9
	10	0+	0+	0+	0+	001	014	061	155	229	172	035	004	0+	10
	11	0+	0+	0+	0+	0+	003	022	085	194	250	114	026	0+	11
	12	0+	0+	0+	0+	0+	001	006	032	113	250	257	123	008	12
	13	0+	0+	0+	0+	0+	0+	001	007	041	154	356	359	123	13
	14	0+	0+	0+	0+	0+	0+	0+	001	007	044	229	488	869	14
15	0	860	463	206	035	005	0+	0+	0+	0+	0+	0+	0+	0+	0
	1	130	366	343	132	031	005	0+	0+	0+	0+	0+	0+	0+	1
	2	009	135	260	231	092	022	003	0+	0+	0+	0+	0+	0+	2
	3	0+	031	129	250	170	063	014	002	0+	0+	0+	0+	0+	3
	4	0+	005	043	188	219	127	042	007	001	0+	0+	0+	0+	4
	5	0+	001	010	103	206	186	092	024	003	0+	0+	0+	0+	5
	6	0+	0+	002	043	147	207	153	061	012	001	0+	0+	0+	6
	7	0+	0+	0+	014	081	177	196	118	035	003	0+	0+	0+	7
	8	0+	0+	0+	003	035	118	196	177	081	014	0+	0+	0+	8
	9	0+	0+	0+	001	012	061	153	207	147	043	002	0+	0+	9

Reprinted from Federick Mosteller, Robert E. K. Rourke, and George B. Thomas, Jr., *Probability with Statistical Applications,* Addison-Wesley Publishing Company, 1970, pp. 475–477.

Table IV: Random Numbers

10 09 73 25 33	76 52 01 35 86	34 67 35 48 76	80 95 90 91 17	39 29 27 49 45
37 54 20 48 05	64 89 47 42 96	24 80 52 40 37	20 63 61 04 02	00 82 29 16 65
08 42 26 89 53	19 64 50 93 03	23 20 90 25 60	15 95 33 47 64	35 08 03 36 06
99 01 90 25 29	09 37 67 07 15	38 31 13 11 65	88 67 67 43 97	04 43 62 76 59
12 80 79 99 70	80 15 73 61 47	64 03 23 66 53	98 95 11 68 77	12 17 17 68 33
66 06 57 47 17	34 07 27 68 50	36 69 73 61 70	65 81 33 98 85	11 19 92 91 70
31 06 01 08 05	45 57 18 24 06	35 30 34 26 14	86 79 90 74 39	23 40 30 97 32
85 26 97 76 02	02 05 16 56 92	68 66 57 48 18	73 05 38 52 47	18 62 38 85 79
63 57 33 21 35	05 32 54 70 48	90 55 35 75 48	28 46 82 87 09	83 49 12 56 24
73 79 64 57 53	03 52 96 47 78	35 80 83 42 82	60 93 52 03 44	35 27 38 84 35
98 52 01 77 67	14 90 56 86 07	22 10 94 05 58	60 97 09 34 33	50 50 07 39 98
11 80 50 54 31	39 80 82 77 32	50 72 56 82 48	29 40 52 42 01	52 77 56 78 51
83 45 29 96 34	06 28 89 80 83	13 74 67 00 78	18 47 54 06 10	68 71 17 78 17
88 68 54 02 00	86 50 75 84 01	36 76 66 79 51	90 36 47 64 93	29 60 91 10 62
99 59 46 73 48	87 51 76 49 69	91 82 60 89 28	93 78 56 13 68	23 47 83 41 13
65 48 11 76 74	17 46 85 09 50	58 04 77 69 74	73 03 95 71 86	40 21 81 65 44
80 12 43 56 35	17 72 70 80 15	45 31 82 23 74	21 11 57 82 53	14 38 55 37 63
74 35 09 98 17	77 40 27 72 14	43 23 60 02 10	45 52 16 42 37	96 28 60 26 55
69 91 62 68 03	66 25 22 91 48	36 93 68 72 03	76 62 11 39 90	94 40 05 64 18
09 89 32 05 05	14 22 56 85 14	46 42 75 67 88	96 29 77 88 22	54 38 21 45 98
91 49 91 45 23	68 47 92 76 86	46 16 28 35 54	94 75 08 99 23	37 08 92 00 48
80 33 69 45 93	26 94 03 68 58	70 29 73 41 35	53 14 03 33 40	42 05 08 23 41
44 10 48 19 49	85 15 74 79 54	32 97 92 65 75	57 60 04 08 81	22 22 20 64 13
12 55 07 37 42	11 10 00 20 40	12 86 07 46 97	96 64 48 94 39	28 70 72 58 15
63 60 64 93 29	16 50 53 44 84	40 21 95 25 63	43 65 17 70 82	07 20 73 17 90
61 19 69 04 46	26 45 74 77 74	51 92 43 37 29	65 39 45 95 93	42 58 26 05 27
15 47 44 52 66	95 27 07 99 53	59 36 78 38 48	82 39 61 01 18	33 21 15 94 66
94 55 72 85 73	67 89 75 43 87	54 62 24 44 31	91 19 04 25 92	92 92 74 59 73
42 48 11 62 13	97 34 40 87 21	16 86 84 87 67	03 07 11 20 59	25 70 14 66 70
23 52 37 83 17	73 20 88 98 37	68 93 59 14 16	26 25 22 96 63	05 52 28 25 62
04 49 35 24 94	75 24 63 38 24	45 86 25 10 25	61 96 27 93 35	65 33 71 24 72
00 54 99 76 54	64 05 18 81 59	96 11 96 38 96	54 69 28 23 91	23 28 72 95 29
35 96 31 53 07	26 89 80 93 54	33 35 13 54 62	77 97 45 00 24	90 10 33 93 33
59 80 80 83 91	45 42 72 68 42	83 60 94 97 00	13 02 12 48 92	78 56 52 01 06
46 05 88 52 36	01 39 09 22 86	77 28 14 40 77	93 91 08 36 47	70 61 74 29 41
32 17 90 05 97	87 37 92 52 41	05 56 70 70 07	86 74 31 71 57	85 39 41 18 38
69 23 46 14 06	20 11 74 52 04	15 95 66 00 00	18 74 39 24 23	97 11 89 63 38
19 56 54 14 30	01 75 87 53 79	40 41 92 15 85	66 67 43 68 06	84 96 28 52 07
45 15 51 49 38	19 47 60 72 46	43 66 79 45 43	59 04 79 00 33	20 82 66 95 41
94 86 43 19 94	36 16 81 08 51	34 88 88 15 53	01 54 03 54 56	05 01 45 11 76
98 08 62 48 26	45 24 02 84 04	44 99 90 88 96	39 09 47 34 07	35 44 13 18 80
33 18 51 62 32	41 94 15 09 49	89 43 54 85 81	88 69 54 19 94	37 54 87 30 43
80 95 10 04 06	96 38 27 07 74	20 15 12 33 87	25 01 62 52 98	94 62 46 11 71
79 75 24 91 40	71 96 12 82 96	69 86 10 25 91	74 85 22 05 39	00 38 75 95 79
18 63 33 25 37	98 14 50 65 71	31 01 02 46 74	05 45 56 14 27	77 93 89 19 36
74 02 94 39 02	77 55 73 22 70	97 79 01 71 19	52 52 75 80 21	80 81 45 17 48
54 17 84 56 11	80 99 33 71 43	05 33 51 29 69	56 12 71 92 55	36 04 09 03 24
11 66 44 98 83	52 07 98 48 27	59 38 17 15 39	09 97 33 34 40	88 46 12 33 56
48 32 47 79 28	31 24 96 47 10	02 29 53 68 70	32 30 75 75 46	15 02 00 99 94
69 07 49 41 38	87 63 79 19 76	35 58 40 44 01	10 51 82 16 15	01 84 87 69 38

From tables of the RAND Corporation. Reprinted from Wilfred J. Dixon and Frank J. Massey, Jr., *Introduction to Statistical Analysis*, 3rd ed., McGraw-Hill Book Company, New York, 1969, p. 446.

USE OF TABLE V

How to find the square root of a number not between 1 and 10. Consider, for example,

$$\sqrt{25{,}600}$$

1. "Block off" the number in two-digit blocks left of the decimal point.
2. Above the first block on the left, write the number that is closest to the square root of the digits in that block. Also locate a decimal point directly above the decimal point in the number.
3. Now take the first three digits in the number and locate them in the "n column" in Table V (neglect all decimal points in Table V). You will see:

2 56 00.

1

2 56 00.

n	n^2	$\sqrt{n}$	$\sqrt{10n}$
2.56	6.5536	1.60000	5.05964

(Remember, neglect decimal points here.)

The answer to $\sqrt{25{,}600}$ will be either or . The correct answer will be the sequence of digits that begins with the digit

1 6 0.0

2 56 00.

Notice that you place one digit over each two-digit block.

Hence, $\sqrt{25{,}600} = 160$.

Example 1. Find $\sqrt{356{,}000}$.

Solution: First write

35 60 00.

Next,

5

35 60 00.

Now look in Table V.

n	n^2	$\sqrt{n}$	$\sqrt{10n}$
3.56	12.6736	1.88680	5.96657

Choose this one because our answer must begin with a 5.

So we write:

5 9 6.6

35 60 00.

Thus, $\sqrt{35{,}6000} \approx 597$.

Example 2. Find $\sqrt{894}$.

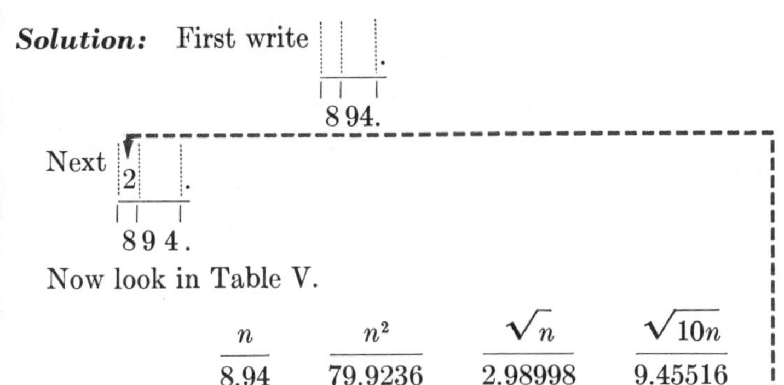

Solution: First write

Now look in Table V.

n	n^2	$\sqrt{n}$	$\sqrt{10n}$
8.94	79.9236	2.98998	9.45516

So 894 ≈ $\sqrt{29.9}$. Choose this one because our answer must begin with a 2.

When the number is less than one, we begin in the same way. Find $\sqrt{.847}$.

1. "Block off" the number in two-digit pairs to the right of the decimal point.
2. Above the first block, write the number that is closest to the square root of the digits in that block. Also, locate a decimal point directly above the decimal point in the number.
3. Now find 847 in the "n column" of Table V.

.84 7

. 9
. 84 7

You will see:

n	n^2	$\sqrt{n}$	$\sqrt{10n}$
8.47	71.7409	2.91033	9.20326

Choose this one because our answer must begin with a 9.

So, $\sqrt{.847} \approx .92$.

For a situation where there are zeros to the right of the decimal point, such as $\sqrt{.00000549}$, do everything as in the illustration above except first place a single zero over each two-digit block of zeros to the right of the decimal point.

. 0 0 2
.00 00 05 49

In Table V you will see:

n	n^2	$\sqrt{n}$	$\sqrt{10n}$
5.49	30.1401	2.34521	7.41620

We need this sequence of digits.

Hence $\sqrt{.00000549} \approx .0023$.

Table V: Squares and Square Roots

n	n^2	$\sqrt{n}$	$\sqrt{10n}$	n	n^2	$\sqrt{n}$	$\sqrt{10n}$
1.00	1.0000	1.00000	3.16228	1.50	2.2500	1.22474	3.87298
1.01	1.0201	1.00499	3.17805	1.51	2.2801	1.22882	3.88587
1.02	1.0404	1.00995	3.19374	1.52	2.3104	1.23288	3.89872
1.03	1.0609	1.01489	3.20936	1.53	2.3409	1.23693	3.91152
1.04	1.0816	1.01980	3.22490	1.54	2.3716	1.24097	3.92428
1.05	1.1025	1.02470	3.24037	1.55	2.4025	1.24499	3.93700
1.06	1.1236	1.02956	3.25576	1.56	2.4336	1.24900	3.94968
1.07	1.1449	1.03441	3.27109	1.57	2.4649	1.25300	3.96232
1.08	1.1664	1.03923	3.28634	1.58	2.4964	1.25698	3.97492
1.09	1.1881	1.04403	3.30151	1.59	2.5281	1.26095	3.98748
1.10	1.2100	1.04881	3.31662	1.60	2.5600	1.26491	4.00000
1.11	1.2321	1.05357	3.33167	1.61	2.5921	1.26886	4.01248
1.12	1.2544	1.05830	3.34664	1.62	2.6244	1.27279	4.02492
1.13	1.2769	1.06301	3.36155	1.63	2.6569	1.27671	4.03733
1.14	1.2996	1.06771	3.37639	1.64	2.6896	1.28062	4.04969
1.15	1.3225	1.07238	3.39116	1.65	2.7225	1.28452	4.06202
1.16	1.3456	1.07703	3.40588	1.66	2.7556	1.28841	4.07431
1.17	1.3689	1.08167	3.42053	1.67	2.7889	1.29228	4.08656
1.18	1.3924	1.08628	3.43511	1.68	2.8224	1.29615	4.09878
1.19	1.4161	1.09087	3.44964	1.69	2.8561	1.30000	4.11096
1.20	1.4400	1.09545	3.46410	1.70	2.8900	1.30384	4.12311
1.21	1.4641	1.10000	3.47851	1.71	2.9241	1.30767	4.13521
1.22	1.4884	1.10454	3.49285	1.72	2.9584	1.31149	4.14729
1.23	1.5129	1.10905	3.50714	1.73	2.9929	1.31529	4.15933
1.24	1.5376	1.11355	3.52136	1.74	3.0276	1.31909	4.17133
1.25	1.5625	1.11803	3.53553	1.75	3.0625	1.32288	4.18330
1.26	1.5876	1.12250	3.54965	1.76	3.0976	1.32665	4.19524
1.27	1.6129	1.12694	3.56371	1.77	3.1329	1.33041	4.20714
1.28	1.6384	1.13137	3.57771	1.78	3.1684	1.33417	4.21900
1.29	1.6641	1.13578	3.59166	1.79	3.2041	1.33791	4.23084
1.30	1.6900	1.14018	3.60555	1.80	3.2400	1.34164	4.24264
1.31	1.7161	1.14455	3.61939	1.81	3.2761	1.34536	4.25441
1.32	1.7424	1.14891	3.63318	1.82	3.3124	1.34907	4.26615
1.33	1.7689	1.15326	3.64692	1.83	3.3489	1.35277	4.27785
1.34	1.7956	1.15758	3.66060	1.84	3.3856	1.35647	4.28952
1.35	1.8225	1.16190	3.67423	1.85	3.4225	1.36015	4.30116
1.36	1.8496	1.16619	3.68782	1.86	3.4596	1.36382	4.31277
1.37	1.8769	1.17047	3.70135	1.87	3.4969	1.36748	4.32435
1.38	1.9044	1.17473	3.71484	1.88	3.5344	1.37113	4.33590
1.39	1.9321	1.17898	3.72827	1.89	3.5721	1.37477	4.34741
1.40	1.9600	1.18322	3.74166	1.90	3.6100	1.37840	4.35890
1.41	1.9881	1.18743	3.75500	1.91	3.6481	1.38203	4.37035
1.42	2.0164	1.19164	3.76829	1.92	3.6864	1.38564	4.38178
1.43	2.0449	1.19583	3.78153	1.93	3.7249	1.38924	4.39318
1.44	2.0736	1.20000	3.79473	1.94	3.7636	1.39284	4.40454
1.45	2.1025	1.20416	3.80789	1.95	3.8025	1.39642	4.41588
1.46	2.1316	1.20830	3.82099	1.96	3.8416	1.40000	4.42719
1.47	2.1609	1.21244	3.83406	1.97	3.8809	1.40357	4.43847
1.48	2.1904	1.21655	3.84708	1.98	3.9204	1.40712	4.44972
1.49	2.2201	1.22066	3.86005	1.99	3.9601	1.41067	4.46094
1.50	2.2500	1.22474	3.87298	2.00	4.0000	1.41421	4.47214
n	n^2	$\sqrt{n}$	$\sqrt{10n}$	n	n^2	$\sqrt{n}$	$\sqrt{10n}$

Table V: (*Continued*)

n	n^2	$\sqrt{n}$	$\sqrt{10n}$
2.00	4.0000	1.41421	4.47214
2.01	4.0401	1.41774	4.48330
2.02	4.0804	1.42127	4.49444
2.03	4.1209	1.42478	4.50555
2.04	4.1616	1.42829	4.51664
2.05	4.2025	1.43178	4.52769
2.06	4.2436	1.43527	4.53872
2.07	4.2849	1.43875	4.54973
2.08	4.3264	1.44222	4.56070
2.09	4.3681	1.44568	4.57165
2.10	4.4100	1.44914	4.58258
2.11	4.4521	1.45258	4.59347
2.12	4.4944	1.45602	4.60435
2.13	4.5369	1.45945	4.61519
2.14	4.5796	1.46287	4.62601
2.15	4.6225	1.46629	4.63681
2.16	4.6656	1.46969	4.64758
2.17	4.7089	1.47309	4.65833
2.18	4.7524	1.47648	4.66905
2.19	4.7961	1.47986	4.67974
2.20	4.8400	1.48324	4.69042
2.21	4.8841	1.48661	4.70106
2.22	4.9284	1.48997	4.71169
2.23	4.9729	1.49332	4.72229
2.24	5.0176	1.49666	4.73286
2.25	5.0625	1.50000	4.74342
2.26	5.1076	1.50333	4.75395
2.27	5.1129	1.50665	4.76445
2.28	5.1984	1.50997	4.77493
2.29	5.2441	1.51327	4.78539
2.30	5.2900	1.51658	4.79583
2.31	5.3361	1.51987	4.80625
2.32	5.3824	1.52315	4.81664
2.33	5.4289	1.52643	4.82701
2.34	5.4756	1.52971	4.83735
2.35	5.5225	1.53297	4.84768
2.36	5.5696	1.53623	4.85798
2.37	5.6169	1.53948	4.86826
2.38	5.6644	1.54272	4.87852
2.39	5.7121	1.54596	4.88876
2.40	5.7600	1.54919	4.89898
2.41	5.8081	1.55242	4.90918
2.42	5.8564	1.55563	4.91935
2.43	5.9049	1.55885	4.92950
2.44	5.9536	1.56205	4.93964
2.45	6.0025	1.56525	4.94975
2.46	6.0516	1.56844	4.95984
2.47	6.1009	1.57162	4.96991
2.48	6.1504	1.57480	4.97996
2.49	6.2001	1.57797	4.98999
2.50	6.2500	1.58114	5.00000
n	n^2	$\sqrt{n}$	$\sqrt{10n}$

n	n^2	$\sqrt{n}$	$\sqrt{10n}$
2.50	6.2500	1.58114	5.00000
2.51	6.3001	1.58430	5.00999
2.52	6.3504	1.58745	5.01996
2.53	6.4009	1.59060	5.02991
2.54	6.4516	1.59374	5.03984
2.55	6.5025	1.59687	5.04975
2.56	6.5536	1.60000	5.05964
2.57	6.6049	1.60312	5.06952
2.58	6.6564	1.60624	5.07937
2.59	6.7081	1.60935	5.08920
2.60	6.7600	1.61245	5.09902
2.61	6.8121	1.61555	5.10882
2.62	6.8644	1.61864	5.11859
2.63	6.9169	1.62173	5.12835
2.64	6.9696	1.62481	5.13809
2.65	7.0225	1.62788	5.14782
2.66	7.0756	1.63095	5.15752
2.67	7.1289	1.63401	5.16720
2.68	7.1824	1.63707	5.17687
2.69	7.2361	1.64012	5.18652
2.70	7.2900	1.64317	5.19615
2.71	7.3441	1.64621	5.20577
2.72	7.3984	1.64924	5.21536
2.73	7.4529	1.65227	5.22494
2.74	7.5076	1.65529	5.23450
2.75	7.5625	1.65831	5.24404
2.76	7.6176	1.66132	5.25357
2.77	7.6729	1.66433	5.26308
2.78	7.7284	1.66733	5.27257
2.79	7.7841	1.67033	5.28205
2.80	7.8400	1.67332	5.29150
2.81	7.8961	1.67631	5.30094
2.82	7.9524	1.67929	5.31037
2.83	8.0089	1.68226	5.31977
2.84	8.0656	1.68523	5.32917
2.85	8.1225	1.68819	5.33854
2.86	8.1796	1.69115	5.34790
2.87	8.2369	1.69411	5.35724
2.88	8.2944	1.69706	5.36656
2.89	8.3521	1.70000	5.37587
2.90	8.4100	1.70294	5.38516
2.91	8.4681	1.70587	5.39444
2.92	8.5264	1.70880	5.40370
2.93	8.5849	1.71172	5.41295
2.94	8.6436	1.71464	5.42218
2.95	8.7025	1.71756	5.43139
2.96	8.7616	1.72047	5.44059
2.97	8.8209	1.72337	5.44977
2.98	8.8804	1.72627	5.45894
2.99	8.9401	1.72916	5.46809
3.00	9.0000	1.73205	5.47723
n	n^2	$\sqrt{n}$	$\sqrt{10n}$

Table V: (*Continued*)

n	n^2	$\sqrt{n}$	$\sqrt{10n}$
3.00	9.0000	1.73205	5.47723
3.01	9.0601	1.73494	5.48635
3.02	9.1204	1.73781	5.49545
3.03	9.1809	1.74069	5.50454
3.04	9.2416	1.74356	5.51362
3.05	9.3025	1.74642	5.52268
3.06	9.3636	1.74929	5.53173
3.07	9.4249	1.75214	5.54076
3.08	9.4864	1.75499	5.54977
3.09	9.5481	1.75784	5.55878
3.10	9.6100	1.76068	5.56776
3.11	9.6721	1.76352	5.57674
3.12	9.7344	1.76635	5.58570
3.13	9.7969	1.76918	5.59464
3.14	9.8596	1.77200	5.60357
3.15	9.9225	1.77482	5.61249
3.16	9.9856	1.77764	5.62139
3.17	10.0489	1.78045	5.63028
3.18	10.1124	1.78326	5.63915
3.19	10.1761	1.78606	5.64801
3.20	10.2400	1.78885	5.65685
3.21	10.3041	1.79165	5.66569
3.22	10.3684	1.79444	5.67450
3.23	10.4329	1.79722	5.68331
3.24	10.4976	1.80000	5.69210
3.25	10.5625	1.80278	5.70088
3.26	10.6276	1.80555	5.70964
3.27	10.6929	1.80831	5.71839
3.28	10.7584	1.81108	5.72713
3.29	10.8241	1.81384	5.73585
3.30	10.8900	1.81659	5.74456
3.31	10.9561	1.81934	5.75326
3.32	11.0224	1.82209	5.76194
3.33	11.0889	1.82483	5.77062
3.34	11.1556	1.82757	5.77927
3.35	11.2225	1.83030	5.78792
3.36	11.2896	1.83303	5.79655
3.37	11.3569	1.83576	5.80517
3.38	11.4244	1.83848	5.81378
3.39	11.4921	1.84120	5.82237
3.40	11.5600	1.84391	5.83095
3.41	11.6281	1.84662	5.83952
3.42	11.6964	1.84932	5.84808
3.43	11.7649	1.85203	5.85662
3.44	11.8336	1.85472	5.86515
3.45	11.9025	1.85742	5.87367
3.46	11.9716	1.86011	5.88218
3.47	12.0409	1.86279	5.89067
3.48	12.1104	1.86548	5.89915
3.49	12.1801	1.86815	5.90762
3.50	12.2500	1.87083	5.91608
n	n^2	$\sqrt{n}$	$\sqrt{10n}$

n	n^2	$\sqrt{n}$	$\sqrt{10n}$
3.50	12.2500	1.87083	5.91608
3.51	12.3201	1.87350	5.92453
3.52	12.3904	1.87617	5.93296
3.53	12.4609	1.87883	5.94138
3.54	12.5316	1.88149	5.94979
3.55	12.6025	1.88414	5.95819
3.56	12.6736	1.88680	5.96657
3.57	12.7449	1.88944	5.97495
3.58	12.8164	1.89209	5.98331
3.59	12.8881	1.89473	5.99166
3.60	12.9600	1.89737	6.00000
3.61	13.0321	1.90000	6.00833
3.62	13.1044	1.90263	6.01664
3.63	13.1769	1.90526	6.02495
3.64	13.2496	1.90788	6.03324
3.65	13.3225	1.91050	6.04152
3.66	13.3956	1.91311	6.04979
3.67	13.4689	1.91572	6.05805
3.68	13.5424	1.91833	6.06630
3.69	13.6161	1.92094	6.07454
3.70	13.6900	1.92354	6.08276
3.71	13.7641	1.92614	6.09098
3.72	13.8384	1.92873	6.09918
3.73	13.9129	1.93132	6.10737
3.74	13.9876	1.93391	6.11555
3.75	14.0625	1.93649	6.12372
3.76	14.1376	1.93907	6.13188
3.77	14.2129	1.94165	6.14003
3.78	14.2884	1.94422	6.14817
3.79	14.3641	1.94679	6.15630
3.80	14.4400	1.94936	6.16441
3.81	14.5161	1.95192	6.17252
3.82	14.5924	1.95448	6.18061
3.83	14.6689	1.95704	6.18870
3.84	14.7456	1.95959	6.19677
3.85	14.8225	1.96214	6.20484
3.86	14.8996	1.96469	6.21289
3.87	14.9769	1.96723	6.22093
3.88	15.0544	1.96977	6.22896
3.89	15.1321	1.97231	6.23699
3.90	15.2100	1.97484	6.24500
3.91	15.2881	1.97737	6.25300
3.92	15.3664	1.97990	6.26099
3.93	15.4449	1.98242	6.26897
3.94	15.5236	1.98494	6.27694
3.95	15.6025	1.98746	6.28490
3.96	15.6816	1.98997	6.29285
3.97	15.7609	1.99249	6.30079
3.98	15.8408	1.99499	6.30872
3.99	15.9201	1.99750	6.31664
4.00	16.0000	2.00000	6.32456
n	n^2	$\sqrt{n}$	$\sqrt{10n}$

Table V: (*Continued*)

n	n^2	$\sqrt{n}$	$\sqrt{10n}$
4.00	16.0000	2.00000	6.32456
4.01	16.0801	2.00250	6.33246
4.02	16.1604	2.00499	6.34035
4.03	16.2409	2.00749	6.34823
4.04	16.3216	2.00998	6.35610
4.05	16.4025	2.01246	6.36396
4.06	16.4836	2.01494	6.37181
4.07	16.5649	2.01742	6.37966
4.08	16.6464	2.01990	6.38749
4.09	16.7281	2.02237	6.39531
4.10	16.8100	2.02485	6.40312
4.11	16.8921	2.02731	6.41093
4.12	16.9744	2.02978	6.41872
4.13	17.0569	2.03224	6.42651
4.14	17.1396	2.03470	6.43428
4.15	17.2225	2.03715	6.44205
4.16	17.3056	2.03961	6.44981
4.17	17.3889	2.04206	6.45755
4.18	17.4724	2.04450	6.46529
4.19	17.5561	2.04695	6.47302
4.20	17.6400	2.04939	6.48074
4.21	17.7241	2.05183	6.48845
4.22	17.8084	2.05426	6.49615
4.23	17.8929	2.05670	6.50384
4.24	17.9776	2.05913	6.51153
4.25	18.0625	2.06155	6.51920
4.26	18.1476	2.06398	6.52687
4.27	18.2329	2.06640	6.53452
4.28	18.3184	2.06882	6.54217
4.29	18.4041	2.07123	6.54981
4.30	18.4900	2.07364	6.55744
4.31	18.5761	2.07605	6.56506
4.32	18.6624	2.07846	6.57267
4.33	18.7489	2.08087	6.58027
4.34	18.8356	2.08327	6.58787
4.35	18.9225	2.08567	6.59545
4.36	19.0096	2.08806	6.60303
4.37	19.0969	2.09045	6.61060
4.38	19.1844	2.09284	6.61816
4.39	19.2721	2.09523	6.62571
4.40	19.3600	2.09762	6.63325
4.41	19.4481	2.10000	6.64078
4.42	19.5364	2.10238	6.64831
4.43	19.6249	2.10476	6.65582
4.44	19.7136	2.10713	6.66333
4.45	19.8025	2.10950	6.67083
4.46	19.8916	2.11187	6.67832
4.47	19.9809	2.11424	6.68581
4.48	20.0704	2.11660	6.69328
4.49	20.1601	2.11896	6.70075
4.50	20.2500	2.12132	6.70820
n	n^2	$\sqrt{n}$	$\sqrt{10n}$

n	n^2	$\sqrt{n}$	$\sqrt{10n}$
4.50	20.2500	2.12132	6.70820
4.51	20.3401	2.12368	6.71565
4.52	20.4304	2.12603	6.72309
4.53	20.5209	2.12838	6.73053
4.54	20.6116	2.13073	6.73795
4.55	20.7025	2.13307	6.74537
4.56	20.7936	2.13542	6.75278
4.57	20.8849	2.13776	6.76018
4.58	20.9764	2.14009	6.76757
4.59	21.0681	2.14243	6.77495
4.60	21.1600	2.14476	6.78233
4.61	21.2521	2.14709	6.78970
4.62	21.3444	2.14942	6.79706
4.63	21.4369	2.15174	6.80441
4.64	21.5296	2.15407	6.81175
4.65	21.6225	2.15639	6.81909
4.66	21.7156	2.15870	6.82642
4.67	21.8089	2.16102	6.83374
4.68	21.9024	2.16333	6.84105
4.69	21.9961	2.16564	6.84836
4.70	22.0900	2.16795	6.85565
4.71	22.1841	2.17025	6.86294
4.72	22.2784	2.17256	6.87023
4.73	22.3729	2.17486	6.87750
4.74	22.4676	2.17715	6.88477
4.75	22.5625	2.17945	6.89202
4.76	22.6576	2.18174	6.89928
4.77	22.7529	2.18403	6.90652
4.78	22.8484	2.18632	6.91375
4.79	22.9441	2.18861	6.92098
4.80	23.0400	2.19089	6.92820
4.81	23.1361	2.19317	6.93542
4.82	23.2324	2.19545	6.94262
4.83	23.3289	2.19773	6.94982
4.84	23.4256	2.20000	6.95701
4.85	23.5225	2.20227	6.96419
4.86	23.6196	2.20454	6.97137
4.87	23.7169	2.20681	6.97854
4.88	23.8144	2.20907	6.98570
4.89	23.9121	2.21133	6.99285
4.90	24.0100	2.21359	7.00000
4.91	24.1081	2.21585	7.00714
4.92	24.2064	2.21811	7.01427
4.93	24.3049	2.22036	7.02140
4.94	24.4036	2.22261	7.02851
4.95	24.5025	2.22486	7.03562
4.96	24.6016	2.22711	7.04273
4.97	24.7009	2.22935	7.04982
4.98	24.8004	2.23159	7.05691
4.99	24.9001	2.23383	7.06399
5.00	25.0000	2.23607	7.07107
n	n^2	$\sqrt{n}$	$\sqrt{10n}$

Table V: (*Continued*)

n	n^2	$\sqrt{n}$	$\sqrt{10n}$
5.00	25.0000	2.23607	7.07107
5.01	25.1001	2.23830	7.07814
5.02	25.2004	2.24054	7.08520
5.03	25.3009	2.24277	7.09225
5.04	25.4016	2.24499	7.09930
5.05	25.5025	2.24722	7.10634
5.06	25.6036	2.24944	7.11337
5.07	25.7049	2.25167	7.12039
5.08	25.8064	2.25389	7.12741
5.09	25.9081	2.25610	7.13442
5.10	26.0100	2.25832	7.14143
5.11	26.1121	2.26053	7.14843
5.12	26.2144	2.26274	7.15542
5.13	26.3169	2.26495	7.16240
5.14	26.4196	2.26716	7.16938
5.15	26.5225	2.26936	7.17635
5.16	26.6256	2.27156	7.18331
5.17	26.7289	2.27376	7.19027
5.18	26.8324	2.27596	7.19722
5.19	26.9361	2.27816	7.20417
5.20	27.0400	2.28035	7.21110
5.21	27.1441	2.28254	7.21803
5.22	27.2484	2.28473	7.22496
5.23	27.3529	2.28692	7.23187
5.24	27.4576	2.28910	7.23878
5.25	27.5625	2.29129	7.24569
5.26	27.6676	2.29347	7.25259
5.27	27.7729	2.29565	7.25948
5.28	27.8784	2.29783	7.26636
5.29	27.9841	2.30000	7.27324
5.30	28.0900	2.30217	7.28011
5.31	28.1961	2.30434	7.28697
5.32	28.3024	2.30651	7.29383
5.33	28.4089	2.30868	7.30068
5.34	28.5156	2.31084	7.30753
5.35	28.6225	2.31301	7.31437
5.36	28.7296	2.31517	7.32120
5.37	28.8369	2.31733	7.32803
5.38	28.9444	2.31948	7.33485
5.39	29.0521	2.32164	7.34166
5.40	29.1600	2.32379	7.34847
5.41	29.2781	2.32594	7.35527
5.42	29.3764	2.32809	7.36206
5.43	29.4849	2.33024	7.36885
5.44	29.5936	2.33238	7.37564
5.45	29.7025	2.33452	7.38241
5.46	29.8116	2.33666	7.38918
5.47	29.9209	2.33880	7.39594
5.48	30.0304	2.34094	7.40270
5.49	30.1401	2.34307	7.40945
5.50	30.2500	2.34521	7.41620
n	n^2	$\sqrt{n}$	$\sqrt{10n}$

n	n^2	$\sqrt{n}$	$\sqrt{10n}$
5.50	30.2500	2.34521	7.41620
5.51	30.3601	2.34734	7.42294
5.52	30.4704	2.34947	7.42967
5.53	30.5809	2.35160	7.43640
5.54	30.6916	2.35372	7.44312
5.55	30.8025	2.35584	7.44983
5.56	30.9136	2.35797	7.45654
5.57	31.0249	2.36008	7.46324
5.58	31.1364	2.36220	7.46994
5.59	31.2481	2.36432	7.47663
5.60	31.3600	2.36643	7.48331
5.61	31.4721	2.36854	7.48999
5.62	31.5844	2.37065	7.49667
5.63	31.6969	2.37276	7.50333
5.64	31.8096	2.37487	7.50999
5.65	31.9225	2.37697	7.51665
5.66	32.0356	2.37908	7.52330
5.67	32.1489	2.38118	7.52994
5.68	32.2624	2.38328	7.53658
5.69	32.3761	2.38537	7.54321
5.70	32.4900	2.38747	7.54983
5.71	32.6041	2.38956	7.55645
5.72	32.7184	2.39165	7.56307
5.73	32.8329	2.39374	7.56968
5.74	32.9476	2.39583	7.57628
5.75	33.0625	2.39792	7.58288
5.76	33.1776	2.40000	7.58947
5.77	33.2929	2.40208	7.59605
5.78	33.3084	2.40416	7.60263
5.79	33.5241	2.40624	7.60920
5.80	33.6400	2.40832	7.61577
5.81	33.7561	2.41039	7.62234
5.82	33.8724	2.41247	7.62889
5.83	33.9889	2.41454	7.63544
5.84	34.1056	2.41661	7.64199
5.85	34.2225	2.41868	7.64853
5.86	34.3396	2.42074	7.65505
5.87	34.4569	2.42281	7.66159
5.88	34.5744	2.42487	7.66812
5.89	34.6921	2.42693	7.67463
5.90	34.8100	2.42899	7.68115
5.91	34.9281	2.43105	7.68765
5.92	35.0464	2.43311	7.69415
5.93	35.1649	2.43516	7.70065
5.94	35.2836	2.43721	7.70714
5.95	35.4025	2.43926	7.71362
5.96	35.5216	2.44131	7.72010
5.97	35.6409	2.44336	7.72658
5.98	35.7604	2.44540	7.73305
5.99	35.8801	2.44745	7.73951
6.00	36.0000	2.44949	7.74597
n	n^2	$\sqrt{n}$	$\sqrt{10n}$

Table V: (*Continued*)

n	n^2	$\sqrt{n}$	$\sqrt{10n}$
6.00	36.0000	2.44949	7.74597
6.01	36.1201	2.45153	7.75242
6.02	36.2404	2.45357	7.75887
6.03	36.3609	2.45561	7.76531
6.04	36.4816	2.45764	7.77174
6.05	36.6025	2.45967	7.77817
6.06	36.7236	2.46171	7.78460
6.07	36.8449	2.46374	7.79102
6.08	36.9664	2.46577	7.79744
6.09	37.0881	2.46779	7.80385
6.10	37.2100	2.46982	7.81025
6.11	37.3321	2.47184	7.81665
6.12	37.4544	2.47386	7.82304
6.13	37.5769	2.47588	7.82943
6.14	37.6996	2.47790	7.83582
6.15	37.8225	2.47992	7.84219
6.16	37.9456	2.48193	7.84857
6.17	38.0689	2.48395	7.85493
6.18	38.1924	2.48596	7.86130
6.19	38.3161	2.48797	7.86766
6.20	38.4400	2.48998	7.87401
6.21	38.5641	2.49199	7.88036
6.22	38.6884	2.49399	7.88670
6.23	38.8129	2.49600	7.89303
6.24	38.9376	2.49800	7.89937
6.25	39.0625	2.50000	7.90569
6.26	39.1876	2.50200	7.91202
6.27	39.3129	2.50400	7.91833
6.28	39.4384	2.50599	7.92465
6.29	39.5641	2.50799	7.93095
6.30	39.6900	2.50998	7.93725
6.31	39.8161	2.51197	7.94355
6.32	39.9424	2.51396	7.94984
6.33	40.0689	2.51595	7.95613
6.34	40.1956	2.51794	7.96241
6.35	40.3225	2.51992	7.96869
6.36	40.4496	2.52190	7.97496
6.37	40.5769	2.52389	7.98123
6.38	40.7044	2.52587	7.98749
6.39	40.8321	2.52784	7.99375
6.40	40.9600	2.52982	8.00000
6.41	41.0881	2.53180	8.00625
6.42	41.2164	2.53377	8.01249
6.43	41.3449	2.53574	8.01873
6.44	41.4736	2.53772	8.02496
6.45	41.6025	2.53969	8.03119
6.46	41.7316	2.54165	8.03741
6.47	41.8609	2.54362	8.04363
6.48	41.9904	2.54558	8.04984
6.49	42.1201	2.54755	8.05605
6.50	42.2500	2.54951	8.06226
n	n^2	$\sqrt{n}$	$\sqrt{10n}$

n	n^2	$\sqrt{n}$	$\sqrt{10n}$
6.50	42.2500	2.54951	8.06226
6.51	42.3801	2.55147	8.06846
6.52	42.5104	2.55343	8.07465
6.53	42.6409	2.55539	8.08084
6.54	42.7716	2.55734	8.08703
6.55	42.9025	2.55930	8.09321
6.56	43.0336	2.56125	8.09938
6.57	43.1649	2.56320	8.10555
6.58	43.2964	2.56515	8.11172
6.59	43.4281	2.56710	8.11788
6.60	43.5600	2.56905	8.12404
6.61	43.6921	2.57099	8.13019
6.62	43.8244	2.57294	8.13634
6.63	43.9569	2.57488	8.14248
6.64	44.0896	2.57682	8.14862
6.65	44.2225	2.57876	8.15475
6.66	44.3556	2.48070	8.16088
6.67	44.4889	2.58263	8.16701
6.68	44.6224	2.58457	8.17313
6.69	44.7561	2.58650	8.17924
6.70	44.8900	2.58844	8.18535
6.71	45.0241	2.59037	8.19146
6.72	45.1584	2.59230	8.19756
6.73	45.2929	2.59422	8.20366
6.74	45.4276	2.59615	8.20975
6.75	45.5625	2.59808	8.21584
6.76	45.6976	2.60000	8.22192
6.77	45.8329	2.60192	8.22800
6.78	45.9684	2.60384	8.23408
6.79	46.1041	2.60576	8.24015
6.80	46.2400	2.60768	8.24621
6.81	46.3761	2.60960	8.25227
6.82	46.5124	2.61151	8.25833
6.83	46.6489	2.61343	8.26438
6.84	46.7856	2.61534	8.27043
6.85	46.9225	2.61725	8.27647
6.86	47.0596	2.61916	8.28251
6.87	47.1969	2.62107	8.28855
6.88	47.3344	2.62298	8.29458
6.89	47.4721	2.62488	8.30060
6.90	47.6100	2.62679	8.30662
6.91	47.7481	2.62869	8.31264
6.92	47.8864	2.63059	8.31865
6.93	48.0249	2.63249	8.32466
6.94	48.1636	2.63439	8.33067
6.95	48.3025	2.63629	8.33667
6.96	48.4416	2.63818	8.34266
6.97	48.5809	2.64008	8.34865
6.98	48.7204	2.64197	8.35464
6.99	48.8601	2.64386	8.36062
7.00	49.0000	2.64575	8.36660
n	n^2	$\sqrt{n}$	$\sqrt{10n}$

Table V: (*Continued*)

n	n^2	$\sqrt{n}$	$\sqrt{10n}$
7.00	49.0000	2.64575	8.36660
7.01	49.1401	2.64764	8.37257
7.02	49.2804	2.64953	8.37854
7.03	49.4209	2.65141	8.38451
7.04	49.5616	2.65330	8.39047
7.05	49.7025	2.65518	8.39643
7.06	49.8436	2.65707	8.40238
7.07	49.9849	2.65895	8.40833
7.08	50.1264	2.66083	8.41427
7.09	50.2681	2.66271	8.24021
7.10	50.4100	2.66458	8.42615
7.11	50.5521	2.66646	8.43208
7.12	50.6944	2.66833	8.43801
7.13	50.8369	2.67021	8.44393
7.14	50.9796	2.67208	8.44985
7.15	51.1225	2.67395	8.45577
7.16	51.2656	2.67582	8.46168
7.17	51.4089	2.67769	8.46759
7.18	51.5524	2.67955	8.47349
7.19	51.6961	2.68142	8.47939
7.20	51.8400	2.68328	8.48528
7.21	51.9841	2.68514	8.49117
7.22	52.1284	2.68701	8.49706
7.23	52.2729	2.68887	8.50294
7.24	52.4176	2.69072	8.50882
7.25	52.5625	2.69258	8.51469
7.26	52.7076	2.69444	8.52056
7.27	52.8529	2.69629	8.52643
7.28	52.9984	2.69815	8.53229
7.29	53.1441	2.70000	8.53815
7.30	53.2900	2.70185	8.54400
7.31	53.4361	2.70370	8.54985
7.32	53.5824	2.70555	8.55570
7.33	53.7289	2.70740	8.56154
7.34	53.8756	2.70924	8.56738
7.35	54.0225	2.71109	8.57321
7.36	54.1696	2.71293	8.57904
7.37	54.3169	2.71477	8.58487
7.38	54.4644	2.71662	8.59069
7.39	54.6121	2.71846	8.59651
7.40	54.7600	2.72029	8.60233
7.41	54.9081	2.72213	8.60814
7.42	55.0564	2.72397	8.61394
7.43	55.2049	2.72580	8.61974
7.44	55.3536	2.72764	8.62554
7.45	55.5025	2.72947	8.63134
7.46	55.6516	2.73130	8.63713
7.47	55.8009	2.73313	8.64292
7.48	55.9504	2.73496	8.64870
7.49	56.1001	2.73679	8.65448
7.50	56.2500	2.73861	8.66025
n	n^2	$\sqrt{n}$	$\sqrt{10n}$

n	n^2	$\sqrt{n}$	$\sqrt{10n}$
7.50	56.2500	2.73861	8.66025
7.51	56.4001	2.74044	8.66603
7.52	56.5504	2.74226	8.67179
7.53	56.7009	2.74408	8.67756
7.54	56.8516	2.74591	8.68332
7.55	57.0025	2.74773	8.68907
7.56	57.1536	2.74955	8.69483
7.57	57.3049	2.75136	8.70057
7.58	57.4564	2.75318	8.70632
7.59	57.6081	2.75500	8.71206
7.60	57.7600	2.75681	8.71780
7.61	57.9121	2.75862	8.72353
7.62	58.0644	2.76043	8.72926
7.63	58.2169	2.76225	8.73499
7.64	58.3696	2.76405	8.74071
7.65	58.5225	2.76586	8.74643
7.66	58.6756	2.76767	8.75214
7.67	58.8289	2.76948	8.75785
7.68	58.9824	2.77128	8.76356
7.69	59.1361	2.77308	8.76926
7.70	59.2900	2.77489	8.77496
7.71	59.4441	2.77669	8.78066
7.72	59.5984	2.77849	8.78635
7.73	59.7529	2.78029	8.79204
7.74	59.9076	2.78209	8.79773
7.75	60.0625	2.78388	8.80341
7.76	60.2176	2.78568	8.80909
7.77	60.3729	2.78747	8.81476
7.78	60.5284	2.78927	8.82043
7.79	60.6841	2.79106	8.82610
7.80	60.8400	2.79285	8.83176
7.81	60.9961	2.79464	8.83742
7.82	61.1524	2.79643	8.84308
7.83	61.3089	2.79821	8.84873
7.84	61.4656	2.80000	8.85438
7.85	61.6225	2.80179	8.86002
7.86	61.7796	2.80357	8.86566
7.87	61.9369	2.80535	8.87130
7.88	62.0944	2.80713	8.87694
7.89	62.2521	2.80891	8.88257
7.90	62.4100	2.81069	8.88819
7.91	62.5681	2.81247	8.89382
7.92	62.7264	2.81425	8.89944
7.93	62.8849	2.81603	8.90505
7.94	63.0436	2.81780	8.91067
7.95	53.2025	2.81957	8.91628
7.96	63.3616	2.82135	8.92188
7.97	63.5209	2.82312	8.92749
7.98	63.6804	2.82489	8.93308
7.99	63.8401	2.82666	8.93868
8.00	64.0000	2.82843	8.94427
n	n^2	$\sqrt{n}$	$\sqrt{10n}$

Table V: (*Continued*)

n	n^2	$\sqrt{n}$	$\sqrt{10n}$	n	n^2	$\sqrt{n}$	$\sqrt{10n}$
8.00	64.0000	2.82843	8.94427	8.50	72.2500	2.91548	9.21954
8.01	64.1601	2.83019	8.94986	8.51	72.4201	2.91719	9.22497
8.02	64.3204	2.83196	8.95545	8.52	72.5904	2.91890	9.23038
8.03	64.4809	2.83373	8.96103	8.53	72.7609	2.92062	9.23580
8.04	64.6416	2.83549	8.96660	8.54	72.9316	2.92233	9.24121
8.05	64.8025	2.83725	8.97218	8.55	73.1025	2.92404	9.24662
8.06	64.9636	2.83901	8.97775	8.56	73.2736	2.92575	9.25203
8.07	65.1249	2.84077	8.98332	8.57	73.4449	2.92746	9.25743
8.08	65.2864	2.84253	8.98888	8.58	73.6164	2.92916	9.26283
8.09	65.4481	2.84429	8.99444	8.59	73.7881	2.93087	9.26823
8.10	65.6100	2.84605	9.00000	8.60	73.9600	2.93258	9.27362
8.11	65.7721	2.84781	9.00555	8.61	74.1321	2.93428	9.27901
8.12	65.9344	2.84956	9.01110	8.62	74.3044	2.93598	9.28440
8.13	66.0969	2.85132	9.01665	8.63	74.4769	2.93769	9.28978
8.14	66.2596	2.85307	9.02219	8.64	74.6496	2.93939	9.29516
8.15	66.4225	2.85482	9.02774	8.65	74.8225	2.94109	9.30054
8.16	66.5856	2.85657	9.03327	8.66	74.9956	2.94279	9.30591
8.17	66.7489	2.85832	9.03881	8.67	75.1689	2.94449	9.31128
8.18	66.9124	2.86007	9.04434	8.68	75.3424	2.94618	9.31665
8.19	67.0761	2.86182	9.04986	8.69	75.5161	2.94788	9.32202
8.20	67.2400	2.86356	9.05539	8.70	75.6900	2.94958	9.32738
8.21	67.4041	2.86531	9.06091	8.71	75.8641	2.95127	9.33274
8.22	67.5684	2.86705	9.06642	8.72	76.0384	2.95296	9.33809
8.23	67.7329	2.86880	9.07193	8.73	76.2129	2.95466	9.34345
8.24	67.8976	2.87054	9.07744	8.74	76.3876	2.95635	9.34880
8.25	68.0625	2.87228	9.08295	8.75	76.5625	2.95804	9.35414
8.26	68.2276	2.87402	9.08845	8.76	76.7376	2.95973	9.35949
8.27	68.3929	2.87576	9.09395	8.77	76.9129	2.96142	9.36483
8.28	68.5584	2.87750	9.09945	8.78	77.0884	2.96311	9.37017
8.29	68.7241	2.87924	9.10494	8.79	77.2641	2.96479	9.37550
8.30	68.8900	2.88097	9.11043	8.80	77.4400	2.96648	9.38083
8.31	69.0561	2.88271	9.11592	8.81	77.6161	2.96816	9.38616
8.32	69.2224	2.88444	9.12140	8.82	77.7924	2.96985	9.39149
8.33	69.3889	2.88617	9.12688	8.83	77.9689	2.97153	9.39681
8.34	69.5556	2.88791	9.13236	8.84	78.1456	2.97321	9.40213
8.35	69.7225	2.88964	9.13783	8.85	78.3225	2.97489	9.40744
8.36	69.8896	2.89137	9.14330	8.86	78.4996	2.97658	9.41276
8.37	70.0569	2.89310	9.14877	8.87	78.6769	2.97825	9.41807
8.38	70.2244	2.89482	9.15423	8.88	78.8544	2.97993	9.42338
8.39	70.3921	2.89655	9.15969	8.89	79.0321	2.98161	9.42868
8.40	70.5600	2.89828	9.16515	8.90	79.2100	2.98329	9.43398
8.41	70.7281	2.90000	9.17061	8.91	79.3881	2.98496	9.43928
8.42	70.8964	2.90172	9.17606	8.92	79.5664	2.98664	9.44458
8.43	71.0649	2.90345	9.18150	8.93	79.7449	2.98831	9.44987
8.44	71.2336	2.90517	9.18695	8.94	79.9236	2.98998	9.45516
8.45	71.4025	2.90689	9.19239	8.95	80.1025	2.99166	9.46044
8.46	71.5716	2.90861	9.19783	8.96	80.2816	2.99333	9.46573
8.47	71.7409	2.91033	9.20326	8.97	80.4609	2.99500	9.47101
8.48	71.9104	2.91204	9.20869	8.98	80.6404	2.99666	9.47629
8.49	72.0801	2.91376	9.21412	8.99	80.8201	2.99833	9.48156
8.50	72.2500	2.91548	9.21954	9.00	81.0000	3.00000	9.48683
n	n^2	$\sqrt{n}$	$\sqrt{10n}$	n	n^2	$\sqrt{n}$	$\sqrt{10n}$

Table V: (*Continued*)

n	n^2	$\sqrt{n}$	$\sqrt{10n}$
9.00	81.0000	3.00000	9.48683
9.01	81.1801	3.00167	9.49210
9.02	81.3604	3.00333	9.49737
9.03	81.5409	3.00500	9.50263
9.04	81.7216	3.00666	9.50789
9.05	81.9025	3.00832	9.51315
9.06	82.0836	3.00998	9.51840
9.07	82.2649	3.01164	9.52365
9.08	82.4464	3.01330	9.52890
9.09	82.6281	3.01496	9.53415
9.10	82.8100	3.01662	9.53939
9.11	82.9921	3.01828	9.54463
9.12	83.1744	3.01993	9.54987
9.13	83.3569	3.02159	9.55510
9.14	83.5396	3.02324	9.56033
9.15	83.7225	3.02490	9.56556
9.16	83.9056	3.02655	9.57079
9.17	84.0889	3.02820	9.57601
9.18	84.2724	3.02985	9.58123
9.19	84.4561	3.03150	9.58645
9.20	84.6400	3.03315	9.59166
9.21	84.8241	3.03480	9.59687
9.22	85.0084	3.03645	9.60208
9.23	85.1929	3.03809	9.60729
9.24	85.3776	3.03974	9.61249
9.25	85.5625	3.04138	9.61769
9.26	85.7476	3.04302	9.62289
9.27	85.9329	3.04467	9.62808
9.28	86.1184	3.04631	9.63328
9.29	86.3041	3.04795	9.63846
9.30	86.4900	3.04959	9.64365
9.31	86.6761	3.05123	9.64883
9.32	86.8624	3.05287	9.65401
9.33	87.0489	3.05450	9.65919
9.34	87.2356	3.05614	9.66437
9.35	87.4225	3.05778	9.66954
9.36	87.6096	3.05941	9.67471
9.37	87.7969	3.06105	9.67988
9.38	87.9844	3.06268	9.68504
9.39	88.1721	3.06431	9.69020
9.40	88.3600	3.06594	9.69536
9.41	88.5481	3.06757	9.70052
9.42	88.7364	3.06920	9.70567
9.43	88.9249	3.07083	9.71082
9.44	89.1136	3.07246	9.71597
9.45	89.3025	3.07409	9.72111
9.46	89.4916	3.07571	9.72625
9.47	89.6809	3.07734	9.73139
9.48	89.8704	3.07896	9.73653
9.49	90.0601	3.08058	9.74166
9.50	90.2500	3.08221	9.74679
n	n^2	$\sqrt{n}$	$\sqrt{10n}$

n	n^2	$\sqrt{n}$	$\sqrt{10n}$
9.50	90.2500	3.08221	9.74679
9.51	90.4401	3.08383	9.75192
9.52	90.6304	3.08545	9.75705
9.53	90.8209	3.08707	9.76217
9.54	91.0116	3.08869	9.76729
9.55	91.2025	3.09031	9.77241
9.56	91.3936	3.09192	9.77753
9.57	91.5849	3.09354	9.78264
9.58	91.7764	3.09516	9.78775
9.59	91.9681	3.09677	9.79285
9.60	92.1600	3.09839	9.79796
9.61	92.3521	3.10000	9.80306
9.62	92.5444	3.10161	9.80816
9.63	92.7369	3.10322	9.81326
9.64	92.9296	3.10483	9.81835
9.65	93.1225	3.10644	9.82344
9.66	93.3156	3.10805	9.82853
9.67	93.5089	3.10966	9.83362
9.68	93.7024	3.11127	9.83870
9.69	93.8961	3.11288	9.84378
9.70	94.0900	3.11448	9.84886
9.71	94.2841	3.11609	9.85393
9.72	94.4784	3.11769	9.85901
9.73	94.6729	3.11929	9.86408
9.74	94.8676	3.12090	9.86914
9.75	95.0625	3.12250	9.87421
9.76	95.2576	3.12410	9.87927
9.77	95.4529	3.12570	9.88433
9.78	95.6484	3.12730	9.88939
9.79	95.8441	3.12890	9.89444
9.80	96.0400	3.13050	9.89949
9.81	96.2361	3.13209	9.90454
9.82	96.4324	3.13369	9.90959
9.83	96.6289	3.13528	9.91464
9.84	96.8256	3.13688	9.91968
9.85	97.0225	3.13847	9.92472
9.86	97.2196	3.14006	9.92975
9.87	97.4169	3.14166	9.93479
9.88	97.6144	3.14325	9.93982
9.89	97.7821	3.14484	9.94485
9.90	98.0100	3.14643	9.94987
9.91	98.2081	3.14802	9.95490
9.92	98.4064	3.14960	9.95992
9.93	98.6049	3.15119	9.96494
9.94	98.8036	3.15278	9.96995
9.95	99.0025	3.15436	9.97497
9.96	99.2016	3.15595	9.97998
9.97	99.4009	3.15753	9.98499
9.98	99.6004	3.15911	9.98999
9.99	99.8001	3.16070	9.99500
10.00	100.000	3.16228	10.0000
n	n^2	$\sqrt{n}$	$\sqrt{10n}$

Table VI: Areas of the Standard Normal Distribution

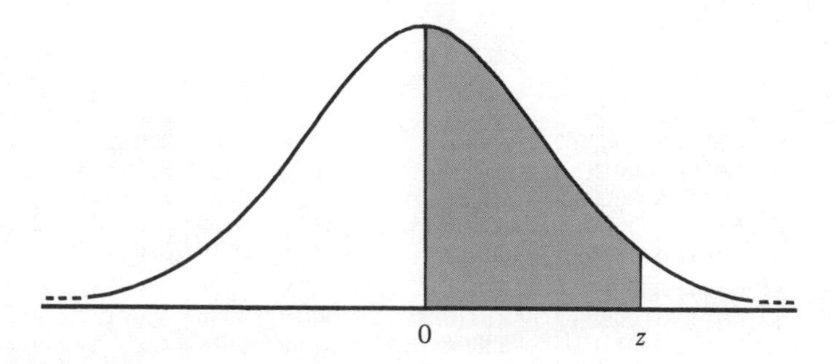

The tabulated values on the next page are areas under the normal distribution for the region shaded above.

Table VI

	Second decimal place in z									
z	.00	.01	.02	.03	.04	.05	.06	.07	.08	.09
.0	.0000	.0040	.0080	.0120	.0160	.0199	.0239	.0279	.0319	.0359
.1	.0398	.0438	.0478	.0517	.0557	.0596	.0636	.0675	.0714	.0753
.2	.0793	.0832	.0871	.0910	.0948	.0987	.1026	.1064	.1103	.1141
.3	.1179	.1217	.1255	.1293	.1331	.1368	.1406	.1443	.1480	.1517
.4	.1554	.1591	.1628	.1664	.1700	.1736	.1772	.1808	.1844	.1879
.5	.1915	.1950	.1985	.2019	.2054	.2088	.2123	.2157	.2190	.2224
.6	.2257	.2291	.2324	.2357	.2389	.2422	.2454	.2486	.2517	.2549
.7	.2580	.2611	.2642	.2673	.2704	.2734	.2764	.2794	.2823	.2852
.8	.2881	.2910	.2939	.2967	.2995	.3023	.3051	.3078	.3106	.3133
.9	.3159	.3186	.3212	.3238	.3264	.3289	.3315	.3340	.3365	.3389
1.0	.3413	.3438	.3461	.3485	.3508	.3531	.3554	.3577	.3599	.3621
1.1	.3643	.3665	.3686	.3708	.3729	.3749	.3770	.3790	.3810	.3830
1.2	.3849	.3869	.3888	.3907	.3925	.3944	.3962	.3980	.3997	.4015
1.3	.4032	.4049	.4066	.4082	.4099	.4115	.4131	.4147	.4162	.4177
1.4	.4192	.4207	.4222	.4236	.4251	.4265	.4279	.4292	.4306	.4319
1.5	.4332	.4345	.4357	.4370	.4382	.4394	.4406	.4418	.4429	.4441
1.6	.4452	.4463	.4474	.4484	.4495	.4505	.4515	.4525	.4535	.4545
1.7	.4554	.4564	.4573	.4582	.4591	.4599	.4608	.4616	.4625	.4633
1.8	.4641	.4649	.4656	.4664	.4671	.4678	.4686	.4693	.4699	.4706
1.9	.4713	.4719	.4726	.4732	.4738	.4744	.4750	.4756	.4761	.4767
2.0	.4772	.4778	.4783	.4788	.4793	.4798	.4803	.4808	.4812	.4817
2.1	.4821	.4826	.4830	.4834	.4838	.4842	.4846	.4850	.4854	.4857
2.2	.4861	.4864	.4868	.4871	.4875	.4878	.4881	.4884	.4887	.4890
2.3	.4893	.4896	.4898	.4901	.4904	.4906	.4909	.4911	.4913	.4916
2.4	.4918	.4920	.4922	.4925	.4927	.4929	.4931	.4932	.4934	.4936
2.5	.4938	.4940	.4941	.4943	.4945	.4946	.4948	.4949	.4951	.4952
2.6	.4953	.4955	.4956	.4957	.4959	.4960	.4961	.4962	.4963	.4964
2.7	.4965	.4966	.4967	.4968	.4969	.4970	.4971	.4972	.4973	.4974
2.8	.4974	.4975	.4976	.4977	.4977	.4978	.4979	.4979	.4980	.4981
2.9	.4981	.4982	.4982	.4983	.4984	.4984	.4985	.4985	.4986	.4986
3.0	.4987	.4987	.4987	.4988	.4988	.4989	.4989	.4989	.4990	.4990
3.1	.4990	.4991	.4991	.4991	.4992	.4992	.4992	.4992	.4993	.4993
3.2	.4993	.4993	.4994	.4994	.4994	.4994	.4994	.4995	.4995	.4995
3.3	.4995	.4995	.4995	.4996	.4996	.4996	.4996	.4996	.4996	.4997
3.4	.4997	.4997	.4997	.4997	.4997	.4997	.4997	.4997	.4997	.4998
3.5	.4998									
4.0	.49997									
4.5	.499997									
5.0	.4999997									

appendix IV
Answers to Odd-Numbered Exercises

CHAPTER ONE

Sections 1.1 and 1.2. **1.** b, c, e **3.** a. F b. F c. F d. T e. T f. T g. F h. T i. F j. T **5.** a. T b. T c. T d. It depends e. F f. T g. F h. F **7.** a. T b. F c. T d. F

Section 1.3. **1.** a. $p \wedge \sim r$ b. $q \rightarrow p$ c. $\sim(q \rightarrow p)$ d. $r \underline{\vee} q$ e. $p \leftrightarrow (r \vee q)$ **3.** a. $\sim(p \rightarrow r)$ b. $\sim q \wedge \sim r$ c. $\sim[(q \vee p) \rightarrow \sim r]$ d. $p \wedge \sim r$ e. $\sim p \leftrightarrow \sim(q \wedge r)$ f. $r \rightarrow p$

Section 1.4. **1.** a. TTTF b. FTFF c. TTTFTTFT d. TFFT e. FFFT f. FFFT g. TFFFFFFF h. FFTFFFFF
3. a. neither b. self-contradiction c. tautology d. neither e. tautology f. self-contradiction g. self-contradiction h. neither i. tautology j. tautology

Section 1.5. **1.** a. neither b. $(p \underline{\vee} q) \Rightarrow (p \vee q)$ c. neither d. $(p \leftrightarrow q) \Rightarrow (p \rightarrow q)$ e. both f. both **5.** all pairs are equivalent
7. a. $\sim(\sim p \vee \sim q)$ b. $\sim[\sim(p \vee q) \vee \sim(\sim p \vee \sim q)]$ c. $\sim p \vee q$ d. $\sim[\sim(\sim p \vee q) \vee \sim(\sim q \vee p)]$ **11.** $\sim(p \vee d)$

Section 1.6. **1.** Converse: If I do not lose, then I played the game fairly. Inverse: If I do not play the game fairly, then I will lose. Contrapositive: If I lost, then I did not play the game fairly.
3. For each of *a–e* the statement is: $q \rightarrow p$.
For each of *a–e* the contrapositive is: $\sim p \rightarrow \sim q$.

Section 1.7. **1.** a. valid b. valid c. valid d. valid e. fallacy f. valid g. fallacy **3.** valid

5. $\left.\begin{array}{l} g \rightarrow h \\ h \leftrightarrow i \\ \underline{\sim g \quad\quad} \\ \therefore h \vee i \end{array}\right\}$ fallacy

9. a. valid b. fallacy c. valid d. valid e. fallacy f. valid

Section 1.8. **3.** a. $(p \wedge q \wedge r) \vee (\sim p \wedge q \wedge \sim r)$
b. $(p \wedge q \wedge r) \vee (\sim p \wedge q \wedge r) \vee (\sim p \wedge \sim q \wedge r)$
c. $\sim[(p \wedge q \wedge r) \vee (\sim p \wedge q \wedge r) \vee (\sim p \wedge \sim q \wedge r)]$
5. a. $(p \wedge q) \vee (\sim p \wedge \sim q)$ b. $p \wedge \sim q$
c. $(p \wedge \sim q) \vee (\sim p \wedge q)$ d. $\sim(p \wedge q)$

Section 1.9. **1.** a. $p \wedge q \wedge (r \vee s)$ b. $p \vee (q \wedge r)$
c. $[p \vee (q \wedge r)] \wedge \sim p$ d. $(p \vee q) \wedge [(q \wedge \sim q) \vee \sim p]$
e. $[p \vee (q \wedge r) \vee \sim r] \wedge q$
f. $[(p \vee q) \wedge r] \vee [p \wedge (\sim q \vee s)] \vee [(a \vee b) \vee c]$
3. T_1——Q——T_2
5.

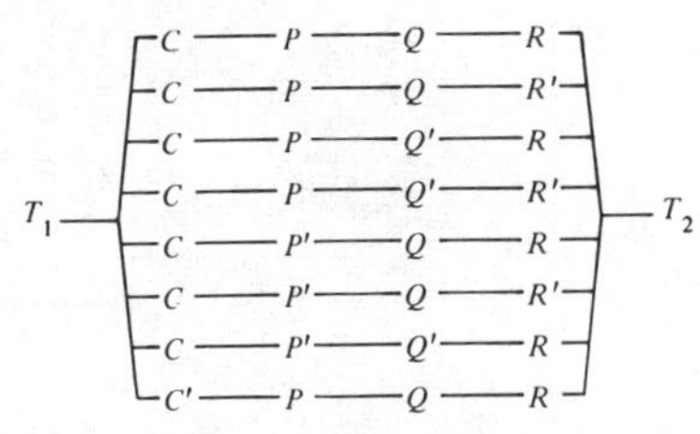

Section 1.10. **1.** Al did it

CHAPTER TWO

Sections 2.1 and 2.2. **1.** a. $\{H1,H3,H5,T1,T3,T5\}$
b. $\{T1,T2,T3,T4,T5,T6\}$ c. $\{H1,H2,H3,H4,H5,H6,T1,T3,T4,T5\}$
d. $\{H2,H4,H6\}$ e. $\{H2,H4,H6,T2,T4,T6\}$ f. $\{T1,T3,T5\}$
g. $\{H2,H4,H6,T4\}$ h. $\varnothing$ i. $\{H2,H4,H6,T2,T6\}$
j. $\{H1,H3,H5,T1,T3,T4,T5\}$ k. $\{T1,T2,T3,T5,T6\}$ l. $\{T1,T3,T5\}$
3. a. $\{H1,H3,H5,T2,T4,T6\}$
b. $\{H1,H2,H3,H4,H5,H6,T2,T6\}$ c. $\{H2,H4,H6,T4\}$
5. a. {Los Angeles, Houston, Dallas} b. {Los Angeles, Houston, Dallas} c. {New York, Chicago, Philadelphia, Detroit, Baltimore, Washington, Cleveland} d. {New York, Chicago, Philadelphia, Detroit, Baltimore, Washington, Cleveland} e. $\varnothing$
7. a. {GE, IBM, ITT, Western Electric} b. {Chrysler} c. {GE}

Section 2.3. **1.** a. no b. no c. yes d. yes e. yes f. yes g. yes (one such pair is A and C) h. $\varnothing$

3. a.

no. of elements in set	1	2	3	4	5	10	n
no. of possible subsets	2	4	8	16	32	1024	2^n

b. $\{a,b,c,d\}$, $\{a,b,c\}$, $\{a,b,d\}$, $\{a,c,d\}$, $\{b,c,d\}$, $\{a,b\}$, $\{a,c\}$, $\{a,d\}$, $\{b,c\}$, $\{b,d\}$, $\{c,d\}$, $\{a\}$, $\{b\}$, $\{c\}$, $\{d\}$, $\varnothing$

5. a. $\{1,5,9,17,18,19,20\}$ b. $\{2,10,13,14,15,16,17,18,19,20,21,22,23,24\}$ c. $\{15\}$ d. $\{22,24\}$ e. $\{1,2,3,4,5,9,10,11,12,13,14,15,16,17,21,22,23,24\}$

Section 2.4. **1.** a. $\{HHH\}$
b. $\{HHH,HHT,HTH,HTT,THH,THT,TTH\}$
c. $\{HHT,HTH,HTT,THH,THT,TTH,TTT\}$ d. $\{TTT\}$
e. $\{HHH\}$ f. $\{HHT,HTH,HTT,THH,THT,TTH\}$ g. $\mathcal{U}$ h. $\mathcal{U}$
i. $\{HHT,HTH,HTT,THH,THT,TTH\}$ j. $\{HHH,TTT\}$
k. $\{HHT,HTH,THH,HTT,THT,TTH\}$ l. $\{TTT\}$ m. $\varnothing$
n. $\{HHH,HHT,HTH,HTT,THH,THT,TTH\}$

5. a.

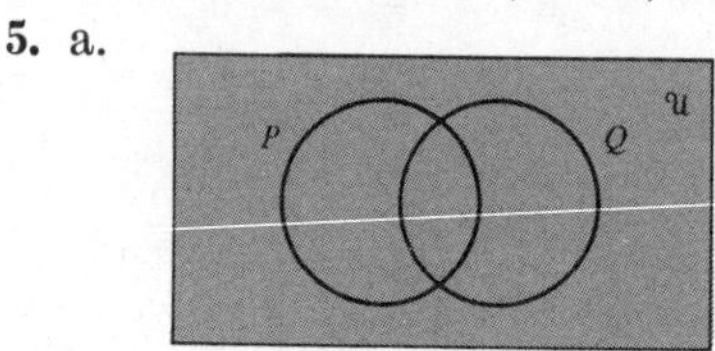

b.

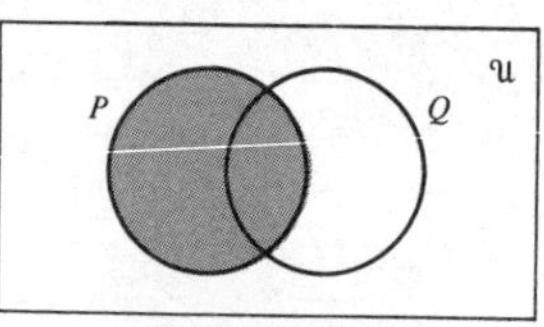

c.

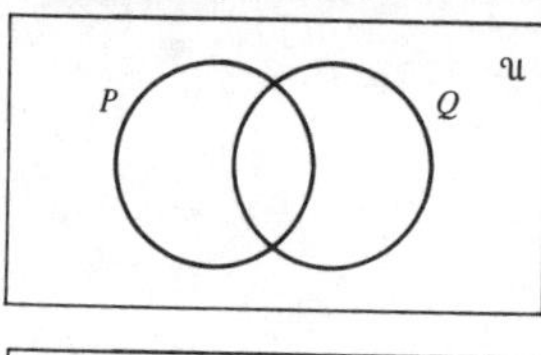

d.

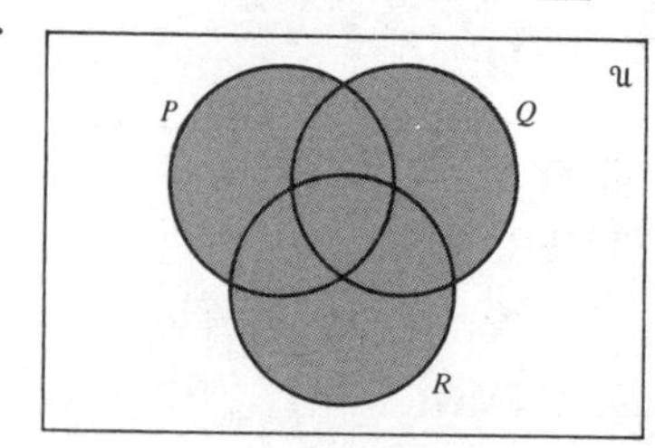

e.

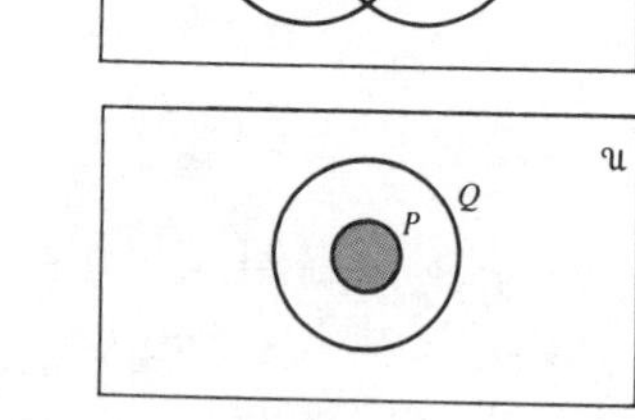

7. (*Refer to diagram at top of facing page*)

1. $P \cap Q \cap R \cap S$
2. $P \cap Q \cap R \cap \tilde{S}$
3. $P \cap Q \cap \tilde{R} \cap S$
4. $P \cap Q \cap \tilde{R} \cap \tilde{S}$
5. $P \cap \tilde{Q} \cap R \cap S$
6. $P \cap \tilde{Q} \cap R \cap \tilde{S}$
7. $P \cap \tilde{Q} \cap \tilde{R} \cap S$
8. $P \cap \tilde{Q} \cap \tilde{R} \cap \tilde{S}$
9. $\tilde{P} \cap Q \cap R \cap S$
10. $\tilde{P} \cap Q \cap R \cap \tilde{S}$
11. $\tilde{P} \cap Q \cap \tilde{R} \cap S$
12. $\tilde{P} \cap Q \cap \tilde{R} \cap \tilde{S}$
13. $\tilde{P} \cap \tilde{Q} \cap R \cap S$
14. $\tilde{P} \cap \tilde{Q} \cap R \cap \tilde{S}$
15. $\tilde{P} \cap \tilde{Q} \cap \tilde{R} \cap S$
16. $\tilde{P} \cap \tilde{Q} \cap \tilde{R} \cap \tilde{S}$

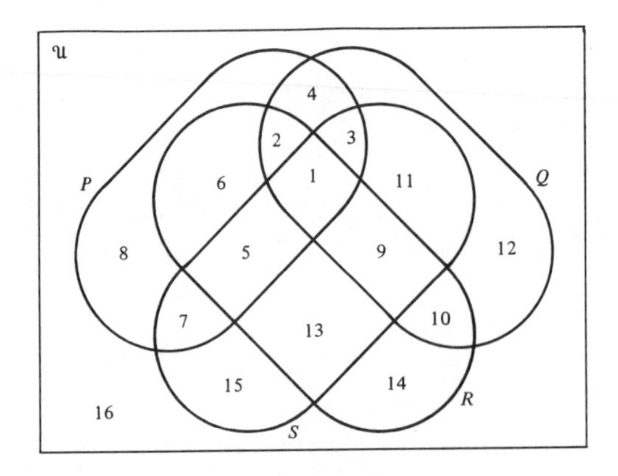

Section 2.5. **1.** a. self-contradiction b. neither c. neither d. neither e. tautology f. neither g. tautology
3. a. yes b. no c. no d. yes
5. a. fallacy b. valid c. valid d. valid e. fallacy
7. a. 1 b. 5 c. 11 **9.** a. 10 b. 25 c. 9 d. 15 e. 68

Section 2.6. **1.** $\mathcal{U} = \{HHHH, HHHT, HHTH, HHTT, HTHH, HTHT, HTTH, HTTT, THHH, THHT, THTH, THTT, TTHH, TTHT, TTTH, TTTT\}$

3. a. *1st die* *2nd die*

1 — 1, 2, 3, 4, 5, 6
2 — 1, 2, 3, 4, 5, 6
3 — 1, 2, 3, 4, 5, 6
4 — 1, 2, 3, 4, 5, 6
5 — 1, 2, 3, 4, 5, 6
6 — 1, 2, 3, 4, 5, 6

b.

1,1	2,1	3,1	4,1	5,1	6,1
1,2	2,2	3,2	4,2	5,2	6,2
1,3	2,3	3,3	4,3	5,3	6,3
1,4	2,4	3,4	4,4	5,4	6,4
1,5	2,5	3,5	4,5	5,5	6,5
1,6	2,6	3,6	4,6	5,6	6,6

c. 6
d. 2
e. 18
f. 0
g. 8
h. 3
i. 21

5. a.

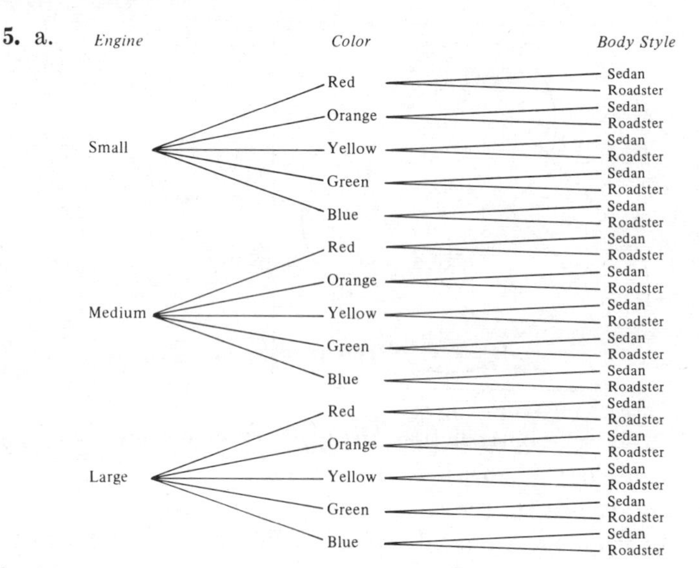

Engine
Color
Body Style
Small
Medium
Large
Red
Orange
Yellow
Green
Blue
Red
Orange
Yellow
Green
Blue
Red
Orange
Yellow
Green
Blue
Sedan
Roadster
Sedan
Roadster
Sedan
Roadster
Sedan
Roadster
Sedan
Roadster
Sedan
Roadster
Sedan
Roadster
Sedan
Roadster
Sedan
Roadster
Sedan
Roadster
Sedan
Roadster
Sedan
Roadster
Sedan
Roadster
Sedan
Roadster
Sedan
Roadster

b. 30

7.

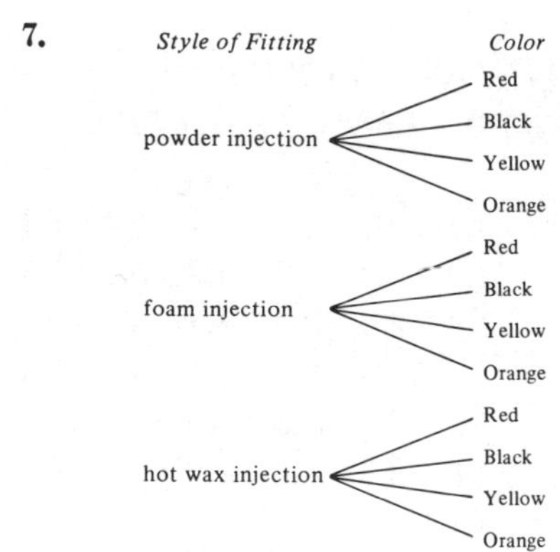

Style of Fitting
Color
powder injection
foam injection
hot wax injection
Red
Black
Yellow
Orange
Red
Black
Yellow
Orange
Red
Black
Yellow
Orange

9.

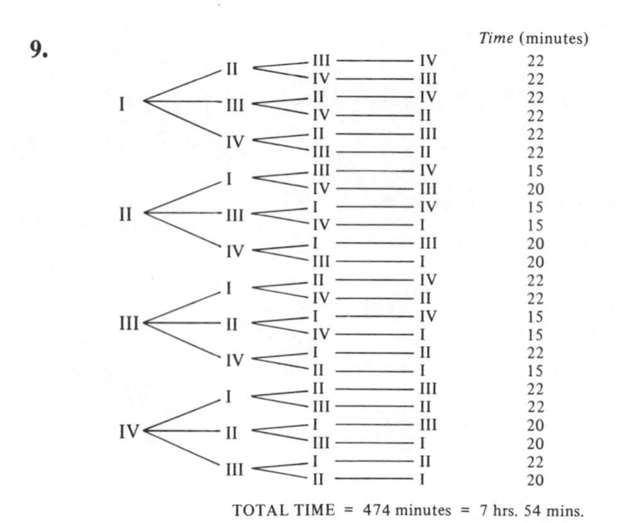

Time (minutes)
I
II
III
IV
II — III — IV 22
II — IV — III 22
III — II — IV 22
III — IV — II 22
IV — II — III 22
IV — III — II 22
I — III — IV 15
I — IV — III 20
III — I — IV 15
III — IV — I 15
IV — I — III 20
IV — III — I 20
I — II — IV 22
I — IV — II 22
II — I — IV 15
II — IV — I 15
IV — I — II 22
IV — II — I 15
I — II — III 22
I — III — II 22
II — I — III 20
II — III — I 20
III — I — II 22
III — II — I 20
TOTAL TIME = 474 minutes = 7 hrs. 54 mins.

Section 2.7. **1.** a. $\{a,b,c,d,e\}$
b. $\{b,c,d,f,g,h,j,k,l,m,n,p,q,r,s,t,u,w,x,y,z\}$ c. $\{b,c,d,e,g,p,t,v,z\}$
d. 130 e. 3380 f. 945 g. 14,196 h. 6,084 i. 17,576 j. 3,380
3. a. $\{(1,3),(2,3)\}$ b. $\{(1,1),(1,2),(1,3),(1,4),(2,1),(2,2),(2,3),(2,4)\}$
c. $\{(2,1),(2,3),(2,4)\}$ d. $\{(1,1),(2,1),(3,1)\}$ e. 4 f. 12
5. $D \times D = \{(1,1),(1,2),(1,3),(1,4),(1,5),(1,6),$
$(2,1),(2,2),(2,3),(2,4),(2,5),(2,6),$
$(3,1),(3,2),(3,3),(3,4),(3,5),(3,6),$
$(4,1),(4,2),(4,3),(4,4),(4,5),(4,6),$
$(5,1),(5,2),(5,3),(5,4),(5,5),(5,6),$
$(6,1),(6,2),(6,3),(6,4),(6,5),(6,6)\}$

CHAPTER THREE

Sections 3.1 and 3.2. **1.** a. 120 b. 625 **3.** a. 10^9 b. $10^9 - 1$
5. a. BARIB 60 MUIBE 120 NOGALS 720 TREVIN 720
b. 5,040 **7.** 4,536 **9.** a. 300 b. 156 **11.** a. 240,240
b. 210,000 **13.** a. $9! = 362{,}880$ b. 2,880 **15.** 24 **17.** 42
19. 64 **21.** 27 **23.** $4^{10}2^{15} = 2^{35}$ **25.** 30

Section 3.3. **1.** a. 21 b. 21 c. 28 d. 28 e. 190 f. 1 g. 576
h. 2 i. 11 j. 4,999,950,000 k. 60 l. 4 m. meaningless
3. a. 24,360 b. 4,060 **5.** $18 + 816 = 834$ **7.** a. 1,326 b. 585
9. $\binom{12}{3} - \binom{8}{3} = 164$

Section 3.4. **1.** 325 **3.** $2(6!)7 = 10{,}080$
5. $\binom{10}{8} + \binom{10}{9} + \binom{10}{10} = 56$ **7.** $\binom{15}{2} = 105$ **9.** 90
11. $\binom{7}{5} \cdot \binom{7}{6} = 147$ **13.** a. $\binom{100}{20}$ b. $\binom{90}{20}$ c. $\binom{10}{2} \cdot \binom{90}{18}$
d. $\binom{10}{1} \cdot \binom{90}{19}$ e. $\binom{100}{20} - \binom{90}{20}$ **15.** a. 45 b. $\binom{5}{3}\binom{5}{4} = 50$
c. $\binom{5}{3}\binom{5}{5} + \binom{5}{4}\binom{5}{4} + \binom{5}{5}\binom{5}{3} = 45$
17. $\binom{3}{1}\binom{2}{1}\binom{3}{1}\binom{7}{2}\binom{7}{3}\binom{18}{1} = 238{,}140$
19. 420 **21.** a. 15 b. 14

Section 3.5. **1.** a. $x^8 + 8x^7y + 28x^6y^2 + 56x^5y^3 + 70x^4y^4 + 56x^3y^5 + 28x^2y^6 + 8xy^7 + y^8$ b. $x^3 - 3x^2 + 3x - 1$
c. $8x^3 - 60x^2y + 150xy^2 - 125y^3$
d. $a^4 - 12a^3b + 54a^2b^2 - 108ab^3 + 81b^4$ e. $-x^3 + 3x^2y - 3xy^2 + y^3$
f. $-(32x^5 + 80x^4y + 80x^3y^2 + 40x^2y^3 + 10xy^4 + y^5)$ **3.** 128

5. $\binom{13}{9} = 715$ **7.** a. 15 b. 10 **9.** 45 **11.** $\binom{100}{59} x^{41}y^{59}$

Section 3.6. **1.** a. $\frac{624}{2{,}598{,}960}$ b. $\frac{3{,}744}{2{,}598{,}960}$ c. $\frac{10{,}200}{2{,}598{,}960}$ d. $\frac{1{,}302{,}540}{2{,}598{,}960}$
3. a. $\binom{52}{7} = 133{,}784{,}560$ b. $4\binom{13}{5}\binom{39}{2} + 4\binom{13}{6}\binom{39}{1} + 4\binom{13}{7}$
c. $13\left(\frac{48 \cdot 44 \cdot 40}{6}\right)$ **5.** 4 **7.** $\binom{10}{2} = 45$

CHAPTER FOUR

Sections 4.1 and 4.2. **1.** a. yes b. no c. no d. yes e. yes f. yes g. yes h. no i. yes j. yes **3.** a. yes b. no c. no d. no e. yes f. no **5.** a. no b. no c. no d. no e. yes f. yes
7. 720 **9.** $\binom{4}{3}\binom{4}{2} = 24$
11. Sunday, Monday, Tuesday, Wednesday, Thursday, Friday, Saturday

Section 4.3. **1.** a. {1,2,3, . . . ,20} b. 1/20 c. 1/2 d. 11/20
3. $P(H) = 4/9$, $P(T) = 4/9$, $P(E) = 1/9$ **5.** a. 1/3 b. 2/3 c. 0
7. a. no b. no c. no d. 4/36 e. 18/36 f. 2/36 g. 20/36
9. a. 29/100 b. 18/100 c. 33/100 d. 71/100

Section 4.4. **1.** a. 1/52 b. 4/52 c. 13/52 d. 16/52 e. 2/52 f. 8/52 g. 26/52 h. 39/52 **3.** a. 2/12 b. 6/12 c. 6/12 d. 10/12 e. 1 to 5 f. 1 to 3 **5.** a. 150/100,000 b. 1,350/100,000 c. 850/100,000 d. 98,650/100,000 e. 99,850/100,000
7. a. 7/12 b. 5/12
9. a.

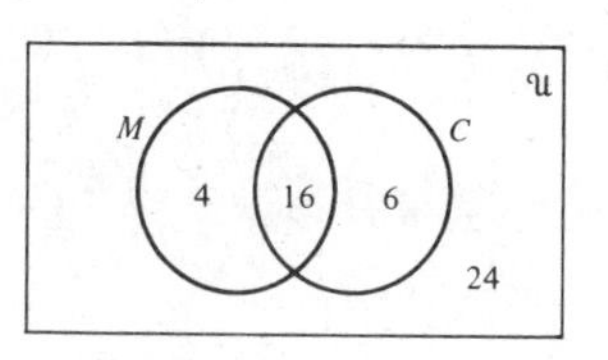

b.

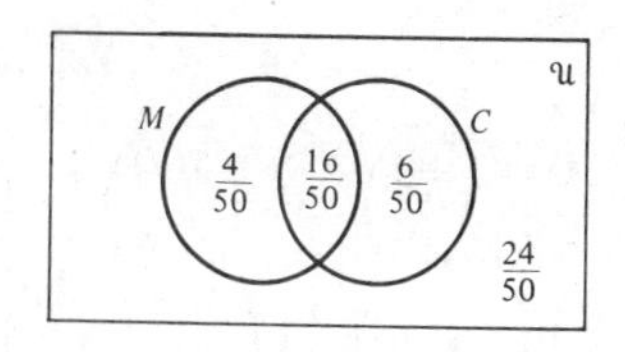

c. .40 d. .44 e. .32 f. .52 g. .08 h. .12 i. .48
11. a. .27 b. .13 c. .04 d. .90 e. .73
13.

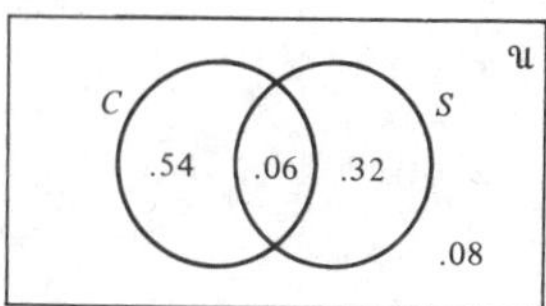

a. .38
b. .08

Section 4.5. **1.** a. 4/36 b. 10/36 c. 6/36 d. 0 e. 0 f. 0 g. 14/36 h. 10/36 i. 16/36 j. 1 to 5 **3.** a. 6/25 b. 3/100

c. 97/100 **5.** a. yes b. yes **7.** 5/18 **9.** a. $\dfrac{\binom{80}{5}}{\binom{100}{5}}$

b. $\dfrac{\binom{20}{5}}{\binom{100}{5}}$ c. $\dfrac{\binom{5}{5}}{\binom{100}{5}}$ d. $\dfrac{\binom{15}{5}}{\binom{100}{5}}$ **11.** $\dfrac{1}{3^5} = \dfrac{1}{243}$

13. $P(\text{crashing}) = .0396$; $P(\text{not crashing}) = .9604$

15. a. $\left(\dfrac{4}{7}\right)^3 = \dfrac{64}{343}$ b. $\dfrac{4}{7}\cdot\dfrac{3}{6}\cdot\dfrac{2}{5} = \dfrac{4}{35}$

17. 11/18 **19.** $\dfrac{\binom{6}{5}+\binom{7}{5}}{\binom{17}{5}}$

21. $\dfrac{40}{\binom{52}{5}} \approx .0000154$ **23.** $\dfrac{10{,}200}{\binom{52}{5}} \approx .0039246$

25. $1 - \dfrac{1{,}302{,}540}{\binom{52}{5}} \approx .4988226$

27. $1 - \dfrac{365\cdot 364\cdot 363\ \cdots\ 341}{(365)^{25}} \approx .569$

Section 4.6. **1.** a.

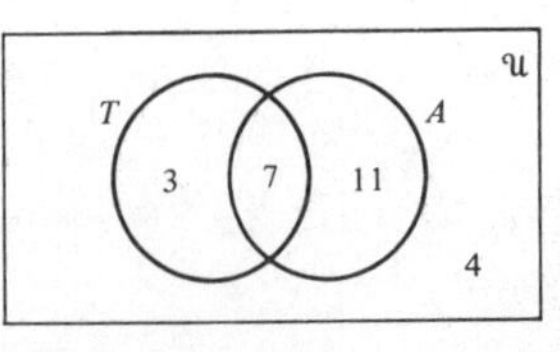

b.

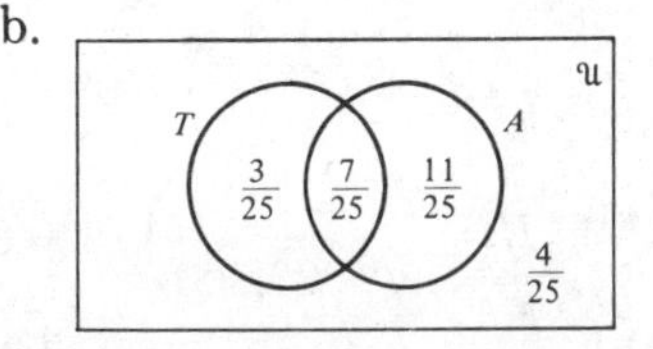

c. 7/10 d. 7/18

3. a. 1/2 b. 7/12 c. 8/55

5. a. 1/4 b. 1/7 **7.** 1/3 **9.** a. 38/100 b. 2/38

11. a. $\dfrac{11{,}100}{33{,}600}$ b. $\dfrac{10{,}000}{31{,}000}$ c. $\dfrac{10{,}000}{11{,}100}$ d. $\dfrac{3{,}500}{3{,}900}$ e. $\dfrac{22{,}500}{26{,}400}$

13. $\dfrac{147}{201}$ **15.** a. .393 b. .164 c. .443

Section 4.7. **1.** a. 1/2 b. 1/8 c. $\left(\dfrac{1}{4}\right)^7$ d. $\left(\dfrac{1}{8}\right)^7$

3. $P(F_1 \text{ Red}) = 1/4$ $P(F_2 \text{ Red}) = 1/4$
$P(F_1 \text{ Roan}) = 1/2$ $P(F_2 \text{ Roan}) = 1/2$
$P(F_1 \text{ White}) = 1/4$ $P(F_2 \text{ White}) = 1/4$

Section 4.8. **1.** a.

Possibility	Arlo (per game)	Arlo (in total)	Bob (per game)	Bob (in total)
HH appears 250 times	loses \$1 each time	loses \$250 in all	wins \$1 each time	wins \$250 in all
HT appears 250 times	wins 30¢ each time	wins \$75 in all	loses 30¢ each time	loses \$75 in all
TH appears 50 times	wins 30¢ each time	wins \$75 in all	loses 30¢ each time	loses \$75 in all
TT appears 50 times	wins 30¢ each time	wins \$75 in all	loses 30¢ each time	loses \$75 in all

Arlo's net losses are \$25

Bob's net winnings are \$25

b. \$25 c. \$25 d. $\frac{\$25}{1000} = 2\tfrac{1}{2}¢$ e. $2\tfrac{1}{2}¢$

3. $E = \left(\frac{1}{1000}\right)(\$199.45) + \left(\frac{1}{1000}\right)(\$149.45) + \left(\frac{1}{1000}\right)(\$99.45) + \left(\frac{1}{1000}\right)(\$99.45) + \left(\frac{996}{1000}\right)(-\$.55) = 0$

5. $E = \left(\frac{1}{1{,}000{,}000}\right)(\$50{,}000) + \left(\frac{1}{1{,}000{,}000}\right)(\$25{,}000)$

a. $E = 7.5¢$ b. no

7. \$18.50 **9.** a. .0079 b. \$79 **11.** \$1300 **13.** $2\tfrac{1}{3}$ **15.** $3\tfrac{1}{2}$

Section 4.9. **1.** $n = 3,\ p = \frac{1}{6},\ q = \frac{5}{6}$

a. $x = 3$: $P(3) = \binom{3}{3}(\frac{1}{6})^3(\frac{5}{6})^0 = \frac{1}{216}$

b. $x = 1$: $P(1) = \binom{3}{1}(\frac{1}{6})^1(\frac{5}{6})^2 = \frac{75}{216}$

3. $n = 4,\ p = \frac{1}{6},\ q = \frac{5}{6}$, success = a 6 occurs. $\therefore x = 2$ and $P(2) = \frac{25}{216}$.

5. a. .246 b. .890 **7.** a. .870 b. $n = 5$ **9.** $\frac{19{,}440}{117{,}649}$

11. .091 **13.** a. .315 b. .599 c. .075

CHAPTER FIVE

Sections 5.1 and 5.2. **1.** $\mathbf{A}$: 3×3, $\mathbf{B}$: 3×4, $\mathbf{C}$: 2×4, $\mathbf{D}$: 6×3, $\mathbf{E}$: $m \times n$ **3.** $\mathbf{A} - \mathbf{B} = (26 \quad 17 \quad 21 \quad 19 \quad 18 \quad 17 \quad 12)$ **5.** 14

7. $\mathbf{N} = (100 \quad 50 \quad 50)$, $\mathbf{P} = \begin{pmatrix} 53.80 \\ 134.20 \\ 1.75 \end{pmatrix}$, $\mathbf{N} \cdot \mathbf{P} = \$12{,}177.50$

9. $(150 \quad 450 \quad 750 \quad 900 \quad 1200 \quad 1500)$

Section 5.3. **1.** a. yes b. yes, $\mathbf{O} = \begin{pmatrix} 0 & 0 & \cdots & 0 \\ 0 & 0 & \cdots & 0 \\ \cdot & \cdot & & \\ \cdot & \cdot & & \\ \cdot & \cdot & & \\ 0 & 0 & \cdots & 0 \end{pmatrix}$

c. yes, $\mathbf{I} = \begin{pmatrix} 1 & 0 & 0 & \cdots & 0 \\ 0 & 1 & 0 & \cdots & 0 \\ \cdot & \cdot & & & \\ \cdot & \cdot & & & \\ \cdot & \cdot & & & \\ 0 & 0 & 0 & \cdots & 1 \end{pmatrix}$ d. no

3. $\begin{pmatrix} 0 & 0 & 0 \\ -\frac{5}{2} & 0 & \frac{5}{2} \\ 0 & 0 & 0 \end{pmatrix}$ **5.** $\begin{pmatrix} 16 & 59 \\ 33 & 95 \\ 115 & -10 \end{pmatrix}$ **7.** not possible

9. $\begin{pmatrix} 37 & -14 \\ 23 & 38 \\ 31 & 54 \end{pmatrix}$ **11.** $\begin{pmatrix} 7 & 10 & 12 \\ 11 & 26 & 26\frac{1}{8} \end{pmatrix}$

13. $\begin{pmatrix} 22 & 11 & 9 \\ 36 & 47 & \frac{5}{4} \end{pmatrix}$ **15.** $\begin{pmatrix} 8 & 13 & 18 \\ 14 & 31 & 32 \end{pmatrix}$ **17.** $\begin{pmatrix} 6 & 7 & 6 \\ 8 & 21 & \frac{81}{4} \end{pmatrix}$

19. a.

	NM	1	2	More
N	.0025	.1500	.0150	.0050
HS	.0400	.1050	.0325	.0120
BA	.0650	.2500	.1125	.0300
MA	.0125	.1175	.0435	.0015
PhD	.0035	.0005	.0005	.0010

b. $\mathbf{Y}$ =

	N	HS	BA	MA	PhD
Monroe	125	5400	100	230	50
L.A.	2,150,000	2,900,000	2,600,000	1,350,000	700
Penfield	31	28	5	4	0

c.

	NM	1	2	More
Monroe	50	638	199	267
L.A.	307,252	1,436,325	477,725	125,576
Penfield	30	9	2	1

Note: Values have been rounded off.

21. $\begin{matrix} & T_1 & T_2 & T_3 \\ S_1 & 180 & 200 & 180 \\ S_2 & 330 & 100 & 90 \end{matrix}$ **23.** b. Replace the 1 in the nth column of an $m \times m$ identity matrix with a k.

25. a. $\begin{pmatrix} 1 & 4 \\ 2 & 5 \\ 3 & 6 \end{pmatrix}$ b. $\begin{pmatrix} 5 \\ 7 \\ -6 \end{pmatrix}$ c. $(1 \quad 2 \quad \frac{1}{2})$

d. $\begin{pmatrix} 1 & 0 & 0 & 0 \\ 0 & 1 & 0 & 0 \\ 0 & 0 & 1 & 0 \\ 0 & 0 & 0 & 1 \end{pmatrix}$ e. $\begin{pmatrix} \frac{1}{2} & 1 & \frac{1}{2} \\ \frac{2}{5} & 0 & \frac{1}{3} \\ \frac{1}{10} & 0 & \frac{1}{6} \end{pmatrix}$

Section 5.4. **1.** a. yes b. no c. yes **3.** $\begin{pmatrix} -4 & 7 \\ 3 & -5 \end{pmatrix}$

5. $\begin{pmatrix} 1 & 0 & 0 \\ -40 & -2 & 3 \\ 97 & 5 & -7 \end{pmatrix}$ **7.** The inverse does not exist.

9. $\begin{pmatrix} 0 & .2 & 0 & 0 \\ 0 & 0 & .2 & 0 \\ 0 & 0 & 0 & .2 \\ .2 & 0 & 0 & 0 \end{pmatrix}$ **11.** $\begin{pmatrix} \frac{4}{7} & -\frac{1}{2} & -\frac{13}{14} \\ -\frac{5}{7} & 1 & \frac{9}{7} \\ \frac{2}{7} & -\frac{1}{2} & -\frac{3}{14} \end{pmatrix}$

Section 5.5. **1.** $x = 7$, $y = 4$ **3.** $x = -1$, $y = -7$, $z = 9$
5. $w = 0$, $x = 0$, $y = 5$, $z = 6$ **7.** no solution
9. $x_1 = 4$, $x_2 = 5$, $x_3 = 0$ **11.** 19 and -1 **13.** \$100 in 6% savings account, \$100 in 7% savings account, \$800 in stock
15. CONZVMGSHD **17.** LOCFKWOUEPVHFYQXXY
19. JESUS WEPT **21.** VD IS NO PICNIC
23. SJIXRUULU **25.** UBVJKOIDEIXVEVVEJY

Section 5.6. **1.** a. $\mathbf{P}^5 = \begin{pmatrix} .34016 & .65984 \\ .32992 & .67008 \end{pmatrix}$ b. $\begin{pmatrix} \frac{1}{3} & \frac{2}{3} \\ \frac{1}{3} & \frac{2}{3} \end{pmatrix}$
c. $\mathbf{v} = (\frac{1}{3} \quad \frac{2}{3})$ d. $\begin{pmatrix} \frac{1}{3} & \frac{2}{3} \\ \frac{1}{3} & \frac{2}{3} \end{pmatrix}$
3. $\mathbf{P}_0 \cdot \mathbf{P}^n = (\frac{1}{3} \quad \frac{1}{3} \quad \frac{1}{3})$; store C ends with 1/3 of the business
5. a. a_1 = pass previous exam, a_2 = pass present exam
b. $\mathbf{L} = \begin{pmatrix} .6 & .4 \\ .9 & .1 \end{pmatrix}$ c. $\mathbf{L}^4 = \begin{pmatrix} .6948 & .3052 \\ .6867 & .3133 \end{pmatrix}$ d. .69075
7. a. $\mathbf{M} = \begin{pmatrix} \frac{1}{2} & \frac{1}{2} \\ \frac{1}{2} & \frac{1}{2} \end{pmatrix}$ b. $v = (\frac{1}{2} \quad \frac{1}{2})$ c. $\frac{1}{2}$

CHAPTER SIX

Section 6.1.

1.

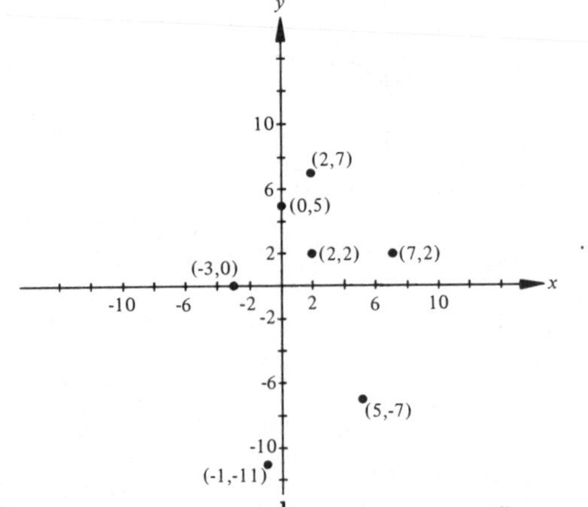

3. a.

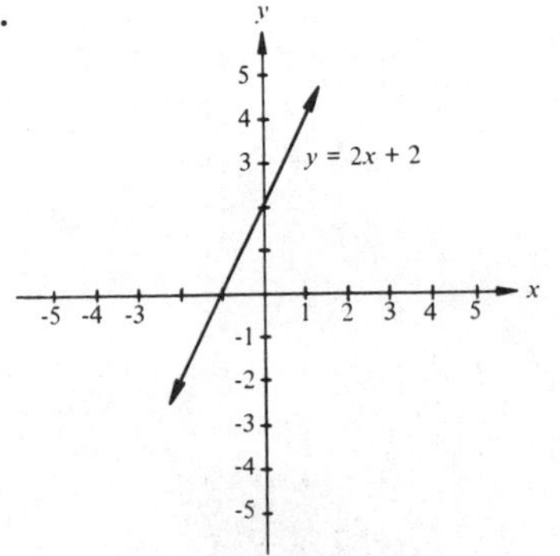

b.

$y = x + \frac{1}{2}$

c. d.

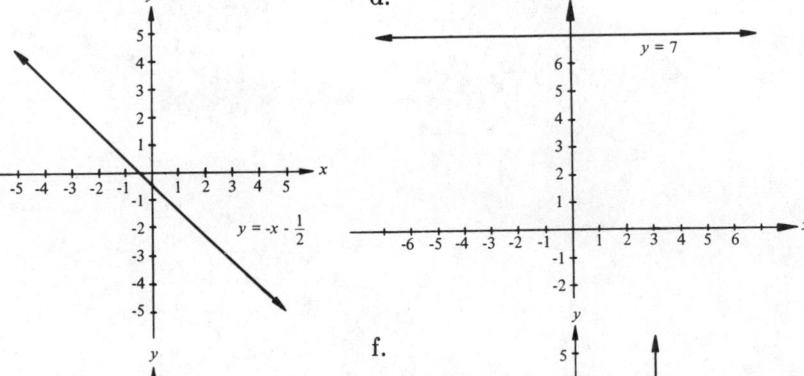

e.

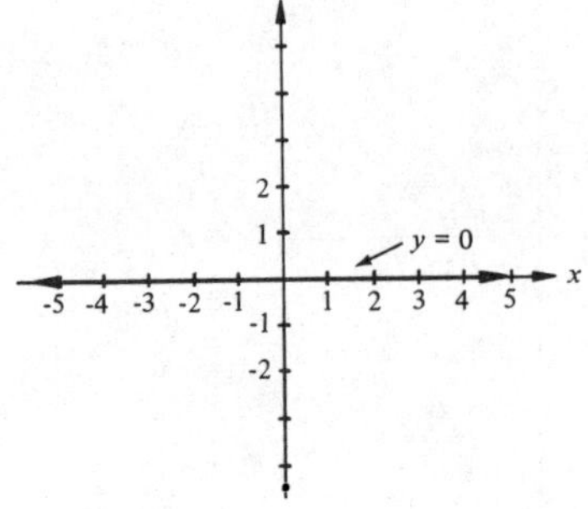

f.

$x = 3$

g. h.

i. j.

5. a. b.

c.

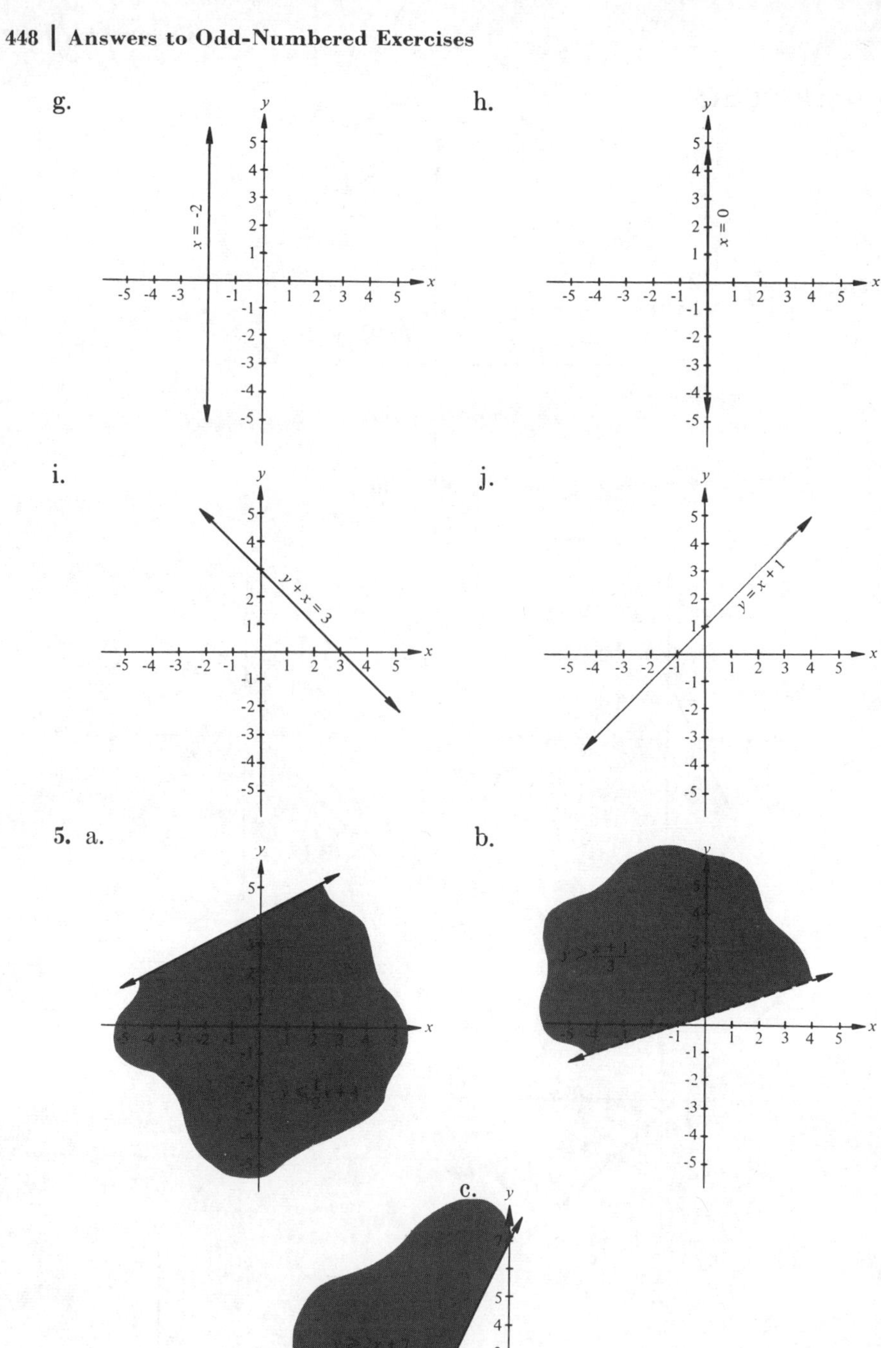

d. same answer as 5c

e.

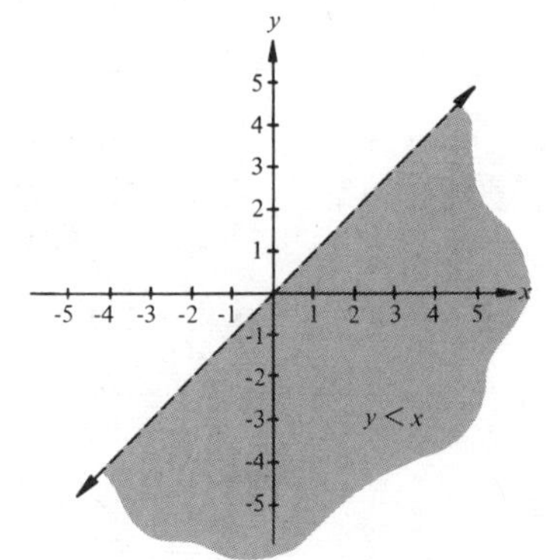

7. a.

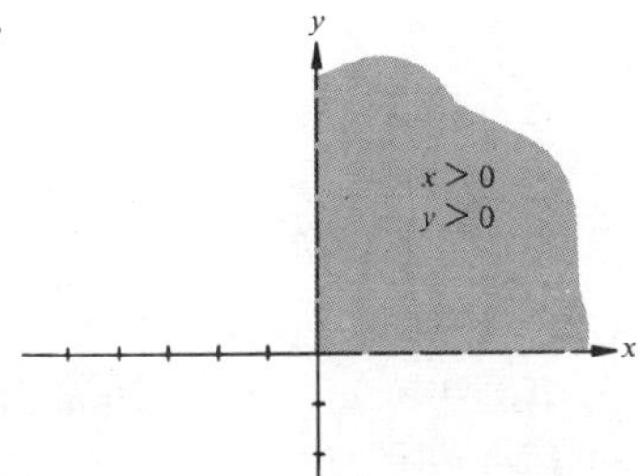

b.

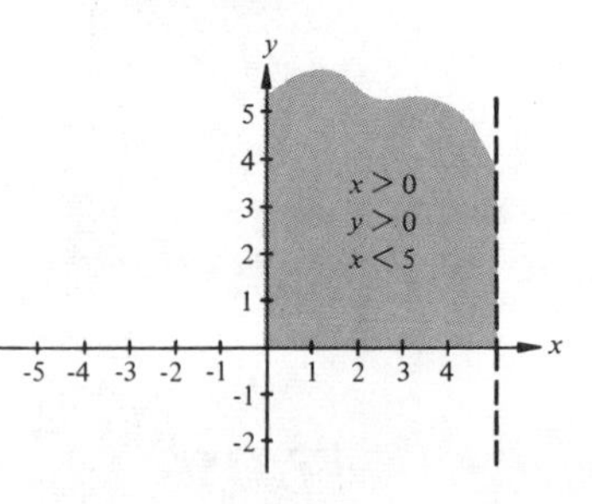

c.

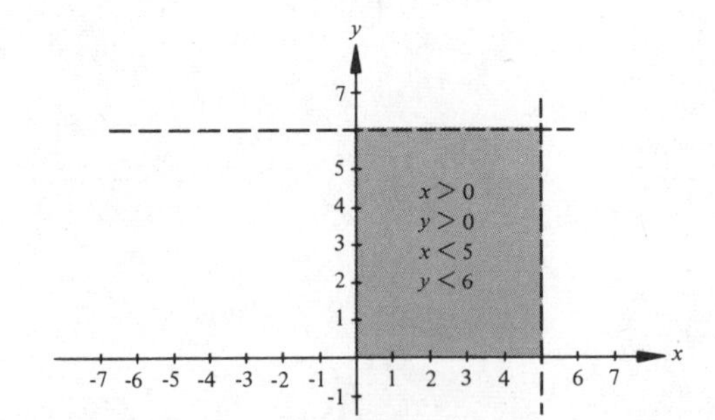

d.

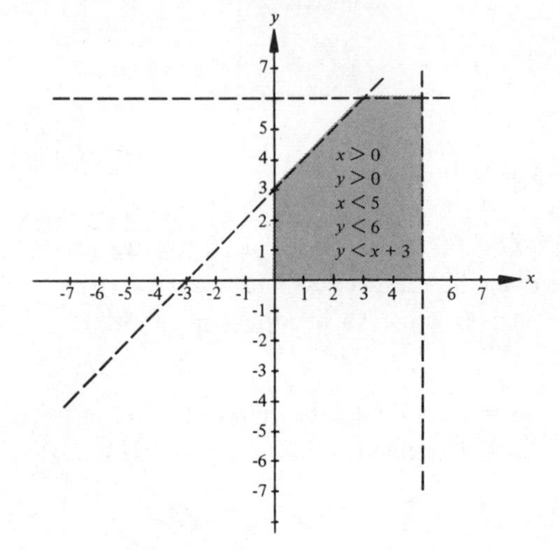

e.

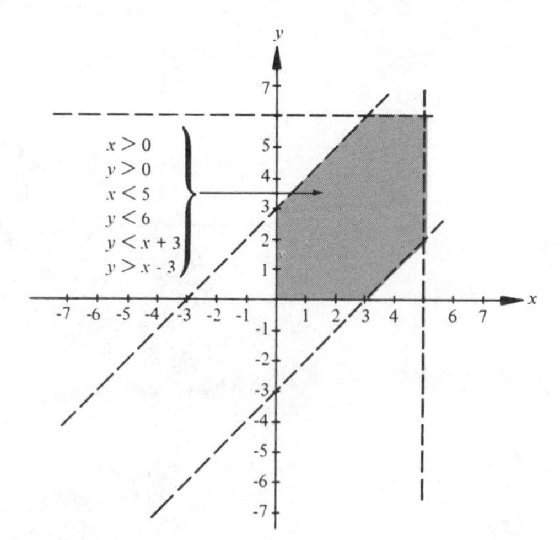

Section 6.2.

1. a.

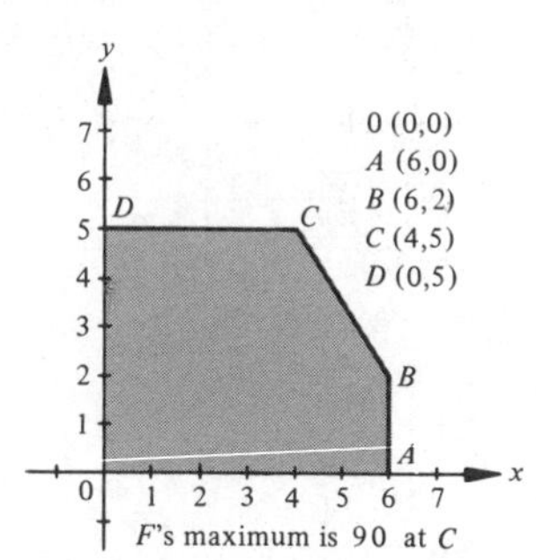

b. F's minimum is 0 at (0,0)
c. G's minimum is -40 at (6,2)
d. $10x + y$ achieves a maximum of 62 at (6,2)
e. $10x - y$ achieves a maximum of 60 at (6,0)
f. $-x + 2y$ achieves a maximum of 10 at (0,5)

3. a.

$$\begin{aligned} x &\leq 40 \\ y &\leq 60 \\ x + y &\leq 80 \\ x &\geq 0 \\ y &\geq 0 \end{aligned}$$

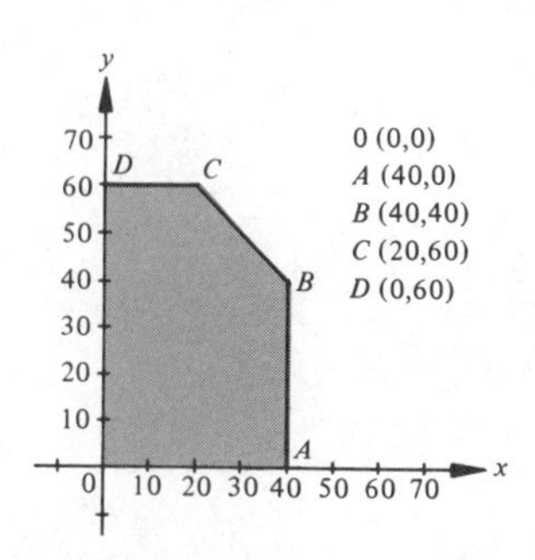

c. The company should produce 20 block sets and 60 toy trains.
5. The company should produce 30 pianos and 5 organs to realize a profit of \$6500. **7.** Joseph's cost is 0 when he produces nothing.

Section 6.3. **9.** $P = 7x + 10y$ achieves its maximum of 176 when $x = 8$, $y = 12$ **11.** $x - y - z$ achieves its minimum value of -18 when $x = 0$, $y = 18$, $z = 0$

Section 6.4.

1. a.

x	y	s_1	s_2	B
5	2	1	0	6
7	3	0	1	20
-10	-5	0	0	0

b.

x	y	s_1	s_2	B
1	3	1	0	30
4	7	0	1	100
-10	-5	0	0	0

c.

x	y	s_1	s_2	s_3	B
1	3	1	0	0	30
4	7	0	1	0	100
1	0	0	0	1	18
-10	-5	0	0	0	0

d.

x_1	x_2	x_3	s_1	s_2	s_3	s_4	B
3	2	0	1	0	0	0	14
1	0	10	0	1	0	0	102
1	1	5	0	0	1	0	56
0	0	1	0	0	0	1	10
-1	-2	-10	0	0	0	0	0

3. P reaches its maximum value of 250 when $x = 25$, $y = 0$
5. P reaches its maximum value of 110 when $x_1 = 2$, $x_2 = 4$, $x_3 = 10$
7. When 20 block sets and 60 toy trains are manufactured, the profit is \$220, a maximum. **9.** The maximum profit of \$85 is obtained when 5 candlesticks, 10 end tables, and 15 bookcases are made.

Section 6.5. **1.** a. Maximize $P = 7x_1 + 17x_2 + 2x_3$
subject to:

$$\begin{aligned} x_1 + 3x_2 &\leq 1 \\ x_1 + x_2 + x_3 &\leq 3 \\ x_1 &\geq 0 \\ x_2 &\geq 0 \\ x_3 &\geq 0 \end{aligned}$$

b. Maximize $P = 21x_1 + 28x_2 + 69x_3 + 6x_4$
subject to:

$$\begin{aligned} x_1 + 6x_2 + x_3 + x_4 &\leq 1 \\ x_1 + 9x_3 &\leq 10 \\ x_1 - x_2 &\leq 20 \\ x_1 &\geq 0 \\ x_2 &\geq 0 \\ x_3 &\geq 0 \\ x_4 &\geq 0 \end{aligned}$$

3. Q's minimum of 17 occurs when $x = \frac{1}{2}$, $y = 4$

CHAPTER SEVEN

Sections 7.1 and 7.2.

1. $A \begin{matrix} H \\ T \end{matrix} \overset{\begin{matrix} & B & \\ H & & T \end{matrix}}{\begin{pmatrix} 25 & -25 \\ -25 & 25 \end{pmatrix}}$ **3.** David $\begin{matrix} 1 \\ 2 \end{matrix} \overset{\begin{matrix} & \text{Joan} & \\ 1 & & 2 \end{matrix}}{\begin{pmatrix} -2 & -3 \\ -3 & -4 \end{pmatrix}}$

5. You $\begin{matrix} \text{Invest} \\ \text{Do not invest} \end{matrix} \overset{\begin{matrix} & \text{Mother Nature} & \\ \text{Good year} & & \text{Recession year} \end{matrix}}{\begin{pmatrix} 8{,}000 & -5{,}000 \\ -2{,}000 & 0 \end{pmatrix}}$

7. a. This is a game of pure strategies. The value of the game is 0. Neither player is favored. The row player's best strategy is row one. The column player's best strategy is column two. b. This is a game of pure strategies. The value of the game is 2. The game favors the row player. The row player's best strategy is row one. The column player's best strategy is column two. c. This is not a game of pure strategies. d. This is a game of pure strategies. The value of the game is -2. The column player is favored. The row player's best strategy is row one. The column player's best strategy is column one. e. This is a game of pure strategies. The value of the game is -3. The game favors the column player. The row player's best strategy is row three. The column player's best strategy is column two. **9.** The value of the game is -1. The column player should always play column two. The row player can choose either row two or row three.

Section 7.3. **1.** a. $\mathbf{P} = (.2 \quad .8)$ b. $\mathbf{Q} = \begin{pmatrix} .4 \\ .6 \end{pmatrix}$ c. \$.80 d. The row player is favored, by 80¢ per game on the average. **5.** -1. **7.** a. 3; row player is favored b. 4/3; row player is favored c. 0; neither player is favored d. 5/4; row player is favored e. $-.09$; column player is favored

Section 7.4. **1.** a. $\mathbf{P} = (\frac{3}{7} \quad \frac{4}{7})$, $\mathbf{Q} = \begin{pmatrix} \frac{4}{7} \\ \frac{3}{7} \end{pmatrix}$, value is 35/49, the row player is favored b. $\mathbf{P} = (\frac{3}{4} \quad \frac{1}{4})$, $\mathbf{Q} = \begin{pmatrix} \frac{1}{2} \\ \frac{1}{2} \end{pmatrix}$, value is 1/2, the row player is favored c. $\mathbf{P} = (1 \quad 0)$, $\mathbf{Q} = \begin{pmatrix} 0 \\ 1 \end{pmatrix}$, value is -2, the column player is favored d. $\mathbf{P} = (\frac{7}{12} \quad \frac{5}{12})$, $\mathbf{Q} = \begin{pmatrix} \frac{7}{12} \\ \frac{5}{12} \end{pmatrix}$, value is $-1/12$, the column player is favored e. $\mathbf{P} = (\frac{5}{7} \quad \frac{2}{7})$, $\mathbf{Q} = \begin{pmatrix} \frac{1}{2} \\ \frac{1}{2} \end{pmatrix}$, value is 0, neither player is favored

3. a. $\mathbf{P} = (\frac{7}{10} \quad \frac{3}{10})$, $\mathbf{Q} = \begin{pmatrix} \frac{6}{10} \\ \frac{4}{10} \end{pmatrix}$ b. .8 c. no **5.** a. $\mathbf{P} = (1 \quad 0)$
b. \$540

Section 7.5. **1.** a. $\mathbf{A}_r = \begin{pmatrix} 2 & 0 \\ -1 & 3 \end{pmatrix}$ b. $\mathbf{A}_r = \begin{pmatrix} 0 & -1 \\ -1 & 4 \end{pmatrix}$

c. $\mathbf{A}_r = \begin{pmatrix} 2 & -3 \\ 1 & 4 \end{pmatrix}$ d. $\mathbf{A}_r = \begin{pmatrix} 2 & 0 & 0 \\ -1 & 3 & 4 \\ 6 & -3 & -4 \end{pmatrix}$ e. does not reduce

f. $\mathbf{A}_r = \begin{pmatrix} 2 & 1 & -4 \\ -4 & -3 & 5 \end{pmatrix}$ g. does not reduce

3. a. $\mathbf{A}_r = \begin{pmatrix} 1 & -4 \\ -3 & 5 \end{pmatrix}$, $\mathbf{P}_r = (\frac{8}{13} \quad \frac{5}{13})$, $\mathbf{Q}_r = \begin{pmatrix} \frac{9}{13} \\ \frac{4}{13} \end{pmatrix}$

row player $\begin{cases} \text{should play row one } 8/13 \text{ of the time} \\ \text{should play row two } 5/13 \text{ of the time} \end{cases}$

column player $\begin{cases} \text{should never play column one} \\ \text{should play column two } 9/13 \text{ of the time} \\ \text{should play column three } 4/13 \text{ of the time} \end{cases}$

b. $-7/13$; the column player is favored.

Section 7.6. **1.** $\mathbf{P} = (\frac{5}{7} \quad \frac{2}{7})$, $\mathbf{Q} = \begin{pmatrix} \frac{4}{7} \\ \frac{3}{7} \\ 0 \end{pmatrix}$, value is $\frac{8}{7}$

3. $\mathbf{P} = (\frac{2}{5} \quad 0 \quad \frac{3}{5})$, $\mathbf{Q} = \begin{pmatrix} \frac{4}{5} \\ \frac{1}{5} \end{pmatrix}$, value is $\frac{7}{5}$ **7.** $\mathbf{P} = (0 \quad 1 \quad 0)$,

$\mathbf{Q} = \begin{pmatrix} 0 \\ 1 \\ 0 \\ 0 \end{pmatrix}$, value is -2 **9.** $\mathbf{P} = (\frac{3}{4} \quad \frac{1}{4} \quad 0)$, $\mathbf{Q} = \begin{pmatrix} 0 \\ \frac{3}{4} \\ \frac{1}{4} \end{pmatrix}$, value is $2\frac{1}{4}$

11. *B* should $\begin{cases} \text{advertise } 50\% \text{ on the radio} \\ \text{never advertise in the newspapers} \end{cases}$
The value is 1.4

CHAPTER EIGHT

Sections 8.1 and 8.2. **1.** New York State voters **9.** a. discrete b. discrete c. continuous d. attribute e. continuous f. discrete g. discrete h. continuous

Section 8.3. **1.** $\bar{x} = 2.4$, $\tilde{x} = 2$, mode $= 2$ **3.** $\bar{x} = 4\frac{3}{4} \approx 4.8$, $\tilde{x} = 4.5$, there is no mode **5.** 1, 3, 4, 6, 6 (there are many more)
7. a. increases b. remains the same c. remains the same
9. 34.6%, 25.8%, 12.8%, 68.1%, 37.3%, 8.7%, 26.3%, 3.6%

Section 8.4. **1.** $R = 4$, $v = 2.00$, $s = 1.4$ **3.** 3.7 **5.** no
7. Criterion 2 is not satisfied; the process is not under control. ($\bar{x} = 5.03$, $s = .08$) **9.** $\bar{x} \pm s$ is from 1.42 to 4.58, $\bar{x} \pm 2s$ is from $-.16$ to 6.16, $\bar{x} \pm 3s$ is from -1.74 to 7.74 **11.** 1.58
13. a. An A is any score greater than 96.1; B is any score from 85.9 to 96.1; C is any score from 65.5 to 85.9; D is any score from 45.1 to 65.5; F is any score less than 45.1 b. 76% c. 100% d. 12%

Section 8.5. **1.**

Class boundaries	*Frequency*
46.5–56.5	1
56.5–66.5	4
66.5–76.5	3
76.5–86.5	3
86.5–96.5	8
96.5–106.5	11
106.5–116.5	10
116.5–126.5	3
126.5–136.5	4
136.5–146.5	2
146.5–156.5	1

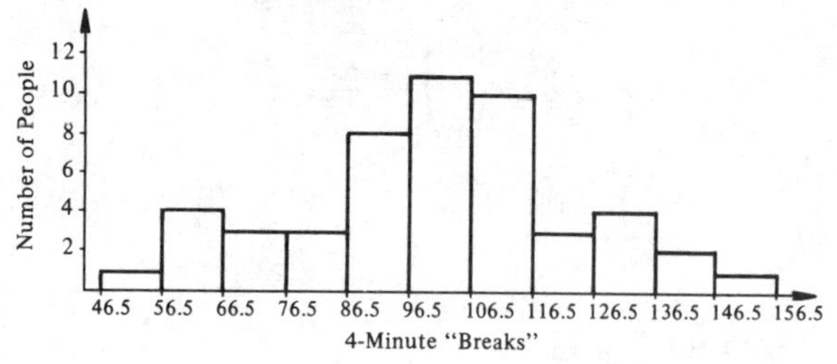

3.

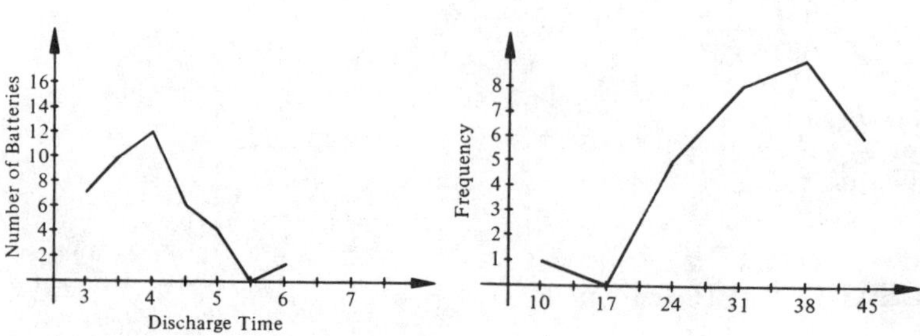

7. a.

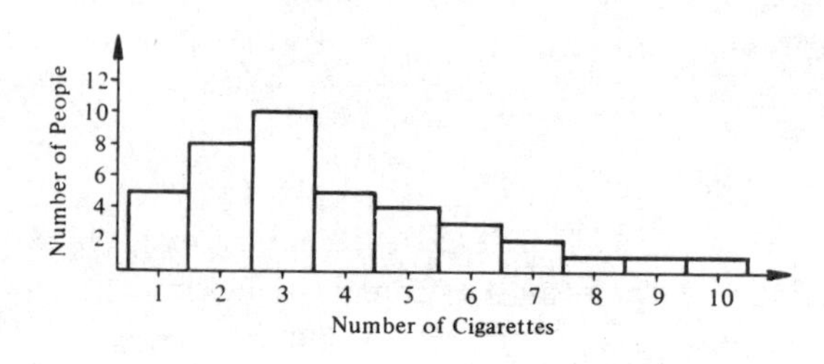

b.

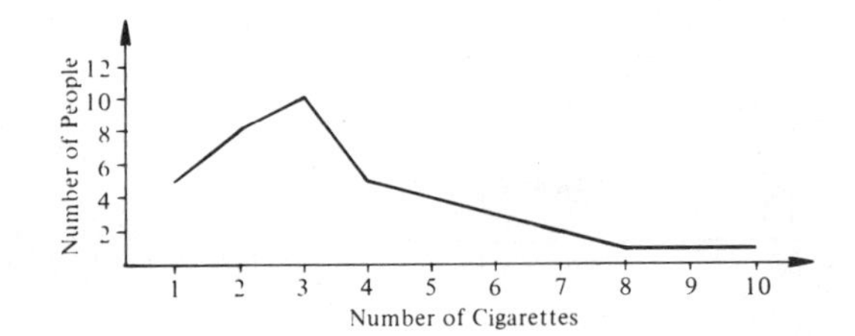

c. The modal class is the one with class mark 3. d. 3
e. $\bar{x} = 3.75 \approx 3.8$

9. $\bar{x} \approx 34.1$, $s \approx 8.5$ a. $\bar{x} \pm s$ is from 25.6 to 42.6, $\bar{x} \pm 2s$ is from 17.1 to 51.1, $\bar{x} \pm 3s$ is from 8.6 to 59.6 b. 59% of the data is in $\bar{x} \pm s$, 97% of the data is in $\bar{x} \pm 2s$, 100% of the data is in $\bar{x} \pm 3s$
c. no

11. a.

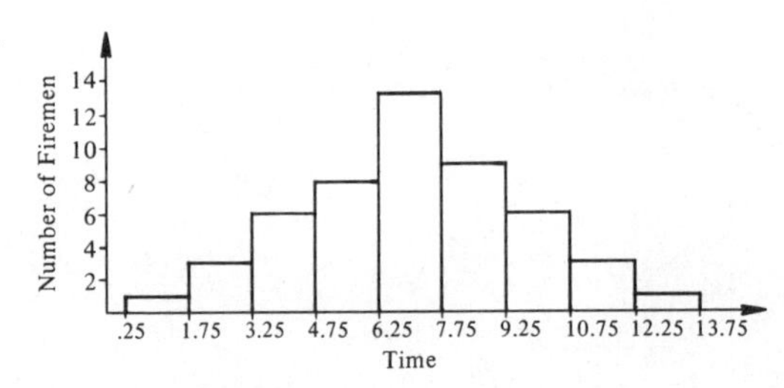

b. $\bar{x} = 7.03$, $s = 2.63$ c. $\bar{x} \pm s$ is from 4.40 to 9.66, $\bar{x} \pm 2s$ is from 1.77 to 12.29, $\bar{x} \pm 3s$ is from −.86 to 14.92 d. 60% lies in $\bar{x} \pm s$, 96% lies in $\bar{x} \pm 2s$, 100% lies in $\bar{x} \pm 3s$ e. no **13.** The percents of change from top to bottom are: −1.9%, +22.2%, +122.9%, +31.4%, −33.3%, +64.3%, −6.9%, +800%, 0%, +345.5%

15.

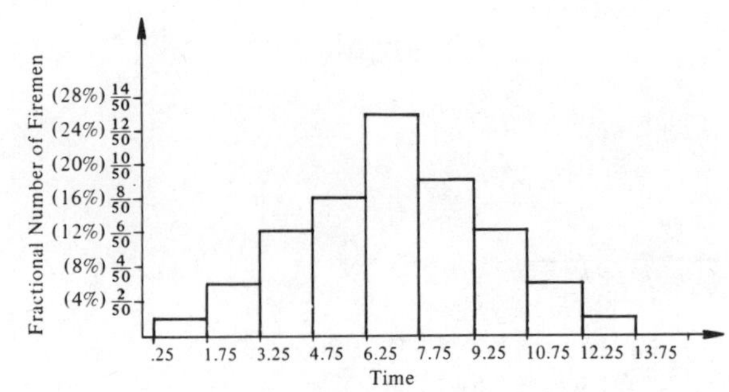

17. a.

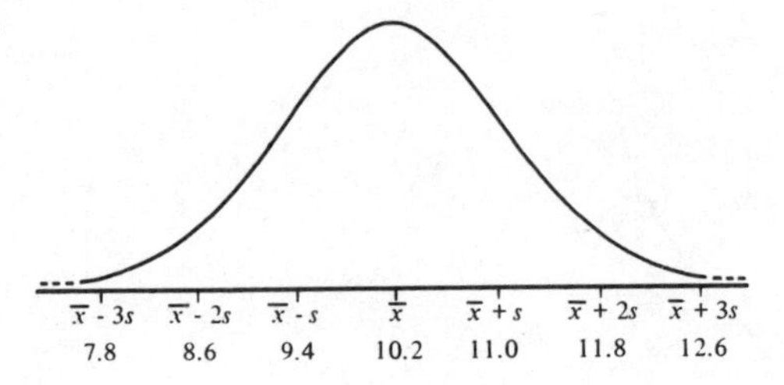

b.

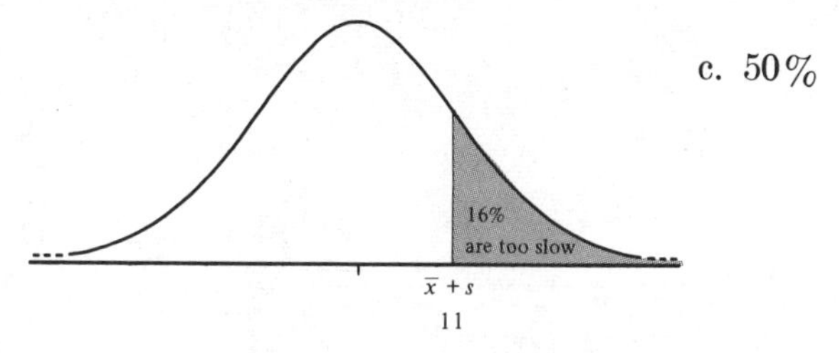

c. 50%

Section 8.6. **1.** a. 18 b. 29 c. 4 d. 2 **3.** $D_1 = 4$, $D_4 = 14$, $D_9 = 28$ **5.** a. 130 b. between 130 and 145

Section 8.7. **1.** a. .9824 b. .0272 c. .1138 d. .0217 e. .9783 f. .8997 g. .0014 **3.** a. .0000003 = .00003% b. 30 **5.** 157.6 **7.** a. 73.24% b. 11.84 ($\therefore$ about 12)

CHAPTER NINE

Section 9.2. **1.** a.

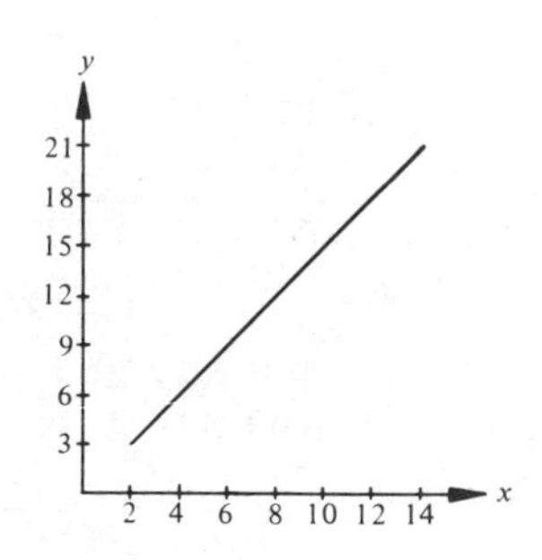

b. yes c. $y = \frac{3}{2}x$

3. a.

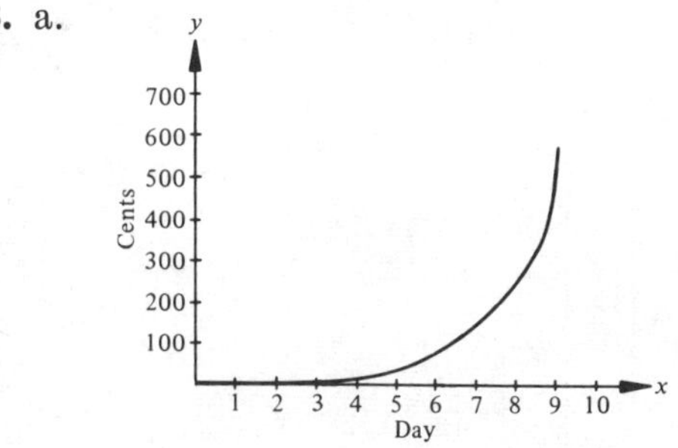

b. $y = 2^x$ c. $2^{10} = 1024¢ = \$10.24$ d. $20.46
e. 2^{29} cents = $5,368,709.12 f. $2^{30} - 2$ cents

5. a.

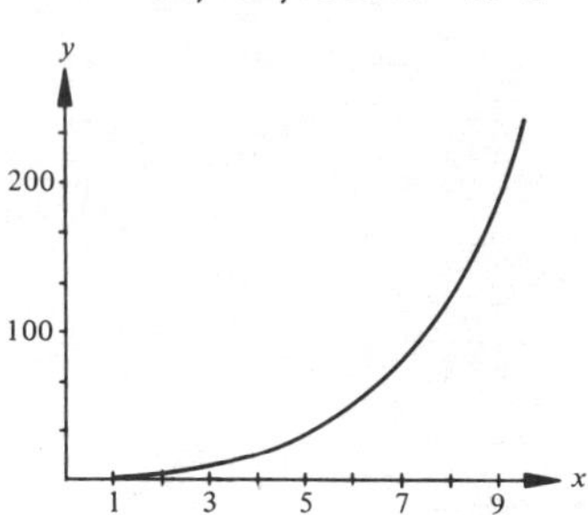

b. exponential c. $y = 3^{\frac{x+1}{2}}$ **7.** .0037 **13.** a. b b. no winner c. b d. b

APPENDIX

Appendix I. Since problems 1, 3, and 5 are very similar, we do problem 3 only. **3.** Step 1 is easily verified: $1^2 = \frac{1(2)(3)}{6} = 1$. Assume the equation holds for $n = k$; show it holds for $n = k + 1$ $\left(\text{that is, show } 1^2 + 2^2 + \cdots + (k+1)^2 = \frac{(k+1)(k+2)(2k+3)}{6}\right)$.

Assuming it true for $n = k$ yields:

$$1^2 + 2^2 + 3^2 + \cdots + k^2 = \frac{k(k+1)(2k+1)}{6}$$

Add $(k+1)^2$ to both sides:

$$\begin{aligned} 1^2 + 2^2 + \cdots + k^2 + (k+1)^2 &= \frac{k(k+1)(2k+1)}{6} + \frac{6(k+1)^2}{6} \\ &= \frac{(k+1)[k(2k+1) + 6(k+1)]}{6} \\ &= \frac{(k+1)[2k^2 + 7k + 6]}{6} \\ &= \frac{(k+1)[(k+2)(2k+3)]}{6} \end{aligned}$$

which was to be shown.

7. a. Fails in step 2, checks in step 1.
b. Fails in step 1, checks in step 2.

Appendix II. **1.** a. $x_1 + x_2 + x_3 + x_4 + x_5$ b. $x_1y_1 + x_2y_2 + x_3y_3$ c. $x_1^2f_1 + x_2^2f_2 + x_3^2f_3 + x_4^2f_4$ d. $(x_1 + y_1) + (x_2 + y_2) + (x_3 + y_3)$
3. a. 7 b. 15 c. 49 **5.** a. 15 b. 45 c. 225

Index

Glossary of Symbols

Symbol	*Meaning*
$p, q, r, \ldots$	simple statements
$\sim p$	negation of p (not p)
$p \wedge q$	conjunction of p with q (p and q)
$p \vee q$	disjunction of p with q (p or q)
$p \underline{\vee} q$	exclusive disjunction of p with q (p or q, but not both)
$p \rightarrow q$	if p, then q (or p conditional q)
$p \leftrightarrow q$	p if and only if q (or p biconditional q)
p iff q	p if and only if q
$a \Rightarrow b$	a implies b
$a \Leftrightarrow b$	a is equivalent to b
$x \mathbf{I} y$	x is inconsistent with y
$\therefore$	therefore
$P, Q, R, \ldots$	sets
$2 \in \{1,2,3\}$	2 is an element of $\{1,2,3\}$
$\mathcal{U}$	the universal set (the set of all logical possibilities)
$P \cap Q$	P intersection Q
$P \cup Q$	P union Q
$A - B$	the difference of A and B
$A \Delta B$	the symmetric difference of A and B
$P \subseteq Q$	P is a subset of Q
$\varnothing$ or $\{\}$	the empty set
$n(A)$	the number of elements in set A
$A = B$	set A equals set B
$G \times H$	the Cartesian product of G with H
$n!$	n factorial
$\binom{n}{j}$	the number of unordered arrangements of n things choosing j at a time
$(x + y)^n$	raising a binomial to the nth power
$P(E)$	the probability of event E
$P(A \mid B)$	the probability of A, knowing B has occurred
$\binom{n}{x} p^x q^{n-x}$	the probability of exactly x successes in n trials of a binomial experiment
$\begin{pmatrix} a_{11} & a_{12} & \cdots & a_{1n} \\ a_{21} & & & \\ \vdots & & & \\ a_{m1} & & \cdots & a_{mn} \end{pmatrix}$	an $m \times n$ matrix